APPLICATIONS OF ION CHROMATOGRAPHY FOR PHARMACEUTICAL AND BIOLOGICAL PRODUCTS

APPLICATIONS OF ION CHROMATOGRAPHY FOR PHARMACEUTICAL AND BIOLOGICAL PRODUCTS

Edited by

Lokesh Bhattacharyya
Jeffrey S. Rohrer

An appendix bibliography listing and select color images from the book can be found at: ftp://ftp.wiley.com/public/sci_tech_med/ion_chromatography.

WILEY

A JOHN WILEY & SONS, INC., PUBLICATION

Published by John Wiley & Sons, Inc., Hoboken, New Jersey
Published simultaneously in Canada

Library of Congress Cataloging-in-Publication Data:

Applications of ion chromatography for pharmaceutical and biological products / edited by Lokesh Bhattacharyya, Jeffrey S. Rohrer.
p. ; cm.
Includes index.
ISBN 978-0-470-46709-1 (cloth)
1. Ion exchange chromatography. 2. Biological products–Analysis. 3. Pharmaceutical industry–Standards. I. Bhattacharyya, Lokesh. II. Rohrer, Jeffrey S.
[DNLM: 1. Chromatography, Ion Exchange–methods. 2. Biological Products–analysis. 3. Drug Industry–standards. 4. Mass Spectrometry. 5. Pharmaceutical Preparations–analysis. QD 79.C453]
QD79.C453A67 2011
660.6′3–dc23

2011022706

10 9 8 7 6 5 4 3 2 1

CONTENTS

CONTRIBUTORS ix

PREFACE xi

PART I PRINCIPLES, MECHANISM, AND INSTRUMENTATION 1

1 ION CHROMATOGRAPHY—PRINCIPLES AND APPLICATIONS 3
Lokesh Bhattacharyya

2 RETENTION PROCESSES IN ION-EXCLUSION CHROMATOGRAPHY: A NEW PERSPECTIVE 23
Milko Novič and Paul R. Haddad

3 PULSED ELECTROCHEMICAL DETECTION IN ION CHROMATOGRAPHY 49
William R. LaCourse

4 SUPPRESSOR DESIGN AND DETECTION FOR ION CHROMATOGRAPHY 91
Kannan Srinivasan, Sheetal Bhardwaj, Rong Lin, and Christopher Pohl

5 MODELLING AND OPTIMIZATION OF ION CHROMATOGRAPHIC SEPARATIONS OF PHARMACEUTICALLY RELEVANT ORGANIC IONS 107
Philip Zakaria, Greg Dicinoski, Melissa Hanna-Brown, and Paul R. Haddad

PART II PHARMACEUTICAL APPLICATIONS 135

6 ION CHROMATOGRAPHY IN PHARMACEUTICAL DRUG ANALYSIS 137
Lokesh Bhattacharyya

7 **ANALYSIS FOR IMPURITIES BY ION CHROMATOGRAPHY** **159**
Shreekant V. Karmarkar

8 **ION CHROMATOGRAPHY ANALYSIS OF AMINOGLYCOSIDE ANTIBIOTICS** **175**
Valoran P. Hanko and Jeffrey S. Rohrer

9 **USE OF CATION-EXCHANGE ION CHROMATOGRAPHY IN THE ANALYSIS OF PHARMACEUTICALS** **193**
Christopher Pohl, Maria Rey, and Rosanne Slingsby

10 **COMPREHENSIVE APPROACHES FOR MEASUREMENT OF ACTIVE PHARMACEUTICAL INGREDIENTS, COUNTER-IONS, AND EXCIPIENTS USING HPLC WITH CHARGED AEROSOL DETECTION** **221**
Christopher Crafts, Bruce Bailey, Paul Gamache, Xiaodong Liu, and Ian Acworth

11 **HIGH PERFORMANCE ION CHROMATOGRAPHIC ANALYSIS OF CHOLINERGIC COMPOUNDS: CARBACHOL AND BETHANECHOL, AND ASSOCIATED DEGRADATION PRODUCTS** **237**
David Trowbridge

12 **ION CHROMATOGRAPHIC ANALYSIS OF PHARMACEUTICALS FOR AUTHENTICITY AND ADULTERATION** **247**
David S. Jackson

13 **ION CHROMATOGRAPHY OF DRUG PRODUCT COUNTER-IONS AND OTHER IONS IN DRUG PRODUCTS** **259**
Brian M. De Borba and Jeffrey S. Rohrer

14 **SAMPLE PREPARATION TECHNIQUES IN ION CHROMATOGRAPHY FOR PHARMACEUTICAL APPLICATIONS** **271**
Jay Gandhi

15 **VALIDATION OF ION CHROMATOGRAPHIC METHODS** **285**
Shreekant V. Karmarkar

16 **PHARMACOPEIAL APPLICATIONS OF ION CHROMATOGRAPHY** **309**
Ravi Ravichandran

PART III BIOTECHNOLOGY APPLICATIONS 321

17 GLYCAN ANALYSIS BY HIGH PERFORMANCE ANION EXCHANGE CHROMATOGRAPHY WITH PULSED AMPEROMETRIC DETECTION 323
Lokesh Bhattacharyya

18 MONOSACCHARIDE ANALYSIS OF GLYCOPROTEINS BY HIGH-PERFORMANCE ANION-EXCHANGE CHROMATOGRAPHY WITH PULSED AMPEROMETRIC DETECTION 339
Jeffrey S. Rohrer

19 TWO-DIMENSIONAL ION CHROMATOGRAPHY FOR SIMULTANEOUS DETERMINATION OF AMINO ACIDS AND CARBOHYDRATES 351
Petr Jandik and Jun Cheng

20 MONITORING PROTEIN *N*-GLYCOSYLATION CONSISTENCY USING HPAEC-PAD 365
Carl Grey

21 OLIGOSACCHARIDE ANALYSIS BY HIGH-PERFORMANCE ANION-EXCHANGE CHROMATOGRAPHY HYPHENATED TO INTEGRATED PULSED AMPEROMETRIC DETECTION AND ON-LINE ION-TRAP MASS SPECTROMETRY 379
Cees Bruggink

PART IV VACCINE APPLICATIONS 393

22 APPLICATIONS OF ION CHROMATOGRAPHY IN BIOLOGICAL PRODUCT ANALYSES 395
Alfred V. Del Grosso and Brandon Duong

23 HIGH PERFORMANCE ANION-EXCHANGE CHROMATOGRAPHY WITH PULSED AMPEROMETRIC DETECTION (HPAEC-PAD) ANALYSIS OF GBS GLYCOCONJUGATE VACCINES: FROM FERMENTATION TO FINAL PRODUCT 423
Stefano Ricci, Barbara Fabbri, Sara Giannini, Silvia Bernardini, Marcello Morelli, and Sandro D'Ascenzi

24 STABILITY STUDIES AND CHARACTERIZATION OF MENINGOCOCCAL A, C, Y, AND W_{135} GLYCOCONJUGATE VACCINE 435
Valeria Carinci, Alberto Bernardini, Cristiana Campa, Claudia Magagnoli, Francesca Beccai, and Sandro D'Ascenzi

PART V APPENDICES 449

Appendix 1 DISSOCIATION CONSTANTS (pK_a) OF ORGANIC ACIDS (AT 20°C) 451

Appendix 2 DISSOCIATION CONSTANTS (pK_a) OF COMMON SUGARS AND ALCOHOLS 455

INDEX 457

CONTRIBUTORS

Ian Acworth, Thermo Fisher Scientific, Chelmsford, Massachusetts, USA

Bruce Bailey, Thermo Fisher Scientific, Chelmsford, Massachusetts, USA

Francesca Beccai, Technology Development Department, Novartis Vaccines & Diagnostics S.r.L., Siena, Italy

Alberto Bernardini, Technology Development, Novartis Vaccines & Diagnostics S.r.L., Siena, Italy

Silvia Bernardini, Technology Development, Novartis Vaccines & Diagnostics, Siena, Italy

Sheetal Bhardwaj, Thermo Fisher Scientific, Sunnyvale, California, USA

Lokesh Bhattacharyya, Division of Biological Standards and Quality Control, Office of Compliance and Biologics Quality, Center for Biologics Evaluation and Research, U.S. Food and Drug Administration, Rockville, Maryland, USA

Cees Bruggink, Thermo Fisher Scientific, Breda, Amsterdam, and Biomolecular Mass Spectrometry Unit, Department of Parasitology, Leiden University Medical Center, RC Leiden, The Netherlands

Cristiana Campa, Technology Development Department, Novartis Vaccines & Diagnostics S.r.L., Siena, Italy

Valeria Carinci, Technology Development, Novartis Vaccines & Diagnostics S.r.L., Siena, Italy

Jun Cheng, Thermo Fisher Scientific, Sunnyvale, California, USA

Christopher Crafts, Thermo Fisher Scientific, Chelmsford, Massachusetts, USA

Brian M. De Borba, Thermo Fisher Scientific, Sunnyvale, California, USA

Sandro D'Ascenzi, Technology Development Department, Novartis Vaccines & Diagnostics S.r.L., Siena, Italy

Alfred V. Del Grosso, Division of Biological Standards and Quality Control, Office of Compliance and Biologics Quality, Center for Biologics Evaluation and Research, U.S. Food and Drug Administration, Rockville, Maryland, USA

Greg Dicinoski, Pfizer Analytical Research Centre, Australian Centre for Research on Separation Science, School of Chemistry, University of Tasmania, Hobart, Tasmania, Australia

Brandon Duong, Division of Biological Standards and Quality Control, Office of Compliance and Biologics Quality, Center for Biologics Evaluation and Research, U.S. Food and Drug Administration, Rockville, Maryland, USA

Barbara Fabbri, Technology Development, Novartis Vaccines & Diagnostics S.r.L., Siena, Italy

Paul Gamache, Thermo Fisher Scientific, Chelmsford, Massachusetts, USA

Jay Gandhi, Metrohm USA, Inc., Riverview, Florida, USA

Sara Giannini, Technology Development, Novartis Vaccines & Diagnostics S.r.L., Siena, Italy

Carl Grey, Department of Biotechnology, Lund University, Lund, Sweden

Paul R. Haddad, Australian Centre for Research On Separation Science (ACROSS), Pfizer Analytical Research Centre, School of Chemistry, Faculty of Science and Engineering, University of Tasmania, Hobart, Tasmania, Australia

Valoran P. Hanko, Thermo Fisher Scientific, Sunnyvale, California, USA

Melissa Hanna-Brown, Pfizer Global Research and Development Laboratories, Sandwich, Kent, UK

David S. Jackson, US Food and Drug Administration, Forensic Chemistry Center, Cincinnati, Ohio, USA

Petr Jandik, Thermo Fisher Scientific, Sunnyvale, California, USA

Shreekant V. Karmarkar, Baxter Healthcare, Round Lake, Illinois, USA

William R. LaCourse, Department of Chemistry and Biochemistry, University of Maryland, Baltimore County, Baltimore, Maryland, USA

Rong Lin, Thermo Fisher Scientific, Sunnyvale, California, USA

Xiaodong Liu, Thermo Fisher Scientific, Chelmsford, Massachusetts, USA

Claudia Magagnoli, Technology Development, Novartis Vaccines & Diagnostics S.r.L., Siena, Italy

Marcello Morelli, Technology Development Department, Novartis Vaccines & Diagnostics S.r.L., Siena, Italy

Milko Novič, Faculty of Chemistry and Chemical Technology, Aškerčeva 5, Ljubljana, Slovenia

Christopher Pohl, Thermo Fisher Scientific, Sunnyvale, California, USA

Ravi Ravichandran, Small Molecules Department, Documentary Standards Division, U.S. Pharmacopeial Convention, Rockville, Maryland, USA

Maria Rey, Thermo Fisher Scientific, Sunnyvale, California, USA

Jeffrey S. Rohrer, Thermo Fisher Scientific, Sunnyvale, California, USA

Stefano Ricci, Technology Development, Novartis Vaccines & Diagnostics, Siena, Italy

Rosanne Slingsby, Thermo Fisher Scientific, Sunnyvale, California, USA

Kannan Srinivasan, Thermo Fisher Scientific, Sunnyvale, California, USA

David Trowbridge, Alcon Laboratories, Fort Worth, Texas, USA

Philip Zakaria, Pfizer Analytical Research Centre, Australian Centre for Research on Separation Science, School of Chemistry, University of Tasmania, Hobart, Tasmania, Australia

PREFACE

Ion chromatography (IC) is a form of HPLC, which involves separation based on ion exchange or ion exclusion followed by detection of analytes using a suitable detector. Typically, electrochemical detectors (ED), such as conductivity or pulsed amperometry detectors, are used, however, conventional detectors, e.g., UV/VIS, RI, or more complex types of detection systems, e.g., mass spectrometry, also can be employed.

Since its introduction in the mid-1970s, IC has developed into an important analytical tool in a number of applications in pharmaceutical and biotechnology industries. Over the last two decades a significant number of IC methods have been successfully developed and validated for the characterization, lot-release, and stability studies of pharmaceutical and biological products. IC has also been used on-line as a tool for process monitoring. The technique has been used widely for the analyses of amino acids, peptides, proteins, glycoproteins, carbohydrates, antibiotics, vaccines, and other products. It has been successfully applied to the analysis of raw materials, bulk active ingredients, counterions, impurities and degradation products, excipients, diluents, and at different stages of the production process, as well as for the analysis of production equipment cleaning solutions, waste streams, container compatibility, and other applications.

With the increasing demand for newer types of therapeutic products (new molecular entities), use of different counter ions to improve stabilities and solubility properties of pharmaceutically active drug molecules, availability of mixed mode columns, and use of detection systems that have higher tolerance for organic solvents, there has been a growing interest in the application of IC in the pharmaceutical industry. Advances in technology have permitted IC to be coupled with mass spectrometry, which is a critical milestone for a wider interest in IC as a product characterization tool. Furthermore, because the principle of operation of IC is different from that of the other forms of chromatography, it can be the method of choice where other forms of chromatography are not suitable, e.g., for ionic and highly polar compounds, inorganic ions, and molecules that do not have suitable chromophores for detection by absorption measurement. The sample preparation often requires minimum pretreatment, generally as simple as dilution with water. When used together with ED, IC requires no pre- or post-column derivatization. In addition, the technique provides the ability to detect analytes with a high degree of selectivity through a combination of on-column resolution and the response selectivity of electrochemical detectors. As an orthogonal method to not only chromatographic and electrophoretic techniques, but also techniques such as AA and ICP, IC can play a critical role in method development and validation. The current regulatory focus on product characterization encourages industry researchers to search for alternative forms of analyses using different types of technologies, which can measure different characteristics of the products, their components, and drug-matrix and drug-container interactions. The interest in the pharmaceutical and biotechnology industries for IC is still growing and is expected to grow steadily in the foreseeable future.

Unlike reversed-phase HPLC, which uses organic solvents, IC generally employs dilute acids, alkalis, or salt solutions as eluents with little or no organic solvent, and as such the eluents are less toxic and less costly to use and dispose of. The recent increase in environmental consciousness together with cost considerations is expected to contribute toward greater interest in IC. Furthermore, with the expiration of key patents a few years ago, manufacturing of IC systems and their components are competitive now. The open competition is expected to result in improved instrumentation and column characteristics, availability of columns suitable for diverse applications, and wider customer support.

The number of USP-NF monographs that include IC-based procedures has grown significantly over the last decade. In 2006, USP-NF has added a new general information chapter on IC (<1065>) and a new general chapter on an IC-based procedure (<345>), illustrating the growing acceptance of IC as one of the methods of choice in the pharmaceutical industry.

No book has been published yet specifically on the application of IC in the analysis of pharmaceutical and biological products and their ingredients. Applications have been presented mostly in peer reviewed journals, scientific review articles, application notes of instrument and column manufacturers, trade publications, and scientific conferences. Although the technology has been around for many years, given the recent interest the time is now appropriate to produce a book on this topic. An edited volume with contributors from academia, pharmaceutical and biotechnology industries, and instrument manufacturers, from North America, Europe, and Australia provides different perspectives, experience, and expertise of the leaders in the field in a comprehensive manner in one place.

The book is designed as an introduction of IC to the beginners and a reference guide for experienced scientists and investigators in the pharmaceutical and biotechnology industry, working in the research, product development, and quality control areas. In addition, this book should be of interest to the student of pharmacy, pharmaceutical and biotechnology science, and academic researchers as a learning tool as well as a comprehensive reference.

We wish to thank all authors and reviewers who have contributed to this book. We recognize that this book would not have been possible but for their hard work. Special thanks go to Dr. Shreekant Karmarkar of Baxter Healthcare for his thoughts and helpful suggestions in the conception and the design of this compilation.

Lokesh Bhattacharyya
Jeffrey S. Rohrer

PART I

PRINCIPLES, MECHANISM, AND INSTRUMENTATION

1

ION CHROMATOGRAPHY—PRINCIPLES AND APPLICATIONS

Lokesh Bhattacharyya

Division of Biological Standards and Quality Control, Office of Compliance and Biologics Quality, Center for Biologics Evaluation and Research, Food and Drug Administration, Rockville, MD

1.1 INTRODUCTION

Ionic methods of separation have been used since the industrial revolution in Europe to reduce hardness of water. In the mid-nineteenth century, British researchers treated various clays with ammonium sulfate or carbonate in solution to release calcium. In the early twentieth century, zeolite columns were used to remove interfering calcium and magnesium ions from solutions to permit determination of sulfate. Ionic separation procedures were used in the Manhattan project to purify and concentrate radioactive materials needed to make atom bombs. Peterson and Sober [1] reported in 1956 a chromatographic method based on ion exchange to separate proteins. However, ion chromatography (IC), in its modern form, was introduced in 1975 by Small et al. [2]. The technique has since gained significant attention for the analysis of a wide variety of analytes in pharmaceutical, biotechnology, environmental, agricultural, and other industries. Several books and chapters on IC have provided a detailed review of its principles and instrumentation [3–5]. In 2000, United States

Applications of Ion Chromatography for Pharmaceutical and Biological Products, First Edition.
Edited by Lokesh Bhattacharyya and Jeffrey S. Rohrer.

Pharmacopeia-National Formulary (USP-NF) had only a few monographs that described test methods involving IC [6] and no general chapter on this technique. However, the number of monographs that include one or more IC-based test procedures has increased dramatically in the last 10 years. In addition, the current USP-NF [7] contains two general chapters on IC (<345> and <1065>) and at least four general chapters that include IC-based test methods (<1045>, <1052>, <1055>, <1086>), indicating its importance as a chromatographic technique for the analysis of pharmaceutical drug substances, products and excipients. In General Chapter <1065>, entitled "Ion Chromatography", USP-NF describes ion chromatography as "a high-performance liquid chromatography (HPLC) instrumental technique used in USP test procedures such as identification tests and assays to measure inorganic anions and cations, organic acids, carbohydrates, sugar alcohols, aminoglycosides, amino acids, proteins, glycoproteins, and potentially other analytes" [7].

This chapter will present an introduction to IC providing an outline of its principles and applications in the analysis of active and inactive ingredients, counter-ions, excipients, degradation products, and impurities relevant to the analysis of pharmaceutical, biologic and biotechnology-derived therapeutic and prophylactic products.

1.2 WHAT IS ION CHROMATOGRAPHY?

Modern IC is a form of HPLC, just as normal phase, reversed-phase and size exclusion chromatographies are different forms of HPLC. The separation in IC is based on ionic (or electrostatic) interactions between ionic and polar analytes, ions present in the eluent, and ionic functional groups derivatized to the chromatographic support. This can lead to two distinct mechanisms of separation—(a) ion exchange due to competitive ionic binding (attraction), and (b) ion exclusion due to repulsion between similarly charged analyte ions and the ions derivatized on the chromatographic support. Separation based on ion exchange has been the predominant form of IC to-date. In addition, chromatographic methods in which the separation due to ion exchange or ion exclusion is modified by the hydrophobic characters of the analyte or the chromatographic support material, by the presence of the organic modifiers in the eluent or due to ion-pair agents, resulting in better resolution that were not achieved otherwise, have gained popularity recently (mixed mode separation).

Numerous studies have been conducted in the last 30 years to understand the details of the mechanisms of ion-exchange and ion-exclusion chromatographies and the effect of different elution parameters, including flow rate, salt concentration, pH, presence of organic solvents, and temperature, on them. The current chapter is not meant to provide a comprehensive review of the studies. Rather, it is meant to provide a general introduction to both types of IC explaining in a qualitative non-mathematical approach how they work, what types of analytes are suitable for separation by ion-exchange and ion-exclusion chromatographies, and the effect of different factors on their performance.

1.3 ION-EXCHANGE CHROMATOGRAPHY

Ion-exchange chromatography involves separation of ionic and polar analytes using chromatographic supports derivatized with ionic functional groups that have charges opposite that of the analyte ions. That is, a column used to separate cations, called a cation-exchange column, contains negatively charged functional groups. Similarly, an anion-exchange column, which separates anions, is derivatized with positively charged functional groups. Ion-exchange chromatography has been widely used in the analysis of anions and cations, including metal ions, mono- and oligosaccharides, alditols and other polyhydroxy compounds, aminoglycosides (antibiotics), amino acids and peptides, organic acids, amines, alcohols, phenols, thiols, nucleotides and nucleosides, and other polar molecules.

The analyte ions and similarly charged ions of the eluent compete to bind to the oppositely charged ionic functional group on the surface of the stationary phase. Assuming that the exchanging ions (analytes and ions in the mobile phase) are cations, the competition can be represented by the following scheme:

$$S - X^-C^+ + M^+ \leftrightarrow S - X^-M^+ + C^+ \tag{1}$$

In this process, the cation M^+ of the eluent exchanges for the analyte cation C^+ bound to the anion X^- derivatized on the surface of the chromatographic support (S). If, on the other hand, the exchanging ions are anions, it is called anion-exchange chromatography and is represented as:

$$S - X^+A^- + B^- \leftrightarrow S - X^+B^- + A^- \tag{2}$$

in which, the anion B^- of the eluent exchanges for the analyte cation A^- bound to the positively charged ion X^+ on the surface of the stationary phase. The adsorption of the analyte to the stationary phase and desorption by the eluent ions is repeated as they travel along the length of the column, resulting in the separation due to ion-exchange [8].

1.3.1 Mechanism

The mechanism of the two processes, cation exchange and anion exchange, are indeed, very similar. In the first step of the process, analyte ions diffuse close to the stationary phase and bind to the oppositely charged ionic sites derivatized on the stationary phase through the Coulombic attraction. The Coulombic force of interaction (f) between the two ions in solution, in its simplified form, is given by the equation,

$$f = q_1q_2/\varepsilon r^2 \tag{3}$$

in which q_1 and q_2 are charges on two ions, ε is the dielectric constant of the medium, and r is the distance between them. In most of the ion chromatographic separations, except when organic solvents are included as modifiers, the medium is water (solutions of acids, alkalis or salts). Therefore, we can consider ε to be a constant. If the charges on both ions are similar (either both positive or both negative), the force is repulsive. Where they are dissimilar (one positive and the other negative), the force is attractive. We need to remember two basic principles of thermodynamics to understand the mechanism. (1) Attractive force between two oppositely charged ions results in decrease in enthalpy (H) and free energy (G). (2) The thermodynamic principles favor the process in which the free energy change is negative.

In a column, the bound analyte ions face competition from similarly charged ions present in the eluent as they compete for binding to the same oppositely charged ionic sites of the stationary phase. For example, the negatively charged analyte ions and the negative ions present in the eluent both compete for the positively charged sites on the stationary phase. Overcoming binding due to the ionic attraction between negatively charged analyte ions and the positively charged ionic site of the stationary phase requires 'work' and leads to an increase in free energy (and enthalpy) of the system and, as such, is not thermodynamically favorable. However, the increase is overwhelmingly compensated by the decrease in free energy (and enthalpy) due to the binding of the negative ions of the eluent because the concentration of the negative ions of the eluent is overwhelmingly greater than that of the analyte ion concentration. To illustrate this with a simple example, the typical concentration of an eluent in IC ranges between 10–100 mM (in some cases, as low as 1 mM or as high as 500 mM). However, the typical concentration of each analyte is in the micromolar to sub-micromolar range. Thus, the concentration of the eluent ion is 10^4–10^5 fold higher than that of the concentration of the analyte ion. The energy input needed to displace an analyte ion from the stationary phase is significantly less than the energy released due to attractive interactions between the stationary phase ion and the overwhelmingly larger number of ions in the eluent resulting in a decrease of free energy and the overall process is thermodynamically favored.

When ionic or polar analytes enter an ion-exchange column, they first bind to the charged sites of the stationary phase in a layer. As different amounts of energy are needed to unbind different analytes from the stationary phase, due to differences in charge density and other factors (see later), the desorption takes place at a different rate and/or requires different concentrations of eluent ions. This leads to separation of the analytes—the analyte requiring lesser energy is desorbed (eluted) earlier from the stationary phase. This adsorption-desorption phenomenon continues from layer to layer as the analytes travel along the length of the chromatographic column, increasing separation between the analytes (Figure 1.1). In an optimized separation procedure, the analytes are resolved when they exit the column.

Equation (3) predicts that the force of attraction between a monovalent analyte ion with one unit of charge (e.g., chloride) and an ionic site on the stationary phase will be lesser than that between a divalent analyte ion (e.g., sulfate), which has two units of charge, and the same stationary phase ionic site. Thus, a higher concentration of eluent ion will be necessary to displace a divalent ion from the stationary phase than that required to displace a monovalent ion, resulting in a separation of the two by IC, and the monovalent ion will be eluted from the column earlier than a divalent ion.

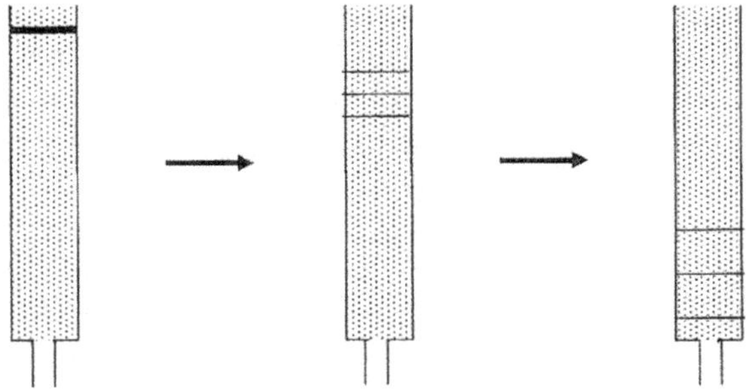

Figure 1.1. A schematic diagram of separation of analytes by ion-exchange chromatography.

Similarly, a trivalent ion will bind the stationary phase more strongly than a divalent ion and will be eluted from the column after the divalent ion.

The above discussion, however, does not explain separation of monovalent ions from an ion exchange column. It is conceivable that we should consider the charge density on the surface of an ion rather than its actual charge, since the ions, particularly those of interest in the analysis of pharmaceutical drugs, are not point masses and the underlying assumption of equation (3) is that the charges are points. A larger monovalent ion (e.g., chloride) will have less charge density than a smaller monovalent ion (e.g., fluoride), since both have a total of one unit of charge. Thus, fluoride ion is expected to bind more strongly on a stationary phase than chloride, require a higher eluent concentration to displace, and elute later from the column. So, when a mixture of fluoride, chloride and bromide is chromatographed on an IC column, bromide is expected to be eluted first (being the largest and therefore having the lowest charge density among the three ions), then chloride and then fluoride. In reality, however, the elution order is found to be reversed. For example, when a mixture of different anions are eluted from an IonPac AS11 column with sodium hydroxide [9], fluoride ion is eluted first, then chloride and then bromide, that is, in the reverse order of what is expected based on the charge density. In fact, the results from the same example show that when a mixture of fluoride, chloride, bromide, nitrate, acetate, and benzoate, all of which are monovalent ions, are eluted from an IonPac AS11 with sodium hydroxide [9], the elution sequence of the ions is,

$$\text{Fluoride} > \text{acetate} > \text{chloride} > \text{bromide} > \text{nitrate} > \text{benzoate} \qquad (4)$$

With the exception of acetate, it appears that a smaller ion is eluted earlier than a larger ion. Similarly, when a mixture of trivalent ions, phosphate and citrate, are eluted from an IonPac AS11 column with sodium hydroxide, the less bulkier phosphate ion is eluted before the bulkier citrate ion [10]. That is, the elution sequence is the reverse of what is expected based on their charge densities.

It is of interest to note that the sequence in which these ions are eluted from the column closely resembles the Hofmeister series (or the lyotropic series) [11]. It is conceivable that the mechanism of separation is somehow related to the mechanism that led to the Hofmeister series [12]. The binding of the analyte ions to the ions on the stationary phase followed by competitive desorption by similar ions present in the eluent, as discussed above, indeed, represent only part of the overall process. Water molecules play a very critical role in the overall process.

An ion in aqueous solution (or for that matter in solution of a polar solvent) does not exist as a free ion. It is hydrated (or generally speaking solvated) with several molecules of water (or solvent). The hydration extends over several layers of water molecules, primarily through coordinate bond formation, formation of hydrogen bonds, and Van der Waals type ion-dipole and dipole-dipole interactions, depending on the nature and charge of the ions, forming a hydration sheath around each ion. The thickness of this sheath is roughly proportional to the charge density of the ion. The water molecules of the sheath interact with the molecules of the bulk water through ion-dipole and dipole-dipole interactions and thereby become part of an overall water structure. Thus, when an eluent ion binds to the stationary phase, it has to free itself from this structure. While free energy (G) is reduced due to the attractive binding between the oppositely charged ions, a considerable amount of free energy is required to break the water structure. However, the ion that was exchanged out of the stationary phase due to the above binding has the same charge as the ion that exchanges in. The former ion immediately forms its own water structure in the solution. While energy needs to be put in to unbind the ion, a significant amount of free energy is released due to the formation of the water structure. Schematically, the overall process can be described as:

Destruction of water structure of the eluent ion	$\longrightarrow$	Increase in G
Binding of the eluent ion to the stationary phase	$\longrightarrow$	Decrease in G
Unbinding of an analyte ion from the stationary phase	$\longrightarrow$	Increase in G
Formation of the water structure around the analyte ion	$\longrightarrow$	Decrease in G

The overall change in free energy is a combination of the free energy changes of the individual steps. A smaller ion will have a high charge density. So, it will be able to form a significantly extended water structure around it resulting in a large decrease in free energy. Thus, a smaller monovalent ion (e.g., fluoride) is eluted from the column earlier than a larger monovalent ion (e.g., chloride) because of a larger reduction of free energy as a result of extended hydration around it. Oxygenated ions such as acetate can form a significantly thicker hydration sheath around it than is expected from its charge density. The oxygen atoms present in these ions can form strong hydrogen bonds with hydrogen atoms of water in the initial layer. Subsequent layers of hydration are formed through hydrogen bonding among the water molecules as well as due to strong ion-dipole and dipole-dipole interactions. Such ions in solution can form a very stable structure permitting a large decrease in the free energy. Thus, even though acetate ion is bulky it is eluted earlier from the column than the chloride and bromide ions, which are smaller than acetate.

1.3.2 Eluent

Typically the eluents used in ion exchange chromatography are acids, alkalis or salt solutions, and do not contain an organic solvent (however, see later). The extremes of pH conditions offered by acids or alkalis help ionize polar molecules into ions. An excellent example is the ionization of neutral sugars and alditols under the high pH conditions, typically 10–500 mM sodium hydroxide, used in High Performance Anion Exchange Chromatography (HPAEC). However, such applications will require analyte molecules to be stable in the acid or alkali used as the eluent. This sometimes limits the application of IC in the analysis of pharmaceutical drugs because the analyte may not be stable under the extreme pH conditions of acids or alkalis. If the analyte molecules are ionic or strongly polarized, elution by salt solutions or buffers of controlled pH conditions, often provide an excellent opportunity for separation by IC. [Using acids or alkalis as eluents has an additional advantage, when suppressed conductivity detection is used. This will be discussed later.]

The elution can be isocratic or with increasing salt concentrations, either by batch or gradient elution, or by altering pH of the eluent. Less tightly bound ions are eluted initially; more tightly bound analytes are eluted either under altered elution conditions (e.g., higher salt concentration or different pH) or simply later, resulting in separation. When gradient elution is used, the peak is expected to be slightly asymmetric and the tailing factor [7] is expected to be greater than 1. As an analyte band travels through the column (Figure 1.1), the eluent behind it has a concentration higher than the concentration at which it is eluted. So, the back of the band cannot bind to the column but can diffuse through the eluent. However, the eluent concentration at the front of a band is lower than the concentration at which it is eluted. It, therefore, binds to the column and its diffusion is restricted.

Changing eluent pH can change the ionic characters of the analytes and/or the functional groups on the chromatographic support. Thus, an anion may become less ionic at a lower pH. However, the actual ionic character depends on the pK_a of the acid containing the anion (A^-), which is the negative logarithm of the equilibrium constant of the following equilibrium:

$$A^- + H^+ \leftrightarrow HA \tag{5}$$

The further the elution pH is from the pK_a, the more ionic it will be. Thus, the anion with a lower pK_a value (more acidic) will be eluted after an anion with a higher pK_a value (less acidic). Similarly, a cation having a lower pK_b value (more basic) will be eluted after a cation with a higher pK_a value (less basic).

1.3.3 Organic Solvents

Sometimes small quantities of organic solvents (organic modifier) are added to IC eluent to achieve better separation, to reduce hydrophobic interaction with the column packings, and for improving chromatographic/peak parameters (e.g., theoretical plate, resolution, peak shape). We now need to consider the ε term used in Equation 3 above to understand the effect of organic modifiers. The dielectric constant of water is around 80 at 20°C. The value of this parameter is below 50 for most of the organic solvents. Thus, when organic solvents are added to an aqueous eluent, the dielectric constant of the medium is decreased. This results in a tighter binding of the analyte and eluent ions to the stationary phase because this term appears in the denominator in Equation 3, which alters the elution pattern.

Inclusion of organic solvents also affects the formation of water structure around an ion by (a) altering the forces of ion-dipole and dipole-dipole interactions and hydrogen bonding due to altered dielectric constant, and (b) interferes with the formation of water structure by inserting itself into the structure. The forces of ion-dipole and dipole-dipole interactions, which, in turn, also affect hydrogen bond formation, are governed by the Coulomb's Law of interaction (Equation 3). The force of such interaction is, thereby, altered by the inclusion of organic solvents. However, the impact will not be significant when a small quantity of organic solvent is used.

The polar organic solvent molecules, particularly those containing oxygen atoms, also enter into the hydration sheath by forming hydrogen bonds. However, they cannot form as extensive a hydrogen bond network as water due to the hydrophobic nature of such molecules and their larger size, thereby weakening the water structure. Thus, less free energy is needed to break such structures as an eluent ion binds to the stationary phase. Similarly, there is a lower reduction of free energy when the analyte ion is released into the eluent.

Inclusion of an organic solvent also reduces the effect of hydrophobic association between the analyte molecules and the stationary phase. In particular, when the analyte has a significant hydrophobic surface, as is the case for many pharmaceutical drugs, it often shows a broad peak in IC due to its interaction with the hydrophobic surface of the chromatographic support. Inclusion of a small quantity of organic solvent often results in sharper peaks thereby improving peak characteristics and other chromatographic parameters (e.g., resolution) by reducing the effect of hydrophobicity.

1.3.4 Other Factors

The dissociation constants of analytes vary with temperature, although the extent of variation is usually small. This does not have any effect on the chromatographic profile, where the analytes are fully ionized under the conditions of chromatography. However, the retention times of analytes that are not fully ionized will vary slightly with temperature. This variation does not pose a significant problem because samples relevant to pharmaceutical applications are usually run with a reference standard. Thus, ion-exchange chromatography is typically run under ambient or near ambient temperatures. Similarly, pressure does not affect elution profiles, as the effects of pressure on dissociation constants are negligible. However, the columns should be operated at their optimum operating pressures (or pressure range) to maintain high performance.

Since ion-exchange chromatography involves binding and unbinding of analyte ions to charges on the surface of the chromatographic support, it is critical that analyte ions are able to diffuse to the chromatographic support to bind to it and diffuse away from the support when desorbed. Therefore, the flow rate must be such as to permit diffusion of the ions. This is usually not a problem for smaller ions, as their diffusion rates are high. Larger ions may need more time. In most cases, a flow rate of 0.5–2.0 mL per minute is sufficient to meet this condition. Anomalies have been observed when higher flow rates are used due to incomplete binding and desorption.

1.4 ION-EXCLUSION CHROMATOGRAPHY

Introduced by Wheaton and Bauman in 1953 [13], Ion-exclusion Chromatography uses strong cation- or anion-exchange chromatographic supports to separate ionic, polar, weakly polar, and apolar analytes, and has been used in the analysis of organic acids, alcohols, glycols and sugars. In contrast to ion-exchange chromatography, the charge on the functional groups on the chromatographic support is the same as the charge on the analyte ion. That is, to separate negatively charged or negatively polarized analytes, the chromatographic supports are derivatized with negatively charged functional groups (typically, sulfonate). Similarly, analytes with positive charge or polarity are separated using a chromatographic support that carries positive charges (most frequently, quaternary ammonium ions).

1.4.1 Mechanism

Although the actual mechanism of separation is not fully understood, it is widely held that the separation is effected by partition of analytes between the stationary phase and the mobile phase across a hypothetical semipermeable Donnan membrane. This theory will be discussed briefly in this chapter. An alternate explanation is presented in Chapter 2 of this book.

Water molecules bind to the ionic functional groups of the chromatographic support through coordination, hydrogen bond, and Van der Waals type ion-dipole interaction forming hydration spheres around the functional groups. Water molecules in this hydration sphere and also those trapped in the interstitial spaces (and pores) of the resins are immobilized around the chromatographic support forming the stationary phase system. As a fully ionized analyte in the mobile phase approaches a stationary phase containing like charges (e.g., chloride ion approaching a stationary phase around a sulfonate-derivatized resin), it is strongly repelled by the similar charge. The repulsion is Coulombic and the repulsive force is given by the Equation 3 above. The repulsive force increases rapidly as the ionic analyte approaches the stationary phase because the Equation 3 contains the r^2 term in the denominator. The repulsion does not permit the ionic analytes to come more than a certain distance from the stationary phase system forming the outer surface of the hypothetical Donnan membrane (Figure 1.2) and such analytes elute from the column without being retained.

When an apolar molecule approaches the same stationary phase, it experiences no repulsion as the q term corresponding to an apolar molecule in Equation 3 is zero. So, it can freely penetrate deep into the immobilized water layer, which permits it to stay longer in the column. Such molecules partition back and forth at different layers as it they travels along the length of the column. Thus, an apolar analyte is

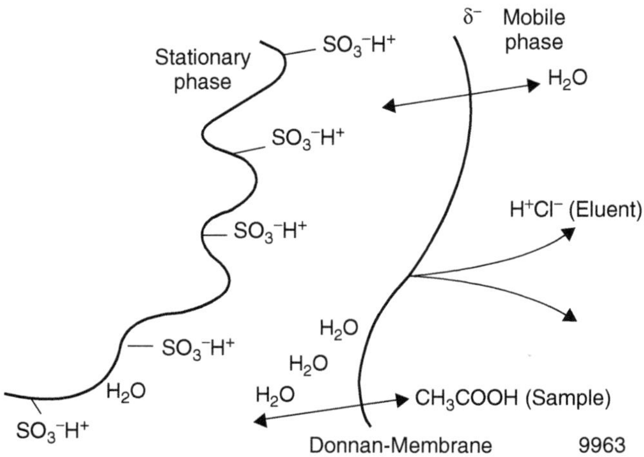

Figure 1.2. A schematic diagram of the formation of the hypothetical Donnan membrane and mechanism of separation by ion-exclusion chromatography. (Reproduced from Application Note 106, with permission from Dionex, Inc.)

eluted from the column well after ionic and polar analytes. A polar analyte, which has partial separation of charges within the molecule (forming a dipole), experiences less repulsion than an ion but more than an apolar molecule. Thus, the degree of penetration of such an analyte is in between an ion and an apolar molecule and it is eluted from the column in between ionic and apolar analytes.

It is also clear from Equation 3 above that the force of repulsion experienced by a polar analyte depends on its dipolar character. An analyte that is more polar has more ionic character, thus, experiences greater repulsion and, therefore, will penetrate less into the stationary phase and will be eluted earlier from the column, compared to a less polar analyte. Thus, less and less polar molecules elute later and later from the column and an apolar molecule elutes at the end resulting in separation.

However, it appears that the partition mechanism does not fully explain many of the separations achieved by ion-exclusion chromatography. Additional mechanisms seem to play some role in the process (see Chapter 2 of this book).

Hydrophobic Properties of analyte molecules play an important role in the separation. Molecules with extended hydrophobic surface are retained longer in the column due to stronger hydrophobic association with the stationary phase system. For example, the elution times of aliphatic carboxylic acids become longer as the length of the alkyl groups increases [14]. The elution order of a mixture of the first three aliphatic carboxylic acids is:

$$\text{formic acid} > \text{acetic acid} > \text{propionic acid.}$$

Calculations based on their pK_a values indicate that these three aliphatic carboxylic acids are strongly ionized in solution (60–97%). Thus, they should come out close to the void volume of the column based on the partition mechanism discussed above. Although formic acid is eluted close to the void volume, the other two are eluted later. Similarly, higher aliphatic amines (e.g., butylamine, pentylamine, diethylamine) show longer elution time due to the hydrophobic character of their long aliphatic chains. The elution times are reduced and the peak shapes are considerably improved when an organic solvent is included in the mobile phase [15].

$\pi-\pi$ interaction also plays a role in the separation by ion exclusion chromatography when the support contains a double bond or an aromatic ring (e.g., polystyrene). For example, acrylic acid, which contains a double bond, elutes after propionic acid. Aromatic acids, which contain a benzene ring show long retention time on the column [14].

Hydrogen bonding is an important factor, particularly in the separation of molecules that contains several hydroxyl groups, e.g., carbohydrates. These molecules are retained longer by the stationary phase, presumably due to hydrogen bonding with the hydration sphere of the stationary phase system.

Steric factors also play a role in ion-exclusion chromatography. Molecules with bulkier groups are excluded earlier. For example, a dicarboxylate (e.g., oxalate) is eluted earlier than a monocarboxylate (e.g., acetate) when eluted with 7.5 mM sulfuric acid. An iso-carboxylic acid (e.g., iso-butyric acid) is eluted earlier than the corresponding normal carboxylic acid [14].

Complexation with the positive counter-ion of the chromatographic support also plays a role in the separation of analytes containing hydroxyl groups (e.g., sugars). Calcium and lead forms of a cation-exchange resin are often used to separate neutral monosaccharides.

1.4.2 Eluent

Based on the partition mechanism discussed above, it is conceivable that deionized water can be used as the eluent during ion-exclusion chromatography. However, several problems have been encountered [14–16]. Although water is found suitable for the resolution of very weak acids, such as carbonic and boric acids, or very weak bases, strong or even moderately strong acids and bases are too ionized in water to be separated. They are not retained sufficiently due to their high degree of ionization and are eluted within the void volume or close to the void volume without adequate resolution. Secondly, the peaks are often fronted, broad, and/or significantly tailed, due to factors other than pure partition mechanism described above. Typically, dilute solutions of strong acids and alkalis are used in the separation of anionic (e.g., carboxylic acids) and cationic (e.g., amines) solutes, respectively, to overcome the problem. Sulfuric, hydrochloric and aliphatic sulfonic acids are widely used. The strong acids suppress ionization of carboxylic acids permitting them to be resolved. Phosphoric acid and perfluorobutyric acid have been used successfully for the separation of weaker acids. Amines are separated using dilute alkalis, such as sodium hydroxide. It is interesting to note that eluents of the same pH, when used with the same stationary phase, produce very similar chromatographic profiles, irrespective of the nature of the acid used as the eluent. The choice of actual acid to be used as eluent, therefore, is often determined by the detection system to be used.

Sometimes, addition of organic solvents to aqueous eluents leads to reduction of run time, sharper peaks and higher resolution because organic solvents minimize the hydrophobic effects. The organic solvent to be used and its concentration are determined by its compatibility with the detection system.

1.4.3 Other Factors

Ion-exclusion chromatography is usually run at ambient temperature, however, higher resolution is obtained at an elevated temperature because the partition rate is increased and the hydrophobic effect is reduced. In some cases, pure water is used as eluent at 60–80°C. [However, note that many analytes, including almost all proteins and some of the pharmaceutical drug molecules, are not stable at such a high temperature.] The efficiency of separation increases with decreased flow rate because it is necessary to permit sufficient time to the analyte molecules to diffuse into the hydration sphere of the stationary phase system to achieve optimal separation. A flow rate in the range of 0.3–0.5 mL/min is recommended for most separations. Ion-exclusion chromatography requires columns that are usually large in size, typically 30 cm, because a considerable volume of chromatographic support material is necessary to provide sufficient occluded liquid to obtain a stationary phase that permits separation of solutes of similar characteristics.

1.5 INSTRUMENTATION

Figure 1.3 shows a schematic of the set up of an IC system. An examination of the figure shows that the set up closely resembles that of a typical HPLC system. The components include an autosampler, a high-pressure pump, an injection valve with sample loop of suitable size (typically, 10–250 μL), a guard column, an analytical column, an optional suppressor or a post-column reagent mixing system, a flow-through detector, and a processing system ranging from a data-processing integrator to a computerized system management unit, which contains software to run the system using pre-programmed method and schedule (sequence) files, perform data acquisition and processing to crunch out the final results.

Since the mobile phase generally contains dilute acids, alkalis or salt solutions, the components in contact with mobile phase are typically made of a completely metal free inert material, such as polyetheretherketone (PEEK). A conventional HPLC system also may be used provided that its components are made of materials that are compatible with the mobile phase. Following suitable preparation, the sample is introduced through the injection valve. After optional chemical suppression or other post column treatment of the effluent, the analyte is detected using a suitable detection system (see later). Because IC typically uses an ionic mobile phase, a suppression of background conductivity of the eluent is often necessary prior to conductometric detection, when such a detector is used, although nonsuppressed conductometric detection has been

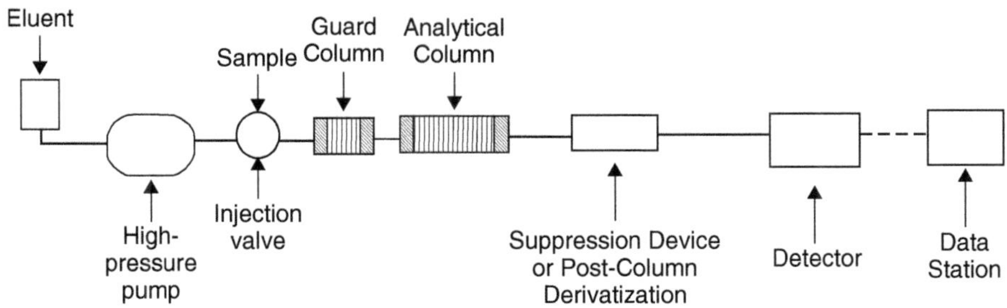

Figure 1.3. A schematic diagram of the set-up and components of a typical IC system. (Adapted from USP-NF General Chapter <1065> with permission.)

used in pharmaceutical analysis, particularly when water, weak acids or weak bases are used as eluents, as is common in ion-exclusion chromatography.

A detailed description of each of the individual components of an IC system is beyond the scope of this chapter. Furthermore, with the exception of the detector system, including the suppressor, and the need to have metal-free components for most IC applications, the components are no different from those used in a traditional HPLC system. A brief discussion on the suppressor and the detectors used in IC is provided below.

1.6 DETECTION

Any suitable detector can be used for the detection and quantitation of analytes by IC. The choice of detector depends upon the nature of the analyte molecules. This may include the universal refractive index (RI) detector, UV detector for analytes that absorb UV, fluorescence detector for analytes that contains fluorophores, or radiochemical detectors, where appropriate [*cf*. 7]. However, traditionally, IC is associated with electrochemical detectors. So, only a discussion of the electrochemical detector systems is included in this chapter. It is not the intention of this chapter to suggest that other types of detectors should not be used with IC. Indeed, they should be, if the application dictates. However, the ability of electrochemical detectors is less appreciated in the pharmaceutical industry, presumably because mechanisms of action of these detectors are less understood compared to those of the traditional photometric detectors mentioned above.

Two types of modern electrochemical detectors are widely used in IC—conductivity (suppressed and nonsuppressed) and pulsed amperometry.

1.6.1 Conductivity Detection

When a constant voltage is applied across two electrodes between which the effluent from a column flows, a current is generated because the effluent contains ions or polar molecules. The strength of the current is proportional to the conductivity of the solution, which, in turn, is proportional to the concentration of ionic species in solution and their ion conductances. The concentration is the number of ions carrying electricity. The ion conductance of an ion determines its ability to carry electricity. The ions present in effluent provide the background (baseline) conductivity of a chromatographic profile. The additional conductivity due to an analyte ion or a polar molecule, when they are present in the effluent, provides the peak, which is proportional to its concentration. Different analytes at the same concentration show different peak areas (or peak heights) due to the difference in their ion conductances.

The problem, however, is that the conductivities of effluent solutions are often significantly higher than the conductivities of the analytes, simply because, as mentioned above, the concentrations of ions in effluent are 10^4–10^5 higher than that of the analytes, particularly in ion-exchange chromatography. Thus, early attempts to apply conductivity measurement to IC had significant limitations.

1.6.1.1 Suppressed and Nonsuppressed Conductivity Detections. This limitation was overcome when Small et al. [2] introduced the concept of *suppressed* IC. Small et al. used a packed-bed suppressor in the hydroxide form to achieve sensitive detection of the ions by chemically modifying the effluent before it enters the conductivity detector. The suppression was achieved by converting the mineral acid eluent to water and thereby obtaining a very low background signal and low noise, while converting the analyte to its base form, which is fully dissociated and actually carries more current than the analyte itself, thereby increasing the sensitivity of the detection (see later). In this system, the effluent containing HA (A being the anion) passes through the suppressor that exchanges A^- for OH^- to produce water, which does not conduct electricity. Noise is proportional to the background signal and elimination of the background electrolyte lowers the noise, provides more stable baseline and improves analyte sensitivity. However, in 1979, Gjerde et al. [17] reported an IC method in which the analytical column is directly linked to a conductivity detector without any suppressor. The methods employed a low capacity analytical column and dilute solutions of weak acids or bases as eluents to achieve low background signals.

The question then is, to suppress or not to suppress. The conductivity of an electrolyte, MX, is given by the following equation:

$$C = c_{MX}\Lambda_{MX} = c_{MX}(\lambda_M + \lambda_X) \quad (6)$$

where C is the conductivity of the electrolyte, c_{MX} is the concentration of MX in Normality (N), Λ_{MX} is the equivalent conductance of the electrolyte MX, and λ_M and λ_X are equivalent ion conductances of M^+ and X^- ions, respectively (including their respective waters of hydrations). The ion conductances of a few common ions are shown in Table 1.1.

To understand suppressed and nonsuppressed detection, let us consider two identical cation-exchange chromatographic runs of the analyte MX using a strong acid, HA, as the eluent (where A is an anion), with the difference that in the first system the effluent first passes through a suppressor before entering the conductivity cell, whereas in the second system the effluent flows directly through the conductivity cell.

TABLE 1.1. Equivalent Ion Conductances of Common Ions

Cation	Eq. Ion Conductance (mho)	Anion	Eq. Ion Conductance (mho)
H^+	349.8	OH^-	198.0
K^+	73.5	Br^-	78.4
Na^+	50.1	Cl^-	76.3
Li^+	38.7	$HCO_3{}^-$	44.5
$NH_4{}^+$	73.4	$SO_4{}^{2-}$	79.8
Mg^{2+}	53.1	Acetate	40.9
Ca^{2+}	59.5	Propionate	35.8

In the second (nonsuppressed) system, the analyte signal is measured by the conductivity difference (ΔC) between MX and HA, given by the following equation:

$$\Delta C = c_M[(\lambda_M - \lambda_H) + (\lambda_X - \lambda_A)] \qquad (7)$$

where c_M is the concentration of M^+ (same as the concentration of MX) in the effluent. The change in conductivity in Equation (7) is around -300 ± 30 times c_{MX} because the equivalent ion conductance values of anions are within approximately ± 30 of each other, except when the anion is the hydroxyl ion (Table 1.1). However, hydroxyl ion cannot be used in a cation-exchange chromatography in which an acid is used as the eluent. The result is, therefore, a negative peak, which can be viewed as a positive peak by reversing the signal polarity of the detector. In the suppressed system, the MX first passes through the suppressor, in which both X^- and A^- are converted to OH^-. Most of the OH^- combines with H^+ of the acidic eluent to form water. Therefore, the analyte is now detected essentially in a background of pure water, resulting in a positive analyte response, given by the equation:

$$\Delta C = c_M(\lambda_M + \lambda_{OH}) \qquad (8)$$

The change in conductivity in Equation (7) is around 250 ± 20 times c_{MX} ($C_M = C_{MX}$) resulting in a positive peak (the equivalent ion conductances of cations are within range of approximately ± 20, except when the cation is H^+).

The above calculation appears to indicate that the nonsuppressed conductivity is about as much or more sensitive than the suppressed conductivity detection. However, we have not taken into consideration the difference in baseline conductivities. In suppressed conductivity, the eluent that passes through the detector is essentially pure water with the baseline conductivity approaching zero, compared to baseline conductivities of 1000–1500 μS when a strong acid is used as the eluent in a nonsuppressed detection system and lower when a weak acid is used because a weak acid is not fully dissociated. Thus, a peak equivalent to the conductivity of 250 times c_{MX} is observed against a background of essentially zero conductivity with suppressed conductivity detection. The same chromatography produces a peak equivalent to around 300 times c_{MX} in a background of 1000–1500 μS when nonsuppressed conductivity is used.

As the baseline conductivity is proportional to the concentration of ions in the eluent, it is critical to use dilute solutions of weak acids and bases when nonsuppressed conductivity detection is employed because they are slightly dissociated, even in dilute solutions. Consequently, it is necessary to use low-capacity ion-exchange columns. However, the capacity is much less of a factor while choosing a column when suppressed conductivity detection is used, because high eluent concentrations may be used without any significant change in the background conductance, as long as the suppressor capacity is not exceeded.

The baseline conductivities (noise) in a typical suppressed conductivity detection is found to be <0.5 nS (using strong acids or alkalis) while the same is ~10 nS (using weak acids or alkalis) with nonsuppressed detection. Thus, considering the above example,

$$\text{Signal-to-noise ratio for the suppressed system} = 250.c_{MX}/0.5 = 500.c_{MX}$$

$$\text{Signal-to-noise ratio for the nonsuppressed system} = 300.c_{MX}/10 = 30.c_{MX}$$

Thus, suppressed conductivity detection provides about an order of magnitude better signal-to-noise ratio than the nonsuppressed system.

Furthermore, Detection Limit and Quantitation Limit are related to the signal-to-noise ratio [*cf*. 18]. Thus, both validation parameters are expected to be an order of magnitude lower when suppressed conductivity detection is used compared to nonsuppressed detection, attributing greater detection and quantitation sensitivity to the former technique.

When gradient elution is used, the baseline changes continuously with nonsuppressed conductivity detection. This makes peak area (or height) measurement less accurate. The baseline does not change when gradient elution is used in conjunction with suppressed conductivity detection.

However, for analytes that form weak bases from the suppressor reaction, such as $NH4^+$, a nonlinear calibration curve has been observed. Thus, a quadratic curve fit is typically required for acceptable correlation of the calibration curve (see Chapter 4 of this book for more details). A linear calibration curve is observed using nonsuppressed conductivity detection.

1.6.1.2 Mechanism of Suppression. Although originally introduced by Small et al. [2], chemical suppressors are seldom used today. The suppressors that are widely used today operate electrolytically. The design and the mode of operation of electrolytic suppressors from different manufacturers vary to some degree in details but the basic mechanism of their operation is essentially the same, which will be discussed here. More recently suppressors have been developed which recycle the eluent back to the eluent delivery chambers, thereby resulting in reduction of the operating cost. These suppressors work only in conjunction with electrolytic eluent generation systems where the feed from the eluent chamber is water. The mechanism of operation of such suppressors is discussed elsewhere in this book (see Chapter 4).

To explain the mechanism of operation of electrolytic autosuppressors, let us consider anion suppression in the effluent from an anion-exchange column (Figure 1.4).

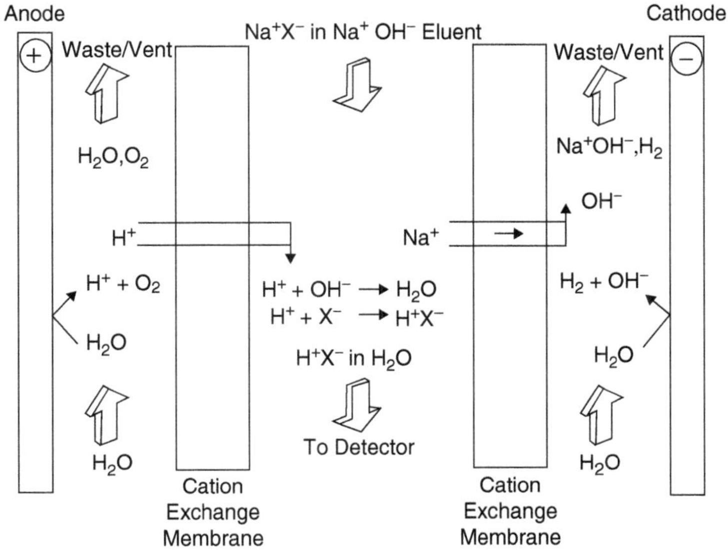

Figure 1.4. A Schematic diagram explaining the mechanism of ion suppression of an effluent from an anion-exchange column by an electrolytic autosuppressor.

The analyte, X^-, having Na^+ as the counter-ion, is eluted from the column by the eluent, sodium hydroxide. As shown in Figure 1.4, the effluent enters into the suppressor through the central chamber enclosed by a "semi-permeable membrane", which permits transfer of positive charges under the influence of an electric field but not transfer of negative charges (for an anion suppressor) nor transfer of material by diffusion. The central chamber has an anode chamber on one side and a cathode chamber on the other. Water is pumped into both the cathode and the anode chambers. When an electric field is applied, water in both chambers undergoes electrolysis. In the anode chamber, the electrolysis generates hydrogen ion and oxygen molecules. Similarly, hydroxyl ion and hydrogen is generated in the cathode chamber. Hydrogen ion travels across the membrane from the anode chamber into the central chamber and sodium ion moves out of the central chamber into the cathode chamber. Hydroxyl ion of the eluent binds to the hydrogen ion in the central chamber to form water. Sodium ion that has moved out of the central chamber is replaced by hydrogen ion that has moved in. The central chamber now contains H^+X^- (instead of Na^+X^-) in pure water, which moves to the detector. Thus, the acid form of the analyte X^- in pure water enters the detector and the eluent is converted to pure water, which provides essentially zero background.

Similarly, when a cation suppressor is used, the "semi-permeable membrane" permits transfer of negative charges only under an electric field and the analyte cation with hydroxyl counter-ion in pure water goes to the detector.

The eluent is converted to pure water only when the eluent is either an acid or an alkali. If a salt is used as an eluent, the anion will combine with the hydrogen ion produced by the electrolysis of water to form the corresponding acid when an anion suppressor is used (e.g., HCl if NaCl is used in the eluent). Similarly, the suppression will produce the hydroxyl form of the cation, if a cation suppressor is used. The suppression will not lead to near zero background under such conditions, however, the background could be still acceptably low if the acid form of the anion is a very weak acid or the hydroxyl form of the cation is a very weak base.

1.6.2 Pulsed Amperometric Detection

Used typically in combination with high-performance anion-exchange chromatography (HPAEC), pulsed amperometric detection (PAD) has proved to be a powerful tool in the detection of mono- and oligosaccharides, alditols, amino acids and peptides without requiring any sample derivatization.

At high pH, the analytes are oxidized at the surface of the gold electrode by the application of pulses of positive potentials. The current (hence amperometry) generated is proportional to the analyte concentration, which therefore can be detected and quantitated. When a single potential is applied to the electrode, oxidation products that deposit on the electrode surface gradually "poison" the electrode surface, resulting in loss of analyte signal. To prevent signal loss, the electrode surface is cleaned by a series of potential pulses that are applied for fixed time periods after the detection potential. Repeated application of a series of potentials designated E_1,E_2,E_3,. . ., over defined time periods t_1,t_2,t_3,. . . constitutes the basis of pulsed amperometry. The series of potentials applied for defined time periods is referred to as a waveform.

The potential E_1 applied over the time period, t_1, is subdivided into two time periods related to two functions. In the initial part, called the delay period (t_d), time is allowed to permit the current to stabilize due to changing potentials so that only stable current from analyte oxidation is measured during the second part of E_1, the detection period (t_{det}), also known as integration time (t_i), as data acquisition takes place to integrate peak areas during this time. The time periods during which each potential is applied, is of the order of a millisecond or less so that data acquisition is continuous for all practical purposes.

Several waveforms are used in the analysis of different molecules. They were developed to increase detection sensitivity, and minimize the sensitivity to dissolved oxygen, the baseline drift, and the loss of electrode surface, when used continuously.

Typically, an Ag/AgCl reference electrode is used as half electrode (in combination with gold electrode) in PAD. The principles and applications of PAD are described in further details in Chapter 3.

REFERENCES

1. Peterson EA, Sober HA. Chromatography of proteins: I. Cellulose ion exchange adsorbents. J. Amer. Chem. Soc. 1956;78:751–755.
2. Small H, Stevens TS, Bauman WC. Novel ion-exchange chromatographic method using conductometric detection. Anal. Chem. 1975;47:1801–1809.

3. Haddad PR, Jackson PE. Ion Chromatography—Principles and Applications. Amsterdam (The Netherlands): ElsivierElsevier; 1990.
4. Fritz, J, Gjerde, DT. Ion Chromatography, 3rd ed. Weinheim (Germany): Wiley-VCH; 2000.
5. Weiss J. Ion Chromatography, 3rd ed. Weinheim (Germany): VCH Verlag; 2004.
6. Bhattacharyya L. Ion chromatography in biological and pharmaceutical drug analysis: USP perspectives, presented at the Intl. IC Symp. Baltimore: September 29–October 2, 2002.
7. USP33-NF28, Rockville:US Pharmacopeial Convention; 2010.
8. Himmelhoch SR. Chromatography of proteins on ion-exchange adsorbents. Methods Enzymol. 1971;22:273–286.
9. Dionex Corporation, Application Note 116: Quantification of anions in pharmaceuticals.
10. DeBorba BM, Rohrer JS, Bhattacharyya L. Development and validation of an assay for citric acid/citrate and phosphate in pharmaceutical dosage forms using ion chromatography with suppressed conductivity detection. J. Pharm. Biomed. Anal. 2004;36:517–524.
11. Hofmeister F. Exp. Pathol. Pharmacol. 1888;24:247–260.
12. Zhang Y, Cremer PS. Interactions between macromolecules and ions: The Hofmeister series. Current Opinion Chem. Biol. 2006;10:658–663.
13. Wheaton RM, Bauman WC. Ion exclusion. Annals of the NY Acad. Sci. 1953;57: 159–176.
14. Harlow GA, Morman DH. Automatic Ion exclusion-partition chromatography of acids. Anal. Chem. 1964;36:2438–2442.
15. Morris J, Fritz, JS. Eluent modifiers for the liquid chromatographic separation of carboxylic acids using conductivity detection. Anal. Chem. 1994;66:2390–2395.
16. Ohta K, Tanaka K, Haddad PR. Ion-exclusion chromatography of aliphatic carboxylic acids on an unmodified silica gel column. J. Chromatogr. A 1996;739:359–365.
17. Gjerde DT, Fritz JS, Schmuckler G. Anion chromatography with low-conductivity eluents. J. Chromatogr. 1979;186:509–519.
18. ICH Harmonised Tripartite Guideline. Validation of analytical procedures: text and methodology, Q2(R1). International conference on harmonisation of technical requirements for registration of pharmaceuticals for human use, November 2005.

2

RETENTION PROCESSES IN ION-EXCLUSION CHROMATOGRAPHY: A NEW PERSPECTIVE

Milko Novič

Faculty of Chemistry and Chemical Technology, Aškerčeva, Ljubljana, Slovenia

Paul R. Haddad

Australian Centre for Research On Separation Science (ACROSS), School of Chemistry, Faculty of Science and Engineering, University of Tasmania, Hobart, Tasmania, Australia

2.1 INTRODUCTION

Ion-exclusion chromatography (IEC) is a type of liquid chromatography that was first introduced by Wheaton and Bauman in 1953 [1]. Since then IEC has attracted the intensive interest of researchers because of its ability to separate biologically interesting species (low fatty acids, amines, alcohols etc.), which are of special importance for the pharmaceutical industry [2–8].

IEC is based on the separation of partially ionized species on strong anion- or strong cation-exchange stationary phases, with Donnan exclusion of the analytes from the charged stationary phase being considered to be the basic separation mechanism. IEC is referred to by a variety of alternative names which reflect the continuous search for the exact separation mechanism of the technique [10]. Examples include: ion-exclusion partition chromatography, Donnan exclusion chromatography, and ion-moderated partition chromatography. It has been demonstrated that the retention of

Applications of Ion Chromatography for Pharmaceutical and Biological Products, First Edition.
Edited by Lokesh Bhattacharyya and Jeffrey S. Rohrer.

an analyte is influenced by a large number of parameters. These include: the degree of ionization of the analyte [11], the molecular size and structure of the analyte [12–14], the eluent concentration and its pH value [15,16], the presence of organic solvents in the eluent [17,18], the ionic strength of the eluent [19,20], the temperature of the column [10,21,22], the material comprising the ion-exchanger used and its hydrophobicity [23], the type of ion-exchange functional group on the stationary phase [20], the degree of cross-linking of the polymer used in the stationary phase [11], the ion-exchange capacity [24], and the ionic form of the resin [14].

According to the currently accepted theory, the mechanism of IEC can be represented schematically as in Figure 2.1. The analytical column used in IEC separations of anionic analytes is usually packed with fully sulfonated (typical total cation-exchange capacity of approx. 5.4 meq/g of dry resin) polystyrene-divinylbenzene (PS-DVB) co-polymer (usually 8% cross-linked) of an average diameter of approximately 7 μm. [Fully sulfonated means that there is one sulfonic group attached to each aromatic ring.] In the case where cationic analytes are to be separated, the resin is usually fully functionalized with quaternary ammonium groups. For simplicity, only IEC of anions is considered further in this chapter. That is, the stationary phase will be assumed to be a fully sulfonated resin.

The current mechanism of IEC proposes that the sulfonate groups are fixed mostly on the surface of the PS-DVB resin and form a negatively charged shield on the polymeric surface, often referred to as the "Donnan membrane". The interior of the resin contains some occluded, or trapped, eluent, which is considered to act as the stationary phase. There is no general agreement regarding the precise morphology of this occluded eluent, but for the retention mechanism to be operative, this eluent liquid must be physically trapped within the polymer network and remain stationary. For convenience, we will refer to this eluent as being contained in "pores", but use of this term does not imply that a physical pore exists in the polymeric structure. For example, the eluent liquid might be trapped within a loose network of polymer chains.

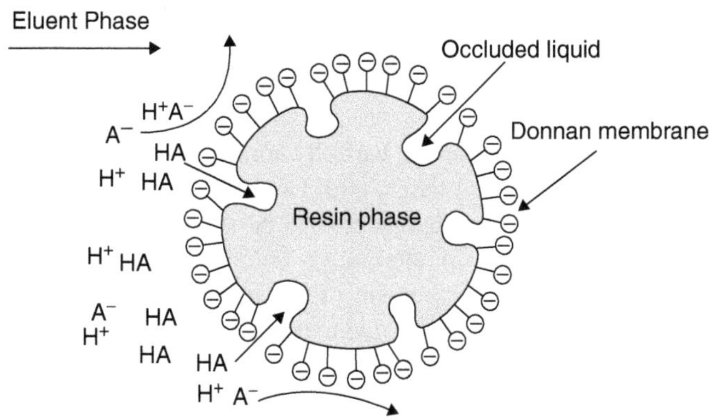

Figure 2.1. Schematic representation of the current ion-exclusion chromatography separation mechanism based on Donnan exclusion effects.

The interstitial fraction of the eluent moves outside the pores and constitutes the carrier stream for the injected analytes. The Donnan membrane separates the moving fraction of the eluent (i.e. the mobile phase) from the static, occluded component of the eluent (i.e. the stationary phase). Once the analytes enter the column, they interact with the sulfonated PS-DVB co-polymer in such a way that the dissociated fraction of the analyte is repelled from the vicinity of the Donnan membrane into the bulk of the interstitial eluent, while the protonated fraction penetrates the membrane and enters the occluded fraction of the eluent, where it may experience additional retention by surface adsorption onto the unfunctionalized parts of the resin [25–28]. The higher the pKa of an individual acid, the higher the protonated fraction and consequently the longer its retention time. Anomalies for analyte acids showing significantly different retention times but having almost identical pKa values have been explained by the increased hydrophobic character of some acids which leads to increased hydrophobic adsorption.

We will critically examine some of these concepts using experimental data based predominantly on conversion of the stationary phase from the eluent form to the analyte form and *vice-versa*. These data are used to indicate some potential shortcomings of the current retention mechanism and will lead to the suggestion of a possible alternative mechanism.

2.2 DEFICIENCIES OF THE CURRENT IEC SEPARATION MECHANISM

In the current IEC mechanism, negative charge originating from sulfonate groups bound covalently to the resin are considered to form a Donnan membrane which acts as a "filter" to resist the passage of negatively charged analytes into the occluded eluent comprising the stationary phase. It is customary to consider only the analyte as carrying an *average* charge determined by the equilibrium existing between the protonated and dissociated forms. The average negative charge on the analyte (given by the relative concentrations of the protonated and deprotonated [dissociated] forms) then determines the extent to which the analyte cloud as a whole is repelled by the Donnan membrane.

Prolonged retention times of some long chain aliphatic carboxylic acids having almost the same pKa values as shorter chained species are explained currently as a consequence of the increased hydrophobic character of the longer chain aliphatic acids and their subsequent hydrophobic interaction with the PS-DVB resin. In the case of aromatic acids, strong $\pi-\pi$ interactions with the PS-DVB resin are also proposed to contribute significantly to retention. For example, benzoic acid (pKa = 4.00) shows a very much longer retention time than acetic acid (pKa = 4.56). In both of the above cases, the analyte is assumed to come into direct contact with the resin, but it is not stated specifically whether this contact occurs inside the pores or elsewhere on the resin. This explanation becomes questionable in terms of two aspects. First, the number of sites for hydrophobic adsorption on a fully sulfonated polymer of high ion-exchange capacity is likely to be small. Second, poor peak shapes (namely, strongly tailed peaks) should be evident for analytes which are retained by hydrophobic adsorption in the fully aqueous eluents used typically in IEC. However, almost all analytes (including aromatic acids) show strong peak *fronting* when water is used as the eluent, while when an aqueous acidic eluent is used, those analytes having strong retention in IEC normally show symmetrical peaks.

The current IEC separation mechanism is based on the penetration of the Donnan membrane by the analyte into the pores of the fully functionalized PS-DVB resin. The mass-transfer for this process is driven only by diffusion resulting from the concentration gradient existing between the two liquid phases and there is no identifiable peak re-focusing mechanism which can counteract the diffusional broadening. This suggests the likelihood of broad peaks, but peaks in IEC generally show good separation efficiencies.

There are some other phenomena occurring in IEC, the origins of which are not readily apparent from current theory. These include the appearance of system peaks, temperature effects on retention, the ability to perform indirect spectrophotometric detection [29–31], and the ability to perform vacancy ion-exclusion chromatography wherein the sample is used as eluent and water is injected as sample [32,33].

2.2.1 Dynamic Column Capacity

The active ingredients in typical acidic IEC eluents can be broadly classified as strong mineral acids (sulfuric acid, hydrochloric acid, etc.) or weaker acids having *Ka* values usually less than 0.01. Interesting behavior occurs when breakthrough experiments are conducted to convert a column from the water form to the acid eluent form and vice-versa [34], as shown in Figures 2.2 and 2.3. Figure 2.2a shows the conversion of the column from the water form into the sulfuric acid form, while Figure 2.2b shows the same data for acetic acid. As can be seen in Figure 2.2a, the time for conversion of the column from the water form to the sulfuric acid form was relatively short (approx. 5 min), reflecting the relatively low dynamic capacity of the column towards sulfuric acid in the tested concentration range. Figure 2.3a shows the reverse interconversion process. The column conversion from the water form to the acetic acid form (Figure 2.2b) and also the reverse process (Figure 2.3b) were significantly longer than in the case of sulfuric acid, indicating that the column showed significantly higher dynamic capacity for acetic acid than for sulfuric acid (by approx. a factor of 3).

The results presented in Figures 2.2 and 2.3 demonstrate some very important characteristics of a fully sulfonated microporous IEC stationary phase:

(i) The total capacity of the stationary phase (measured as the number of meq. of the eluent species retained and able to be replaced by water) increased with the pKa value of the active eluent component.

(ii) The detector response obtained during the complete back-conversion to the water form (e.g., from 0 to 4 min in Figure 2.3a and 0 to 11 min in Figure 2.3b) was constant and was identical to the detector response occurring after the same eluent had been used to equilibrate the column from the water form. That is, the stationary phase fully loaded with the eluent acid released this acid at a concentration identical to that which had been used to load it.

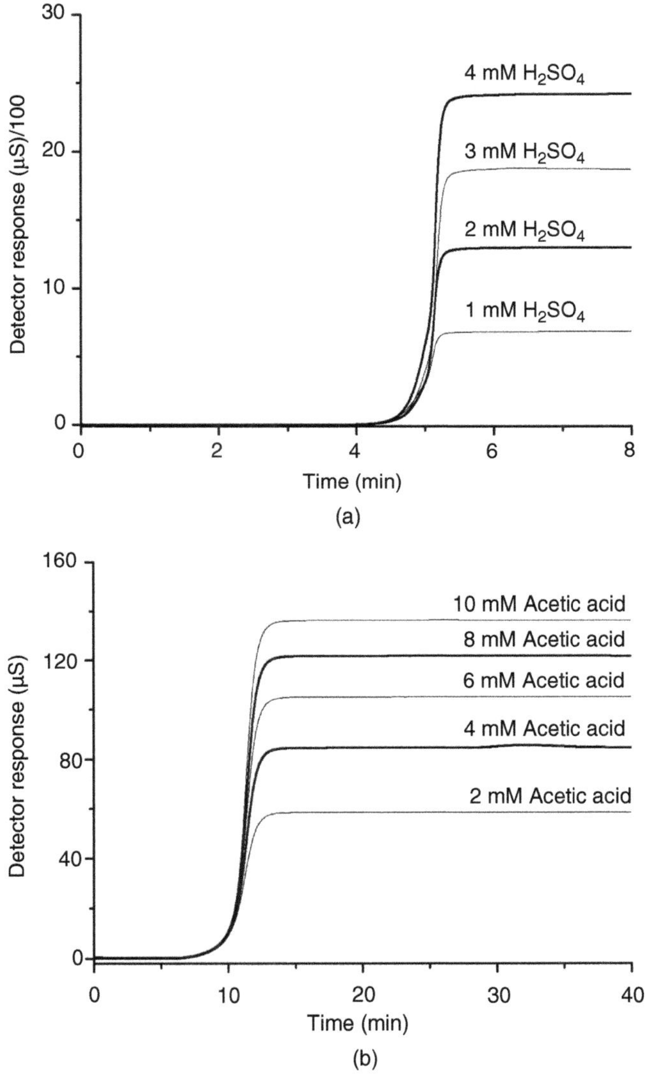

Figure 2.2. Detector responses obtained during conversion of the column from the water form to (a) the sulfuric acid form or (b) the acetic acid form. Experimental conditions: analytical system DX-500 with IonPac ICE-AS1 analytical column, eluent flow rate 1 mL/min, column temperature was 30°C. [Reproduced with permission from Ref. 34.]

(iii) Column conversion from the water form to the acidic form and vice-versa was always very effective, resulting in a sharp change in the detector response indicating either complete loading of the eluent acid or its complete release from the column.

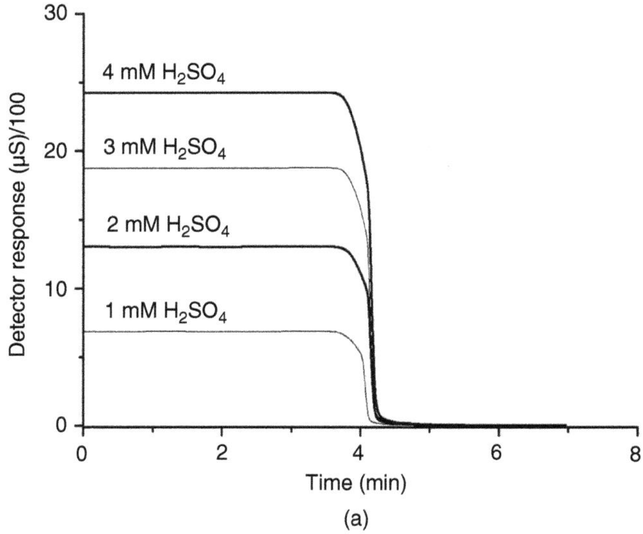

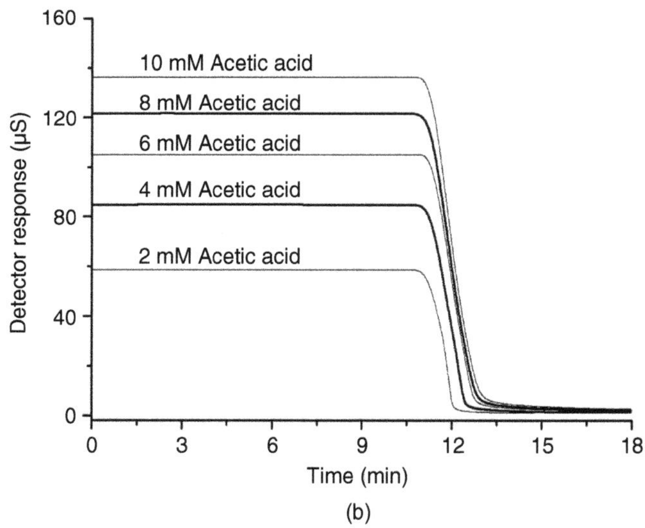

Figure 2.3. Detector responses obtained during conversion of the column from (a) the sulfuric acid form or (b) the acetic acid form to the water form. Experimental conditions: analytical system DX-500 with IonPac ICE-AS1 analytical column, eluent flow rate 1 mL/min, column temperature was 30°C. [Reproduced with permission from Ref. 34.]

(iv) Column capacity, measured as column conversion time, did not change significantly when the concentration of a particular active eluent component was increased over the range of eluent concentrations studied. For example, the breakthrough volumes shown in Figures 2.2 and 2.3 were the same irrespective of the concentrations of sulfuric acid or acetic acid.

The observed detector responses shown in Figures 2.2 and 2.3 cannot be explained adequately by the current Donnan membrane-based mechanism for IEC (i.e. by the partial penetration of acetic acid through the Donnan membrane in comparison to the complete exclusion of the totally dissociated sulfuric acid).

The next interesting characteristic of the detector responses in Figures 2.2 and 2.3 is the sharp change which signals the end of the loading or displacement processes. It should be noted that for all acids used as the active eluent components, a sharp conclusion to the column back-conversion into the water form is observed, while detector responses obtained during column conversion into the specific eluent form differed according to the individual acid tested. The observed behavior could arise under the Donnan membrane mechanism for IEC only if the Peclet number (defined to be the ratio of the rate of advection by the flow to the rate of diffusion) for the system was sufficiently small that there was rapid transfer of the analyte from the occluded eluent to the interstitial eluent.

2.2.2 Retention Behavior and Peak Shapes of Organic Acids

Figure 2.4a shows the retention behavior of acetic acid on the AS1 stationary phase using eluent compositions in the range 0–10 mM sulfuric acid. Figure 2.4b shows the same data for benzoic acid. In both cases it can be noted that retention time increases and peak shape becomes generally more symmetrical as the eluent pH decreases. The conventional explanations for these observations are as follows. First, decreasing the eluent pH in the general proximity of the pKa of the analyte decreases the average negative charge on the analyte and therefore leads to increased retention as a result of diminished repulsion by the Donnan membrane. Second, the fronted peak shape observed in water eluents results from the fact that the zones of lower concentration in the analyte band (i.e. those near the edges of the analyte band) become more dissociated, and are therefore more highly repelled by the Donnan membrane, and thus move more quickly through the column, leading to a fronted peak shape. Use of an acidic eluent minimizes this effect and therefore leads to improved peak shape.

When an acidic eluent (e.g., 1 mM H_2SO_4) is used, increased retention is observed and conventionally this is attributed to increased penetration of the Donnan membrane by the analyte as a result of a decrease in its average negative charge caused by the lowered pH of the eluent. Once the analyte is protonated fully, no further increases in retention are observed (e.g., in the range 1–10 mM H_2SO_4 in Figure 2.4a). When the analyte acids have differing pKa values, they will be separated on the basis of their differing average charge. However, it is well-known in IEC that some analyte acids having very similar pKa values can also be separated, and some analytes show a degree of retention which is much greater than would be predicted on the basis of their pKa

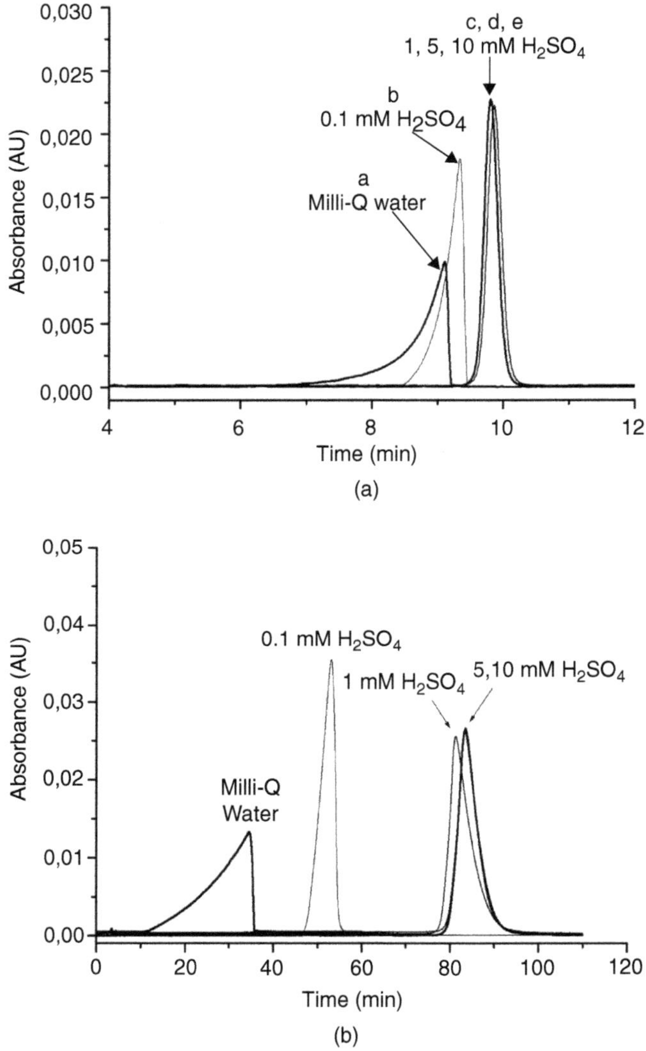

Figure 2.4. Chromatograms for injections of (a) 2 mM acetic acid solution or (b) 2 mM benzoic acid solution obtained using Milli-Q water, and 0.1–10 mM sulfuric acid as eluents. Other experimental conditions: Sample volume 100 μL, analytical column IonPac ICE-AS1 (9 × 250 mm, Dionex), detector UV/Vis (210 nm), eluent flow-rate 1 mL/min, column temperature column temperature 30°C. [Reproduced with permission from Ref. 34.]

value alone (i.e. solely on their degree of ionization). Analytes in the latter category are generally longer-chain aliphatic acids or aromatic acids. The high level of retention of these species is usually explained by the presence of an additional, adsorptive retention mechanism resulting from hydrophobic or π-π interactions between the analyte and the unfunctionalized portions of the stationary phase [35,36]. High pH-dependence of the retention times of these species has been explained as the consequence of the effects of the changing degree of dissociation of the analyte on these adsorption processes. An organic acid having strong affinity towards microporous, fully sulfonated PS-DVB stationary phases is benzoic acid, which shows very strong retention in comparison to aliphatic mono- and di-carboxylic acids having similar pKa values. This is evident from Figure 2.4b where the retention time for benzoic acid in 2 mM H_2SO_4 is about eight times higher than acetic acid, although the latter is the weaker acid. Under the current mechanism for IEC, this increased retention would be attributed to adsorption onto the unfunctionalized portions of the polymer support phase.

Interconversion of the stationary phase between the water and benzoic acid forms (shown in Figure 2.5) provides some important insight into the likelihood of substantial hydrophobic adsorption effects. While the dynamic capacity of the column for benzoic acid was much greater than for acetic acid, the column exhibited a very sharp transition from the benzoic acid form to the water form. Such a sharp transition would not be expected to occur if benzoic acid had been retained on the column by hydrophobic adsorption during the column loading step and was being desorbed (in an aqueous eluent) during the conversion of the column back to the water form. Such hydrophobic desorption would be expected to exhibit a much more gradual desorption profile. This effect can be illustrated by conducting column interconversion studies between the water and benzoic acid forms using a resin which is identical to that used in Figure 2.5, but without any functionalization with sulfonate groups. Figure 2.6a shows the column interconversion plots, and demonstrates that the unfunctionalized resin showed relatively minor total adsorption of benzoic acid (as evidenced by the breakthrough point) and also shows that the transition between the water and benzoic acid forms was more gradual than was observed in Figure 2.5 for the functionalized stationary phase. In Figure 2.6b, the column interconversion plots for a 4× 250 mm column filled with Dionex IonPac NS1 stationary phase are shown. This stationary phase has strong reversed-phase character. In Figure 2.6b, a large NS1 column capacity towards 2 mM benzoic acid is evident, and there is extremely slow desorption of benzoic acid from the stationary phase after the eluent was converted into pure water. These observations support the hypothesis that there is minimal hydrophobic adsorption of benzoic acid on the functionalized resin.

The IEC retention times of very strongly retained analytes, such as pentanoic, hexanoic, and benzoic acids, can be reduced substantially by the addition of an organic solvent (typically methanol or acetonitrile) to the eluent [27,37]. This behavior is generally explained by a reduction in hydrophobic adsorption of analytes caused by the organic solvent, leading to reduced retention. Figure 2.7 shows the conversion of the IEC-AS1 resin column from the benzoic acid form (in 25% aqueous methanol) to the 25% aqueous methanol form. As observed in the aqueous system, there was a sharp transition from the benzoic acid form to the 25% methanol form and the

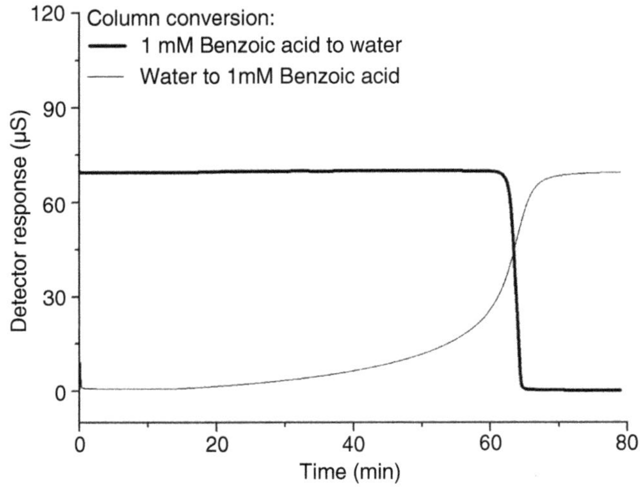

Figure 2.5. Interconversion of the IonPac IEC-AS1 column between benzoic acid and water forms. The thin line shows conversion from water to 1 mM benzoic acid and the thick line shows conversion from 1 mM benzoic acid back to water form. Other experimental conditions: Analytical column ICE-AS1 (9 × 250 mm, Dionex), conductivity detection, eluent flow-rate 1 mL/min, column temperature 30°C. [Reproduced with permission from Ref. 34.]

column capacity for benzoic acid was decreased by almost 50% in the presence of 25% methanol (as evident from a comparison of Figs 2.5 and 2.7).

2.3 RE-EVALUATION OF THE IEC SEPARATION MECHANISM

The experimental data presented above provide motivation for a re-evaluation of the currently accepted mechanism of IEC, especially regarding the role of the Donnan membrane presumed to cover the pores of the stationary phase and the role of hydrophobic adsorption in the retention of some analytes. In searching for an alternative IEC separation mechanism the following theoretical and experimental considerations should also be taken into account:

(i) The substrate beads are composed of 8% cross-linked PS-DVB with an average diameter of approximately 7 μm.
(ii) The PS-DVB resin is fully functionalized, having a total ion-exchange capacity of approx 5 meq/g of dry resin. As mentioned earlier, the complete functionalization means that there is one sulfonic group attached to each aromatic ring.
(iii) There are "pores" on the stationary phase used for IEC separations [38,39]. The estimated pore size in nonfunctionalized microporous polystyrene-divinylbenzene copolymers ranges from 77 Å for 1% divinylbenzene to 13 Å

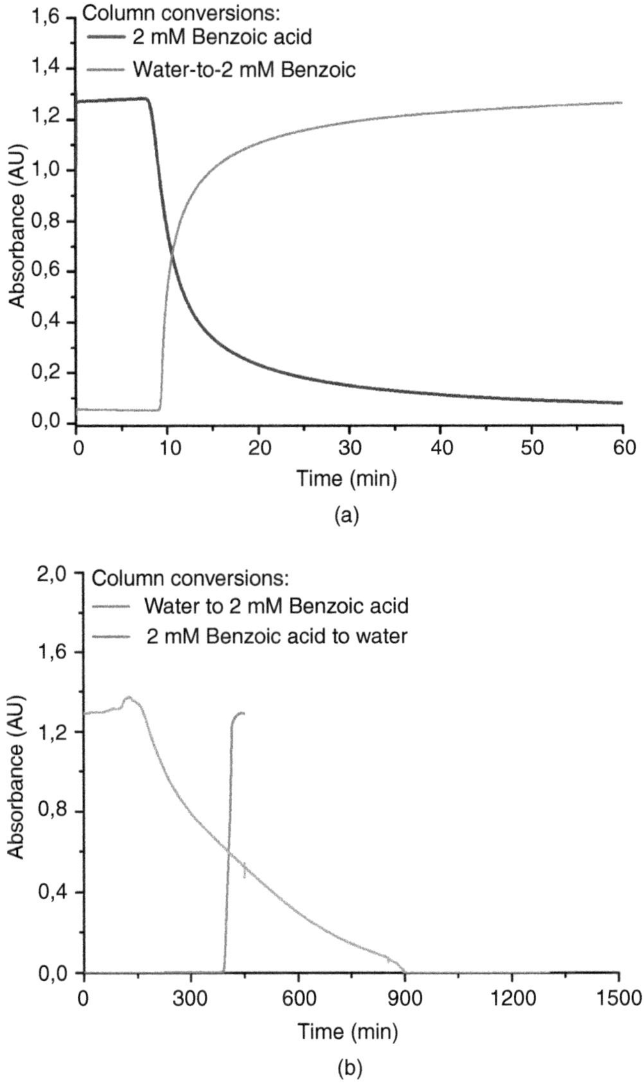

Figure 2.6. Interconversion of (a) a 9 × 250 mm column packed with unfunctionalized AS1 PS-DVB co-polymer between 2 mM benzoic acid and water forms (reproduced with permission from Ref. 34) and (b) a 4 × 250 mm Dionex IonPac NS1 reversed-phase ethylvinylbenzene-divinylbenzene polymer between 2 mM benzoic acid and water forms. The flow-rate was 1 mL/min, detector UV/Vis (210 nm) and the temperature was 30°C in each case.

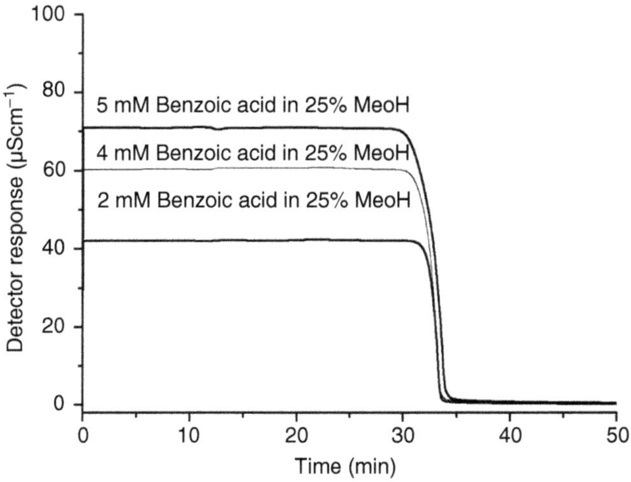

Figure 2.7. Interconversion of the AS1 column between benzoic acid in 25% methanol (v/v) form and 25% methanol (v/v) form. Benzoic acid concentrations are as marked in the Figure. Other experimental conditions: Sample volume 100 μL, analytical column ICE-AS1 (9 × 250 mm, Dionex), conductivity detection, eluent flow-rate 1 mL/min, column temperature 30°C. [Reproduced with permission from Ref. 34.]

for 16% divinylbenzene [39]. In view of the strong electrostatic repulsion of fully dissociated sulfonic functional groups introduced into PS-DVB, the resulting pores should have larger diameters due to swelling of this kind of cation exchanger in water. The interior of the pores represent areas of lower charge density than the rest of the bead and due to the microporous structure, there is assumed to be no flow of eluent through the pores. The structure of such a fully functionalized PS-DVB co-polymer is influenced by the repulsion effects occurring between sulfonate groups and this factor can result in changes to pore sizes when the eluent composition is changed (added organic solvent for example).

(iv) The fixed functional groups on the surface of the substrate do not create a charged Donnan membrane screening the pores, but they create a Donnan membrane on the surface of the functionalized polymer (both on the exterior of the resin and also within the pores).

A suggested alternative retention mechanism for IEC is presented schematically in Figure 2.8. In this figure a pore of the resin is shown and features a sulfonated surface and an interior volume containing occluded liquid. As an initial step, it will be assumed that the eluent used is pure water. In this case, above the fixed monolayer of covalently bound sulfonate groups, there will be a diffuse layer of H_3O^+ cations, according to the Gouy-Chapman model [40,41]. Taking into account the microporous structure of the fully functionalized PS-DVB polymer used in IEC columns and the expected average pore diameter, there should always be some negative electrostatic potential in the

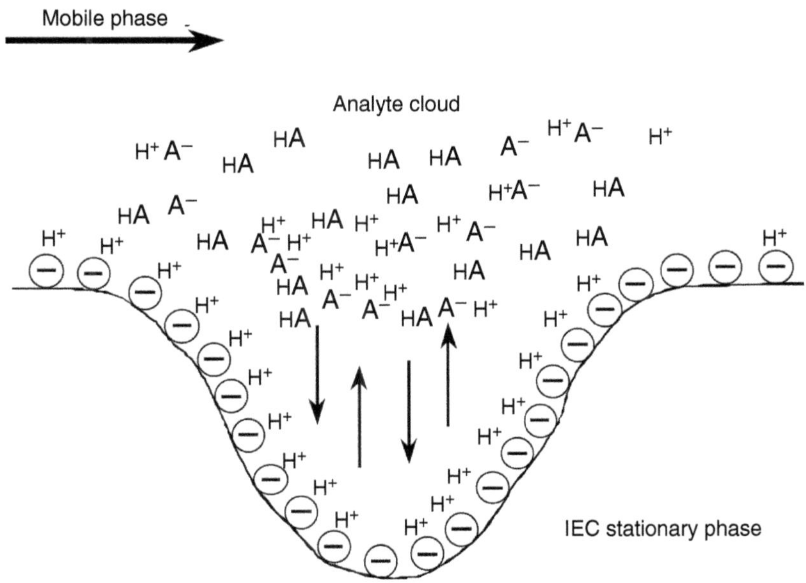

Figure 2.8. Schematic presentation of the suggested alternative retention mechanism for ion-exclusion chromatography. [Reproduced with permission from Ref. 34.]

middle of the pores under most eluent conditions. This negative potential provides a permanent repulsion force for any negatively charged species, so that the fully functionalized microporous PS-DVB polymer bead tends to retain the water form. That is, when a strong electrolyte is used as the eluent component (e.g., H_2SO_4), the electrolyte is present chiefly in the interstitial liquid, while the occluded phase tends to retain its pure water form.

Because of the high concentration of negative charge and consequent high negative electrostatic potential at the outer surface of the resin bead, the presence of a pore is essential for analyte retention because it represents a zone with significantly lower electrostatic repulsion than the outer surface of the bead. A cloud of analyte molecules carrying the same charge polarity as the functional groups on the resin can therefore move between the interstitial eluent and the occluded eluent in the pores, providing a basic retention process. In Figure 2.8 the retention mechanism can be divided into two processes which are occurring simultaneously. The diffusion of the analyte cloud into the areas with lower negative potential (occluded liquid within the pore, the first process) is opposed by the electrostatic repulsion of the charged cloud out of the pore into the interstitial eluent (the second process). The intensity of the diffusion process is controlled by the concentration gradient of the analyte, whilst the degree of electrostatic repulsion depends on the average charge density of the analyte cloud (as determined by the degree of dissociation of the analyte) and the average electrostatic potential across the pore. Because the electrostatic repulsion of the analyte out of the pore is more effective than its diffusion into the pore, peak shapes in IEC are always strongly fronted when water is used as the eluent. In acidic eluents, the dissociation of the analyte is suppressed and the electrostatic surface potential at the surface of the pore is lower, so the analyte can penetrate deeper into the pore and will produce peaks which are symmetrical or tailed in shape, rather than fronted.

2.3.1 Rationalization of the Proposed IEC Retention Model with Observed Retention Data

When an acidic eluent (such as 1 mM H_2SO_4) is used, two on-column liquid phases are established. The first is the interstitial eluent composed of 1 mM H_2SO_4 and the second is the occluded liquid, composed of pure water (or very dilute H_2SO_4) because sulfate ion is repelled from the pores. Analyte dissociation in the interstitial phase is controlled by the H_3O^+ concentration of the eluent and in the case of acetic acid as analyte, dissociation is almost completely suppressed in 1 mM H_2SO_4. The protonated acetic acid diffuses under a concentration gradient into the occluded phase where deprotonation is favored due to the higher pH of the occluded phase. For this reason a maximal concentration gradient of the protonated form of the acetic acid exists, favoring diffusion into the occluded phase and therefore minimizing the processes leading to peak fronting. As a consequence, a more symmetrical peak shape results. Back-diffusion of the analyte from the occluded phase to the interstitial phase is also affected by the degree of acidity in the eluent. The more acidic the eluent, the more protonation occurs as the ionized analyte diffuses from the occluded phase and this in turn increases the propensity for the analyte to re-enter the pore and leads to increased retention of the analyte. In summary, acidic eluents stimulate strong diffusion of the protonated analyte into the pores at the leading edge of the analyte band. Combined with the decreased intensity of its back-repulsion into the interstitial eluent at the tailing edge of the analyte band, this produces both symmetrical peaks and increased retention times.

We turn now to the interconversion plots shown in Figure 2.3, and in particular the fact that the detector signal remains at a constant level during the back-conversion to the water form, before a sharp transition point is reached. During this back-conversion of the column to the water form, when the first water segment reaches the first loaded pore, both the concentration gradient and the electrostatic repulsion act in the same direction to rapidly transfer the eluent components from the occluded liquid into the interstitial eluent phase. This process continues rapidly on the following column segments as the eluent moves down the column until the equilibrium of the eluent acid between the occluded and interstitial phases is attained, giving an effluent with the *same eluent concentration* as that used to load the column pores during the equilibration step. At the point where the loaded stationary phase pore nears exhaustion, the forces of repulsion by the surface potential and the concentration gradient ensure that diffusion of the cloud of eluent molecules into the pore is prevented. That is, the process by which the eluent acid is removed from the pore causes the cloud of eluent acid in the pore to move progressively towards the outer surface of the pore. This can be viewed as an "ion pump" process which counteracts any concentration drop due to hydrodynamic dispersion in the interstitial volume. The final eluent acid is therefore expelled as a discrete band, leading to the observed sharp change in detector signal.

In the case of benzoic acid as analyte, the long retention time and high column loading capacity must be explained without introducing hydrophobic surface adsorption effects since the evidence provided from the column interconversion plots does not support such adsorption. In fact, the capacity of the stationary phase towards benzoic acid can be seen to have *increased* after sulfonation of the base polymer. A secondary retention mechanism must be operative in view of the fact that benzoic acid is eluted at much longer retention time than would be possible even if it fully penetrated the pores of the stationary phase. However, the nature of this additional retention mechanism is not yet clear, although it could involve electrostatic effects which retain the analyte near the internal surface of the pore. This notion is based on the fact that there is a clear correlation between the zeta potential of an analyte (as reflected by its electrophoretic mobility) and its retention time in IEC [34].

2.4 CONSIDERATION OF OTHER PHENOMENA IN IEC

The discussion thus far has focused on the conventional form of IEC which involves the separation of weakly acidic analytes on sulfonated polymeric cation-exchangers using acidic eluents. However, there are ways in which IEC is utilized and other aspects of the technique that need to be explained by the proposed mechanism outlined in Section 3 above. These include an evaluation of the peak shapes observed in IEC, the simultaneous separation of anions and cations by IEC using a carboxylate weak cation-exchange column, and the use of indirect UV-Vis detection in IEC. These areas are discussed below.

2.4.1 Peak Shapes in IEC

As noted earlier, IEC separations performed using water as eluent characteristically exhibit severely fronted peaks, but when the eluent is acidified, peak shapes become close to symmetrical and retention time is generally increased. To date, this behavior has been explained by differences in degree of dissociation of the analyte along the dispersed analyte zone on the column. The lower the analyte concentration, the higher the degree of dissociation and the faster-moving this analyte zone. The dilute parts of the analyte zone therefore overtake the areas of higher concentration, leading to peak fronting.

Under the proposed new mechanism, the capacity of the stationary phase towards the selected analyte is concentration-dependent. This means that the higher the analyte concentration, the more the analyte cloud fraction diffuses into the pore and the slower it moves along the column. Consequently, a fronting analyte distribution profile on the stationary phase is obtained. Moreover, there is certain threshold analyte concentration below which the analyte will not diffuse into pores on the surface of the stationary phase and this fraction of the analyte is eluted out of the column within the system void volume. This is the reason why all peaks obtained in IEC using water as the eluent, tend to start elution within the same time window.

When the eluent is acidified the analyte-stationary phase interactions change. First of all, the analyte dissociation within the interstitial eluent is suppressed. Second, the stationary phase tends to retain its water form due to intensive repulsion of any charged species such as the acid ions. This means that the degree of dissociation of the analyte within the interstitial and occluded liquids is different. The protonated fraction of the analyte in the interstitial liquid tends to diffuse into the pores because it does not experience repulsion effects, whereas the dissociated fraction is repelled strongly out of the pores into the interstitial liquid where it becomes protonated and diffuses back into the pores. The overall result of these two processes is that the analyte cloud spends longer in the pores (leading to increased retention time) and the cloud moves as a symmetrical band down the column (leading to symmetrical peak shapes).

2.4.2 Simultaneous Separation of Anions and Cations in IEC

Thus far, we have discussed IEC performed on high ion-exchange capacity, fully functionalized sulfonated styrene-divinylbenzene stationary phases. The localized surface negative charge on such a stationary phase is greater than the average negative charge in the analyte cloud for typical sample concentrations used in IEC. If the localized surface negative charge is decreased, for example by using a stationary phase containing a mixture of sulfonic acid and carboxylic functionalities, then it would be expected that repulsion of the analyte from the pores would decrease. This can be illustrated by examining the column conversion curves for a Dionex AS1 column (sulfonate) and a Dionex AS6 (sulfonate/carboxylate) shown in Figure 2.9. The interconversion of the AS1 column between the water and sulfuric acid forms is more rapid and shows a sharper profile than for the AS6 column. The AS6 column capacity is higher for 2 mM sulfuric acid than that for AS1, which means that the sulfate anions can penetrate deeper into the pores located on the surface of AS6 stationary phase. On the other hand, the repulsion of sulfate back into the interstitial liquid is less intensive for the AS6 stationary phase, as evidenced by the shape of the end of the column conversion process from 2 mM H_2SO_4 form into the water form. The edge of the curve is much steeper when the AS1 column is used as a result of stronger sulfate repulsion into the interstitial liquid.

Stationary phases having diluted surface negative charge can be used to discriminate between fully ionized analytes, such as the anions of strong mineral acids (e.g., chloride, nitrate, and sulfate). The negative charge on these anions is distributed differently depending on the size and shape of the ions. For example, the negative charge on the nitrate anion is very much dispersed and nitrate can therefore penetrate further into the pores than can chloride, which in turn penetrates further than the doubly charged sulfate ion. These three anions should therefore show weak retention on a carboxylate cation-exchanger, as evident from Figure 2.10 [42]. In this example, an acid rain sample is injected. Sulfate, chloride, and nitrate are separated by IEC, while a series of cations is separated by cation-exchange using a weakly acidic eluent. Two separation mechanisms occur simultaneously on the same stationary phase using the same eluent. Optimization of the simultaneous determination of anions and cations is

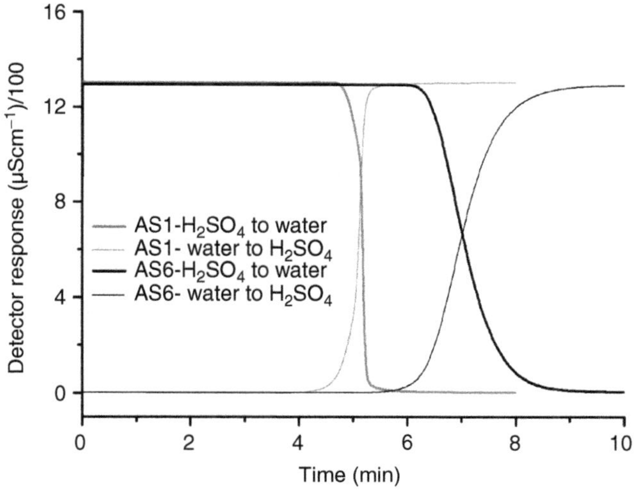

Figure 2.9. Column conversion from water to 2 mM H_2SO_4 form and vice versa. IonPac ICE-AS1 and IonPac ICE-AS6 stationary phases were tested. Experimental conditions: analytical system DX500, eluent flow-rate 1 mL/min, column temperature was 30°C.

based on careful selection of the stationary phase (functional groups and total capacity) and by the concentration of H_3O^+ in the eluent, with the latter parameter affecting both the IEC and cation-exchange mechanisms.

2.4.3 Indirect UV-Vis Detection in IEC

UV/Visible spectrophotometric detection is applied frequently in many forms of chromatography. Conceptually, there are two possibilities: direct and indirect detection. Direct detection involves the measurement of the absorption of the analyte, normally against a background of low absorbance by the eluent. On the other hand, indirect detection involves monitoring the background absorbance of the eluent under conditions where this absorbance will be modified (normally by diminution) by the presence of an analyte. Direct detection is therefore analyte-specific in that only analytes which absorb light at the detection wavelength will be detected, whereas indirect detection is usually general in nature in that all analytes will be detected.

Indirect UV/Vis detection has been utilized widely in ion-exchange chromatography. The principle of operation here is that an eluent containing a UV-absorbing competing ion is used for the ion-exchange separation and the absorbance is normally measured at the wavelength of maximal absorbance of the competing ion. When an analyte is eluted from the column, considerations of electroneutrality decree that there must be a local decrease in the concentration of the competing ion in the vicinity of the analyte band. This causes a decrease in absorbance when an analyte is eluted, leading to negative peaks. In IEC, indirect detection should not be applicable because the eluent components do not participate directly in the retention process, so their

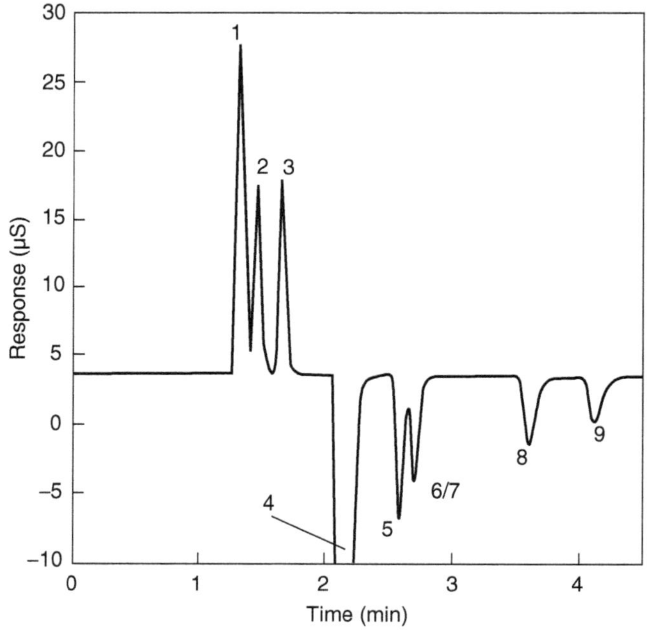

Figure 2.10. Simultaneous IEC/cation-exchange chromatograms of anions and cations by elution with 15 mM tartaric acid (pH 2.44). Experimental conditions: TSKgel Super-IC-A/C column (15 cm x 6 mm I.D.), column temperature 40°C, injection volume 30 μL, eluent flow-rate 1.5 mL/min. Peak identities (mM) 1 = sulfate (0.2), 2 = chloride (0.1), 3 = nitrate (0.2), 4 = elution dip, 5 = sodium (0.1), 6 = ammonium (0.1), 7 = potassium (0.1), 8 = magnesium (0.1), 9 = calcium (0.1). [Reproduced with permission from Ref. 42.]

concentration outside and inside a chromatographic peak should be constant, giving no physical background for the indirect detection of an analyte. However, some reports have been published describing the applicability of indirect UV detection of UV-transparent analytes in IEC [30,43]. It was proposed that the observed peak height is proportional to the solute dissociation constant and inversely proportional to the dissociation constant and concentration of aromatic acids used as competing ions. We suggest here an alternative explanation that is not based on displacement of an eluent ion by an analyte ion, but rather on a detection signal that arises from a change in the degree of dissociation of the eluent ion, induced by the presence of an analyte.

If one proceeds on the basis that there can be no direct displacement of the eluent ion by the analyte ion (i.e. that the mechanism which applies in ion-exchange chromatography does not operate in IEC), the observed indirect detection signal arises as shown in Figure 2.11. This figure shows that the eluent concentration remains constant, even in the presence of an analyte band, but the detector exhibits a negative, indirect signal which corresponds to the eluted analyte band. The maximum detector response at the desired wavelength is the difference between the background absorbance (e.g., at position T in Figure 2.11) and that at position P (i.e. where the analyte concentration is at a maximum). The detected peak signal ΔDR_{λ} can be expressed as:

$$\Delta DR_{\lambda} = DR_{\lambda,P} - DR_{\lambda,T} \tag{1}$$

Remembering that both the analyte and eluent are normally weak acids, represented by HA and HE, respectively, then both the eluent and analyte will contain both protonated and deprotonated forms (i.e. HA, A^-, HE and E^-, assuming that both the eluent and analyte are monoprotic acids). Equation (1) can be expanded to give:

$$\begin{aligned}\Delta DR_{\lambda} = &((DR_{E^{y-},P} + DR_{HE,P}) + (DR_{A^{x-},P} + DR_{HA,P})) \\ &- ((DR_{E^{y-},T} + DR_{HE,T}) + (DR_{A^{x-},T} + DR_{HA,T}))\end{aligned} \tag{2}$$

If it is assumed that the detection wavelength has been chosen so that the eluent absorbs light but the analyte does not (which is the normal case for indirect spectrophotometric detection), Equation (2) can be rewritten as:

$$\Delta DR_{\lambda} = (DR_{E^{y-},P} + DR_{HE,P}) - (DR_{E^{y-},T} + DR_{HE,T}) \tag{3}$$

By applying the Beer–Lambert Law, Equation (3) takes the form:

$$\Delta DR_{\lambda} = l((\varepsilon_{E-,\lambda}([E_P^-] - [E_T^-]) + \varepsilon_{HE,\lambda}([HE_P] - [HE_T])) \tag{4}$$

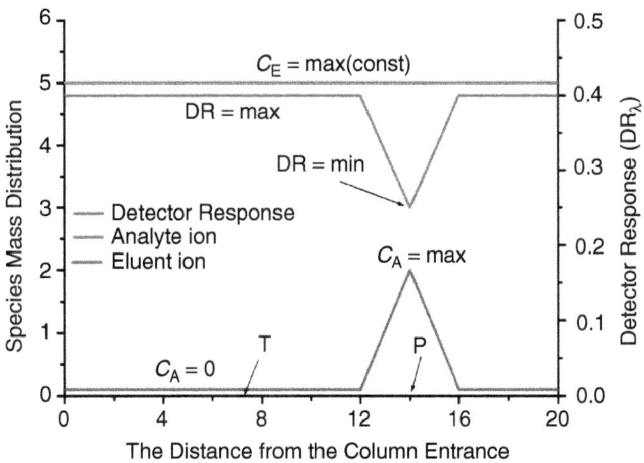

Figure 2.11. Schematic diagram of indirect UV/Vis detection in IEC. The analyte and eluent mass-distribution is shown on the left *y*-axis and the corresponding UV/Vis absorbance is shown on the right *y*-axis. The eluent concentration is not affected by the presence of the analyte peak due to nonactive eluent participation in the IEC separation mechanism.

Where l is the path length and ε is the absorptivity. If the degree of eluent dissociation in zones P and T differ, then the form is:

$$([E_P^-] - [E_T^-]) = -([HE_P] - [HE_T]) \tag{5}$$

and finally:

$$\Delta DR_\lambda = l([HE_P] - [HE_T])(\varepsilon_{HE,\lambda} - \varepsilon_{E-,\lambda}) \tag{6}$$

A nonzero value is expected forΔDR_λ only if $[HE_P] - [HE_T] \neq 0$. and $\varepsilon_{HE,\lambda} - \varepsilon_{E-\lambda} \neq 0$ This means that a signal can be recorded only if the degree of eluent dissociation within the analyte band differs from that outside the band and the absorptivity for the dissociated and protonated forms of the eluent are different. In such a case, a negative or positive indirect UV/Vis signal can be expected if $\varepsilon_{HE,\lambda} - \varepsilon_{E-\lambda} < 0$ or $\varepsilon_{HE,\lambda} - \varepsilon_{E-,\lambda} > 0$ respectively.

We can now consider published data for indirect detection in IEC using an aromatic acid as eluent, aliphatic acids as analytes, and a detection wavelength of 290 nm (at which the analytes show very little absorbance) [43]. Figure 2.12 shows the detector responses obtained by both conductivity and spectrophotometric detection with pyridinedicarboxylic acid (PDCA) as eluent. PDCA is a dicarboxylic acid having $pKa_1 = 2.16$ and $pKa_2 = 6.92$. The pKa values of the analytes used are shown in Table 2.1

Inspection of Figure 2.12 and Table 2.1 shows that pKa_1 values and retention times of the analyte acids are strongly correlated, in accordance with the IEC retention mechanism discussed in this chapter. Even more pronounced is the effect of the pKa_1 value of the analyte acid on the indirect UV/Vis signal shown in Figure 2.12a. The lower the pKa_1, the more intensive the negative signal. This effect can be illustrated by injecting a plug of an analyte acid (such as acetic acid or sulfuric acid) into a flowing stream of PDCA and monitoring the absorbance (Figure 2.13). As can be seen in Figure 2.13, sulfuric acid causes a much more intensive reduction in the detector response than does an equimolar concentration of acetic acid. Sulfuric acid is a fully dissociated mineral acid and can suppress the PDCA dissociation much more effectively than a weak acid such as acetic acid. It should also be remembered that when the protonated form of the eluent has a higher absorptivity than the dissociated form, the indirect detection signal will be positive rather than negative.

2.5 CONCLUSIONS

Despite being a long established analytical technique, complete understanding of some aspects of the mechanism of IEC has remained elusive. While the current Donnan exclusion interpretation of retention in IEC provides a satisfactory explanation of many aspects of IEC, there are also some inconsistencies. The alternative retention mechanism described here is based on an electrostatic repulsion process in a manner similar to the Donnan exclusion mechanism. However, additional interpretations

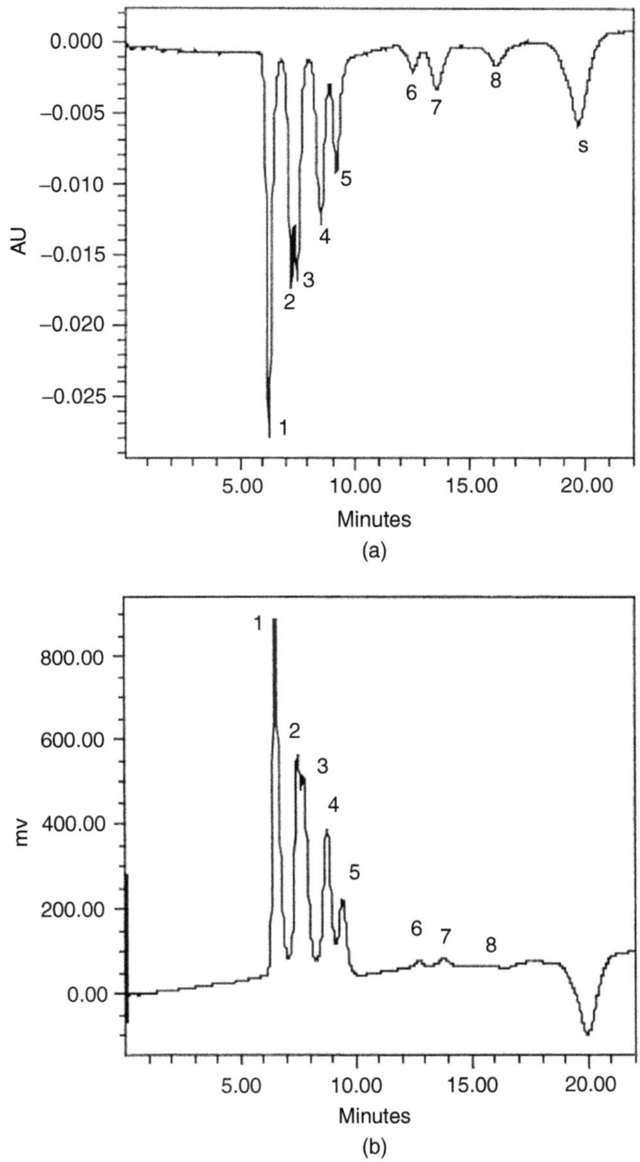

Figure 2.12. Ion exclusion chromatography of aliphatic carboxylic acids using 1 mM PDCA as the eluent: (a.) indirect UV detection at 290 nm, (b) conductivity detection. Separation conditions: analytical column HPX-87, eluent PDCA (pyridine dicarboxylic acid), 1 mM, eluent flow rate 0.6 mL/min, sample volume 10 μ L and column temperature 40°C. On both chromatograms the analytes are marked as follows: 1, oxalic acid (0.5 mM); 2, pyruvic acid (0.25 mM); 3, citric acid (0.25 mM); 4, T-aconitic acid (0.5 mM); 5, malic acid (0.5 mM); 6, formic acid (0.5 mM); 7, succinic acid (0.5 mM); 8, acetic acid (0.5 mM); s, system peak. [Reproduced with permission from Ref. 43.]

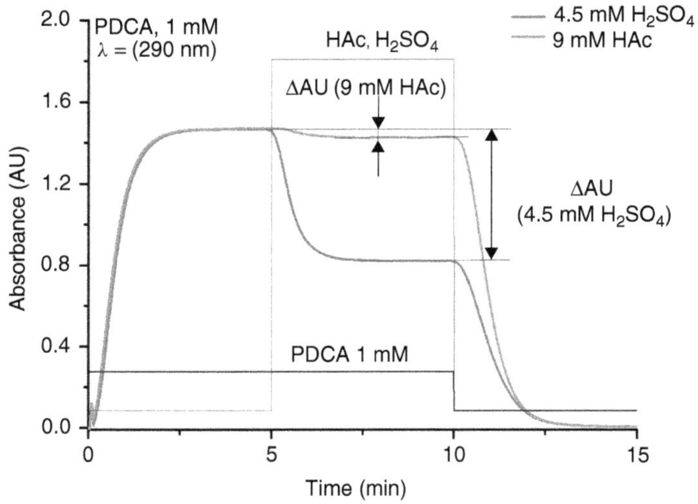

Figure 2.13. Absorbance measurements of 1 mM PDCA injected with a plug of 9 mM H_2SO_4 (dark gray line) or 9 mM acetic acid (HAc, dark gray line) in a continuous flow analyzer. The total flow-rate through the detector was 1.0 mL/min, absorbance was measured at 290 nm. The PDCA gradient is marked by thicker gray line, while sulfuric or acetic acids gradients are marked by thinner gray lines.

TABLE 2.1. Dissociation Constants of the Acids for Which IEC Chromatograms Are shown in Figure 2.12.

Peak number (acid, Fig. 2.4a,b)	pKa_1	pKa_2	pKa_3
1. Oxalic acid	1.27	4.19	
2. Pyruvic acid	2.39		
3. Citric acid	3.13	4.77	6.40
4. T-aconitic acid	2.80	4.46	
5. Malic acid	3.40		
6. Formic acid	3.75		
7. Succinic acid	4.21	4.61	
8. Acetic acid	4.76		

of the retention mechanism are introduced, based predominantly on the existence of "pores" on the surface of the stationary phase. Analytes diffuse into these "pores" containing occluded water under the effects of concentration gradients, while simultaneously undergoing repulsion effects from the fixed charges inside the pores. These two processes determine the degree of retention of an analyte. IEC is seen to be based on a permeation retention process, but differs from other such processes (such as size-exclusion chromatography) in that the suggested mechanism introduces variable permeation volume, combined with electrostatic repulsion out of the permeation volume. For analytes showing retention times longer than that which can result from total permeation of the analyte into the pores, there is an additional retention mechanism which is not the result of hydrophobic adsorption of the analyte onto the polymeric resin. This retention process has not been identified but there are indications that electrostatic interactions are involved. The proposed retention mechanism has been shown to be consistent with the peak shapes observed in IEC, the simultaneous separation of anions and cations on weak cation-exchange stationary phases, and the use of indirect spectrophotometric detection in IEC.

REFERENCES

1. Wheaton RM, Bauman WC. Ion exclusion, a unit operation utilizing ion exchange materials. Ind. Eng. Chem. 1953;45:228–233.
2. Jenke DR. Development and Validation of an Ion-Exclusion Chromatographic Method for the Quantitation of Organic Acids in Complex Pharmaceutical Products. J.Chromatogr. Sci. 1998;36(4):179–186.
3. Iwinski G, Jenke DR. Determination of alcohols in pharmaceuticals by ion-exclusion chromatography, J. Chromatogr. A 1987;392:397–405
4. Warnock M, Nair L, Sadain S, Jenke DR. Evaluation of an ion exclusion/direct conductivity method for quantitating acetic and lactic acids in pharmaceutical LVP base formulations. J. Liq. Chromatogr. 2002;25(4):541–560.
5. Allan GA, Gedge JI, Nedderman ANR, Roffey SJ, Small HF, Webster R. Pharmacokinetics and metabolism of UK-383,367 in rats and dogs: A rationale for long-lived plasma radioactivity. Xenobiotica 2006;36(5):399–418.
6. Claes PJ, Compernolle F, Vanderhaeghe H. Chromatographic analysis of neomycin isolation and identification of minor components. J. Antibiotics 1974;27(1):931–942.
7. Karmarkar S, Koberda M, Momani J, Kotecki D, Garber R. Validated ion-exclusion chromatographic method for citrate and acetate in medical fluids. J. Chromatogr. A 2004;1039:147–153.
8. Schiller M, von der Heydt H, Maerz F, Schmidt PC. Quantification of sugars and organic acids in hygroscopic, pharmaceutical herbal dry extracts. J. Chromatogr. A 2002;968:101–111.
9. Yang L, Liu L, Olsen BA, Nussbaum MA. The determination of oxalic acid, oxamic acid, and oxamide in a drug substance by ion-exclusion chromatography. J. Pharm. Biomed. Anal. 2000;22:487–493.
10. Haddad PR, Jackson PE. Ion Chromatography: Principles and Applications, Amsterdam: Elsevier; 1990. p 195.
11. Harlow GA, Morman DH. Automatic Ion Exclusion-Partition Chromatography of Acids. Anal. Chem. 1964;36:2438–2442.
12. Medved AL, Ivanov AA, Shpigun OA. Ion-exclusion chromatography of organic acids: Regularities in the retention of aliphatic carboxylic acids. J. Anal. Chem. 1996;51:964–971.
13. Waki H, Tokunaga Y. Donnan exclusion chromatography: I. Theory and application to the separation of phosphorus oxoanions or metal cations. J. Chromatogr. 1980;201:259–264.

14. Kihara K, Rokushika S, Hatano H. Elution behavior of aliphatic carboxylic acids on a strong cation-exchange resin column. J. Chromatogr. 1987;410:103–110.
15. Turkelson VT, Richards M. Separation of the citric acid cycle acids by liquid chromatography. Anal. Chem. 1978;50:1420–1423.
16. Albarran G, Collins CH. Ion-moderated partition chromatographic determination of $Ca^{14}CO_3$ and $Ba^{14}CO_3$ self-radiolysis products. J. Chromatogr. 1987;395:623–629.
17. Tanaka K, Ishizuka T. Ion-exclusion chromatography of condensed phosphates on a cation-exchange resin. J. Chromatogr. 1980;190:77–83.
18. Tanaka K,. Fritz JS. Separation of aliphatic carboxylic acids by ion-exclusion chromatography using a weak-acid eluent. J. Chromatogr. A 1986;361:151–160.
19. Klampfl CW, Buchberger W, Rieder G,. Bonn GK. Retention behavior of carboxylic acids on highly cross-linked poly(styrene-divinylbenzene)-based and silica-based cation exchangers. J. Chromatogr. A 1997;770:23–28.
20. O'Reilly J, Doble P, Tanaka K, Haddad PR. Retention behaviour of strong acid anions in ion-exclusion chromatography on sulfonate and carboxylate stationary phases J. Chromatogr. A 2000;884:61–74.
21. Papp E, Keresztes P. Retention behaviour of mono- and dicarboxylic acids, carbohydrates and alcohols in ion-exclusion chromatography. J. Chromatogr. A 1990;506:157–167.
22. Chen ZL, Adams MA. Retention behavior and simultaneous separation of carboxylic and aromatic acids using ion-exclusion chromatography. J. Liq. Chromatogr. Rel. Technol. 1999;22:2595–2611.
23. Ohta K, Tanaka K, Haddad PR. Ion-exclusion chromatography of benzenecarboxylic acids on an unmodified silica-gel column. J. Chromatogr. A 1997;782:33–40.
24. Lee DP, Lord AD. A high performance phase for the organic acids. LC-GC 1987;5: 261–266.
25. Papadoyannis IN, Samanidou VF. Encyclopedia of Chromatography, Marcel, New York: Dekker; 2001. p 455–457.
26. Tanaka K, Haddad PR. Ion exclusion chromatography: liquid chromatography. In: Wilson ID, Adlard ER, Cooke M, Poole CF, editors. Encyclopedia of Separation Science, San Diego: Academic Press; 2000. p 3193–3201.
27. Ng KL, Paul B, Haddad PR, Tanaka K. Retention modeling of electrostatic and adsorption effects of aliphatic and aromatic carboxylic acids in ion-exclusion chromatography. J. Chromatogr. A 1999;850:17–27.
28. Glód BK. Principles and Applications of ion-exclusion chromatography. Acta Chromatographica 1997;7:72–87.
29. Z. Liang Chen Z, Tang C, Xu J. Non-suppressed conductivity and indirect UV detection of carboxylic acids in environmental samples by ion-exclusion chromatography using 2,6-pyridinedicarboxylic acidic eluent. J. Chromatogr. A, 1999;859:173–181.
30. Chen ZL, Glod BK, Adams MA. Indirect photometric detection of aliphatic acids separated by ion-exclusion chromatography using aromatic acidic eluents. J. Chromatogr. A 1998;818:61–68.
31. Novic M, Haddad PR. Indirect UV spectrophotometric detection in Ion Exclusion Chromatography—Insights into retention processes, presented at the International Ion Chromatography Symposium, 2003 September 21-24 2003, paper B-4; San Diego, CA.

32. Tanaka K, Ding MY, Helaleh MIH, Taoda H, Takahashi H,. Hu WZ, Hasebe K, Haddad PR, Fritz JS, Sarzanini C. Vacancy ion-exclusion chromatography of carboxylic acids on a weakly acidic cation-exchange resin. J. Chromatogr. A 2002;956(1–2):209–214.
33. Tanaka K, Ding MY, Takahashi H, Helaleh MIH, Taoda H, Hu WZ, Hasebe K, Haddad PR, Mori M, Fritz JS, Sarzanini C. Vacancy ion-exclusion chromatography of carboxylic acids on a strongly acidic cation-exchange resin. Anal. Chim. Acta 2002;474(1–2):31–35.
34. Novic M, Haddad PR. Analyte-stationary phase interactions in ion-exclusion chromatography. J. Chromatogr. A 2006;1118:19–28.
35. Haddad PR, Jackson PE, Ion Chromatography: Principles and Applications, Amsterdam: Elsevier; 1990. p 207.
36. Weiss J. Ion Chromatography, 2nd ed. Weinheim: VCH, 1995.
37. Tanaka K, Fritz JS. Separation of aliphatic carboxylic acids by ion-exclusion chromatography using a weak-acid eluent. J. Chromatogr. 1986;361:151–160.
38. Albrighjt RL, Yarnell PA. Rohm and Haas Company, Ion Exchangers, Encyclopedia of Polymer Science and Engineering, Volume 8, New York: Wiley; 1987. p 359.
39. Freeman DH, Schram SB, Characterization of microporous polystyrene-divinylbenzene copolymer gels by inverse gel permeation chromatography. Anal. Chem. 1981;53(8):1235–1238.
40. Gouy G. Sur la constitution de la charge électrique a la surface d'un electrolyte. J. Phys. 1910;9:457–468.
41. Chapman DL. A contribution to the theory of electrocapillarity, Philos. Mag. 1913;25:475–480.
42. Mori M, Tanaka K, Helaleh MIH, Xu Q, Ikedo M, Ogura Y, Sato S, Hu W, Hasebe K, Haddad PR. High-speed simultaneous ion-exclusion/cation-exchange chromatography of anions and cations on a weakly acidic cation-exchange resin column. J. Chromatogr. A 2003;997(1):219–224.
43. Chen ZL, Tang C, Xu J. Non-suppressed conductivity and indirect UV detection of carboxylic acids in environmental samples by ion-exclusion chromatography using 2,6-pyridinedicarboxylic acidic eluent. J. Chromatogr. A 1999;859:173–181.

3

PULSED ELECTROCHEMICAL DETECTION IN ION CHROMATOGRAPHY

William R. LaCourse

Department of Chemistry and Biochemistry, University of Maryland, Baltimore County, MD

3.1 INTRODUCTION

The first commercial electrochemical detector (EC) for high performance liquid chromatography (HPLC) was introduced in 1974, and over the past several decades its popularity, breadth, and application have grown significantly. The reasons for its success can be directly attributed to its high sensitivity, often high selectivity, and wide linear range. Also, EC detectors are fairly simple in construction, relatively inexpensive, and rugged. Of importance more recently, electrochemical cells are easily miniaturized due to the response being dependent on electrode area and not path length as in optical absorbance methods. This facilitates their adaptability to miniaturized chromatographic and electrophoretic separation techniques and lab-on-a-chip devices.

Combining EC with a high efficiency separation technique lends itself to the analysis of relatively complex samples. Some of the earliest applications were directed toward the determination of neurotransmitters in complex biological samples (i.e., brain extracts), drugs and their metabolites in physiological fluids, and contaminants in

Applications of Ion Chromatography for Pharmaceutical and Biological Products, First Edition.
Edited by Lokesh Bhattacharyya and Jeffrey S. Rohrer.

environmental samples. Hence, electrochemical detection following chromatographic separation has proven to be a powerful analytical technique for the determination of compounds that undergo oxidation and, less commonly, reduction. A closer look reveals that the vast majority of these compounds (e.g., phenols, aminophenols, catecholamines, and other metabolic amines) are aromatic in nature. Such compounds tend to be readily oxidized at inert electrodes (e.g., Au, Pt, and carbon) by applying a constant potential, which is commonly denoted as dc amperometry [1,2]. Electronic resonance, or π-resonance, in molecules with extended conjugation stabilizes the free-radical intermediates formed during the anodic oxidation process, and, as a consequence, the activation barrier for the electrochemical reaction is lowered significantly. Hence, these compounds are considered to be self-stabilized, and the electrode, which only functions to accept or donate electrons in the detection mechanism, should ideally be inert. Any electrochemical involvement of the electrode (i.e., adsorption of reactant and/or products) is detrimental to the detection process and leads to electrode deactivation. Thus, glassy carbon electrodes are popular in dc amperometry due to their resistance to fouling. It is important to note that resistant does not imply impervious, and fouling does occur. Exposure of the electrode to the analyte, as the chromatographic band, is much less relative to the amount of mobile phase passing over the electrode, and deactivation phenomena are much less pronounced in hydrodynamic systems (e.g., HPLC). Fouling often occurs over a period of hours or days, and it is generally classified as long-term, which complements the short-term nature of most chromatographic experiments. Hence, in DC amperometry, the electrode surface must be mechanically polished or some other method of electrode reactivation must be undertaken on a daily basis to maintain response sensitivity and reproducibility.

In contrast to the mechanism described above, self-stabilization of the radical intermediates is not available to polar *aliphatic* molecules (e.g., carbohydrates and biogenic amines), and the rate of oxidation is either very slow or nil even though oxidation of the molecule may be favored thermodynamically [3]. Efficient oxidation of polar aliphatic compounds can be accomplished via the stabilization of reaction intermediates by their adsorption at "clean" noble metal electrodes. Faradaic processes that benefit from electrode surface interactions are described as "electrocatalytic".

Unfortunately, exploiting adsorption to the electrode to assist in the electrochemical detection process dramatically increases the rate of electrode deactivation, or fouling, which can lead to complete passivation of the 'working' electrode within seconds. Thus, the historical perspective of non-reactivity for polar aliphatic organic compounds at noble metals can be attributed to rapid fouling of the electrode surface, and the irony is that this fouling is a result of high, but transient, electrocatalytic activity. In order to perform analytically relevant detections following a separation, the electrode must be continuously reactivated, or "cleaned", on-line; and of all the possible cleaning pretreatments, only electrochemical reactivation can be performed effectively.

3.2 HISTORICAL PERSPECTIVE OF PULSED ELECTROCHEMICAL DETECTION

The application of alternate positive and negative potential pulses as an effective means of reactivating noble metal electrodes has been employed since 1924 [4]. Over the next four decades, the use of pulsed potential waveforms for reactivation of fouled noble metal electrodes was used extensively in research directed towards the development of hydrocarbon fuel cells [5]. The application of pulsed potential electrode pretreatments in electrocatalytic systems is attributable to a seminal paper by Kolthoff and Tanaka [6], in which they studied the polarization curves of platinum electrodes in different supporting electrolytes. The benefit of electrode reactivation by alternated positive and negative potential excursions was clearly established by a plethora of researchers [7–13], and the introduction and rapid growth of HPLC in the 1960s and 1970s also stimulated its application to electrochemical detectors in chromatographic systems.

The epitome of pulsed potential cleaning and reactivation of solid anodes is represented by the advent of Pulsed Electrochemical Detection (PED). Its invention can be attributed to Professor Dennis C. Johnson, an academic descendant of Kolthoff, who had been studying anodic oxygen-transfer reactions at noble metal electrodes for a number of years. The Johnson group initiated research on reactivating noble metal electrodes using pulsed potential waveforms in 1981, and in that same year Pulsed Amperometric Detection (PAD) was introduced for the determination of simple alcohols at Pt electrodes in flow injection systems [14,15].

Interestingly, PAD owes much of its popularity and success to Ion Chromatography (IC). Around the time PAD was being introduced to the scientific community by the Johnson group, the Dionex Corporation (Sunnyvale, CA) was working on alkaline-tolerant polymer-based columns to resolve carbohydrates as anions. Prior to 1981, the only approaches to chromatographic detection for carbohydrates without prior derivatization were refractive index and low-wavelength UV absorption both of which suffered significant drawbacks especially in regards to sensitivity. Fortunately PAD was able to detect alcohols and, therefore, carbohydrates under alkaline conditions. In a collaborative effort, the Johnson group and Dionex worked together to put forth the first commercial electrochemical detector for carbohydrates following High Performance Anion Exchange Chromatography (HPAEC) [16,17].

Initially, PAD was characterized by the application of three-step potential waveforms at noble metal electrodes that measured the average current over a fixed period [18]. Modifications of the waveform led to the following techniques:

- *Pulsed Coulometric Detection (PCD)*—PCD integrated the current during the detection step and output the signal in coulombs. The benefit was reduced noise by integrating the signal over a longer period.
- *Reversed-PAD (RPAD)*—RPAD reversed the positive and negative potential pulsed in PAD in an effort to reduce the large background signal from surface oxide formation.

- *Activated-PAD (APAD)*—APAD added a fourth potential pulse to the PAD waveform in an effort to overcome the large background signal from surface oxide formation by freezing the growth of oxide on the electrode surface.
- *Potential Sweep Pulsed Coulometric Detection (PSPCD)*—PSPCD expanded PCD to incorporate a triangular potential sweep in the detection step. This waveform was successful at overcoming the effects of the deleterious background signal by electronically integrating the surface oxide formation and reduction signals.
- *Integrated Pulsed Amperometric Detection (IPAD)*—PSPCD was renamed to IPAD and was expanded to utilize either a triangular, saw tooth, or square wave in the detection step.
- *Pulsed Voltammetric Detection (PVD)*—PVD incorporates a voltammetric scan as part of the detection step, which allows for chronovoltammograms that show current versus potential data at each chromatographic time point, in other words, three-dimensional pulsed electrochemical detection.

All these techniques now use the generic title of Pulsed Electrochemical Detection (PED), which includes the most popular techniques of PAD and IPAD, for all detection strategies based on the application of multistep waveforms regardless of the specific form of signal measurement [19,20]. An in-depth review of all these techniques has been published in a book by LaCourse [21]. The analytical significance of this technique is probably best reflected in that it continues to be the leading detection technique for carbohydrates, as well as being available from many commercial manufacturers.

3.3 ELECTROCATALYSIS AT NOBLE METAL ELECTRODES

3.3.1 Carbohydrates

The surfaces of noble metals (i.e., Pt and Au), which are commonly considered to be inert as solid anodes, are electrochemically active when a voltage is applied in the presence of a supporting electrolyte. Fig. 3.1a shows the current-potential (i-E) plot for an Au rotating disk electrode (RDE) in 0.1 M NaOH with (. . ..) and without (—) dissolved O_2 present. The figure is oriented such that the current from oxidation and reduction processes are up and down, respectively. The observed waves and peaks are attributed as follows:

- *Region 'a'*. This wave, which is due to the formation of surface oxide on the positive scan, consists of several maxima that correspond to the different crystal faces being populated with surface oxide.
- *Region 'b'*. At the upper limit of the voltammetric scan, water is electrolytically decomposed, which results in the evolution of oxygen.
- *Region 'c'*. After the potential direction is reversed, the surface oxide formed on the forward scan is reduced, which results in a clean and renewed electrode surface.

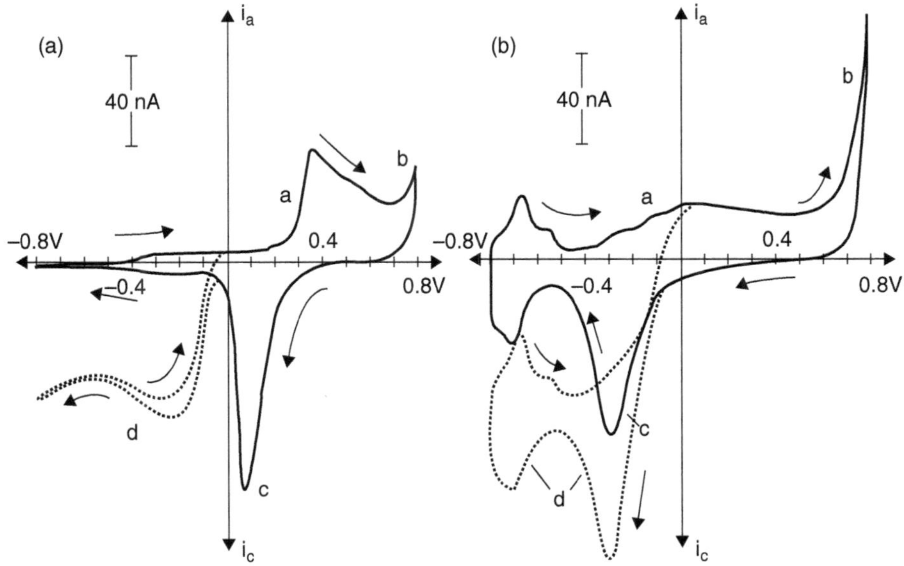

Figure 3.1. Cyclic voltammetric response (i-E) for (a) Au and (b) Pt rotating disk electrode. Conditions: rotation speed, 900 rpm; scan rate, 200 mV/s; Ag/AgCl reference electrode. Solutions: (———) 0.1 M NaOH, de-aerated; (…) 0.1 M NaOH.

- *Region 'd'*. If dissolved O_2 is present, its reduction is observed as a wave during the positive and negative scans due to its mass-transport dependent nature.

The same basic features of Au electrodes are found for Pt electrodes under similar experimental conditions, except the Pt electrode also exhibits hydrogen adsorption and reduction waves at the negative limits of voltammetric scan as shown in Fig. 3.1b. Also and more importantly, the O_2 reduction signal is well-resolved from the surface oxide formation signal for an Au electrode; whereas, there is significant overlap of these same two processes for a Pt electrode. This 100–200 mV potential window is essentially free of any background signal interference, which accounts for the general preference of Au over Pt electrodes for the majority of PED applications. It is important to note that the potentials and intensity of processes for the Au and Pt electrodes described above are dependent on the composition of the supporting electrolyte (e.g., pH, ionic strength, solution composition, and temperature).

The i-E plot for 200 μM glucose (- - - -) at an Au electrode in 0.1 M NaOH is shown in Fig. 3.2. Glucose, a reducing sugar, undergoes irreversible oxidation as noted by the following two signals:

- *Region 'e'*. Beginning at ca. −0.6 V on the positive scan, this signal corresponds to oxidation of the aldehyde group to the carboxylate anion in alkaline media. This process occurs concomitantly with the signal in region *e*.

- *Region 'f'*. From ca. −0.2 to +0.4 V, the combined currents from the alcohol and aldehyde groups produce a large oxidation signal that is attenuated during the positive scan with the onset of surface oxide formation (region a).

As the Au electrode surface becomes covered with oxide, the signal for glucose diminishes and electrode activity for the oxidation of glucose ceases. At this point the electrode is considered to be inert and no signal for the oxidation of glucose is observed on the negative scan in the region of ca. +0.8 to +0.2 V. Following the reductive dissolution of the oxide (region *c*) on the negative scan, glucose oxidation returns and signal is again observed for oxidation of alcohol and aldehyde groups of glucose. The signals in regions *e* and *f* are observed to increase in intensity with increase in glucose concentration.

The i-E plots in Figs. 3.1 and 3.2 were generated using cyclic voltammetry (CV). Although CV is the tool of choice to study electrochemical mechanisms, it does not accurately reflect the voltammetric response of electrolytic oxidations. The signal incorporates a degree of electrode fouling as it scans through its triangular potential-time waveform as is typified by CV. Aside from inappropriate potential profiles, CV is virtually useless for observing the effects of time parameters of a PED waveform. On the other hand, *pulsed voltammetry* (PV) applies a PAD waveform with small incremental changes in one of the parameters of the waveform (e.g., the detection

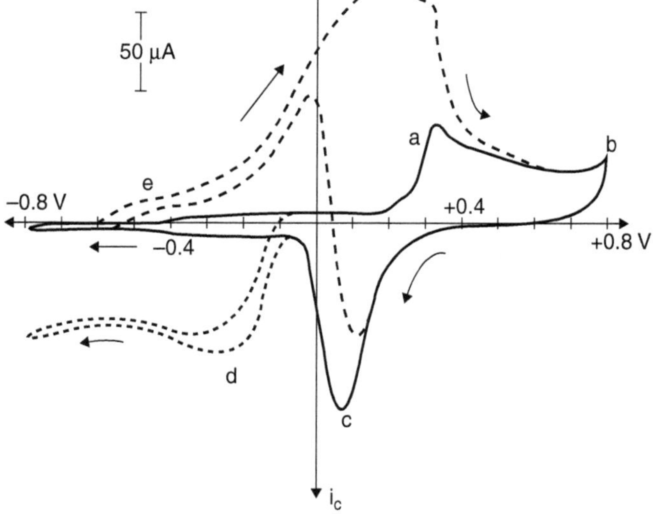

Figure 3.2. Cyclic voltammogram of glucose at gold electrode. Conditions: 900 rpm rotation speed, 200 mV/s scan rate. Solutions: (———) 0.1 M NaOH, de-aerated; (- - -) 0.1 M NaOH; and (- - - -) 0.4 mM glucose.

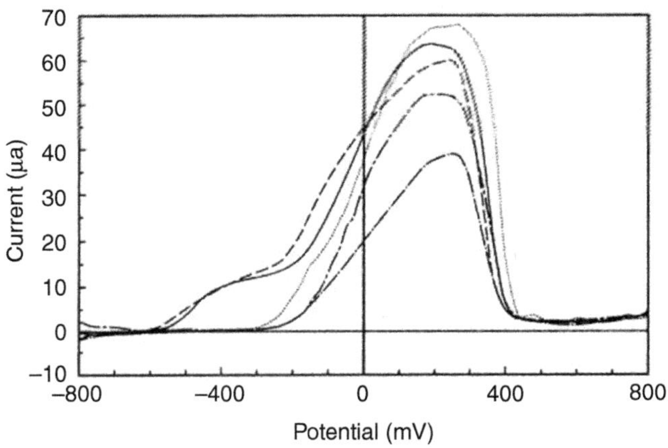

Figure 3.3. PV response for five carbohydrates at the Au RDE in 0.1 M NaOH. Solutions (0.1 mM): (••••) sorbitol; (———) glucose; (_-_-_) fructose; (-.-.-) sucrose; (------) maltose.

potential or E_{det}) for each cycle of a multistep waveform to a hydrodynamically controlled electrode (e.g., RDE). PV has proven to be the definitive method for studying and optimizing PAD waveforms, especially for Mode I detections (see section 3.4 for the details on Mode I detection). The "background-corrected" i-E_{det} response (positive scan direction) for equimolar concentrations of several carbohydrates in 0.1 M NaOH at an Au electrode is shown in Fig. 3.3. Beginning at ca. −600 mV, the signal corresponding to the oxidation of the aldehyde group of any reducing sugars (glucose and maltose) is clearly evident. The peak signal in the region of −200 to +400 mV corresponds to the oxidation of the alcohol groups of each of the carbohydrates and aldehyde oxidations of any reducing sugars. The dramatic attenuation of the carbohydrates occurs abruptly at ca. +400 mV as a consequence of inhibition by surface oxide formation. Since these PVs are 'background-corrected', which is another advantage of this technique, the oxide formation is not shown. Because the electrocatalytic oxidation process is dependent on the surface state of the electrode, similar voltammetric plots are obtained for virtually all other carbohydrates except that non-reducing carbohydrates will not show a wave commencing at -0.6 V, which corresponds to the oxidation of the aldehyde group. The optimal response is obtained at ca. +180 mV for these experimental conditions. The application of this potential is universal and results in the highly sensitive detection of virtually all carbohydrates. Also, pH-dependent shifts of the onset potentials of the oxidations processes mimic that of the surface oxide formation processes. A detailed description of PV has been published [20].

3.3.2 Amine- and Sulfur-Containing Compounds

In addition to the alcohol group of polar aliphatic molecules, electrocatalytic oxidations can also be extended to compounds with amine and/or sulfur moieties (e.g., aminoalcohols, amines, thiocompounds). When compared to alcohols, compounds with these functional groups exhibit much stronger adsorption characteristics at noble metal electrodes. The increased adsorption can be exploited to increase their preferential adsorption, and subsequent oxidation, relative to mobile phase and matrix constituents. Unfortunately, their oxidation signals are often concomitant with surface oxide formation processes, which is deleterious to their sensitive detection.

The voltammetric response for 0.1 M NaOH at an Au RDE (Fig. 3.4A) is essentially the same as that which was described earlier including the reduction of dissolved O_2 during both the positive and negative scans. This voltammogram is oriented in the North American convention with the negative potential polarity plotted to the right, which corresponds to the more negative potentials corresponding to higher electron energies. In the presence of 10 μM lysine (—), an anodic signal is observed on the positive scan beginning at ca. +0.1 V (region *E*). This signal corresponds to the direct oxidation of the amine functional group to a hydroxylamine, and as mentioned above, it occurs only during the positive potential scan in the region where surface oxide is being generated. On the reverse scan, no signal is observed, which suggests that dc amperometric detection in the range of ca. +0.6 to +0.2 V is not expected to produce a useful analytical signal at an Au electrode when covered by inert oxide.

The pulsed voltammograms of lysine at various concentrations in 0.1 M NaOH at an Au RDE (Fig. 3.4B) are in general agreement with the CV of lysine. The major portion of the analytical signal is in the region of ca. +180 mV to +500 mV, which are the potentials at which the kinetics of amine adsorption and the rate of oxide formation compete most effectively, is attributable to the oxidation of the amine groups on lysine. Above +500 mV, the signal is dominated by the formation of surface oxidation, which is very rapid at higher potentials. Also, the amine oxidation signal is observed to increase with an increase in lysine concentration. As with carbohydrates, amino acids and other polar aliphatic amines display similar voltammetric behavior.

Sulfur-containing compounds can also be electrocatalytically oxidized on Au and Pt electrodes. Sulfur moieties that are not in their highest oxidation state must also undergo adsorption to the electrode as a prerequisite to detection, and therefore at least one nonbonded electron pair must reside on the S-atom. Alcohol and amine groups can be oxidized directly at Pt [22] and Au [23,24] electrodes only under alkaline conditions, but S-compounds can be electrocatalytically oxidized across the entire pH range at both electrode materials. Hence, the oxidation of sulfur compounds under mildly acidic conditions is highly selective. As mentioned previously, the Au electrode is preferred over the Pt electrode to minimize interference from dissolved O_2.

The i-E plots in Fig. 3.5 are voltammograms with (———) and without (----) penicillamine in 250 mM acetate buffer (pH = 3.75) at an Au RDE. A concentration-dependent signal is observed on the positive scan at ca. +300 mV to ca. +1500 mV for (a) 50, (b) 20, and (c) 10 mM penicillamine, which can be attributed to the oxidation of the thiol and, to a limited degree, the amine moiety. The adsorption of S-containing compounds is so strong that the onset of oxide formation is pushed to higher potentials, and the background signal for oxide formation is different in the presence and absence of penicillamine. Since the true background in the presence of the analyte is not known, background correction via subtraction of the unperturbed analyte-free response (. . . .) from the analyte response with an altered background results in the

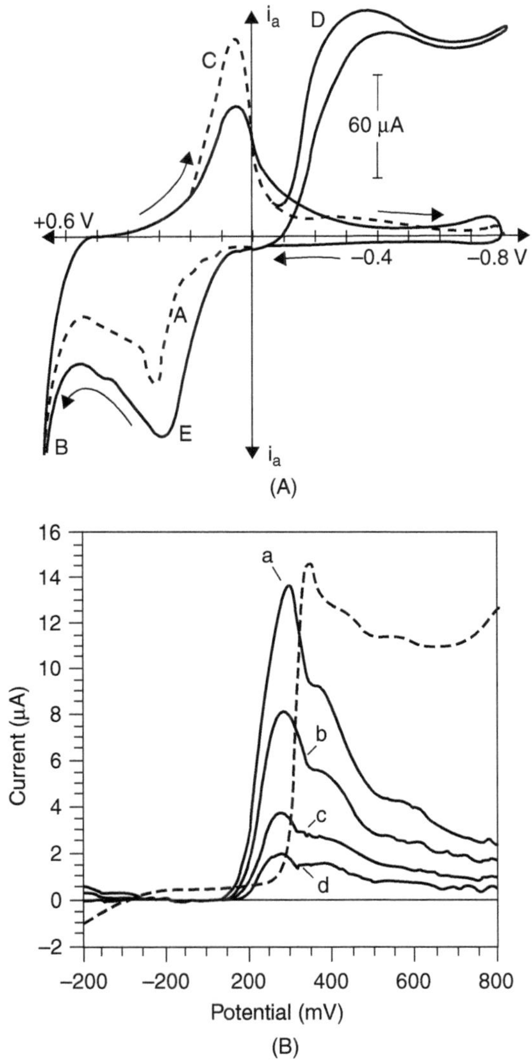

Figure 3.4. Current-potential (i E) response curves for lysine at a Au RDE by (A) cyclic and (B) pulsed voltammetry. Conditions: rotation speed, 1000 rpm; potential scan, 200 mV s^{-1}; and Ag/AgCl reference electrode. (A) Solutions: (.....) 0.1 M NaOH with dissolved O_2; (------) 0.1 M NaOH, deaerated; and (———) 0.2 mM lysine, deaerated. (B) Pulsed voltammetric responses of lysine are background-corrected. Conditions: rotation speed, 900 rpm. Waveform: $E_{\rm det}$, variable; $t_{\rm det}$, 450 ms; $t_{\rm del}$, 240 ms; $t_{\rm int}$, 200 ms; $E_{\rm oxd}$, +800 mV; $E_{\rm oxd}$, 180 ms; $E_{\rm red}$, −800 mV, $t_{\rm red}$, 360 ms. Solutions: (———) lysine, (a) 100 μM, (b) 50 μM, (c) 20 μM, (d) 10 μM; and (----) 0.1 M NaOH, de-aerated.

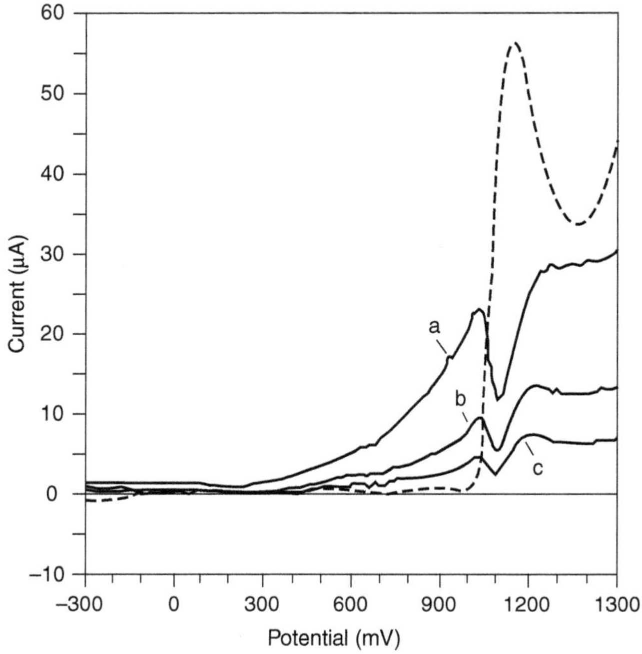

Figure 3.5. Pulsed voltammetric response of penicillamine at a Au RDE in 250 mM acetate buffer, pH 3.75. This plot is 'background-corrected'. Conditions: rotation speed, 900 rpm. Waveform: E_{det}, variable; t_{det}, 450 ms; t_{del}, 240 ms; t_{int}, 200 ms; E_{oxd}, +1500 mV; t_{oxd}, 180 ms; E_{red}, −300 mV, t_{red}, 360 ms. Solutions: (———) penicillamine, (a) 50 μM, (b) 20 μM, and (c) 10 μM; (----) residual, de-aerated.

negative dip at ca. +1050 mV. Hence, the negative dip in the analyte signal is an artifact. This phenomenon can also lead to negative post-peak dips when using PAD waveforms. In general, all sulfur compounds give very similar voltammetric profiles, except for the degree of perturbation of the residual response.

A deleterious consequence of electrocatalytic detection mechanisms at noble metal electrodes is the accumulation of adsorbed carbonaceous materials on the electrode surface, which eventually passivates the electrode surface resulting in a loss of activity. If a constant potential was applied to the Au electrode as in dc amperometry, the highly transient catalytic activity for carbohydrate oxidation would rapidly diminish due to fouling of the electrode surface. Because the practice of disassembling the electrochemical cell and mechanically polishing the working electrode, as in dc amperometry, is not possible for electrocatalytic detections, the only other option is to combine electrochemical detection with "on-line" cleaning. *Pulsed electrochemical detection* (PED) is designed to maintain uniform and reproducible electrode activity at noble metal electrodes for the oxidation of polar aliphatic compounds via the incorporation of on-line pulsed potential cleaning and reactivation of the electrode [14]. A complete description of PED and its application has been published [21].

3.4 FUNDAMENTALS OF PULSED ELECTROCHEMICAL DETECTION

The electrocatalytic detection process exploits the oxide-free, or "clean", surface of noble metal electrodes to adsorb organic compounds and their radical intermediates, which facilitates their oxidation via anodic oxygen-transfer from H_2O by transient, intermediate products in the surface-oxide formation mechanism (i.e., AuOH and PtOH). The on-line electrode cleaning and reactivation of PED is premised on the reversible cycling of the formation and dissolution of surface oxide by the application of alternating positive and negative potentials, respectively. Any fouling, which results as a consequence of the catalytic detection process or adsorbed compounds, is 'cleaned' from the electrode surface by pulsing to a more positive potential, which results in the formation of stable surface oxide (i.e., AuO and PtO) that forms at the expense of any other surface adsorbed species. A negative potential excursion is needed to reductively remove the surface oxide and may induce further cathodic cleaning of the electrode surface. When properly applied, the alternating positive and negative potential steps restore the native reactivity of the oxide-free metal surface, and gives PED its name.

The basis for all PED techniques at noble metal electrodes can be summarized by the application of some form of the following three steps: (1) electrocatalytic oxidation for detection, (2) oxide formation to "clean" the electrode surface, and (3) oxide removal to reactivate the electrode, further clean the electrode, and possibly pre-adsorb analyte prior to the next detection cycle. The steps allow for three modes of detection in PED, which are as follows:

- Mode I: Direct Detection at 'Oxide-Free' Surfaces. At potentials between oxide formation and dissolved O_2 reduction, electrocatalytic oxidation occurs with little or no concurrent formation of surface oxide. The surface-stabilized oxidation results in a product, which may leave the diffusion-layer, re-adsorb for further oxidation, or foul the electrode surface. The baseline signal originates primarily from double-layer charging, which decays quickly to a virtual zero value. All alcohol-containing compounds (e.g., carbohydrates) are detected by Mode I at Au electrodes in alkaline solutions and Pt electrodes in acidic solutions [21].
- Mode II: Direct 'Oxide-Catalyzed' Detection. In Mode II, the analyte is often adsorbed to the electrode surface in a prior potential step. During the detection step, electrocatalyzed oxidation of pre-adsorbed analyte is the primary contributor to the analytical signal; however, simultaneous catalytic oxidation of analyte in the diffusion layer is not excluded. In contrast to Mode I detections, compounds that occur by Mode II detections occur concurrently with the formation of surface oxide. The oxidation products may leave the diffusion layer, re-adsorb to the surface for further oxidation, or foul the electrode surface. The background signal, resulting from anodic formation of surface oxide, is large and has a deleterious effect on quantitation, which has been the driving force behind the development of numerous alternative waveforms. The most successful waveform is that of integrated pulsed amperometric detection (IPAD), which is virtually required for successful Mode II detections [18]. Aliphatic

amines and amino acids are detected by Mode II at Au and Pt electrodes under alkaline conditions. Numerous sulfur compounds are also detected by Mode II at Au and Pt electrodes over a wide range of pH conditions.

- Mode III: 'Indirect Detections' at Oxide Surfaces. Essential to Mode I and Mode II detections is the preadsorption of the analyte at the oxide-free surface of the electrode prior to its subsequent electrocatalytic oxidation. Species which adsorb strongly to the electrode surface and are electroinactive can also be visualized by their effect on the oxide formation process. Preadsorbed species reduce the effective area of the electrode surface, and the analyte signal originates from a suppression of oxide formation. Since the baseline signal results from anodic currents from surface oxide formation at a "clean" electrode surface of larger area, a negative peak results. Detection as a result of the suppression of surface oxide formation is known as Mode III detection. Because Mode III does not involve the electrocatalytic oxidation of the analyte, it is only 'indirectly' a PED technique. Sulfur-containing and inorganic compounds have been detected by Mode III [21,25,26].

A requirement of all modes of detection in PED is adsorption of the analyte to the noble metal electrode. This key feature has lead to the development of several approaches to indirect detection in PED. Mode III detections were first investigated by Polta and Johnson [27] and later Welch and coworkers [28–31] . Unfortunately, the extent of oxide formation is highly sensitive to numerous system parameters (e.g., pH, ionic strength, organic modifiers, temperature), which is deleterious to system performance and reproducibility. In addition, many analytes have a Faradaic response at the potentials of oxide-formation, which further complicates accurate quantitation of even well-separated compounds.

Another form of indirect detection has been investigated that uses the suppression of response by mobile phase constituents. Since detection occurs in the oxide-free region of the electrode, these approaches, strictly speaking, are considered to be derivatives of Mode I. Doscotch et al. [32] tried to circumvent the problems of Mode III detection by exploiting the suppression of the large cathodic signal resulting from the electrochemical reduction of dissolved oxygen in the mobile phase. This approach is called indirect adsorption detection (IAD), and it was found to be more tolerant to changes in pH than indirect detections based on Mode III. Among other problems, IAD suffered from poor interday reproducibility due to the inability to accurately control the amount of dissolved oxygen in the system. LaCourse [26] presented an alternate mode of indirect detection using PED based on the controllable addition of a PED-active reagent (e.g., polyols or carbohydrates) to the mobile phase to generate a stable anodic background signal. This approach exploits the well-understood anodic response of polyols or carbohydrates, which occurs at potentials where oxide formation is minimal. Olson et al. [33] applied this approach to the determination of amino acids and proteins with detection limits in the low picomole and low femtomole levels, respectively. This method, called indirect pulsed electrochemical detection (InPED), allows the use of a single, optimized waveform to detect a wide variety of compounds and depends on an easily controlled background signal.

Independent of the Mode, electrocatalysis-based detection of analytes within a class of compounds is controlled primarily by the dependence of the catalytic surface state on the electrode potential rather than by the redox potentials (E^o) of the reactants. All compounds within a class will give very similar I-E plots, and, as a consequence, voltammetric resolution of complex mixtures is mostly futile. Therefore, general selectivity is achieved via chromatographic separation prior to electrocatalytic detection. This conclusion does not preclude limited selectivity from judicious control of detection parameters.

3.5 WAVEFORM DESIGN AND RECENT ADVANCES

PAD applied to Au and Pt electrodes originally consisted of a three-step potential-time waveform, see Fig. 3.6a [18,34,35]. Waveform labels can be found in Fig. 3.10, *vide infra*. In this waveform, detection occurs at potential E_{det}, during the period t_{det}, with sampling of the faradaic signal over the time period t_{int} after a delay of t_{del} to overcome capacitance currents. The output signal is either the average of the current (amperes or coulombs per second) or the integrated charge (coulombs) over t_{int}. A sampling period of 16.7 ms or a multiple thereof reduces 60-Hz noise. Thereafter, oxidative cleaning of the electrode occurs at potential E_{oxd} during the period t_{oxd}, followed by reductive reactivation at E_{red} during the period t_{red}. Typically, the total waveform cycle ($t_{\mathrm{total}} = t_{\mathrm{det}} + t_{\mathrm{oxd}} + t_{\mathrm{red}}$) is ca. 1 s with a frequency of ca. 1 Hz.

Although the original PAD waveform was optimized using PV for high sensitivity, a four-potential step waveform that was termed the "quadruple-potential" waveform (Fig. 3.6b), was developed to improve long-term reproducibility [36]. During the detection step of a PAD waveform (Fig. 3.6A), soluble Au species are generated, which result in recession of the electrode over time. The quadruple-potential pulse waveform applies a large negative potential step, E_{red}, for a brief period, t_{red}, after the detection step in order to reduce any partially solvated Au species back to metallic Au. In addition, this step invokes reductive "cleaning" of the electrode surface. The negative potential step is followed by a brief positive potential step (E_{oxd}, t_{oxd}) to activate the electrode surface, which is subsequently followed by a negative potential pulse to effect adsorption and preconcentration of the analyte on the electrode surface (E_{ads}, t_{ads}).

Detection of amine and S-containing analytes occurs with concomitant formation of surface oxides that are needed to catalyze the anodic oxygen transfer process. The use of either of the PED waveforms discussed above typically results in a large background signal, which compromises the measurement of analyte response, is sensitive to variations in mobile phase conditions, and leads to post-peak dipping. Fig. 3.6c shows the original IPAD waveform, which was designed to coulometrically reject the large background signals originating from surface oxide formation [24,37,39]. Waveform labels can be found in Fig. 3.10, *vide infra*. In the IPAD waveform, the potential is scanned in a rapid cycle fashion between the values E_{dst} to E_{dmx} to E_{dnd} with a concurrent and continuous electronic integration of the amperometric signal. The values of E_{dst}, E_{dmx}, and E_{dnd} are chosen to correspond to

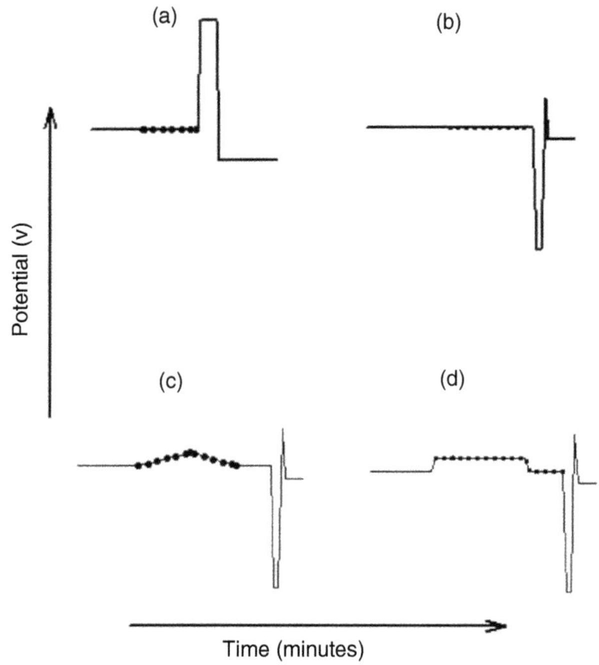

Figure 3.6. Generic potential vs. time waveforms used in PED. (a) Three-potential pulse waveform, PAD; (b) quadruple pulse waveform, PAD; (c) cyclic scan IPAD or PVD waveform; and (d) square wave IPAD waveform. Dots indicate integration periods.

potential regions before oxide formation occurs, maximal mass-transport dependent signal from the analyte oxidation, and post cathodic dissolution of the surface oxide, respectively. Hence, in theory, the anodic formation of the oxide required to catalytically stimulate the detection process does not contribute to the total integration of the amperometric signal. In other words, analyte and oxide formation signal minus oxide dissolution signal gives the response only for the analyte, which makes the detection appear to be independent of the oxide formation process. Because the anodic reactions of aliphatic compounds are highly irreversible, there is no cathodic contribution to the total integral from reduction of the detection products during the scan from E_{dmx} to E_{dnd}. Fig. 3.6d is a variation on the IPAD waveform that incorporates a square wave in place of a triangular wave as part of the detection step.

The IPAD waveform can be modified to record the current as a function of the applied scanning potential to generate three-dimensional data in the form of an i-E plot at every chromatographic time. This PED technique, which allows for "on-the-fly" voltammetry, is known as *Pulsed Voltammetric Detection* (PVD) [30]. Fig. 3.7 shows a surface plot of current vs. potential vs. time for the separation of lysine, galactosamine, serine, and sucrose using HPLC-PVD [21]. These plots are background-corrected "on-the-fly". Note that the amine-containing compounds (i.e., lysine, galactosamine, and

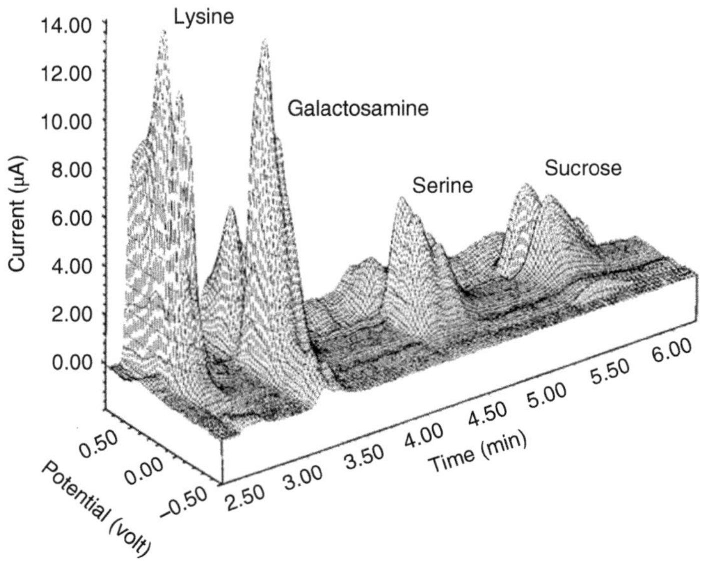

Figure 3.7. HPLC-PVD of various mixtures depicted in a surface plot for bioactive compounds. Note the individual signatures in the contour plot for each of the compounds.

serine) have signal at high potentials; whereas, the compounds with hydroxyl groups (i.e., galactosamine, serine, and sucrose) have signals in the oxide-free region of the electrode. At any potential a chromatogram can be extracted to afford greater selectivity, and at any time point a voltammogram can be extracted to identify/characterize the analyte/peak. The results shown here illustrate the feasibility and doubtless importance of PVD to enhance selectivity and compound characterization, afford a limited degree of functional group identification with peak purity, and allow for quantitation of the compound of interest. The benefits of on-line voltammetric detection in chromatography have been previously reviewed [40,41].

In 2005, Dionex (Sunnyvale, CA) introduced 3D Amperometry [42], which essentially delivers current versus time plots at every chromatographic time point. This information-rich 3D data output allows for signal optimization and compound fingerprinting. The data is viewed in wireframe or isoamperometric displays to study reaction characteristics of analytes. Integration periods for pulsed amperometry are easily optimized post-run without having to reinject samples. This capability is especially useful for waveform optimization.

3.6 MICROELECTRODE CONSIDERATIONS

The trend for most separation techniques including IC is miniaturization. Electrochemical detection, including PED, is ideally suited to microseparation systems because detection is based on a reaction at an electrode surface (i.e., $i = n{\bullet}F{\bullet}A{\bullet}D{\bullet}C^{\mathrm{b}}/\delta$), where i, n,A,D,C^{b}, and δ are current, number of electrons, electrode surface area, diffusion constant, bulk solution concentration, and hydrodynamic diffusion layer thickness, respectively. The response in optical detection methods are based on Beer's Law (i.e., $A = abC^{\mathrm{b}}$), which is dependent on the path length of the detector cell. The loss of response upon miniaturization in optical detection systems is further exacerbated by compounds with poor optical detection properties. Electrochemical detection following microchromatographic and capillary-based separation techniques offers increased mass sensitivity, higher chromatographic efficiencies, less solvent consumption, and, in particular, the ability to analyze samples of limited quantity. PED affords these same advantages to virtually all polar aliphatic compounds, especially those with poor optical detection properties (e.g., carbohydrates). Because PED maintains electrode activity on-line, the electrodes *do not* require daily polishing, which is often difficult in miniaturized detection formats.

Microelectrodes with dimensions of 0.2–50 μm diameters are used extensively for microseparation techniques. In addition to dimensional compatibility, microelectrodes have properties of enhanced mass transfer to the electrode, low iR-drop, and low electrochemical cell time constant. The latter two properties allow for fast response of the electrode to a potential pulse or to a potential ramp; scan rates up to 100,000 V/s^{-1} have been feasible with gold 7-μm discs. Fig. 3.8 shows the results from cyclic voltammetry scans (at 10 V/s^{-1}) with 5 mm, 3 mm, 1 mm RDE's, and a 50 μm × 1 mm Au fiber electrode. In order to present all results for widely differing electrode areas on the same scale, current was normalized by dividing by the anodic peak current. It is evident that the start of anodic activity due to AuO formation is at nearly the same potential for all electrodes, but the shape of the peak is changed greatly with electrode size, showing widening and smearing effects, to the extent that a peak is not developed for the 5 mm and 3 mm electrodes. These effects are due to the high iR drop and large cell time constant that exist for the large electrodes. Even more exaggerated effects are noted for the AuO reduction peaks. Similar comparisons were made by LaCourse and coworkers (unpublished data) when they exploited the advantages of microelectrodes to perform "in-cell" pulsed voltammetric detection (PVD) scans in a microseparation system.

3.7 IMPLEMENTATION OF IC-PED: MODE I AND II REVISITED

The great majority of applications of PED in IC are for Mode I, or oxide-free, detections, which are implemented with a three- or four-step potential-time waveform at a frequency of ca. 0.5–2 Hz. This frequency is appropriate for maintaining chromatographic peak integrity in HPAEC. Optimization of waveform parameters in PAD is best accomplished using *pulsed voltammetry* (PV), but most authors choose to optimize detection either through literature-based waveforms or experientially. Fig. 3.9 summarizes the entire process of chromatographic peak formation in HPAEC-PED for a three-step potential waveform. The development of (a) positive peaks for carbohydrates is illustrated by the (b) chronoamperometric (i-t) response curves generated following the (c) potential step from E_{red} to E_{det} in the PAD waveform. The residual

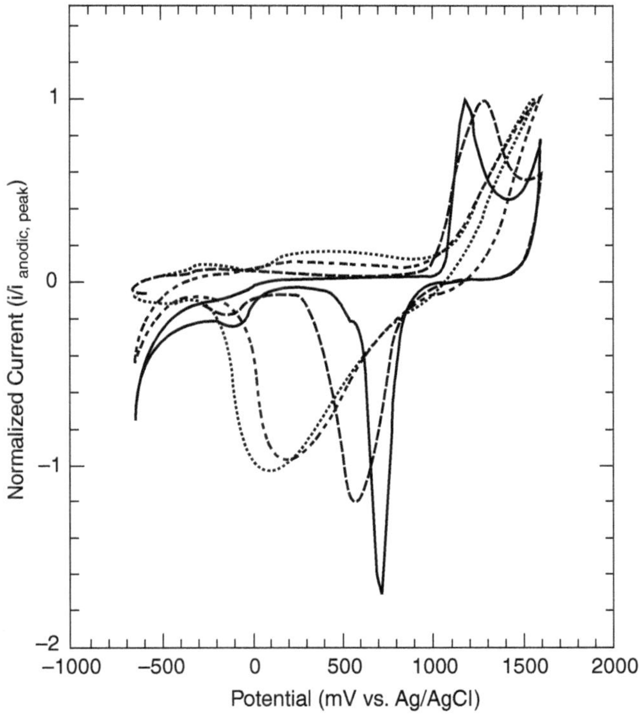

Figure 3.8. (A) Voltammetric response of Au electrodes as a function of electrode size. Conditions: supporting electrolyte, 95% 100 mM phosphate buffer (pH 3)/95% CH_3CN; rotation rate for rotating disk electrodes (RDEs), 900 RPM; scan rate: 10,000 mV/s^{-1}. Curves: (•••••) 5-mm RDE; (----) 3-mm RDE; (– – – –) 1-mm RDE; and (———) 50 μm fiber.

current decays quickly, and the baseline signal in IC-PAD is minimal for t_{del} > ca. 100 ms for 1.0 mm Au electrode in 0.1 M NaOH. The transient *i-t* response for the presence of the carbohydrate is dependent on its concentration in the electrochemical cell, and the peak shown is representative of the corresponding anodic signal expected in IC-PAD for the value of t_{del} indicated. Compound selectivity is achieved primarily via chromatographic separation.

PED at Pt and Au electrodes is also capable of producing sensitive and reproducible detection responses for aliphatic amines and sulfur-containing compounds. For these compounds, detection occurs simultaneously with the formation of surface oxide, and the anodic mechanisms are believed to be catalyzed by the formation of intermediate oxide species (AuOH and PtOH). These detection mechanisms (Mode II) are believed to involve prior adsorption of compounds at the oxide-free electrode surface and, hence, compounds detected are those with at least one pair of non-bonded electrons on nitrogen and/or sulfur atoms. Therefore, quaternary aliphatic amines, sulfones, and sulfonates are not detected. A large baseline signal is encountered for the oxide-catalyzed detections of amine- and sulfur-based compounds (Mode ll). Mode II

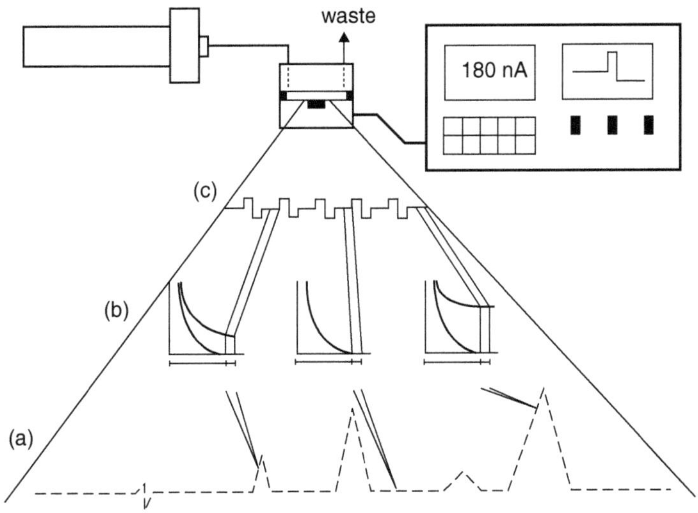

Figure 3.9. Schematic drawing of a typical HPLC-PED set-up showing how a (a) chromatogram is produced from the (b) sampled current response derived from the application of a (c) potential vs. time waveform to the working electrode of an electrochemical cell. The current is integrated typically for a fixed interval of time (t_{int}) after a delay period (t_{del}), which allows capacitive current to dissipate.

detections performed with PAD are subject to a number of disadvantages due to the formation of surface oxide, which is required, and concomitant with the detection of amine- and sulfur-based compounds.

Baseline offset and drift can be significantly diminished by use of the IPAD waveform. Fig. 3.10 schematically compares PAD (left) versus IPAD (right) for an oxide-catalyzed detection. Using the IPAD waveform, the electrode current is integrated throughout a rapid cyclic scan during the detection potential (E_{det}) step within a pulsed waveform. The potential scan proceeds into (positive scan) and backs out (negative scan) of the region of the oxide-catalyzed reaction for detection by Mode II. The anodic charge for oxide formed on the positive sweep during the detection period tends to be compensated by the corresponding cathodic charge (opposite polarity) for dissolution of the oxide on the negative sweep. Hence the 'background signal' on the electronic integrator at the end of the detection period can be virtually zero and is relatively unaffected by the gradual change of electrode area. More importantly, the majority of the analyte, which is preadsorbed to the electrode surface during E_{red}, is oxidized within the first 100 ms. If a typical delay time of 200–300 ms is used as in PAD, the majority of analytical signal is lost. In IPAD, the sweep starts prior to oxidation of the analyte, sweeps into oxide formation with the concomitant Mode II detection of the analyte, and proceeds back to the original starting potential. Hence, the majority of analyte signal is conserved while electronically rejecting the oxide

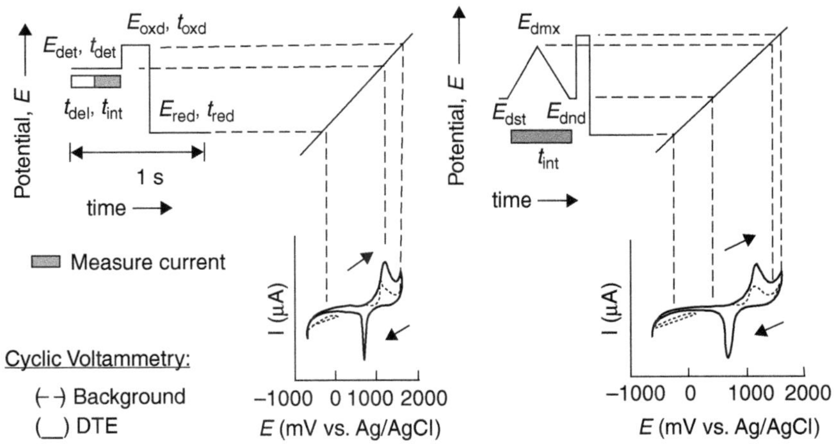

Figure 3.10. Comparison of the PAD and IPAD waveforms as translated to cyclic voltammetric response of supporting electrolyte and dithioerythritol at a Au RDE electrode.

background. The detection procedure based on the IPAD waveform in Fig. 3.10 combines cyclic voltammetry with potential pulse cleaning to maintain uniform electrode activity. The waveform was originally called potential sweep (PS)-PCD; however, the name integrated pulsed amperometric detection (IPAD) is now preferred.

The advantage of IPAD compared with PAD relates to maximization of the analyte signal, minimization of baseline magnitude, and the near elimination of drift for oxide-catalyzed detections (Mode II). There is no significant difference in signal-to-noise (S/N) between PAD and IPAD for carbohydrates, or Mode I detections.

3.8 IC-PED FOR PHARMACEUTICALS AND BIOLOGICAL SAMPLES

PED following ion chromatographic separation of carbohydrates using HPAEC is now well established for the direct detection of monosaccharides, disaccharides, oligosaccharides, polysaccharides, and sugar alcohols. The focus of this chapter up to this point has been on the mechanism of detection, but the popularity of PED is reflected more "accurately" in how well it works for "real-world" applications. The number of papers involving PED has increased significantly since its inception, and many of these have been summarized in review papers, chapters, and a book [21]. Rather than attempting to review the individual papers, I shall focus on a brief review of basic applications that highlight applications for pharmaceuticals and biological samples in relation to classes of compounds (e.g., carbohydrates, amine-based compounds, and thiocompounds). The intent is to enable the reader to better apply HPAEC-PED to one's own application.

3.8.1 Carbohydrates

All aldehyde- and alcohol-containing compounds (e.g., carbohydrates) can be detected by PED at Au and Pt electrodes in alkaline media (pH > 12). However, use of Au electrodes has the distinct advantage that detection can be achieved without simultaneous reduction of oxygen. The maximum anodic signal for polyalcohols and carbohydrates in 0.1 M NaOH is obtained at $E_{det} = +100$ to $+200$ mV vs. Ag/AgCl, and a value in this range is chosen for E_{det} in the PAD waveform.

Carbohydrates are weak acids. Hence, chromatographic separations of complex mixtures can be achieved using HPAEC with alkaline mobile phases. Paskach et al. [43] determined the retention times for 93 sugars and sugar alcohols with one to four residues using HPAEC under isocratic elution with 0.1 M NaOH. Capacity factors for monosaccharides are lowest for sugar alcohols, and higher for analogous aldoses and ketoses. In general, capacity factors increased with an increasing number of carbon atoms. Present technology has pushed the limits of detection for monosaccharides to 0.1 to 1 picomole.

Separations of complex mixtures of carbohydrates often require gradient elution chromatography. One approach is to use a NaOH gradient [44]. Unfortunately, the baseline shift is observed corresponding to the gradient profile as a consequence of the pH dependent oxide formation mechanism in PED. The following approaches have been used to overcome the shift in baseline due to NaOH-gradients:

- The addition of concentrated NaOH (e.g., >600 mM) post-separation/post-column can mitigate the pH change from a difference of 2 units to under 0.1 pH unit. Although the baseline shift is greatly reduced, the effects of dilution and the still changing background signal hamper high sensitivity applications.
- The most technically accurate approach is to substitute a pH reference electrode for the more common fixed-potential Ag/AgCl reference electrode. Since the potential of the pH reference shifts similarly to the pH dependence of the oxide formation mechanism (i.e., -60 mV pH^{-1}), the detection potential relative to the potential of the onset of oxide formation remains essentially constant. The overall result is almost complete alleviation of the background signal due to the changing pH of the gradient elution profile.
- An alternate approach is to perform the gradient separations of complex mixtures of carbohydrates using a PED-inert anionic species while holding the NaOH concentration in the mobile phase constant. Acetate ion gradients can readily be substituted for hydroxide ion gradients with excellent results. This approach allows the use of the Ag/AgCl reference electrode, which is more stable and less noisy then the pH reference electrode. The use of acetate ion gradients for complex mixtures of carbohydrates is the most popular.

The application of the greatest impact to the pharmaceutical and biotechnology industries of HPAEC-PAD is its application to the separation and quantitation of neutral and amino monosaccharides, as well as positional isomers of glycopeptides [45,46], derived from glycoproteins. The introduction of HPAEC-PAD facilitated the

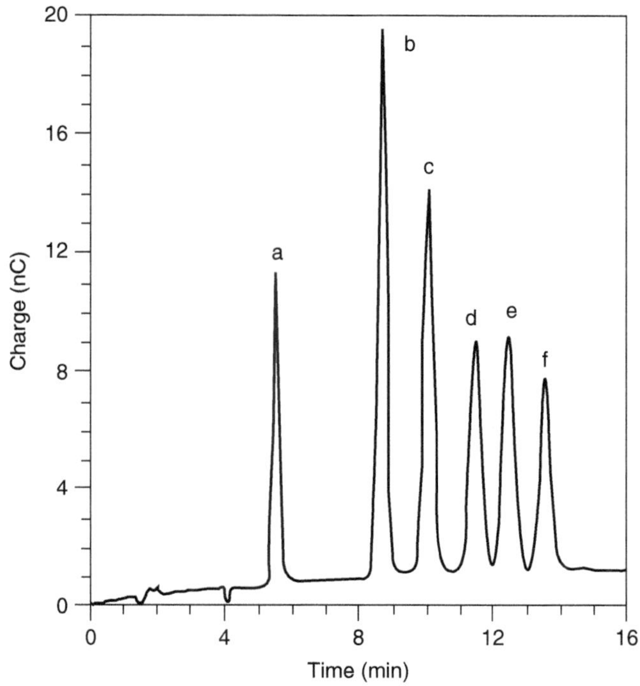

Figure 3.11. Anion-exchange separation of common monosaccharides found typically in mammalian glycoproteins. Peaks (100 pmole each): (a) fucose; (b) galactosamine; (c) glucosamine; (d) galactose; (e) glucose; and (f) mannose. Column: CarboPac PA-10 (Dionex); Mobile phase: 16 mMNaOH (isocratic); Flow rate: 1.0 mL/min; Electrode: Au. Waveform:$E_1 = 100$ mV, $E_2 = -2000$ mV, $E_3 = +600$ mV, $E_4 = -100$ mV, $t_1 = 0$–400 ms, $t_{\rm int} = 200$–400 ms, $t_2 = 410 - 420$ ms, $t_3 = 430$ ms, $t_4 = 450 - 950$ ms (vs. Ag/AgCl).

elucidation of the complex antennary structures of glycoconjugates without the need of any pre-injection or post-column derivatization chemistry. Under weak alkaline conditions (e.g., 10–20 mM NaOH), the separation of isomeric monosaccharides, such as glucose and galactose, is possible, and Fig. 3.11 shows the separation of monosaccharides that typically are found in the hydrolyzates of mammalian glycoproteins [47]. It should be noted that 2-deoxyglucose is often added as an internal standard as it is not endogenous in mammalian systems. Amino sugars, or basic sugars, typically elute earlier then their non-substituted analogues. This assay allows for the compositional analysis of a glycoconjugate. Additionally, the efficient separation of HPAEC combined with PED allowed for the assay and characterization of complex oligosaccharides and polysaccharides.

High-mannose, hybrid, and complex oligosaccharides, as well as polysaccharides are readily separated and detected using HPAEC-PAD. Undoubtedly, the most impressive aspect of HPAEC is its ability to resolve closely related structures of glycopeptides. Fig. 3.12a shows the structures of two asialylated fetuin glycopeptides that

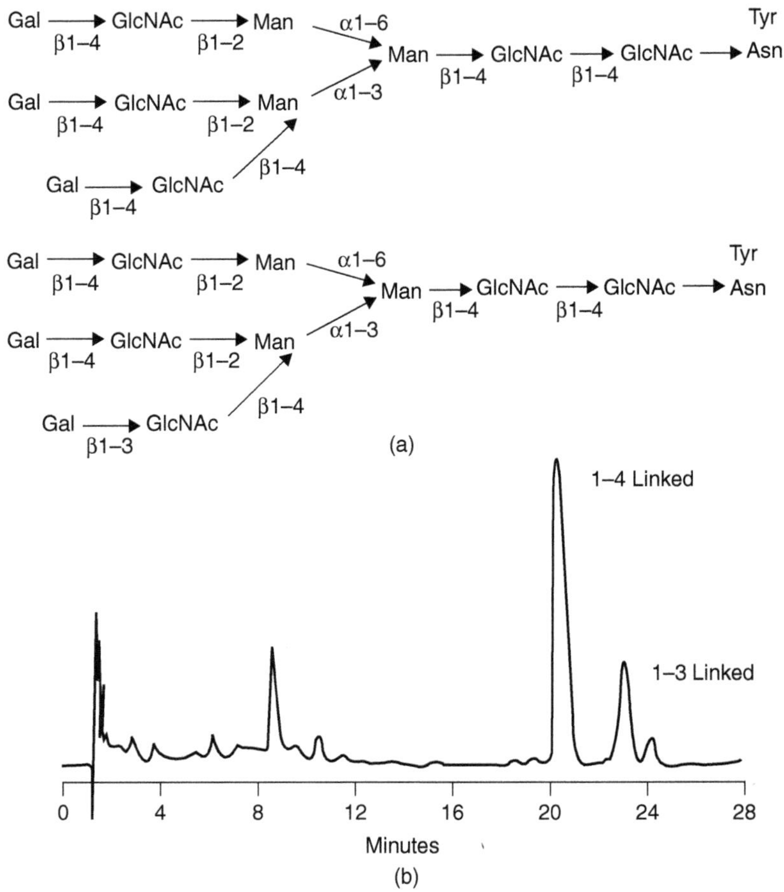

Figure 3.12. (a) Structures and (b) analysis of asialylated fetuing lycopeptides. Run conditions can be found in Technical Note 42 (1997) from Dionex (Sunnyvale, CA).

differ by only a single bond linkage. Still, the two compounds are baseline separated using HPAEC, as shown in Fig. 3.12b. The application of HPAEC-PAD to the characterization of glycoproteins has been reviewed by Townsend [48,49] and is discussed in Chapters 17–21 of this book.

Bioapplications include the determination of xylitol in serum and saliva [50], sugars in red blood cells [51,52], bile acids [53], and polysaccharide and glycoconjugate vaccines [54–58]. Applications of IC in the analyses and characterization of oligo- and polysaccharides in vaccines have been discussed in more details in Chapters 22–24 of this book. HPAEC-PAD has been used for the assessment of carbohydrates in HIV patients [59] and as a diagnostic tool for lysosomal diseases [60] and diabetes [61]. Application of HPAEC-PAD to carbohydrate analysis in "real-world" samples has been extensive, which reflects the analytical utility of this technique.

3.8.2 Alcohols and Polyols

As expected, PAD using either Pt or Au working electrodes under acidic and alkaline mobile phase conditions, respectively, can be used for the determination of underivatized aliphatic alcohols and polyalcohols. A mixture of thirteen alcohols and glycols has been separated using ion exclusion chromatography under isocratic elution with 50 mM $HClO_4$ [62]. Since alcohols and polyalcohols are not ionizable, the retention of smaller alcohols (e.g., ethanol) is based mainly on hydrogen bonding and size exclusion; whereas, larger alcohols are retained by adsorptive interactions with the resin polymer. The limit of detection for ethylene glycol by this method is ca. 10 ppb or ca. 1 pmol in a 50 μl injection. This method was applied to the determination of alcohols in liquid cold formulas and brandy [62]. In the same paper, the authors used reversed-phase chromatography to detect 1,6 hexanediol, 1,4 cyclohexanediol, and cefazolin using gradient chromatography with post column addition of NaOH.

3.8.3 Aminoalcohols, Amino Sugars, and Aminoglycosides

The mechanism of PAD requires adsorption of analyte prior to its detection via electrocatalytic oxidation at the electrode surface. Hence, the detection under Mode I of the alcohol group of aminoalcohols, amino sugars, and aminoglycosides is enhanced by the amine functional group, which is more strongly adsorbed to the electrode surface. The enhanced sensitivity is attributable to increased residence time of the analyte at the electrode surface via the adsorption of the amine group(s). This is the reason that peaks (b) galactosamine and (c) glucosamine in Figure 3.11 have a larger response than the others in the chromatogram. Since the amine group is able to effectively compete with acetonitrile for adsorption sites on the electrode surface, virtually no interference from acetonitrile is observed for the detection of these multi-functional compounds. On the other hand, carbohydrates in the presence of acetonitrile in the mobile phase are blocked from adsorbing to the electrode surface and their signals are dramatically reduced.

Aminoalcohols (i.e., alkanolamines) are important in the chemical and pharmaceutical industries for production of emulsifying agents, corrosion inhibitors, laundry additives, dyes, and for purifying gases. The high sensitivity and selectivity of PED contributes to decreased time for sample preparation and simplified chromatograms. These compounds can be separated using either pH tolerant, mixed-mode (i.e., ion-exchange and reversed-phase) or ion-pairing chromatography. Since the detection of these compounds using a PAD waveform requires alkaline conditions, post-column addition of concentrated NaOH is required. Ethanolamine, triethanolamine, and aminomethylpropanol in cold tablets, hand lotions, and hair spray formulations have been detected using PAD [63].

The ability to quantitate aminoglycosides in a variety of matrices (e.g., fermentation broths, formulations, foodstuffs, and biological matrices) has been hindered by poor optical detection properties, and problems with chemical derivatization as a consequence of their multifunctionality (i.e., forms multiple product species). On the other hand, the multi-functional nature of aminoglycosides, which also results in enhanced sensitivity, is ideally-suited to PED. The earliest application of aminoglycosides by PAD was by Polta et al. [64]. They focused their efforts on nebramycin factors, which are produced by fermentation of *Streptomyces tenebrarius*. Tobramycin and apramycin were determined in fermentation broth and spiked blood serum by LC-PAD using this technique. Detection limits were ca. 80 ppb for tobramycin. HPAEC-PAD was also used for the determination of tobramycin by Statler [65], who achieved limits of detection of 0.2 ppb. McLaughlin and Henion [66] used PAD with ion-spray mass spectrometry (MS) detection for spectinomycin, hygromycin B, streptomycin, and dihydrostreptomycin. They found that PAD detection was compatible with the ion-pairing agent and more versatile than MS detection. Other antibiotics applications include spectinomycin and its degradants [67], the four major components of gentamicin to aid in the identification of the source of pharmaceutical preparations of gentamicin [68], and neomycin A, B, and C in topical lotions [69]. The determination of aminoglycoside antibiotics by IC-PED is of significant research interest, and a review of this topic has been published [70]. Also a chapter in this book (Chapter 8) describes recent advances in HPAEC-PAD assays of aminoglycoside antiobiotics.

3.8.4 Amines, Amino Acids, Peptides, and Proteins

Numerous amine-based compounds can be detected directly at Pt and Au electrodes in alkaline solutions by the oxide catalyzed mechanism of Mode II using PED. The criterion for detection by PED is the existence of a nonbonded electron pair on the N-atom of the amine to facilitate adsorption at the electrode surfaces. Therefore, primary, secondary, and tertiary amines are detected, and quaternary amines are not.

Several studies have focused on the detection of aliphatic monoamines and diamines by PAD [71–73] waveforms, but amine detections are oxide-catalyzed and are best performed using IPAD waveforms. A fine example is by Draisic et al. [71]. He used cation-exchange chromatography followed by IPAD to determine biogenic amines (i.e., putrescine, histidine, cadaverine, histamine, and spermidine) in spoiled canned herring.

Simple mixtures of amino acids can be separated isocratically using anion-exchange chromatography [74]. For complex mixtures, it is essential to perform separations with gradient elution chromatography. Fig. 3.13 shows the chromatogram of 17 amino acids using AAA-Direct™ system (Dionex), which uses an anion-exchange column with a quaternary gradient that incorporates both a pH and an organic modifier gradient [75]. Presently, detection limits for amino acids are typically 1–50 pmole injected with comparable sensitivities for primary and secondary amino acids. Amino acid analysis using ion chromatography is also discussed in Chapter 19 of this book.

Other than the glycopeptide assays discussed earlier in this chapter, IC-IPAD for peptides has been used for the determination of bovine fetuin hydrolyzates [26], tyrosine and tryptophan-containing peptides [76], and sulfur-containing peptides [76,77]. IPAD tends to be more sensitive than UV-absorption detection at 214 nm for small peptides.

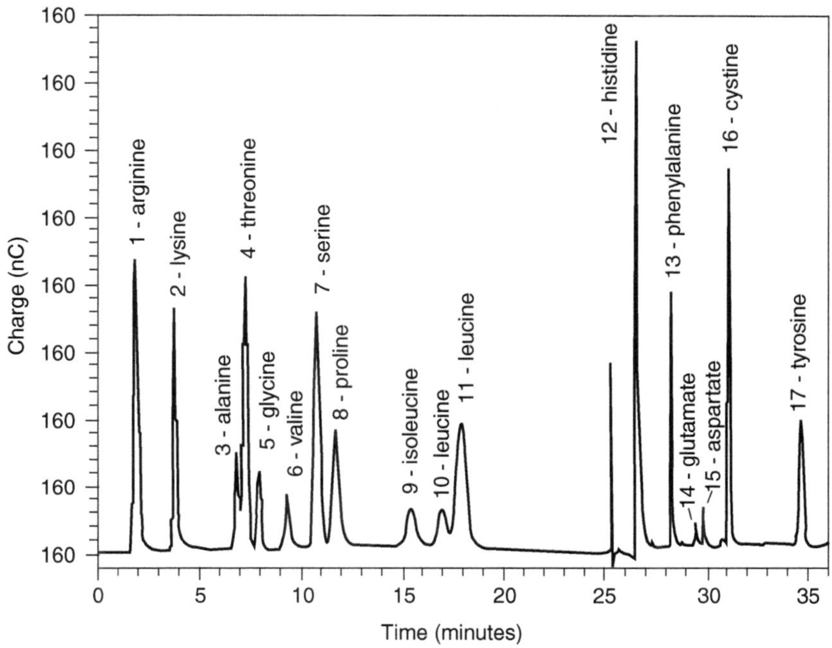

Figure 3.13. HPAEC-PED results for a mixture of seventeen amino acids (AA S-18 standard from Sigma) at 500 pmole each. Run conditions can be found in Technical Note 50 (2001) from Dionex (Sunnyvale, CA).

3.8.5 Sulfur Compounds

As with amine containing compounds, sulfur compounds are strongly adsorbed at the oxide free surfaces of Au and Pt electrodes, and as a consequence, sulfur-containing compounds are best detected using IPAD. Electrode fouling is more prevalent when detecting sulfur compounds than either alcohol or amine compounds, which often requires the use of more aggressive waveforms [21]. Since sulfur-containing compounds are amenable to sensitive and selective detection under mildly acidic conditions, almost all papers published about IPAD have used it following reversed-phase separations. The IPAD waveform typically gives limits of detection of the order of 1 pmole injected. Studies by LaCourse [78–80] have focused on the analysis of sulfur-containing antibiotics. IPAD offers almost universal detection of penicillins and cephalosporins.

3.9 ADVANCED APPLICATIONS OF IC-PED

3.9.1 Fingerprinting of Bioproducts

HPAEC followed by PED has been applied to the direct detection of sugar alcohols, monosaccharides, oligosaccharides, aminoglycosides, amino alcohols, amino acids, and numerous thiocompounds. Many of these compounds are the building blocks of cells, which means that IC-PED may be used to chemically profile or "fingerprint" biomaterials and products.

3.9.2 Carbohydrate Analysis of Bacterial Polysaccharides

In bacterial infections, the difference in virulence between strains within one species is often associated with the structure and composition of the bacterial polysaccharide. Composition of the carbohydrate is readily accomplished using HPAEC-PED. Critical to complete structural characterization is the determination of the absolute configuration of the carbohydrates. Interestingly within regards to bacteria, both enantiomers (i.e., D- and L- configurations) of a particular monosaccharide occur with regular frequency and, occasionally, within the same polysaccharide [81–83].

By combining HPAEC-PED with in-line laser polarimetry (OR) both the composition and the enantiomeric configuration of component sugars [84] can be determined in a single chromatographic run. The response of 500 ppm of (a) L- and (b) D-fucose by PAD and OR following retention using HPAEC is shown in Fig. 3.14. Since the sensitivity of PAD is far greater than the on-line polarimeter, the PAD signal is nearly overwhelmed by the high concentration of fucose. Also, PAD is completely insensitive to the absolute configuration of the carbohydrate. On the other hand, OR clearly shows the configuration of the fucose with a negative peak for the L-configuration and a positive response for the D-configuration. Sensitivity and peak direction are directly related to the magnitude and sign, respectively, of the specific rotations of each monosaccharide. This approach has now been applied to the analysis of the capsular polysaccharides of several Gram positive and Gram negative pathogenic bacteria with success [84]. Fig. 3.14c shows the chromatogram for the monosaccharide mixture derived by hydrolysis of the capsular polysaccharide of *Streptococcus pneumoniae* type 12F. The monosaccharides are identified, in order of increasing retention, as L-fucose, D-galactosamine, D-galactose, and D-glucose. The combination of HPAEC-OR-PAD offers a new paradigm for "carbo-typing" bacteria, which is presently limited by the poor sensitivity of the OR detector.

3.9.3 In-vitro Microdialysis for Carbohydrate Systems

Microdialysis (MD) is an on-line, continuous sampling technique that uses a semi-permeable membrane with a specific molecular weight cut-off (MWCO) [85] to size differentiate between small analyte(s) of interest and larger molecules in a complex sample matrix. This approach, which is more commonly used in vivo, has opened up research in the area of pharmacokinetics and biological responses to various stimuli [86,87]. Recovery of the analytes across the membrane depends upon numerous factors, which include the nature of the analyte(s), matrix properties, temperature, mass transport, perfusion flow rate, membrane type, area, geometry, and thickness. Many of these parameters are either optimized or standardized to a set value or condition. Because the individual recovery value for different analytes may vary, many of the

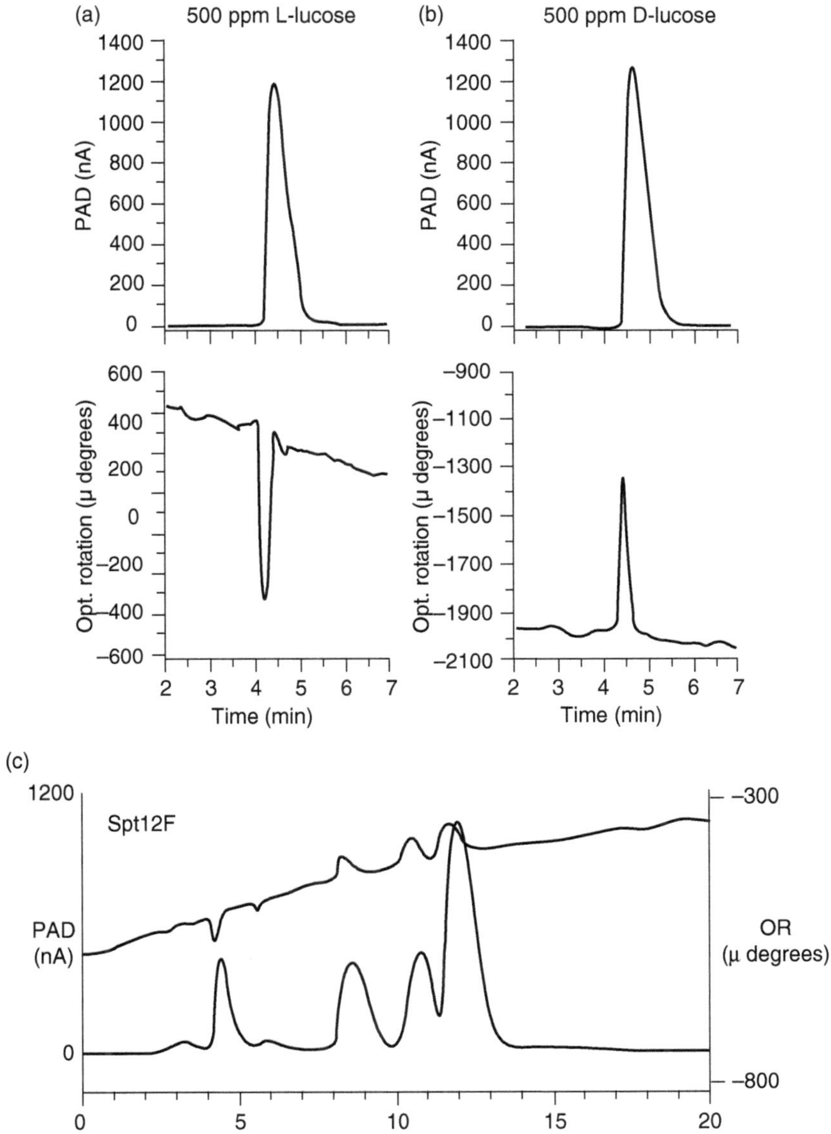

Figure 3.14. Chromatograms of 500 ppm (a) L-fucose and (b) D-fucose. Upper and Lower plots are PAD and OR, respectively. HPAEC-OR-PAD of (c) *Streptococcus pneumoniae* type 12 F. Upper and lower traces are OR and PAD, respectively. Column: CarboPac PA-1 (Dionex); Mobile phase: (a) 18 mM, (b) 16 mMNaOH (isocratic); Flow rate: 0.90 mL/min; Electrode: Au; Optical wavelength of polarimeter: 670 nm.

in vitro MD studies that have been carried out are qualitative, as opposed to providing quantitative results [88].

Combining MD sampling with a chromatographic technique, researchers are able to sample reaction mixtures continuously to obtain a dynamic or kinetic profile [89,90]. This approach is uniquely suited to the characterization of enzyme systems, where the separation of substrate from product is a necessity, thereby allowing kinetics of each to be observed individually by UV absorbance or other forms of detection.

3.9.4 Enzyme Characterization

Enzymes are the reaction catalysts of biological systems. They have extraordinary catalytic power, often a high degree of substrate specificity, and can greatly accelerate specific chemical reactions [91]. Current methods for determining Michaelis-Menten enzyme kinetic parameters most often involve direct or indirect spectrophotometric methods [92], which are not amenable to carbohydrate-enzyme systems due to the poor optical detection properties of the substrates and products.

The first report of on-line microdialysis sampling coupled to HPAEC and PED for carbohydrate analysis came in 1995 [93]. The study involved the determination of different oligosaccharides produced during the hydrolysis of 0.25% ivory nut mannan by endomannanase from *A. niger*. This was a preliminary study and provided *no quantitative results*, but showed that a small-scale bioprocess (hydrolysis) could be monitored continuously for a period of 32 h.

Zook and LaCourse [94] used microdialysis sampling combined with HPAEC-PED to quantitatively monitor the hydrolysis of lactose in skim and whole milk into glucose and galactose via commercially sold Lactaid™ drops containing β-glucosidase/lactase. Quantitative results were obtained by adding an internal standard (e.g., deoxyglucose) to correct for variances in recovery. Limits of detection were found to be 1–3 ng and recovery data in milk matrix showed an average RSD of 5.0%. Standard first order exponential decay curves were used to determine the observed rate constants, which were found to correlate well with the manufacturer's values. Zook and LaCourse used the same system to characterize several other enzyme systems, including the glucose oxidase reaction [85] and the digestion of amylopectin by the enzyme isoamylase [95]. The Torto group [96,97] carried out similar work on the enzymatic hydrolysis on potato and corn starches to determine information about molecular structure. The chain length distribution patterns of the hydrolyzed starch of waxy, normal, and two types of high amylose maize were determined, which highlighted the potential of MD-HPAEC-PAD for fingerprinting of different plant species [97,98].

The LaCourse group [99] extended their earlier work for the accurate determination of K_m values using PAD-active, chromophoric substrates, whose enzymatic parameters could be confirmed using standard spectrophotometric assays and literature values. Microdialysis was used to monitor the enzymatic hydrolysis of carbohydrate substrates by almond β-glucosidase [100,101] to obtain Michaelis-Menten enzyme constants. The enzyme catalyzes the hydrolysis of a broad array of substrates including the model nitrophenyl glycosides. Fig. 3.15 shows the decrease

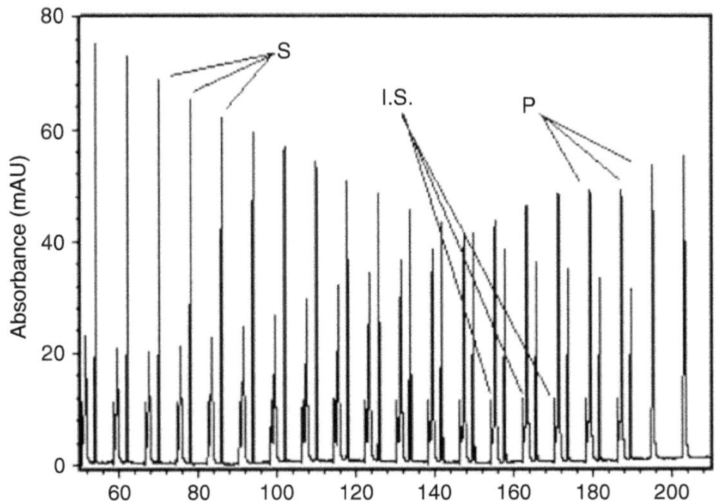

Figure 3.15. Chromatogram of enzymatic solution, consecutive injections, duration 100 minutes. Initial concentrations 12.5 μg/mL of enzyme in 30 mL of 50 ppm internal standard (IS). 100 ppm 4-nitrophenyl-β-D-glucoside(S) added at $t = 0$. Growth of product, 4-nitrophenol (P) and decrease in substrate as reaction progresses. Conditions: Phenomenex 5 micron Luna C18 column 250 × 4.6 mm; Mobile phase: MeOAc buffer (0.05 M, pH 3.00)/ACN (80/20, v/v); Flow rate: 1.0 mL/min; Injection Volume: 20 μL; Temp: 30°C, 350 nm wavelength.

of 4-nitrophenyl-β-D-glucoside and the increase of 4-nitrophenol in the presence of 12.5 μg/mL β-glucosidase. For three different substrates under initial velocity (zero-order) conditions, the K_m and V_{max} values obtained by microdialysis were found to correlate well with literature values.

The analytical utility of in vitro microdialysis was further demonstrated by its ability to monitor carbohydrate reactions in complex matrices. An application is shown for monitoring the glycoside salicin and its hydrolysis product saligenin in a commercially available willow bark product that is used for making tea [99]. Salicin is an analgesic that can be found in many dietary supplements and nutraceutical products, which are sold commercially [102]. The K_m value of salicin was reported for the first time by this new method of in vitro microdialysis.

3.9.5 Toxicological Applications—Glucuronides

Alcohol is the most commonly abused substance and the need to determine its usage pre- or postmortem is an important analytical challenge. Clinical applications may include monitoring alcohol abuse for contraindication of medicines and challenge testing of alcohol abusers. Differentiating between alcohol consumption prior to death or from postmortem decomposition is a common practice in many forensic cases [103]. All these applications are pursued by monitoring biomarkers of alcohol consumption. A highly specific and sensitive biomarker of alcohol consumption is ethyl glucuronide

(EtG). Its widespread adoption is facilitated with the development of simpler and less expensive methods of analysis, and it has been reviewed [104].

Kaushik et al. [104] used LC-PED to determine EtG in post-mortem urine samples. After isocratically separating EtG and methyl glucuronide (MetG), which serves as an internal standard, using reversed-phase chromatography and subsequent addition of post-column NaOH, the glucuronides were detected using PED at an Au working electrode. EtG was found to have a limit of detection of 0.03 μg/mL (7 pmole; 50 μL injection volume) and repeatability at the concentration at the limit of quantitation of 1.7% RSD (relative standard deviation). The forensic utility of this method was further validated by the analysis of 29 postmortem urine specimens, whose results agreed strongly with certified determinations. An improved method for EtG in human urine was later introduced by the LaCourse group [105]. The improved method resulted in higher sensitivity, lower detection limits, better chromatographic performance, and the use of propyl glucuronide (PG) as a superior internal standard. This improved method showed that LC-PED is a sensitive, selective, and direct method for the determination of EtG, a biomarker of alcohol consumption.

Additional biomarkers for alcohol consumption include ethyl sulfate (EtS) and ethyl phosphate (EtP). The use of multiple markers for alcohol would provide further validation of alcohol abuse determinations. Fig. 3.16 shows the separation of all three potential analytes by IC using conductivity detection. Detection of EtG can be improved by adapting PED to this separation. It should be noted that virtually all glucuronides are detectable by PED. The LaCourse group has recently separated a suite of seven different drug glucuronides in a single chromatographic run [106]. The detector showed good analytical sensitivity toward these analytes, as shown by their limit of detection values (e.g., MetG, EtG, M3G, M6G, phenyl-glucuronide, acetaminophen-glucuronide and codeine-6-glucuronide were 10, 10, 30, 40, 50, 100, and 100 ng/mL, respectively). This system has great potential for use in the toxicological and forensic

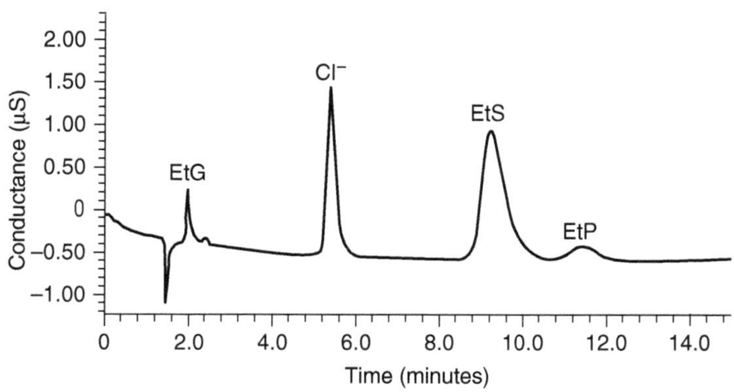

Figure 3.16. Chromatogram of EtG, Cl^-, EtS, and EtP following their separation by IC. Concentrations: 100 ppm each; Column: AG-10 & AS11 (Dionex); Mobile phase: NaOH; Gradient: 0–10 min at 5 mM, 10–14 min at 40 mM, 14–20 min at 5 mM; Flow rate: 1.0 mL/min; ASRS setting: 300 mA.

realm, allowing one to selectively screen for the presence of a variety of substances with high sensitivity, even after the parent compound is eliminated from the body.

3.9.6 Indirect Detection

A prerequisite of PED is adsorption of the analyte to the electrode surface prior to detection. Hence, surface-active species that adsorb more strongly can interfere in the detection mechanism of weakly adsorbed molecules. A consequence of this phenomenon is the attenuation of the analytical signal for carbohydrates when acetonitrile is added to the mobile phase. The effect is attributed to the favored adsorption of the cyano functional group over that of the alcohol group.

This adsorption requirement can be exploited as an indirect detection paradigm for non-PED active compounds by adding a weakly-adsorbing, PAD-active reagent to the mobile phase either pre- or post-column. The background signal from the reagent is then suppressed by the eluting analyte, which is preferentially adsorbed to the electrode surface. LaCourse first demonstrated this technique by adding 1 mM sorbitol to the mobile phase of an IC system [26]. Lysine (Lys) was demonstrated to give an indirect response that was greater in magnitude than its direct PED response. More recently, Olson et al. [33] have extended this work as a means to detect amino acids and proteins. Fig. 3.17 shows the detection of four amino acids at 600 pmol each using indirect-PED (InPED) following their separation using anion-exchange chromatography. Detection limits were on the order of 30 pmol or less. This work has now been extended to carboxylate-containing drugs. Fig. 3.18 shows the separation and detection of lysine, vigabatrin, methionine, gabapentin, and biotin using InPED following HPAEC using a mobile phase of 0.1 M NaOH plus 25 ppm sorbitol. InPED has the following advantages:

- The ability to detect weakly electroactive or electroinactive analytes.

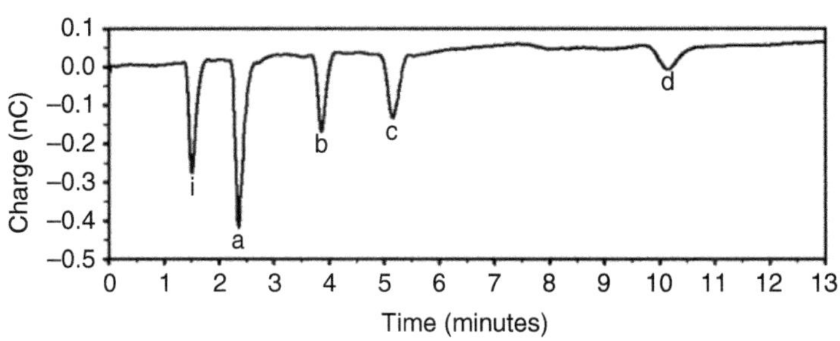

Figure 3.17. Anion-exchange separation of 600 pmol each of (a) Gln, (b) Gly, (c) Met, (d) Phe, and (i) solvent front using a quadruple-potential InPED waveform at Au electrode. Waveform: $E_1 = 0.0$ mV, $E_2 = -2000$ mV, $E_3 = +700$ mV, $E_4 = -100$ mV, $t_1 = 0$–400 ms, $t_{int} = 200$–400 ms, $t_2 = 410 - 420$ ms, $t_3 = 430$ ms, $t_4 = 450 - 950$ ms (vs. Ag/AgCl). Column: Dionex CarboPac-PA1; Mobile phase: 0.1 M NaOH/180 μM gluconic acid (isocratic); Flow rate: 1.0 mL/min.

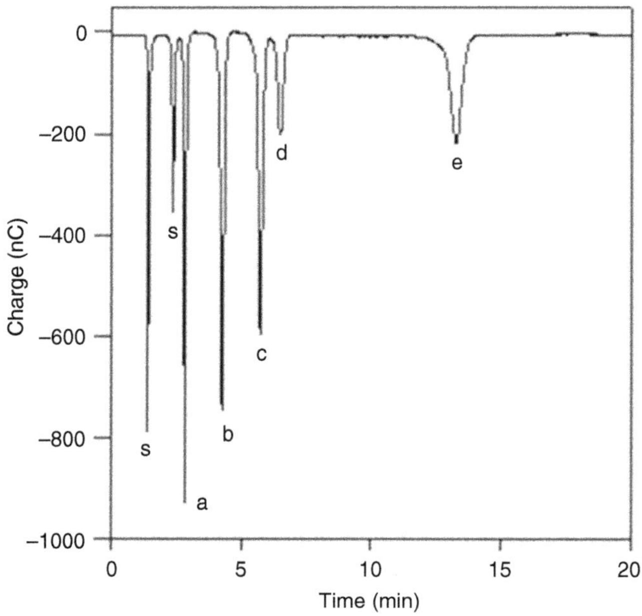

Figure 3.18. The separation and detection of (a) lysine, (b) vigabatrin, (c) methionine, (d) gabapentin, and (e) biotin. Peaks (s) are system peaks and peaks. InPED at a Au electrode following HPAEC using a mobile phase of 0.1 M NaOH plus 25 ppm sorbitol (isocratic). Waveform: $E_1 = 0.0$ mV, $E_2 = -1500$ mV, $E_3 = +700$ mV, $E_4 = -100$ mV, $t_1 = 0$–400 ms, $t_{\text{int}} =$ 200–400 ms, $t_2 = 410$–420 ms, $t_3 = 430$ ms, $t_4 = 450$–950 ms (vs. Ag/AgCl). Column: Dionex CarboPac-PA1; flow rate: 1.0 mL/min.

- Only one optimized waveform is needed for a wide range of compounds, because the mobile phase reagent is detected and not the individual analytes.
- Complex, integrated waveforms are unnecessary because detection takes place in the oxide-free region of the electrode and is unaffected by surface oxide formation.
- An increase in the indirect response was observed with increasing molecular weight that is dependent on the coverage of the electrode surface by the analyte.

3.9.7 Microelectrode Applications in PED

As discussed earlier in this chapter, PED is adaptable to microchromatographic and electrophoretic separations. Microelectrode applications of PED in IC have been limited by the lack of commercial technology to perform the separation. The recent introduction of a capillary IC system will provide the opportunity to take advantage of the highly sensitive detection capabilities of PED for carbohydrates, amines, and thiocompounds. As an example, LaCourse and Owens [107,108] were the first to apply

PED following microbore (i.e., 1-mm i.d. column) and capillary (i.e., 180-μm i.d. column) liquid chromatography to the determination of thiocompounds (e.g., methionine, cystamine, homocysteine, and coenzyme A and derivatives) under typical reversed-phase conditions. Interestingly, PED enables the direct determination of thio redox couples [i.e., -SH/-S-S-] and numerous other sulfur moieties at a single Au electrode without derivatization. Limits of detection were 0.2–0.5 pmole injected except for methionine, which was 2 pmole.

Capillary electrophoresis (CE)-PAD is ideally suited to the separation and detection of charged carbohydrates. The first application of CE-PAD was for glucose in blood [109], see Fig. 3.19. The level of glucose in blood was determined to be 4.25 ± 0.13 mM, in good agreement with that reported in the literature [110]. The PAD response for glucose was determined to be linear over the range of 10–1000 μM, and the detection limit was determined to be ca. 20 fmol per injection. LaCourse and Owens [111] extended the application of PED to the direct detection of many polar aliphatic compounds over a wide range of pH conditions following CE. Limits of detection are typically 10 fmol or less per injection. Lunte and coworkers [112,113] focused efforts on the characterization of glycopeptides derived from recombinant proteins. They determined that CE-PAD was a useful alternative to UV detection in the CE analysis of tryptic digests. Numerous reviews [114–117] have been published that focus on the detection of carbohydrates in CE all of which have a section devoted to PED.

The first report of PAD on an electrophoretic chip was presented by Fanguy and Henry [118]. Using a hybrid poly-(dimethylsiloxane)/glass device coupled with a Pt

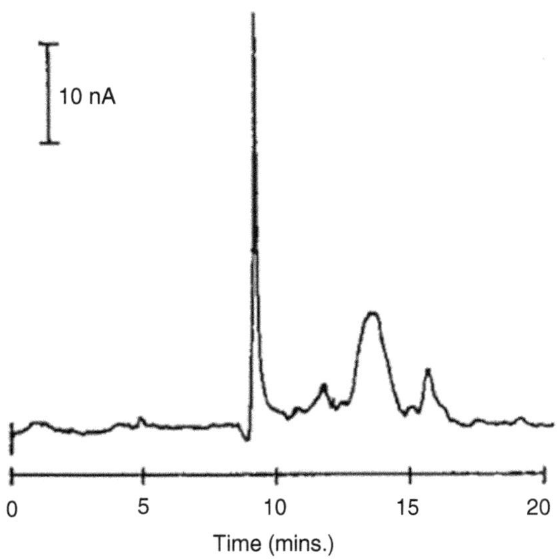

Figure 3.19. Electropherogram of human blood. Peak at ca. 9 min corresponds to 85 μM glucose. (Reprinted with permission from Ref. 105.) Copyright 1993 American Chemical Society. (Reprinted with permission from Ref. 109.)

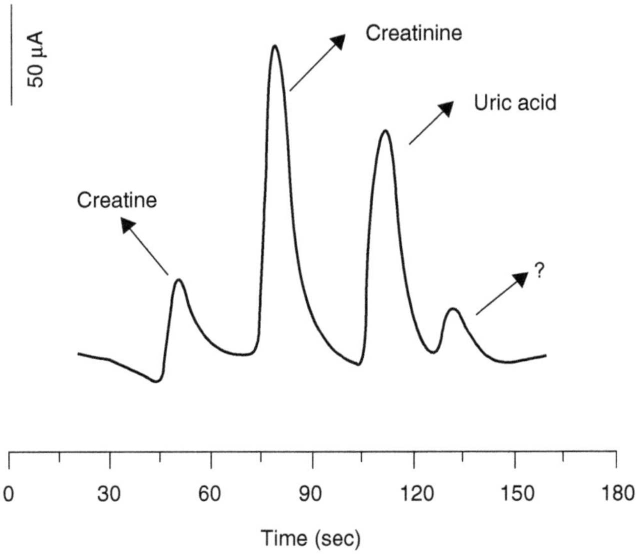

Figure 3.20. Baseline separation of creatine, creatinine, and uric acid on a CE chip using 30 mM borate buffer (pH = 9.4) in less than 200 s. (Ref. 121—Reproduced by permission of The Royal Society of Chemistry.)

working electrode, they were able to detect glucose, maltose, and xylose. Glucose was found to respond linearly from 20 to 500 μM with a measured detection limit of 20 μM. With the addition of sodium dodecyl sulfate (SDS) to the separation buffer, the same group was able to stabilize flow rates (i.e., reduce baseline drift) and decrease migration time. When combined with increasing the pH of the solution in the detection reservoir to pH 11, they were able to significantly improve the sensitivity of the system resulting in a detection limit of approximately 1 μM glucose [119].

Garcia and Henry [120] extended their work to underivatized amino acids and sulfur-containing antibiotics using a CE chip with the electrode position at the end of the separation channel. Detection limits ranging from 6 fmol (5 μM) per injection for penicillin and ampicillin to 455 fmol (350 μM) per injection for histidine were obtained. The best example of microchip CE with PED was produced by Garcia and Henry [121] for the direct detection of renal function markers. Fig. 3.20 shows the baseline separation of creatine, creatinine, and uric acid using 30 mM borate buffer (pH = 9.4) in less than 200 s. Linear calibration curves were obtained with limits of detection of 80, 250, and 270 μM for the compounds, respectively. The analysis of a real urine sample was presented with validation of creatinine concentration using a clinical assay kit.

3.10 CONCLUSIONS

Analytical systems comprised of ion chromatography with pulsed electrochemical detection (IC-PED) have proven reliable for the separation and sensitive detection of numerous polar aliphatic compounds, but most importantly carbohydrates. The use of polymeric ion-exchange phases to achieve separations in IC-PED of complex mixtures of carbohydrates, primarily because because of their stability at pH extremes, is highly complementary to the use of highly alkaline mobile phases required for maximum sensitivity in PED at Au electrodes. Hence, PED owes much of its popularity to its unique compatibility with IC.

The analytical significance of IC-PED in the majority of applications described in this chapter is the result of the fact that these compounds (e.g., carbohydrates) are not detected with comparable sensitivity by conventional direct photometric methodology, nor are these compounds amenable to anodic detection at constant applied potentials at conventional anodes (i.e., Au, Pt and C). PED, which combines the electrocatalytic detection of compounds with pulsed potential cleaning of the electrode, provides a means to the direct, sensitive, and reproducible detection of numerous compounds, including alditols, carbohydrates, simple alcohols, aminoalcohols, amines, amino acids, and organosulfur compounds.

The maturity of PED over the past three decades is reflected in the use of PED in advanced applications, such as 'fingerprinting' of bioproducts, enzyme characterization, and monitoring of biomarkers. These front-end applications strongly support the rugged and reliable nature of PED following a chromatographic separation, which includes virtually all aqueous-based separations (e.g., ion-exchange, ion-pairing, ion-exclusion, and reversed-phase chromatography). PED is now being applied to forensic and toxicological problems of interest, which include the determination of drug metabolites as glucuronides in physiological fluids.

PED offers many advantages over alternate detection schemes for liquid chromatography. Because electrochemical detection relies on a reaction at the electrode surface, detector cells can be miniaturized without sacrificing sensitivity. This advantage makes them especially suited for microbore and capillary techniques. Pulsed potential cleaning eliminates the need for daily polishing of the electrode which renders PED more convenient experimentally than dc amperometry. The sensitivity and selectivity (e.g., sulfur-based compounds under typical reversed-phase conditions) of PED for specific functional groups on the analyte simplifies the analysis of complex (e.g., biological) matrices.

Significant advancements in PED have been made for the detection of amine and sulfur compounds with an emphasis on advanced waveforms (e.g., IPAD, PVD, and 3D Amperometry). The impressive accomplishments in HPLC/IC-PED, thus far, have only accentuated the need for a deeper understanding of PED, its limits, and application of this technology to "real-world" bioanalytical problems of critical significance.

3.11 ACKNOWLEDGEMENTS

The author would like to gratefully acknowledge the many discussions and interactions with Professor Dennis C. Johnson and the financial support from Dionex Corp. (Sunnyvale, CA).

REFERENCES

1. Adams RN. Electrochemistry at Solid Electrodes. New York: Marcel Dekker;1969.
2. Kissinger PT. Laboratory Techniques in Electroanalytical Chemistry. New York: Marcel Dekker;1984.
3. Gilman S. Electroanalytical Chemistry, Vol. 2. New York: Marcel Dekker;1967.
4. Hammett LP. The velocity of the hydrogen electrode reaction on platinum catalysts. JACS. 1924;46:7–19.
5. Gilman S. In: Bard AJ, Editor, Electroanalytical Chemistry, Vol. 2. New York: Marcel Dekker;1967, p 111–192.
6. Kolthoff IM, Tanakan N. Rotated and stationary platinum wire electrodes. Anal. Chem. 1954;26:632–636.
7. Gilman S. A study of the mechanism of carbon monoxide adsorption on platinum by a new electrochemical procedure. J. Phys. Chem. 1963;67:78–84.
8. Breiter MW. Comparative voltammetric study of methanol oxidation and adsorption on noble metal electrodes in perchloric acid solutions. Electrochim. Acta 1963;8: 973–983.
9. Giner J. The anodic oxidation of methanol and formic acid and the reductive adsorption of CO. Electrochim. Acta 1964;9:63–77.
10. Woods R. Electroanalytical Chemistry. New York: Marcel Dekker;1976, p 20–27.
11. Clark D, Fleishman M, Pletcher DJ. Partial anodic oxidation of propylene in acetonitrile. J. Electroanal. Chem. Interfacial Electrochem. 1972;36:137–146.
12. MacDonald A, Duke PD. Small-volume solid-electrode flow-through electrochemical cells: Preliminary evaluation using pulse polarographic techniques. J. Chromatogr. 1973;83:331–342.
13. Stulik W, Hora V. Continuous voltammetric measurements with solid electrodes: Part I. A flow-through cell with tubular electrodes employing pulse polarization of the electrode system. J. Electroanal. Chem. 1976;70:253–263.
14. Hughes S, Meschi PL, Johnson DC. Amperometric detection of simple alcohols in aqueous solutions by application of a triple-pulse potential waveform at platinum electrodes. Anal. Chim. Acta 1981;132:1–10.
15. Hughes S, Johnson DC. Amperometric detection of simple carbohydrates at platinum electrodes in alkaline solutions by application of a triple-pulse potential waveform. Anal. Chim. Acta 1981;132:11–22.
16. Edwards P, Haak KK. A pulsed amperometric detector for ion chromatography. Am. Lab. 1983;15:78–87.
17. Rocklin RD, Pohl CA. Determination of carbohydrates by anion exchange chromatography with pulsed amperometric detection. J. Liq. Chromatogr. 1983;6:1577–1590.

18. Johnson DC, LaCourse WR. Liquid chromatography with pulsed electrochemical detection at gold and platinum electrodes. Anal. Chem. 1990;62:589A–597A.
19. Johnson DC, LaCourse WR. Pulsed electrochemical detection at noble metal electrodes in liquid chromatography. Electroanalysis 1992;4:367–380.
20. LaCourse WR, Johnson DC. Optimization of waveforms for pulsed amperometric detection of carbohydrates based on pulsed voltammetry. Anal. Chem. 1993;65:50–55.
21. LaCourse WR. Pulsed Electrochemical Detection in High Performance Liquid Chromatography. New York: Wiley;1997, p 324.
22. Johnson DC. Carbohydrate detection gains potential. Nature 1986;321:451–452.
23. Dobberpuhl DA, Johnson DC. Pulsed electrochemical detection at the ring of a ring-disk electrode applied to a study of amine adsorption at gold electrodes. Anal. Chem. 1995;67:1254–1258.
24. Welch LE, LaCourse WR, Mead Jr DA, Johnson DC, Hu T. Comparison of pulsed coulometric detection and potential-sweep-pulsed coulometric detection for underivatized amino acids in liquid chromatography. Anal. Chem. 1989;61:555–559.
25. Owens GS. [PhD Dissertation]. Pulsed electrochemical detection of organic sulfur compounds following separation by microbore high performance chromatography or capillary electrophoresis. 1997; University of Maryland, Baltimore County, Baltimore, MD.
26. LaCourse WR. Pulsed electrochemical detection at noble metal electrodes in high-performance liquid chromatography. Analusis 1993;21:181–195.
27. Polta JA, Johnson DC. Pulsed amperometric detection of electroinactive adsorbates at platinum electrodes in a flow injection system. Anal. Chem. 1985;57:1373–1376.
28. Koprowski L, Kirchmann E, Welch LE. The electrochemical oxidation of penicillins on gold electrodes. Electroanalysis 1993;5:473–482.
29. Kirchmann E, Welch LE. High-performance liquid chromatographic separation and electrochemical detection of penicillins. J. Chromatogr. 1993;633:111–118.
30. Kirchmann E, Earley RL, Welch LE. The electrochemical detection of penicillins in milk. J. Liq. Chromatogr. 1994;17:1755–1772.
31. Altunata S, Earley RL, Mossman DM, Welch LE. Pulsed electrochemical detection of penicillins using three and four step waveforms. Talanta 1995;42:17–25.
32. Doscotch MA, Jones JA, Welch LE. Indirect adsorption detection: an alternative pulsed electrochemical detection waveform. Anal. Chim. Acta 1997;344:55–64.
33. Olson MP, Keating LR, LaCourse WR. Indirect pulsed electrochemical detection of amino acids and proteins following high-performance liquid chromatography. Anal. Chim. Acta 2009;652:198–204.
34. Andrews RW, King RM. Selection of potentials for pulsed amperometric detection of carbohydrates at gold electrodes. Anal. Chem. 1990;62:2130–2134.
35. LaCourse WR, Johnson DC. Optimization of waveforms for pulsed amperometric detection (p.a.d.) of carbohydrates following separation by liquid chromatography. Carbohydr. Res. 1991;215:159–178.
36. Rocklin RD, Clarke AP, Weitzhandler M. Improved long-term reproducibility for pulsed amperometric detection of carbohydrates via a new quadruple-potential waveform. Anal. Chem. 1998;70:1496–1501.
37. Neuburger GG, Johnson DC. Pulsed coulometric detection with automatic rejection of background signal in surface-oxide-catalyzed anodic detections at gold electrodes in flow-through cells. Anal. Chem. 1988;60:2288–2293.

38. Clarke AP, Jandik P, Rocklin RD, Liu Y. Avdalovic N. An integrated amperometry waveform for the direct, sensitive detection of amino acids and amino sugars following anion-exchange chromatography. Anal. Chem. 1999;71:2774–2781.
39. Kennedy RT, Jorgenson JW. Quantitative analysis of individual neurons by open tubular liquid chromatography with voltammetric detection. Anal. Chem. 1989;61:436–441.
40. Jandik P, Haddad PR, Sturrock PE. Electrochemical detectors for ion chromatographic analysis: a critical review. Critical Reviews in Analytical Chemistry 1988;20:1–74.
41. Haddad PR, Jandik P. Detection methods in ion chromatography. Chromatographic Science Series 1987;37:87–156.
42. Dionex Corporation, Application Note 179: Carbohydrate and amino acid analysis using 3-D amperometry.
43. Paskach TJ, Lieker HP, Reilley PJ, Thielecke K. High-performance anion-exchange chromatography of sugars and sugar alcohols on quaternary ammonium resins under alkaline conditions. Carbohydr. Res. 1991;215:1–14.
44. LaCourse WR, Mead Jr DA, Johnson DC. Anion-exchange separation of carbohydrates with pulsed amperometric detection using a pH-selective reference electrode. Anal. Chem. 1990;62:220–224.
45. Hardy MR, Townsend RR. Separation of positional isomers of oligosaccharides and glycopeptides by high-performance anion-exchange chromatography with pulsed amperometric detection. PNAS USA 1988;85:3289–3293.
46. Townsend RR, Hardy MR, Hindsgaul O, Lee CE. High-performance anion-exchange chromatography of oligosaccharides using pellicular resins and pulsed amperometric detection. Anal. Biochem. 1988;174:459–470.
47. Hardy MR, Townsend RR, Lee YC. Monosaccharide analysis of glycoconjugates by anion exchange chromatography with pulsed amperometric detection. Anal. Biochem. 1988;170:54–62.
48. Townsend RR. In: El Rassi A, editor. Carbohydrate Analysis: High Performance Liquid Chromatography and Capillary Electrophoresis. Amsterdam: Elsevier;1995, p 181.
49. Townsend RR. Chapter 5: Analysis of glycoconjugates using high-pH anion-exchange chromatography. J. Chromatogr. Libr. 1995;58:181–209.
50. Ohsawa K, Yoshimura Y, Watanabe S, Tanaka H, Yokota A, Tamura K, Imaeda K. Determination of xylitol in the human serum and saliva by ion chromatography with pulsed amperometric detection. Anal. Sciences 1986;2:165–168.
51. Kuroda N, Taka S, Kajikawa T, Niimi M, Ishida T, Kawanishi K. Determination of sugars and polyols in red cells by high-performance liquid chromatography with pulsed amperometric detection; 1. Red cell sorbitol. Tonyobyo (Tokyo) 1994;37:695–698.
52. Kuroda N, Tada S, Kajikawa T, Niimi M, Ishida T, Kawanishi K. Determination of sugars and polyols in red blood cells by high-performance liquid chromatography with pulsed amperometric detection; 2. Red blood cell myo-inositol. Tonyobyo (Tokyo) 1995;38:979–983.
53. Chaplin MF. Analysis of bile acids and their conjugates using high-pH anion-exchange chromatography with pulsed amperometric detection. J. Chromatogr. 1995;664:431–434.
54. Tsai CM, Gu XX, Byrd RA. Quantification of polysaccharide in Haemophilus influenza type b conjugate and polysaccharide vaccines by high-performance anion-exchange chromatography with pulsed amperometric detection. Vaccine 1993;12:700–706.

55. Talaga P, Vialle S, Moreau M. Development of a high-performance anion-exchange chromatography with pulsed-amperometric detection based quantification assay for pneumococcal polysaccharides and conjugates. Vaccine 2002;20:2474–2484.
56. Lei QP, Shannon AG, Heller RK, Lamb DH. Quantification of Free Polysaccharide in Meningococcal Polysaccharide-Diphtheria Toxoid Conjugate Vaccines. In: Brown F, Corbel M, Griffiths E, editors. Developmental Biology, Volume 103: Physico-Chemical Procedures for the Characterization of Vaccines. Basel: Karger;2000. p 259–264.
57. Talaga P, Moreau M. Quantification of C-polysaccharide in Streptococcus pneumoniae polysaccharides by high-performance anion exchange chromatography with pulsed amperometric detection. In: Brown F, Corbel M, Griffiths E, editors. Developmental Biology, Volume 103: Physico-Chemical Procedures for the Characterization of Vaccines. Basel: Karger; 2000. p 27–34.
58. Lei QP, Lamb DH, Heller R, Pietrobon P. Quantitation of low-level unconjugated polysaccharide in tetanus toxoid-conjugate vaccine by HPAEC-PAD following rapid separation by deoxycholate-HCl. J. Pharm. Biomed. Anal. 2000;21:1087–1091.
59. Fleming SC, Kynaston JA, Laker MF, Pearson ADJ, Kapembwa MS, Griffin GE. Analysis of multiple sugar probes in urine and plasma by high-performance anion-exchange chromatography with pulsed electrochemical detection: Application in the assessment of intestinal permeability in human immunodeficiency virus infection. J. Chromatogr. 1993;640:293–297.
60. Blom HJ, Andersson HC, Krasnewich DM, Gahl WA. Pulsed amperometric detection of carbohydrates in lysosomal storage disease fibroblasts: A new screening technique for carbohydrate storage diseases. J. Chromatogr. 1990;533:11–21.
61. Tanaka S, Nakamori K, Akanuma H, Yabuuchi M. High performance liquid chromatographic determination of 1,5-anhydroglucitol in human plasma for diagnosis of diabetes mellitus. Biomed. Chromatogr. 1992;6:63–66.
62. LaCourse WR, Johnson DC, Rey MA, Slingsby RW. Pulsed amperometric detection of aliphatic alcohols in liquid chromatography. Anal. Chem. 1991;63:134–139.
63. LaCourse WR, Jackson Jr WA, Johnson DC. Pulsed amperometric detection of alkanolamines following ion-pair chromatography. Anal. Chem. 1989;61:2466–2471.
64. Polta JA, Johnson DC, Merkel KE. Liquid chromatographic separation of aminoglycosides with pulsed amperometric detection. J. Chromatogr. 1985;324:407–414.
65. Statler JA. Determination of tobramycin using high-performance liquid chromatography with pulsed amperometric detection. J. Chromatogr. 1990;527:244–246.
66. McLaughlin LG, Henion JD. Determination of aminoglycoside antibiotics by reversed-phase ion-pair high-performance liquid chromatography coupled with pulsed amperometry and ion spray mass spectrometry. J. Chromatogr. 1992;591:195–206.
67. Phillips JG, Simmonds C. Determination of spectinomycin using cation-exchange chromatography with pulsed amperometric detection. J. Chromatogr. 1994;675:123–128.
68. Kaine LA, Wolnik KA. Forensic investigation of gentamicin sulfates by anion-exchange ion chromatography with pulsed electrochemical detection. J. Chromatogr. 1994;674:255–261.
69. Dionex Corporation, Application Note 66: Neomycin in Topical Lotions.
70. LaCourse WR, Dasenbrock CO. In: Advances in Chromatography. New York: Marcel Dekker;1998. p 189–232.

71. Draisci R, Cavalli S, Lucentini L, Stacchini A. Ion exchange separation and pulsed amperometric detection for determination of biogenic amines in fish products. Chromatographia 1993;35:9–12.
72. Dobberpuhl DA, Johnson DC. This research contribution is in commemoration of the life and science of I. M. Kolthoff (1894–1993). Pulsed electrochemical detection at the ring of a ring-disk electrode applied to a study of amine adsorption at gold electrodes. Anal. Chem. 1995;67:1254–1258.
73. Dobberpuhl DA, Hoekstra JC, Johnson DC. Pulsed electrochemical detection at old electrodes applied to monoamines and diamines following their chromatographic separation. Anal. Chim. Acta 1996;322:55–62.
74. Welch LE, LaCourse WR, Mead Jr DA, Johnson DC. A comparison of pulsed amperometric detection and conductivity detection of underivatized amino acids in liquid chromatography. Talanta 1990;37:377–380.
75. LaCourse WR. Chapter 25. Pulsed electrochemical detection of carbohydrates at noble metal electrodes following liquid chromatographic and electrophoretic separation. In: El Rassi Z, editor. Carbohydrate analysis by modern chromatography and electrophoresis; J. Chromatogr. Lib. Amsterdam: Elsevier;2002;66: 905–946.
76. van Riel JA, Olieman C. Selective detection in RP-HPLC of Tyr-, Trp-, and sulfur-containing peptides by pulsed amperometry at platinum. Anal. Chem. 1995;67: 3911–3915.
77. Donaldson MJ, Adlard MW. Analysis of δ-L-α-aminoadipyl-L-cysteinyl-D-valine by ion chromatography and pulsed amperometric detection. J. Chromatogr. 1990;509:347–356.
78. Dasenbrock CO, LaCourse WR. Measuring traces of antibiotics in milk. Chemtech 1998;28:26–33.
79. LaCourse WR, Dasenbrock CO. Pulsed electrochemical detection of sulfur-containing antibiotics following high performance liquid chromatography. J. Pharm. Biomed. Anal. 1999;19:239–252.
80. Dasenbrock CO, LaCourse WR. Assay for cephapirin and ampicillin in raw milk by high-performance liquid chromatography-integrated pulsed amperometric detection. Anal. Chem. 1998;70:2415–2420.
81. Horton DR, Rodemeyer G, Rodemeyer R. Characterization of 2-amino-2,6-dideoxy-D-glucose as a constituent of the lipopolysaccharide antigen of Pseudomonas aeruginosa immunotype 4. Carbohydr. Res. 1977;56:129–138.
82. Dmitriev BA, Knirel YA, Kochetkov NA, Stanislavsky NK, Mashilova GM. Somatic antigens of Pseudomonas aeruginosa. The structure of the polysaccharide chain of Ps. aeruginosa O-serogroup 7 (Lanyi) lipopolysacharide. Eur. J. Biochem. 1980;106:643–651.
83. Baumann H, Tzianabos AO, Brisson JR, Kasper DL, Jennings HJ. Structural elucidation of two capsular polysaccharides from one strain of Bacteroides fragilis using high-resolution NMR spectroscopy. Biochemistry 1992;31:4081–4089.
84. Stroop CJM, Bush CA, Marple RL, LaCourse WR. Carbohydrate analysis of bacterial polysaccharides by high-pH-anion-exchange chromatography and online polarimetric determination of absolute configuration. Anal. Biochem. 2002;303:176–185.
85. Zook CM, LaCourse WR. Pulsed amperometric detection of microdialyzates from the glucose oxidase reaction. Anal. Chem. 1998;70:801–806.
86. Hansen DK, Davies MI, Lunte SM, Lunte CE. Pharmacokinetic and metabolism studies using microdialysis sampling. J. Pharm. Sci. 1999;88:14–27.

87. Morrison PF, Bungay PM, Hsaio JK, Mefford IN, Dykstra KH, Dedrick RL. In: Terry E. Robinson TE, Justice JB, editors. Microdialysis in the Neurosciences. New York: Elsevier;1991, p 47–70.
88. Snyder KL, Nathan CE, Lee A, Stenken JA. Diffusion and calibration properties of microdialysis sampling membranes in biological media. Analyst 2001;126:1261–1268.
89. Stenken JA, Puckett DL, Lunte SM, Lunte CE. Detection of N-acetylcysteine, cysteine and their disulfides in urine by liquid chromatography with a dual-electrode amperometric detector. J. Pharm. Biomed. Anal. 1990;8:85–89.
90. Torto N, Gordon L. Technical issues of in vitro microdialysis sampling in bioprocess monitoring. Trends Anal. Chem. 1999;18:252–260.
91. Marangoni AG. Enzyme kinetics: a modern approach. Hoboken NJ: Wiley Interscience;2003.
92. Mandenius CF, Bulow L, Danielsson B. Determination of amygdalin and cyanide in industrial food samples using enzymic methods. Acta Chem. Scand. B. 1983;37:739–742.
93. Torto N, Buttler T, Gorton L, Marko-Varga G, Stalbrand H, Tjerneld F. Monitoring of enzymatic hydrolysis of ivory nut mannan using on-line microdialysis sampling and anion-exchange chromatography with integrated pulsed electrochemical detection. Anal. Chim. Acta 1995;313:15–24.
94. Zook CM, LaCourse WR. Monitoring in vitro enzymic digestion of lactose in milk using microdialysis with pulsed amperometric detection. Curr. Sep. 1998;17:41–45.
95. Zook CM. [PhD Dissertation]. Microdialysis sampling coupled to high performance anion-exchange chromatography-pulsed amperometric detection for analysis of carbohydrates. 1998; University of Maryland, Baltimore County, Baltimore, MD.
96. Torto N, Gorton L, Marko-Varga G, Emneus J, Akerberg C, Zacchi G, Laurell T. Monitoring of enzymatic hydrolysis of starch by microdialysis sampling coupled on-line to anion exchange chromatography and integrated pulsed electrochemical detection using post-column switching. Biotechnol. Bioengin. 1997;56:546–554.
97. Nilsson GS, Richardson S, Huber A, Torto N, Laurell T, Gorton L. Microdialysis clean-up and sampling in enzyme-based methods for the characterization of starch. Carbohydr. Polymers 2001;46:59–68.
98. Okatch H, Torto N. Profiling of carbohydrate polymers in biotechnology using microdialysis sampling, high performance anion exchange chromatography with integrated pulsed electrochemical detection/mass spectrometry. Afr. J. Biotech. 2003;2:636–644.
99. Modi SW, LaCourse WR. Monitoring carbohydrate enzymatic reactions by quantitative in vitro microdialysis. J. Chromatogr. A 2006;1118:125–133.
100. Schwartz J, Sloan J, Lee YC. Mannosidase, glucosidase, and galactosidase in sweet almond emulsion. Arch. Biochem. Biophys. 1970;137:122–127.
101. Dale MP, Ensley HE, Kern J, Sastry KAR, Byers LD. Reversible inhibitors of β-glucosidase. Biochemistry 1985;24:3530–3539.
102. Li LS, Huang WD, He Q, Ye S. Determination of salicin in extract of willow bark by high performance liquid chromatography. Chin. J. Chromatogr. 2001;19:446–448.
103. Wurst FM, Kempter C, Seidl S, Alt A. Ethyl glucuronide–a marker of alcohol consumption and a relapse marker with clinical and forensic implications. Alcohol and Alcoholism 1999;34:71–77.
104. Kaushik R, Levine B, LaCourse WR. A brief review: HPLC methods to directly detect drug glucuronides in biological matrices (Part 1). Anal. Chim. Acta 2006;556:255–266.

105. Shah R, LaCourse WR. An improved method to detect ethyl glucuronide in urine using reversed-phase liquid chromatography and pulsed electrochemical detection. Anal. Chim. Acta 2006;576:239–245.
106. Kaushik R. [MS Thesis] Glucuronide metabolite monitoring in physiological samples by reversed-phase high performance liquid chromatography (HPLC) coupled with pulsed electrochemical detection. 2006; University of Maryland, Baltimore County, Baltimore, MD.
107. LaCourse WR, Owens GS. Pulsed electrochemical detection of thiocompounds following microchromatographic separations. Anal. Chim. Acta 1995;307:301–319.
108. Owens GS, LaCourse WR. Pulsed electrochemical detection of sulfur-containing compounds following microbore liquid chromatography. Curr. Sep. 1996;14:82–88.
109. O'Shea TJ, Lunte SM, LaCourse WR. Detection of carbohydrates by capillary electrophoresis with pulsed amperometric detection. Anal. Chem. 1993;65:948–951.
110. The Merck Manual 15th Ed. Rahway NJ: Merck Sharp and Dohme Research Laboratories;1987, p 2413.
111. LaCourse WR, Owens GS. Pulsed electrochemical detection of nonchromophoric compounds following capillary electrophoresis. Electrophoresis 1996;17:310–318.
112. Weber PL, Kornfelt T, Klausen NK, Lunte SM. Characterization of glycopeptides from recombinant coagulation factor VIIa by high-performance liquid chromatography and capillary zone electrophoresis using ultraviolet and pulsed electrochemical detection. Anal. Biochem. 1995;225:135–142.
113. Weber PL, Lunte SM. Capillary electrophoresis with pulsed amperometric detection of carbohydrates and glycopeptides. Electrophoresis 1996;17:302–309.
114. Lunte SM, O'Shea TJ. Pharmaceutical and biomedical applications of capillary electrophoresis/electrochemistry. Electrophoresis 1994;15:79–86.
115. Paulus A, Klockow A. Detection of carbohydrates in capillary electrophoresis. J. Chromatogr., A 1996;720:353–376.
116. Baldwin, R. P. (2000). Recent advances in electrochemical detection in capillary electrophoresis. Electrophoresis, 21, 4017–4028.
117. LaCourse WR. Pulsed electrochemical detection of carbohydrates at noble metal electrodes following liquid chromatographic and electrophoretic separation. Journal of Chromatography Library 2002;66:905–946.
118. Fanguy JC, Henry CS. Pulsed amperometric detection of carbohydrates on an electrophoretic microchip. Analyst 2002;127:1021–1023.
119. Garcia CD, Henry CS. Enhanced determination of glucose by microchip electrophoresis with pulsed amperometric detection. Anal. Chim. Acta 2004;508:1–9.
120. Garcia CD, Henry CS. Direct Determination of Carbohydrates, Amino Acids, and Antibiotics by Microchip Electrophoresis with Pulsed Amperometric Detection. Anal. Chem. 2003;75:4778–4783.
121. Garcia CD, Henry CS. Direct detection of renal function markers using microchip CE with pulsed electrochemical detection. Analyst 2004;129:579–584.

4

SUPPRESSOR DESIGN AND DETECTION FOR ION CHROMATOGRAPHY

Kannan Srinivasan, Sheetal Bhardwaj, Rong Lin, and Christopher Pohl

Thermo Fisher Scientific, Sunnyvale, California, CA

4.1 INTRODUCTION

Suppressors have become an integral part of ion chromatography and today the term suppressed ion chromatography is synonymous with ion chromatography (IC). The role of the suppressor is to remove the contribution of the eluent before detection. In the suppression process the background from the eluent is converted to weakly dissociated species, while the analyte signals are enhanced for fully dissociated species thereby increasing the overall signal to noise ratio for conductivity detection. The suppressor as originally described by Small et al. [1] was a column packed with ion-exchange resin that was termed a stripper column. The stripper column however had to be regenerated off-line, had issues with analysis of weakly dissociated species, and had too large a delay volume to have sufficient capacity for suppression. Since the original publication, the suppressor design has undergone significant changes with the objective to address some of the limitations of the original packed bed suppressor as well as related suppressor products. For example to facilitate a continuous operation of the ion chromatograph, the packed bed suppressor design was replaced with a continuously

Applications of Ion Chromatography for Pharmaceutical and Biological Products, First Edition.
Edited by Lokesh Bhattacharyya and Jeffrey S. Rohrer.

regenerated suppressor made from a fiber-based ion-exchange material [2]. The fiber was packed to reduce the overall band dispersion and allowed for excellent chromatographic performance [3]. The fiber suppressor was a chemically regenerated suppressor. The fiber design addressed some of the limitations of the packed bed suppressor. However, the device was fragile and had limited suppression capacity. Researchers also explored several suppressor designs using the annular geometry [4–6]. The fiber design was subsequently replaced with a continuously regenerated flat ion-exchange membrane based suppressor termed micromembrane suppressor or MMS [7], which was also operated with a chemical regenerant. Research in the use of electrolysis or an electrical potential to drive suppression was pursued by several groups [8–11]. A commercial electrolytic suppressor based on a flat membrane design was introduced [12–14] and to date this design continues as the "work horse" suppressor for IC.

4.2 MEMBRANE SUPPRESSOR

The flat membrane suppressor supports continuous operation of the ion chromatograph and can be regenerated either chemically with a chemical regenerant or electrolytically by the electrolysis of water. The suppressor could easily suppress 200 μeqv/min of eluent (200 mM at 1 mL/min) while the small delay volume in the device allowed for excellent peak shapes. Unlike packed bed suppressors, there was no compromise between suppression capacity and delay volume of the device, due to the thin eluent channel. The suppressor reactions can be described for anion analysis using a sodium carbonate eluent as follows,

$$Na_2CO_3 \text{ (Eluent)} + RSO_3H\text{(IES)} \Rightarrow H_2CO_3 \text{ (Suppressed Eluent)} + RSO3Na \text{ (IES)} \quad (1)$$

Sodium ion from the eluent is exchanged with hydronium ion from the ion exchange surface (IES) during anion analysis, thereby converting the conductive sodium carbonate eluent to weakly ionized carbonic acid. The analyte sodium chloride, a fully dissociated salt, is converted to a conductive acid as outlined below

$$NaX \text{ (Analyte)} + RSO_3H \text{ (IES)} \Rightarrow HX \text{ (Analyte)} + RSO_3Na \text{ (IES)} \quad (2)$$

In addition to eluent counter-ions, the suppressor also removes or exchanges the counter-ions to the analyte ion. Thus detection of chloride as a conductive acid in a background of a weakly conductive eluent is possible with a suppressor. The lower background leads to lower noise and hence suppressed IC provides a higher S/N ratio and lower detection limits in comparison to a direct conductivity measurement (also see Chapter 1 of this book).

The IES in the original Small suppressor [1] was the surface of an ion-exchange resin, more specifically a cation-exchange resin during anion analysis. The IES in the modern membrane suppressor is the surface of an ion-exchange membrane or screen. It can be seen that in order to sustain the above reactions in the continuously regenerated suppressor the eluent counter-ion, sodium in the above case, should also be continuously and effectively removed from the suppressor and replaced by hydronium ions, so that the IES is always in a regenerated state ready for further reactions.

4.3 MEMBRANE SUPPRESSOR DESIGN

The membrane suppressor itself comprises of three channels defined by two ion-exchange membranes as shown in Figure 4.1. The central channel is the eluent channel and is flanked by two opposing regenerant channels. The fluidic pathway in the suppressor is defined by three gasketed ion-exchange screens. The gaskets provide a fluidic seal around the perimeter when compressed between two polymer plates. The regenerant flow in the suppressor is split internally to divert flow to the two regenerant channels. In the chemical mode of operation the chemical regenerant is supplied to the regenerant channels via a regenerant inlet port. In this mode the eluent counter-ions and sample counter-ions are removed via the two ion-exchange membranes on both sides of the regenerant channel and exit the suppressor via the regenerant outlet port along with any excess regenerant ions. The original MMS suppressor had ion-exchange screens of high capacity in all three channels [7]. It was realized recently that eluent counter-ions and multivalent sample counter-ions are retained by the regenerant screens due to the high capacity. This retention of counter-ions affected the overall capacity of the suppressor device. By redesigning the suppressor with screens that had no retention of the eluent counter-ions in the regenerant channel, the MMS 300 suppressors provided a much higher dynamic capacity relative to the MMS suppressors, as well as a more robust platform for pursuing suppression, particularly in the presence of solvents [15].

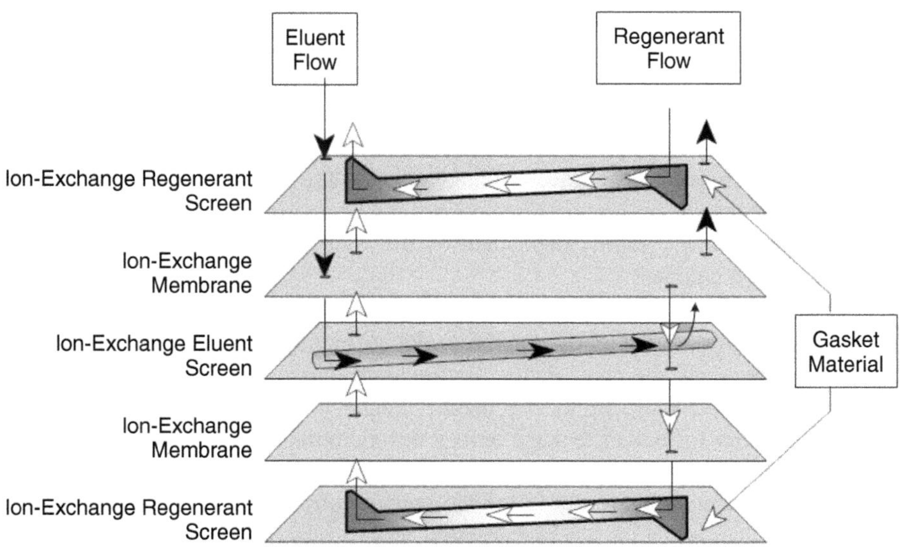

Figure 4.1. A flat membrane based suppressor design schematic.

In the electrolytic suppressor device or self regenerating suppressor (SRS), two electrodes made from a noble metal are placed in the regenerant channel for pursuing electrolysis of water. In the electrolytic mode of operation the regenerant (typically water) is supplied to the regenerant channel via the recycle mode of operation or from an external source reservoir of water termed as the external water mode of operation. A self subsisting source of regenerant or water was obtained in the recycle mode of operation by using the cell effluent and thus contributing to significant ease of use from an operator's perspective [14,16]. In the electrolytic mode of operation, the polarity of the applied potential along with the type of the device dictated the direction of the supply of regenerant ions and the removal of counter-ions. One membrane directed the regenerant ions into the eluent channel while the other membrane directed the eluent counter-ions out of the eluent channel. Thus the SRS has directionality in contrast to the MMS suppressor.

4.4 MMS OPERATION

The MMS suppressor requires a continuous supply of chemical regenerant. The function of the regenerant is to supply the regenerant ions required for suppression or regeneration of the IES and to remove the eluent and analyte counter-ions. For example, during anion analysis sulfuric acid is used as the regenerant and it supplies the hydronium ions required for suppression. The ion-exchange membrane in this case is a cation-exchange membrane that facilitates the exchange of the eluent counter-ion with the regenerant ion on the IES. The regenerant further exchanges the eluent counter-ions with the regenerant ions, thereby continuously regenerating the IES as long as a supply of regenerant is established. It should be noted that the suppressor does have some static ion-exchange capacity that allows the device to operate without regenerant for a short duration, similar to a packed bed suppressor, until the capacity of the suppressor device is consumed. The dynamic capacity of the suppressor is established when the regenerant ion is supplied continuously to the IES. Although sulfate anion in the above example would be repelled by the Donnan potential on the membrane (developed by the fixed anion charges on the cation-exchange membrane) there is some transport of the sulfate anion, termed Donnan forbidden transport, across the membrane which is highly dependent on the concentration of the regenerant ion in the regenerant channel.

The concentration of the regenerant is therefore kept relatively low to minimize the "leakage" of regenerant across the ion-exchange membranes. By adjusting the flow rate of the regenerant, the equivalents of regenerant ions required for the suppression reaction could be satisfied. Bulky regenerant counter-ions are usually chosen to minimize the Donnan forbidden leakage across the membranes [17]. The thickness and charge density of the membrane are key parameters that affect leakage. The thicker the ion-exchange membrane, the lower the leakage observed. As the charge density increases, Donnan potential works favorably and results in Donnan rejection of the regenerant resulting in minimal leakage across the membrane. The implication of the leakage is that the background is no longer low thus increasing the noise. In some cases the leakage can affect the peak signal by changing the equilibrium dissociation of the background eluent or the analyte. For example in an acid environment the dissociation of weak acids is decreased resulting in lower response.

There are two modes of operation for the MMS suppressor. The conventional chemical mode of operation that uses an external reservoir of regenerant that is supplied to the suppressor typically via pneumatic means, and a displacement chemical regeneration (DCR) mode of operation that uses a displacement approach as a regenerant dispensing mechanism [18]. In the conventional chemical mode of operation, the regenerant is dispensed by pressurizing the reservoir using an inert gas such as nitrogen and usually at an applied pressure of about 10 psi or above. The typical flow rate is in the 5 mL/min regime for the 4 mm suppressor. A pump could also be used for dispensing the regenerant however pump pulsations can impact the background signal, affecting the noise pattern. These pulsations can be minimized or eliminated by dampening the pump flow. In this mode the concentration and flow rate of the regenerant is optimized to achieve roughly a seven-fold excess of regenerant equivalent concentration over the eluent equivalent concentration during anion analysis.

The second mode of operation is the DCR mode. In this mode the suppressed cell effluent is routed to a closed fully filled regenerant reservoir and the regenerant is displaced from the reservoir. Since the cell effluent is used to dispense the regenerant, the regenerant flow is the same as the eluent and the concentration is the only term that is optimized. Typically a twofold excess is needed, calculated from the number of equivalents represented by the eluent, to suppress a given eluent concentration during anion analysis. The DCR mode of operation is schematically illustrated in Figure 4.2. The density of the cell effluent when compared to the regenerant dictates if the cell effluent is supplied to the top or bottom of the reservoir. For carbonate based eluents, carbonic acid is less dense than the sulfuric acid regenerant hence the cell effluent is returned to the top of the container while the acid is displaced from the bottom of the container. For hydroxide eluents the scenario is similar. For cation applications with acid eluents, the cell effluent is water and is denser than tetrabutylammonium hydroxide (TBAOH) regenerant. In this case, the cell effluent is returned to the bottom of the regenerant container while the TBAOH regenerant is displaced from the top of the container. For cation analysis, with acetonitrile present in the eluent, however, the cell effluent, which is acetonitrile and water, is less dense than TBAOH and hence is

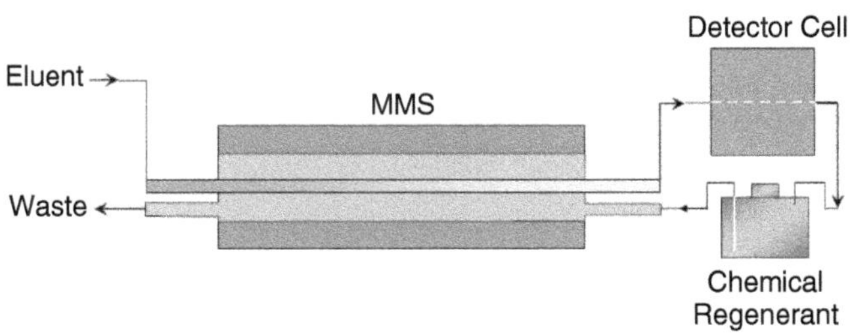

Figure 4.2. DCR setup with a Micromembrane suppressor and detector cell.

returned to the reservoir at the top while dispensing the TBAOH regenerant from the bottom of the container.

The mixing of the regenerant with the cell effluent is minimal hence the entire reservoir content can be displaced [18]. In fact, when the regenerant is completely mixed with the cell effluent excellent suppression performance was shown to be feasible (Figure 4.3). Another means of dispensing the regenerant was by using a bag-in-a-bottle concept. Here the regenerant is poured into a bag that is placed in a 2L container. The cell effluent is returned to the bottle to the outside of the bag; this action pushes the regenerant from the container and is routed to the regenerant channel. In this case there is clear separation of the regenerant from the cell effluent by the bag [18].

In the DCR mode the eluent and regenerant container volumes are the same and it is recommended to refill the regenerant when the eluent container is replenished. In Figure 4.3 the regenerant would be replenished after about 27.7 hours. Hence some mixing does not pose any issues in this mode. From an operation perspective, if the concentration of the regenerant is twofold higher than the eluent concentration, then a small leakage of regenerant is expected, which would increase the background. In comparison to the conventional chemical mode of operation, the DCR mode provides the ease of use as there is no need to have an external pump or pneumatic means for dispensing the regenerant. Since the analytical pump is used to dispense the regenerant there is no need for dampening the pump pulsations.

Another advantage of the DCR mode is that, unlike the conventional chemical mode of operation, supply of eluent into the suppressor always ensures supply of regenerant. This aspect ensures that the suppressor capacity is always maintained. In

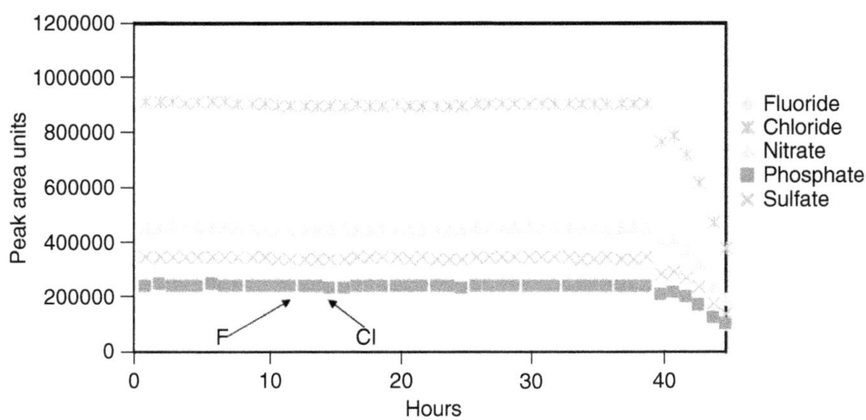

Figure 4.3. Peak area response versus time (hours) for a DCR setup where the 2L regenerant reservoir is actively mixed using a stirrer. Column: IonPac® AS15 (4 × 250 mm); Eluent: NaOH 38 mM at 1.2 mL/min; Regenerant: 75 mN Sulfuric acid. Standards: A 10x dilution of a 5 anion standard. (Dionex Corporation P/N 037157).

contrast, in the conventional mode, supply of eluent alone without regenerant supply would deplete the static capacity in the suppressor.

4.5 SRS OPERATION

The SRS suppressor relies on the electrolytic water splitting reactions to supply regenerant ions. When a DC potential (greater than ~1.5 V) is applied between the two electrodes in the SRS, at the anode

$$H_2O \Rightarrow 2H^+ + \frac{1}{2}O_2 + 2e^- \quad (3)$$

and at the cathode

$$2H_2O + 2e^- \Rightarrow 2OH^- + H_2 \quad (4)$$

The above reactions provide the electrolysis-derived ions required for the suppression process. During anion analysis hydronium ions generated at the anode as per Equation (3) above, are driven towards the cathode, which traverse through the ion-exchange membranes as depicted in Figure 4.4. The anion suppressor has cation-exchange functionality across the suppressor channels and selectively allows transport

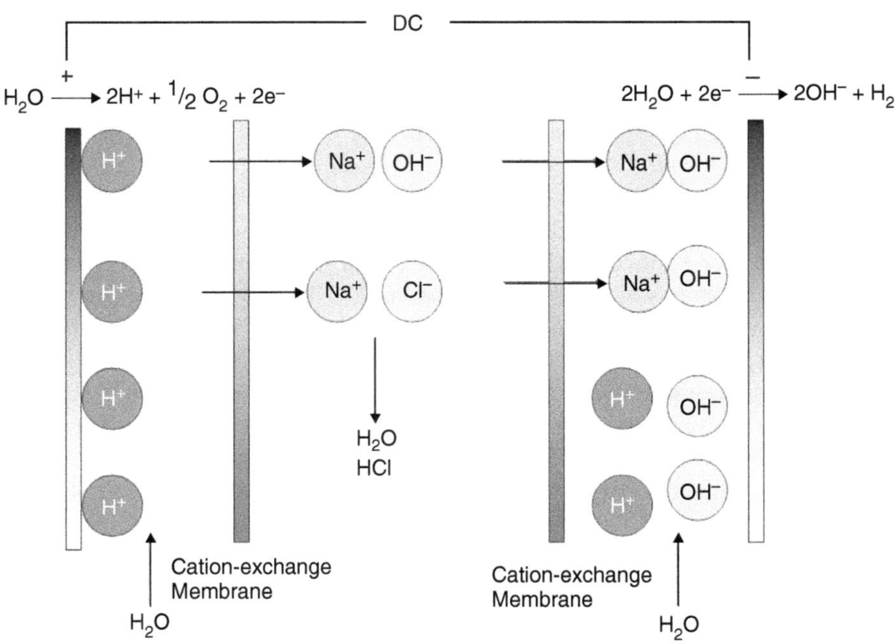

Figure 4.4. SRS suppressor operational schematic.

of cations via the ion-exchange membranes (screens). The hydronium ions along with eluent counter-ions such as sodium or potassium are driven from the eluent channel across the membrane towards the cathode. The cations combine with the hydroxide at the cathode to form a base or water and are removed from the suppressor. The transport of hydronium ions regenerates the suppressor; however, excess transport of hydronium ions would result in lowering of the current efficiency of the device since the applied current would be wasted in the formation of water. The current I in mA, required to suppress a given eluent strength with 100% current efficiency can be easily calculated from the relationship

$$I = FCV/60 \tag{5}$$

where F is Faraday's constant (Coulomb/eqv), C is the eluent concentration in normality, V is the eluent flow rate in mL/min. As the capacity in the suppressor eluent channel increases, the current efficiency of the device suffers [19]. Other means of improving current efficiency of suppressor products are discussed elsewhere [20,21]. The benefit of improved current efficiency is lower wattage and lower noise.

From a suppressor operation perspective the recycle mode of operation is the simplest operation mode since it uses the cell effluent for supplying the water needed for the electrolytic reactions. However, when solvents are present in the eluent, recycling the cell effluent would result in potential oxidation of the solvent at the electrode surface hence the external water mode is recommended. In the external water mode, in the regenerant channel relative to the eluent channel, the transport of the solvent from the eluent channel to the electrode and associated reactions at the electrode surface are significantly reduced. While the external water mode was advocated for achieving low noise with previous-generation suppressors [14], with the introduction of the SRS 300 suppressors, the motivation to pursue the former mode has significantly diminished since the latter suppressors are capable of achieving low noise in the recycle mode of operation.

Another mode of operation is the gas assisted mode of operation. In this mode the velocity in the regenerant channel is significantly enhanced by pumping a gas stream (nitrogen or purified air) in conjunction with the supply of water from either the recycle or external water mode of operation [22]. This would allow one to use a reduced flow rate of the aqueous stream thereby allowing a regenerant container to last for a longer time relative to the standard external water mode of operation. With the current suppressor products, the benefit of this mode can also be realized only in the presence of solvent and in conjunction with the external water mode of operation.

The membrane suppressor SRS discussed above has a dynamic capacity of about 200 μeqv/min. Although this level of capacity is adequate for suppressing most eluents used in IC, this capacity is not adequate for suppressing some samples that contain a high level of analyte counter-ions. For example, when analyzing trace anions in concentrated base, the suppressor dynamic capacity is exceeded when a large volume of sample is injected. Under these conditions it was found that one way to suppress was to pass the sample through the suppressor eluent channel multiple times, thereby compensating for the limited capacity [23]. A recent approach used a "park and neutralize" scheme to pursue this application. The sample of interest is parked within the eluent channel of the suppressor device. By parking the sample and supplying adequate current, the suppressor was capable of suppressing or neutralizing the base sample at the expense of time [24].

4.6 OTHER SUPPRESSOR DESIGNS

Other suppressor designs have evolved over the years [25]. For example a suppressor design called Metrohm Suppressor Module (MSM), based on three packed bed suppressor units that are switched so that when one is being used for suppression, the other two are being either regenerated with a regenerant or equilibrated with water [26]. A pseudo continuous mode of operation was established by switching the three suppressor units. Another suppressor design involved the use of ion-exchange resin that was filled into a groove and used for suppression [27]. After each run, the capacity depleted resin was discarded and replaced with a fresh aliquot of resin. Thus, a pseudo continuous mode of operation was established.

Another continuously regenerated suppressor called Atlas electrolytic suppressor (AES) based on a monolithic phase and flow distributor disks was designed to overcome the limitations of packed bed suppressors [28,29]. This suppressor was capable of suppressing 25 mN eluent at 1.0 mL/min continuously and provided excellent noise performance.

4.7 DETECTION MODES

Detection approaches in IC can be classified into two major groups, namely electrochemical methods or light-based detection methods. Electrochemical methods would include conductivity, D.C. amperometry, pulsed and integrated amperometry, and potentiometry. Light-based detection methods would include UV and visible absorbance based detection, indirect photometry, fluorescence, chemiluminescence, and refractive index detection. For basic information on various detectors and instrumentation setup the reader is referred to various books and articles as outlined in the reference section [30–33]. The most common detection approaches are discussed below.

From the first publication on suppressed IC to date, conductivity detection remains the most sensitive system for the detection of ionic species (both inorganic and organic). Conductivity involves applying an alternating voltage to a pair of electrodes in a cell and measuring the resulting current, which is directly proportional to the conductance of the solution being measured within the cell. The lowest noise possible with conductivity detection is in the order of 0.2 nS/cm. Typical signals for common ions found in drinking water samples are in the μS/cm range, thus yielding high S/N ratios. With preconcentration techniques and suppressed conductivity detection it is also feasible to detect the ppt (part per trillion) level of ions and is particularly suited for ultra pure water analysis in the semiconductor industry.

The suppressor enhances the detection of fully dissociated species with conductivity detection. However, for weakly dissociated species the response is dependent on the pK_a or pK_b and concentration of the analyte ion. For example ammonium ion is converted to ammonium hydroxide after suppression and produces a nonlinear response with concentration. The response against concentration plot has to be fitted to a quadratic fit curve during analysis with suppressed conductivity detection. While from a quantitative perspective this approach is valid, several researchers looked at ways of producing a linear fit curve by converting the weakly dissociated species to suitable salt forms. In one membrane-based approach, the suppressed cation analytes were converted to the methanesulfonate form in a background of methanesulfonic acid (MSA). This was achieved by using a concurrent flow of the regenerant in a CSRS suppressor that had an optimized electrode length that facilitated transport of MSA back into the suppressor eluent channel. The cation analytes were replaced with hydronium ions in the next stage in a converter section thus forming a highly conductive MSA for every cation detected, as depicted in Figure 4.5 [34,35]. This approach not only enhanced the signal for weakly dissociated species, it also enhanced the detection of strongly dissociated species and linear relationships between responses and concentrations were achieved. The salt converter device however was optimized for isocratic operation only. In another approach, a microelectrodialytic eluent generator was used to generate and add sodium hydroxide reagent to the post suppression stream to convert weak acids to the salt form. Detection of weak acids such as silicate, arsenite, and cyanide was feasible by this approach [36]. Though in this approach all analytes are converted to the salt form, fully dissociated species would give a lower response after conversion to the salt form.

Another reason for non-linearity of the suppressed conductivity signal stems from the background and its role in affecting the response. An effect called "spongy platform" was described [30], which reduced the response of acids due to the weak carbonic acid background from carbonate and/or bicarbonate eluent. The weakly dissociated background gets into the neutral form due to the presence of the acid peak and gets depressed at the location of the peak thereby displaying a lower response for an acid analyte. This issue resulted in a nonlinear response at low concentrations. Some researchers worked on removing the carbon dioxide from the suppressed eluent using gas permeable tubings. The net effect of removing the carbon dioxide from the

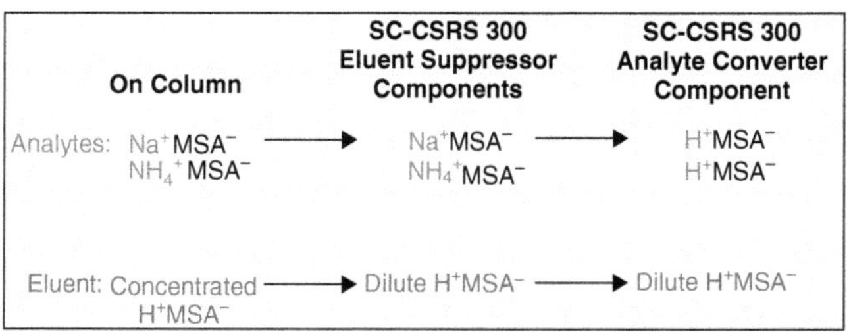

Figure 4.5. Salt converter net reaction schematic.

carbonic acid background was to drive the equilibrium to further formation of carbon dioxide and upon further removal the background was similar to water as described in the equation below.

$$H_2CO_3 \Longleftrightarrow CO_2(\text{removed}) + H_2O \tag{6}$$

This removal of the carbonic acid background or "spongy platform" resulted in improved response and linearity of response with conductivity detection. The use of gas permeable membrane based CO_2 removal devices in conjunction with a suppressor allowed the performance of carbonate/bicarbonate eluents to be similar to hydroxide eluents with suppressed conductivity detection [37]. The approach worked well under isocratic conditions. Under gradient conditions however significant changes were observed in the background.

DC amperometry involves measurement of a current resulting from the oxidation or reduction of analytes on the surface of a working electrode, in response to a single applied potential. DC amperometry is applicable for analysis of catecholamines and other biogenic neurotransmitters and organic amines using a glassy carbon electrode [38].

Analysis of cyanide, sulfide, and halides on a silver electrode is another established application of DC amperometry. Integrated amperometry involves measuring a current during a portion of an applied potential waveform. The purpose of the waveform is to clean the electrode surface and make the clean surface available for subsequent analysis [39,40]. The key applications of this detection method are for analysis of carbohydrates, alcohols, glycols, aldehydes, amines, and thiols. A quadrapole waveform was shown to provide reproducible analysis of carbohydrate samples. Electrode recession was minimized and electrode to electrode response was improved [41].

From an electrode material perspective, while the early work was focused on carbon electrodes, newer electrode materials, such as platinum, gold and silver, are more commonly reported in the literature. Disposable electrodes were introduced to provide a fresh electrode surface for the detection because it was observed that standard electrodes would get fouled over time and require reconditioning. The reconditioning steps were time consuming and cumbersome. For example, when pursuing DC amperometry for the detection of sulfide and cyanide analysis with a silver electrode, a more robust detection method was established by switching to a pulsed waveform and disposable silver electrodes. Disposable electrodes had several advantages over standard electrodes, namely, faster equilibration, electrode to electrode reproducibility, consistent and stable response for a stipulated time period along with the advantage of not needing polishing or activation waveforms [42]. Electrochemical detection is described in more detail elsewhere in this book (see Chapter 3).

With UV detection, direct and indirect detection methods have been used for the detection of ions. Most inorganic ions lack the chromophores and only absorb in the low UV region. On the other hand, aromatic and heterocyclic ions (usually acids and amines) have strong UV-absorbing chromophores and are routinely monitored by UV. However this may be advantageous during UV detection of organic ions because inorganic ions will not interfere with the detection. For example, detection of nitrate, nitrite, bromide, and iodide is feasible in the presence of large concentrations of chloride from a sea water sample. Similarly analyte ions that have chromophores such as in pharmaceutical samples can be selectively detected by the appropriate choice of the wavelength. Post column derivatization and subsequent UV or visible detection is also commonly employed. The most common application is detection of metal ions or lanthanides derivatized with 4-(2-pyridylazo) resorcinol (PAR). The PAR forms a visible light absorbing complex with most transition metals with λ_{max} at 520 nm.

A recent approach with post-column reaction with 4-(2-pyridylazo) resorcinol (PAR) and absorbance detection at 510 nm using a pair of light sensitive diodes with one acting as a light source and the other as a detector was used for analyzing transition metal samples. Comparable detection limits were reported for this LED detection approach versus a standard absorbance detector [43].

Another strategy to overcome the limitation of not having chromophores is to use a UV-absorbing eluent and pursue detection of common inorganic ions in the indirect mode [30]. In this case the UV absorbing eluent elutes the non-chromophoric sample ion off the column, and since the sample occupies a certain equivalent of the eluent a negative signal proportional to the sample concentration is seen at the location. The response of the sample ions is universal provided the sample does not absorb at the chosen detection wavelength. Simultaneous measurement of anions and cations was shown feasible by monitoring at a single wavelength or by monitoring at multiple wavelengths chosen specifically for the anion or cation analysis [30].

Although there are some examples in the literature of both direct and indirect fluorescence detection [44], it is not a popular detection approach with IC because most common ions do not fluoresce. However, fluorescence with post-column derivatization for detection of a primary amine after derivatization with o-phthalaldehyde and 2-mercaptoethanol was a primary application of this detection approach [45].

While the role of suppressor is tailored to increase the S/N ratio of conductivity detection, due to the reduction of background signal, post column techniques are feasible that selectively enhance the analyte signals. For example post column reagents could be tailored to react selectively with the analyte of interest while having little background signal. In practice however there is some background and noise associated with the introduction of the post column reagent in the post-suppressor configuration. Detection of ions such as bromate in the presence of high levels of matrix ions is feasible in drinking water by the above approach. Another possibility was the ability to concentrate the species of interest selectively from the post-suppressor stream for further analysis. Sensitive analysis of species such as bromate or perchlorate in the presence of high concentration of matrix ions was demonstrated for drinking water samples. Again this was feasible due to the weakly dissociated suppressor background (water-like) with hydroxide eluents [46,47].

4.8 CONCLUSION

Suppressed IC with conductivity detection remains the preferred mode for pursuing ion analysis. The flat membrane design continues to be the preferred design due to its low delay volume, high suppression capacity, and versatile performance under both aqueous and solvent conditions. Conductivity detection continues to dominate the detection arena for ions due to its simplicity and its applicability to a wide range of samples and concentrations.

REFERENCES

1. Small H, Stevens TS, Bauman WC. Novel ion exchange chromatographic method using conductimetric detection. Anal. Chem. 1975;47:1801–1809.
2. Stevens TS, Davis JC, Small H. Hollow fiber ion-exchange suppressor for ion chromatography. Anal. Chem. 1981;53:1488–1492.
3. Stevens TS, Jewett GL, Bredeweg RA. Packed hollow fiber suppressors for ion chromatography. Anal. Chem. 1982;54:1206–1208.
4. Dasgupta PK. Linear and helical flow in a perfluorosulfonate membrane of annular geometry as a continuous cation exchanger. Anal. Chem. 1984;56:96–103.
5. Dasgupta PK. Annular helical suppressor for ion chromatography. Anal. Chem. 1984;56:103–105.
6. Dasgupta PK, Bligh RQ, Mercurio MA. Dual membrane annular helical suppressors in ion chromatography. Anal. Chem. 1985;57:484–489.
7. Stillian, JR. An improved suppressor for ion chromatography. LC Liq. Chromatogr. 1985;3:802–806.
8. Ban T, Murayama T, Muramoto S, Hanaoka Y. inventors; Yokogawa Hokushin Electric Works, assignee. Method and apparatus for analysis of ionic species. US patent 4403039. 1983 Sep 6.
9. Jansen; KH, Fischer KH, Wolf B. inventors; Process and apparatus for the quantitative determination of cations or anions by ion chromatography US patent 4459357. 1984 July 10.
10. Tian ZW, Hu RZ, Lin HS, WU. JT. High-Performance Electrochemical suppressor for ion chromatography, J. Chromatogr. A 1988;439:159–163.
11. Strong DL, Dasgupta PK. Electrodialytic membrane suppressor for ion chromatography. Anal. Chem. 1989;61:939–945.
12. Pohl C, Slingsby RW, Stillian JR, Gajek R. inventors; Dionex Corporation, assignee. Modified membrane suppressor and method for use, US patent 4999098. 1991 Mar 12.
13. Henshall A, Rabin S, Statler J, Stillian J. Recent development in ion chromatography detection: the self- regenerating suppressor, Am. Lab. 1992;24:20R.
14. Rabin S, Stillian J, Barreto V, Friedman K, Toofan M. New membrane-based electrolytic suppressor device for suppressed conductivity detection in ion chromatography. J. Chromatogr. 1993;640:97–109.
15. Srinivasan K, Pohl C, inventors; Dionex Corporation, assignee. Chemical suppressors and method of use. US patent 7517696. 2009 Apr 14.
16. Stillian JR, Barreto VM, Friedman KA, Rabin SB, Toofan M. inventors; Dionex Corporation, assignee. Ion chromatography system using electrochemical suppression and detector effluent recycle US patent 5248426. 1993 Sep 28.
17. Dasgupta PK, Bligh RQ, Lee J, D'Agostino V. Ion penetration through tubular ion exchange membranes. Anal. Chem. 1985;57:253–257.

18. Srinivasan K, Pohl C, inventors; Dionex Corporation, assignee. Displacement chemical regeneration method and apparatus. US patent 7524457. 2009 Apr 28.
19. Srinivasan K, Saini, S, Avdalovic, N. Recent advances in continuously regenerated suppressor devices, Pittcon 2001, Pittsburg Conference, New Orleans, LA, March 2001, Abstract 136
20. Srinivasan K, Barreto VB, Pohl CA, Thayer JR, Avdalovic N. inventors; Dionex Corporation, assignee. Current-efficient suppressors and method of use. US patent 6077434. 2000 June 20.
21. Srinivasan K, Barreto VB, Pohl CA, Thayer JR, Avdalovic N. inventors; Dionex Corporation, assignee. Current-efficient suppressors. US patent 6328885. 2001 Dec 11.
22. Srinivasan K, Pohl C. inventors; Dionex Corporation, assignee. Method and apparatus for gas-assisted suppressed chromatography. US patent 6425284. 2002 July 30.
23. Siriraks A, Stillian J. Determination of anions and cations in concentrated bases and acids by ion chromatography: Electrolytic sample pretreatment. J. Chromatogr. 640, 1993:151–160.
24. Srinivasan K, Lin R. inventors; Dionex Corporation, assignee. Parking a sample stream and suppressing the sample US patent 7399415. 2008 July 15.
25. Haddad PR, Jackson PE, Shaw MJ. Developments in suppressor technology for inorganic ion analysis by ion chromatography using conductivity detection. J. Chromatogr. 2003;1000:725–742.
26. Suppressor Modules Operation. Available at http://www.metrohm.com/com/Support/FAQ/ic/0Suppressor%20Modules; Accessed 2010 May 12.
27. Sato S, Ogura Y, Miyanaga A, Sugimoto T, Tanaka K, Moriyama H. Ion chromatographic system with a novel switching suppression device J. Chromatogr. 2002;956:53–58.
28. Small H, Liu Y, Riviello J, Avdalovic N, Srinivasan K. inventors; Dionex Corporation, assignee. Continuously regenerated packed bed suppressor for ion chromatograph, US patent. 6325976. 2001 Dec 4.
29. Liu Y, Srinivasan K, Pohl C, Avdalovic N. Recent developments in electrolytic devices for ion chromatography. J. Biochem. Biophys. Methods 2004;60:205–232.
30. Small H. Ion chromatography, Plenum Press: New York; 1989, 139–210.
31. Weiss J. Handbook of ion chromatography, Wiley-VCH; 2005, 461–547.
32. Dasgupta PK. Postcolumn techniques: A critical perspective for ion chromatography. J. Chromatogr. Sci. 1989;27:422–431.
33. Dasgupta PK. Ion chromatography, the state of the art. Anal. Chem. 1992;64: 775A–783A.
34. Srinivasan K, Saini S, Avdalovic N. inventors; Dionex Corporation, assignee. Suppressed chromatography and salt conversion system US patent 6752927. 2004 June 22.
35. Srinivasan K, Saini S, Avdalovic N. inventors; Dionex Corporation, assignee. Suppressed chromatography and salt conversion system US patent 7618535. 2009 Nov 17.
36. Sjögren A, Dasgupta PK. A planar microelectrodialytic NaOH generator for eluite conversion after suppressed conductometric detection in ion chromatography. Anal. Chimica. Acta. 1999;384:135–141.
37. Ullah SM, Adams RL, Srinivasan K, Dasgupta PK. Asymmetric membrane fiber-based carbon dioxide removal devices for ion chromatography. Anal. Chem. 2004;76:7084–7093.
38. Kissinger PT, Riggin RM, Alcorn RL, Rau LD. Estimation of catecholamines in urine by high performance liquid chromatography with electrochemical detection. Bio. Med. 1975;13:299–306.

39. LaCourse WR. Pulsed Electrochemical Detection in High Performance Liquid Chromatography. John Wiley & Sons: New York; 1997.
40. Polta JA, Johnson DC The Direct Electrochemical Detection of Amino Acids at a Platinum Electrode in an Alkaline Chromatographic Effluent J. Liq. Chromatogr. 1983;6:1727–1743.
41. Rocklin RD, Clarke AP, Weitzhandler M. Improved long-term reproducibility of pulsed amperometric detection of carbohydrates via a new quadrapole-potential waveform. Anal. Chem. 1998;70:1496–1501.
42. Cheng J, Jandik P, Liu X, Pohl C. Pulsed amperometric detection waveform with disposable thin-film platinum working electrodes in high performance liquid chromatography. J. of Elec. Anal. Chem. 2007;608:117–124.
43. Barrona L, O'Toolea M, Diamond D, Nesterenko PN, Paull B. Separation of transition metals on a poly-iminodiacetic acid grafted polymeric resin column with post-column reaction detection utilising a paired emitter–detector diode system. J. Chromatogr. A 2008;1214:31–36.
44. Haddad PR, Jackson PE, Ion chromatography: principles and applications. J. of Chromatography Library, volume 46, New York: Elsevier; 1990.
45. Roth M. Anal. Chem. Fluorescence reaction for amino acids. 1971;43:880–882.
46. Wagner HP, Pepich BV, Later D, Pohl C, Srinivasan K, Borba BD, Lin R, Munch DJ. Determination of perchlorate in drinking water using two-dimensional ion chromatography with suppressed conductivity detection. U. S. EPA Method 314.2, May 2008.
47. Wagner HP, Pepich BV, Later D, Pohl C, Srinivasan K, Borba BD, Lin R, Munch DJ. Determination of bromate in drinking water using two-dimensional ion chromatography with suppressed conductivity detection. U. S. EPA Method 302, Sep 2009.

5

MODELLING AND OPTIMIZATION OF ION CHROMATOGRAPHIC SEPARATIONS OF PHARMACEUTICALLY RELEVANT ORGANIC IONS

Philip Zakaria and Greg Dicinoski

Pfizer Analytical Research Centre, Australian Centre for Research on Separation Science, School of Chemistry, University of Tasmania, Tasmania, Australia

Melissa Hanna-Brown

Pfizer Global Research and Development Laboratories, Sandwich, UK

Paul R. Haddad

Pfizer Analytical Research Centre, Australian Centre for Research on Separation Science, School of Chemistry, University of Tasmania, Tasmania, Australia

5.1 INTRODUCTION

Ion chromatography (IC) is a well-established and robust liquid chromatographic technique for the separation of charged species, especially inorganic and small organic anions and cations. Even today this remains the major focus of IC, with numerous columns and established methods being available for the efficient separation of a large variety of inorganic ions. IC also has potential to be applied to the routine separation of organic ions, such as organic acids and bases relevant to the pharmaceutical

Applications of Ion Chromatography for Pharmaceutical and Biological Products, First Edition.
Edited by Lokesh Bhattacharyya and Jeffrey S. Rohrer.

industry. Indeed, this is an attractive option because it provides chromatographic separations having a different separation selectivity to reversed-phase high performance liquid chromatography, which is the dominant separation technique used in the pharmaceutical industry. However, IC has not been applied routinely to the separation of pharmaceutically relevant compounds largely due to the perceived poor separation efficiencies that result from secondary interactions between the hydrophobic portions of these organic compounds and the backbone of the packing material used in the column.

Despite this limitation, the use of IC for the separation of organic compounds has received considerable attention and some illustrative examples are provided in the following discussion. An early example was demonstrated by Horvath and Lipsky in 1966 who separated a series of small organic compounds under acidic and basic conditions [1]. Mixtures of four compounds were separated in 60–90 min, although relatively broad peaks were exhibited. Megoulas and Koupparis used IC for the determination of carbocisteine in pharmaceuticals in both anion-exchange and cation-exchange modes with nonsuppressed conductometric detection [2]. Best results were obtained using the cation-exchange system with an acidic mobile phase of pH 3–4. This system provided sufficient separation of carbocisteine from other amino acids, though the nonsuppressed conductivity detection exhibited limited sensitivity. IC has been applied to the separation of artificial sweeteners (aspartame, sodium cyclamate, acesulfame-K, and sodium saccharin) by several research groups [3–5]. The separations were all performed in the anion-exchange mode using carbonate [3] or hydroxide [4,5] eluents. The most recent report by Zhu et al. [5] made use of the Dionex Eluent Generator to electrolytically generate hydroxide eluents and combined this with suppressed conductivity detection. The resultant system exhibited good separation and detection for the artificial sweeteners in a range of real samples.

When IC is applied for the analysis of inorganic ions, aqueous eluents are generally used since solubility of these species is generally not a problem. However, in the case of pharmaceutically related compounds, solubility is a major concern since the majority of the compounds of interest have limited solubility in aqueous solutions. To address these problems most IC systems dealing with pharmaceutically relevant compounds include one or more organic modifiers in the eluent. The addition of an organic modifier acts to both improve solubility for the compounds of interest and also to reduce or modify the extent of reversed-phase (RP) interactions between the organic component of the solute and the backbone of the stationary phase (SP). This is further discussed below. Knox and Jurand separated morphine alkaloids using both strong anion- and cation-exchange columns [6]. It was found that apart from morphine, the main alkaloids in opium were strongly retained on the strong cation-exchange column used in the study. However, the addition of a small amount of acetonitrile and n-propanol led to the elution of these components within 50 min. Murgia et al. used a 4% cross-linked cation-exchange resin as a SP for the separation of xanthenes, analgesic drugs and coffee [7]. Although it was found that electrolyte-free 25% ethanol could be used as the eluent, weakly acidic solutes were greatly affected by the acidity of the stationary phase and in these cases the use of buffered eluents (with their associated competing ions) was beneficial to the separations. Perez and Bello investigated the determination of paracetamol and salicylic acid in dosage forms using anion-exchange IC with UV detection [8]. The eluent consisted of a small amount of acetonitrile (5%), presumably to improve peak shapes although this was not specifically mentioned. Jin et al. used anion-exchange IC for the determination of five 4-hydroxycoumarin rodenticides in animal liver tissues using a Dionex Eluent Generator with reagent free ion chromatography (RFIC) and fluorescence detection. These compounds are relatively large and it was found that the addition of acetonitrile reduced analysis time, dramatically improved peak shapes, and minimized peak broadening, with the result that the five target compounds were baseline resolved within 8 min, albeit under gradient conditions.

5.2 RETENTION MODELLING IN ION CHROMATOGRAPHY

The conditions used in the IC separations discussed above were determined by a relatively inefficient procedure whereby each experimental parameter was optimized individually. The use of an experimental design and a defined experimental space over which the system can be investigated provides a more thorough method for examining a system. However, to be statistically valid this approach often requires considerable experimentation and is therefore usually a time-consuming and laborious process. A more attractive and efficient alternative is to employ computer modelling of analyte retention in order to simulate separations and select the optimal separation conditions. In most cases, this can be undertaken with relatively few experiments.

The development of a computer-based optimization routine can be achieved in a number of ways, including simplex, factorial design, and interpretive methods. The interpretive approach will be the focus of this chapter due to its practicality, ease of use, and wide applicability to a range of IC systems. Generally, an interpretive method utilizes a retention model based on a suitable mathematical equation relating the retention factor of an analyte to one or more experimental variables within the separation system. Two types of retention model can be used: theoretical or empirical. The theoretical approach uses the derivation of a retention model from underlying theory and requires a detailed knowledge of the system parameters (relating to the analytes, stationary phase, and eluent) to calculate the retention factor of a desired analyte. Once these parameters have been determined (or estimated) the model allows for the prediction of elution time of the analyte at any condition within a defined experimental space. An added advantage of the theoretical approach is that the parameters used within the model have a physical meaning and can be used to either classify various components of the system or explain observed trends in retention behavior. The empirical approach focuses on describing the manner in which retention changes as parameters within the system are varied, typically through changes in the eluent composition. A major difference between the empirical and theoretical approaches is that in the former case there is no requirement to provide a theoretical basis to the observed trend, but rather to simply describe these trends by some kind of mathematical expression or artificial neural network.

Madden and Haddad [9] have provided a comparison between a developed empirical model and seven theoretical IC models selected from the literature for the isocratic separation of 15 inorganic anions on three different IC columns. The simplest of these models was the Linear Solvent Strength Model (LSSM) which is based on the following equilibrium

$$yA_m^{x-} + xE_r^{y-} \rightleftharpoons yA_r^{x-} + xE_m^{y-} \tag{1}$$

where x and y represent the charge on the anions A and E, respectively, and the subscripts m and r represent the anion either free in the eluent or bound in the resin phase, respectively.

From this equilibrium the LSSM can be derived:

$$\log k_A = \frac{1}{y}\log(K_{A,E}) + \frac{x}{y}\log\left(\frac{Q}{y}\right) + \log\left(\frac{w}{V_m}\right) - \frac{x}{y}\log[E_m^{y-}] \tag{2}$$

where k_A is the retention factor of the analyte, $K_{A,E}$ is the ion-exchange selectivity constant between the analyte and eluent competing ion, Q is the effective ion-exchange capacity of the SP, w is the mass of the stationary phase and V_m is the volume of the eluent phase. For a given system a plot of $\log K_A$ vs. $\log[E_m^{y-}]$ will yield the straight line:

$$\log k_A = C - \frac{x}{y}\log[E_m^{y-}] \tag{3}$$

where C is a constant related to the system. This can also be expressed as:

$$\log k_i = a_i - b_i \log[E_m^{y-}] \tag{4}$$

where $a_i = C$ and $b_i = x/y$.

Each of the models investigated by Madden and Haddad was evaluated using eluents of varying concentrations and pH, with the empirical model being the only model able to predict retention consistently and accurately for all analytes and columns investigated. This was attributed to the nature of the empirical model which was based on the LSSM, but with a modification to allow empirical measurements of a_i and b_i in Eq. (4). Values of a_i and b_i are dependent on the parameters $x, y, K_{A,E}, Q, w$, and V_m from Eq. (2), which in practice can often be difficult to quantify. This approach overcomes one of the major limitations of theoretical models whereby either the parameters needed in the model are difficult to quantify or any competing retention mechanisms (for example, hydrophobic adsorption effects) not specifically considered by the model can be accommodated. This work was extended by the same authors [10] to the separation of 24 inorganic anions by suppressed IC using carbonate eluents. Again the empirical model was consistently the best performer, especially for analytes of higher charge. Figure 5.1 shows the correlation coefficients between the predicted and the experimental data for each individual analyte and each of the models tested. The major disadvantage noted for the empirical model when compared to the theoretical models was that, with the exception of one of the theoretical models, it required the greatest number of initial experiments for two-dimensional optimizations (i.e. those involving eluent concentration and eluent pH).

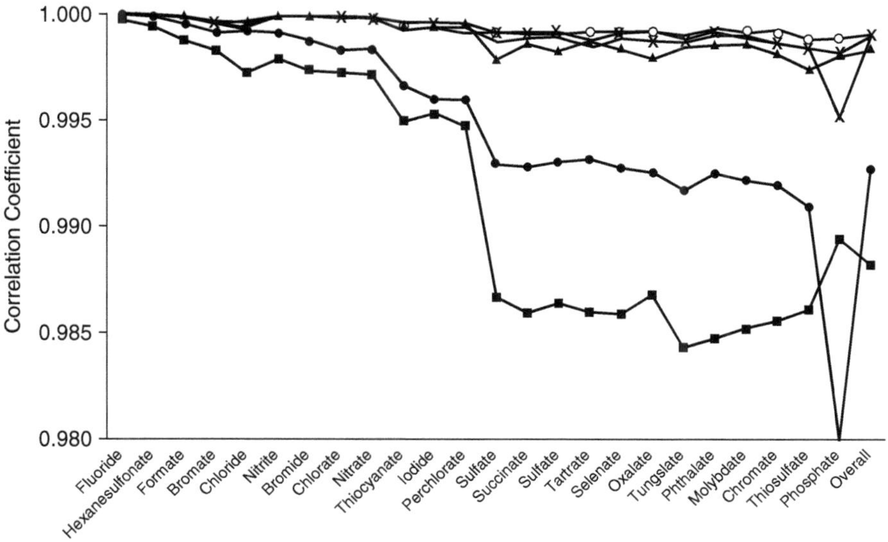

Figure 5.1. Correlation coefficients between predicted and experimental data using a Dionex IonPac AS4A-SC column. ♦ = Dominant equilibrium approach; ■ = effective charge approach; ▲ = dual eluent species model; × = Kuwamoto model; ◊ = extended multiple eluent species model; □ = multiple species eluent/analyte approach; Δ = empirical end-points model. (Taken from [10].)

Madden et al. [11] used the same empirical modelling process in the development of the Virtual Column® software which allowed for the simulation and optimization of the separation of inorganic anions by IC [11]. Figure 5.2 shows the Virtual Column® interface for the separation of inorganic anions using a Dionex IonPac AS20 anion-exchange column. Again, an empirical approach to the LSSM was utilized to predict analyte retention for eluents comprising one or two competing ions. Extensive evaluation of Virtual Column 2 was undertaken using up to 33 analytes and several different columns with both carbonate-bicarbonate and hydroxide eluents. Differences between predicted and observed retention times were generally less than 5%.

Although use of the empirical version of the LSSM has been shown to accurately describe isocratic separations of inorganic anions, its utility is restricted due to the fundamental limitations of performing IC under isocratic conditions. An obvious extension of this work is to include gradient elution in the modelling process. Shellie et al. [12] used a stepwise approach to consider eluent profiles with up to five isocratic and linear gradient steps. Isocratic steps were treated using the empirical LSSM described above, while the linear gradient steps were treated using the gradient elution model proposed by Jandera and Churacek [13]:

$$t_g = \left(\frac{1}{u}\right)\left\{\left(\frac{1}{B}\right)\left[(zb_i + 1)Ba_i t_0 u + C_s^{(zb_i+1)}\right]^{1/(zb_i+1)} - \frac{C_s^{1/z}}{B}\right\} + t_0 \quad (5)$$

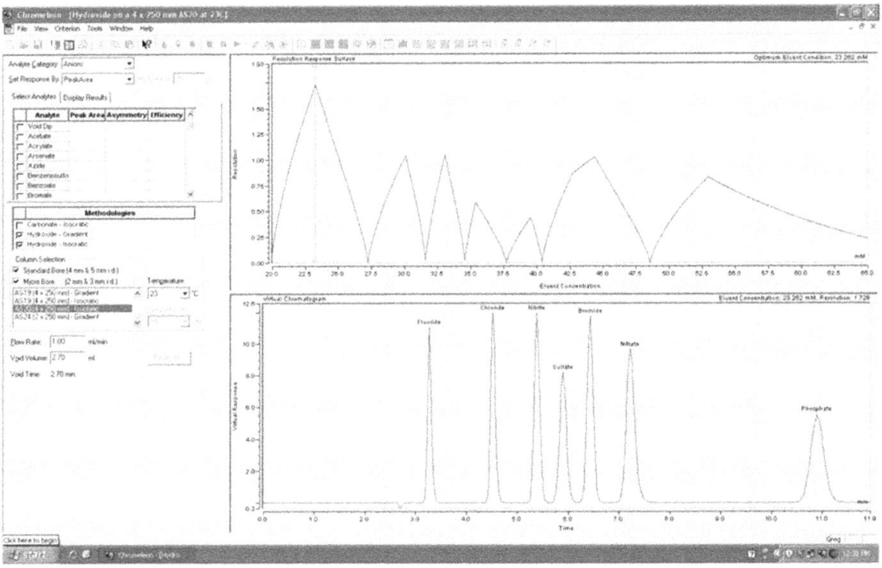

Figure 5.2. Virtual Column interface for the isocratic separation of inorganic anions using a Dionex IonPac AS20 column.

where t_g is the analyte retention time under gradient conditions, u is the mobile phase flow-rate, B is the normalized gradient ramp ($B = R/u$ in mM/mL where R is the gradient ramp rate in mM/min), z is an adjustable parameter used to describe the shape of the gradient profile with $z = 1$ representing a linear gradient, C_s is the starting concentration of the gradient and t_0 is the void time. The constants a_i and b_i in Eq. (5) are the same as those found in the simple LSSM equation shown in Eq. (4).

Figure 5.3 shows a schematic outlining the method adopted by Shellie et al. Using this method it is possible to calculate retention for both isocratic and gradient steps using the same isocratic data that was used in the empirical LSSM. The retention of the analytes under multi step conditions is calculated by treating each step individually using either Eq. (4) or (5) depending on whether the step is isocratic or gradient. At the completion of each step the position of the analyte on the column is calculated and is used as the starting point for the next step. Using this method, along with a large database of previously acquired isocratic retention data, retention prediction for a wide selection of inorganic and small organic anions and cations was achieved over isocratic, linear gradient and multi step gradient (hydroxide or methanesulfonic acid) profiles. Resultant correlation coefficients between observed and predicted retention times were 0.987 and 0.997 for anions and cations, respectively. Shellie et al. [12] also proposed a simple empirical expression for peak width under gradient conditions which comprised a band broadening term as well as a peak compression term. This peak compression term was derived empirically and was used to describe the observed peak narrowing encountered in gradient separations.

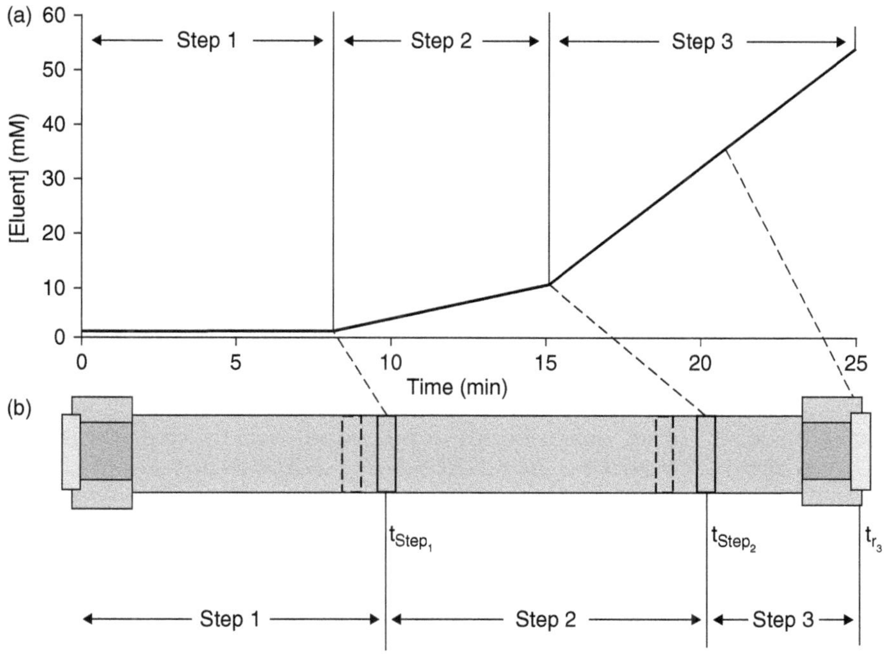

Figure 5.3. Schematic showing stepwise approach to modeling multi step gradient separations. (Taken from [12].)

The major advantage of the empirical LSSM approach to IC modelling arises from the experimental determination of a_i and b_i in Eq. (4). As mentioned previously this process takes into account any additional interactions that may be present. By doing this the practical utility of the model is improved when compared to purely theoretically derived models where all interactions affecting the separation need to be explicitly considered. This adaptability means the approach lends itself perfectly to modelling the IC separation of pharmaceutically relevant compounds that are likely to exhibit secondary interactions with the stationary phase. Such a system was investigated by Zakaria et al. [14] who applied it to 24 anionic and 3 neutral pharmaceutically relevant compounds. The same approach as outlined above was used to model the system with values of a_i and b_i in Eq. (4) being determined experimentally. Because the compounds being investigated consisted of small to medium sized organic molecules, it can be reasonably assumed that hydrophobic adsorption interactions will be present to some degree. As noted in previous work, the inclusion of an organic modifier in the eluent can minimize these interactions so that excessive retention is reduced and peak shapes are improved. However, the use of organic modifiers can cause compatibility issues with various components within an IC system. This is especially true when using a modern IC system utilizing membrane-based devices such as electrolytic eluent generators and membrane suppressors. Two approaches can be used to overcome this issue, namely minimizing the amount of organic modifier to a level that is compatible with the instrumentation (or replacing the non compatible instrumentation with a more robust option), or mixing the organic modifier into the eluent stream via the use of a simple T-piece, thereby by-passing the eluent generator. This latter approach was used by Jin and Zhu [15] for the IC separation of trace chlorophenols in clam tissues coupled to mass spectrometry. It was found that for low concentrations of hydroxide (used as the competing ion) or when organic modifier was absent from the eluent, broad peaks were observed. This was attributed to the interaction of the phenol group of the analytes with the polymeric stationary phase. The addition of 15% (v/v) acetonitrile greatly decreased this broadening for all 14 chlorophenols investigated.

A similar setup was used by Zakaria et al. [14] for the separation of pharmaceutically relevant anionic compounds (Figure 5.4). Similar to the findings of Jin and Zhu [15], peak shapes were improved on addition of methanol to the eluent. Furthermore, while selectivity changes were clearly evident at higher concentrations of methanol (around 50% (v/v)), these changes were not as apparent for concentrations up to 30% where the majority of the improvements in peak shape were observed (Figure 5.5). This study also demonstrated the use of two eluent generators (Elu-Gen units) plumbed in series as a means of increasing the maximum hydroxide concentration above the limit of 100 mM at 1 mL/min achievable by an individual unit. Although the described system worked well, software issues needed to be overcome and the second Elu-Gen unit needed to be connected to another pump contained within the system. This process, although logistically cumbersome, succeeded in raising the maximum achievable

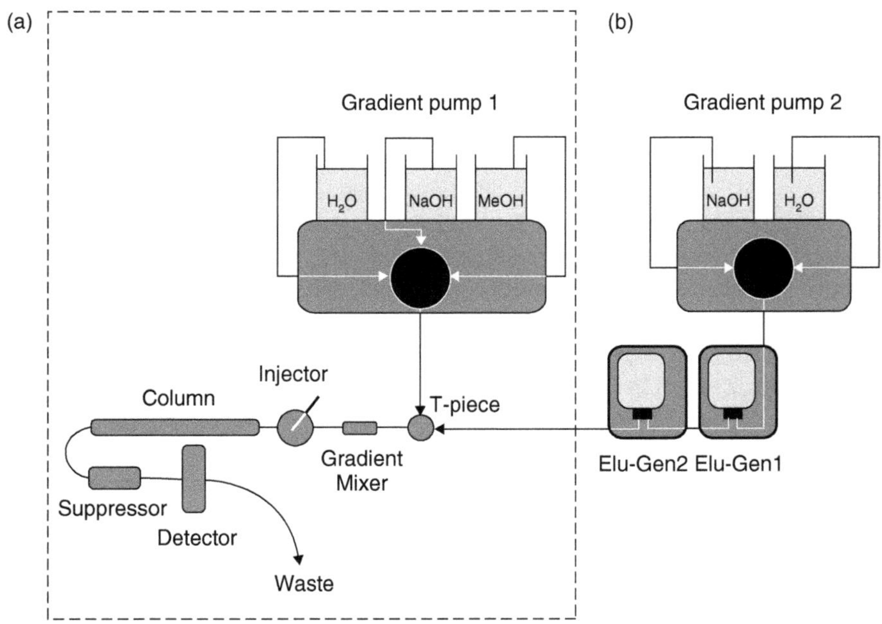

Figure 5.4. Schematic of pump layout used by Zakaria et al. to facilitate use of methanol with non compatible IC components. Layout (a) avoids methanol compatibility issues by removing the Elu-Gen modules and using a manually prepared hydroxide stock solution. Layout (b) avoids passing methanol through the Elu-Gen modules by mixing it with the eluent stream subsequent to hydroxide generation. (Taken from [14].)

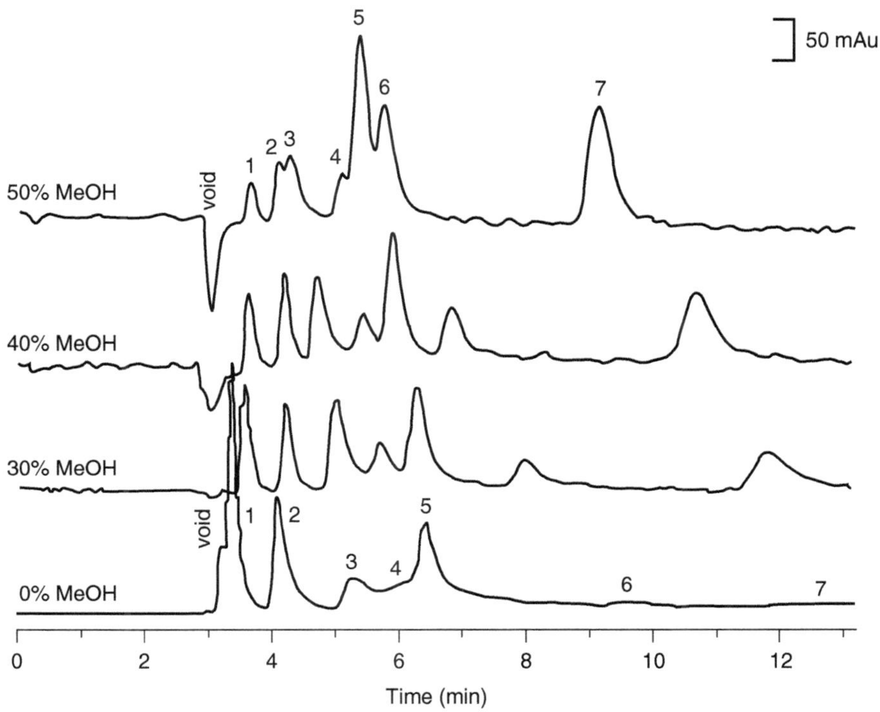

Figure 5.5. Effect of methanol on the separation of a selection of target pharmaceutically relevant anions. (Taken from [14].)

hydroxide concentration and maintained the advantages of convenience and reduced eluent preparation time afforded by the Elu-Gen system. In the same work the use of the empirical LSSM method was shown to be applicable to the 27 pharmaceutically relevant anionic compounds chosen. Furthermore, it was also found that the b_i values obtained for the compounds investigated did indeed deviate from those predicted by theory, with the largest deviation corresponding to those compounds with larger log P values. Despite this deviation, linearity of Eq. (4) was maintained for all compounds investigated, suggesting that although it is likely a component of the observed retention was hydrophobic in nature, this effect was independent of the ion-exchange mechanism resulting in the success of the empirical LSSM. Peak width was predicted using the model proposed by Shellie et al. [12] and despite the broader nature of the peaks, the model performed adequately. Furthermore, the predictive capability of the model was comparable to that obtained for inorganic ions, with errors generally less than 15% between observed and predicted values.

Zakaria et al. also compared the separation of a subset of 10 of the pharmaceutically relevant anionic compounds (chosen as a representative group of the original compound list) on six different Dionex IC columns. It was found that although the columns varied in ion-exchange capacity, the selectivity exhibited by each column was remarkably similar suggesting that, at least for the compounds investigated, the choice of column was not of major importance provided the compounds could be eluted in a reasonable time. This result suggested that the mechanism of separation at constant organic modifier concentration was predominantly the same for each column.

In subsequent work, the same chromatographic system was applied to the separation of pharmaceutically relevant cations [16]. Again, the system involved the post-Elu-Gen addition of methanol to the eluent, but in the case of cations the observed retention was far greater than that observed when separating anions, leading to excessive elution times and broad peaks. Furthermore, the reduction in retention derived from increasing the concentration of the competing ion was less pronounced, thereby limiting the ability of high methanesulfonic acid concentrations to fully counter this increased retention. As a result, the cation separation system was run at higher concentrations of methanol (37.5–50% (v/v)) to reduce retention and facilitate reasonable separation times. The stronger hydrophobic character of the cation system also showed up in the deviation of the experimentally derived b_i values from those predicted by theory. Values ranged from 0.13–0.43 for monovalent cations and 0.43–1.55 for divalent cations, deviating strongly from the theoretical values 1 and 2, respectively. This is a greater deviation than that observed for the equivalent anion system described above and suggests that secondary interactions play a larger role in the separation of cations. The differences in retention of the two cation-exchange columns investigated (Dionex IonPac CS14 and CS17 columns) was attributed to a difference in the relative hydrophobicities of the two columns since the stated capacities were very similar (0.325 and 0.363 meq/column for the CS14 and CS17 columns, respectively).

5.3 EFFECT OF ORGANIC MODIFIERS

Organic modifiers have been routinely used in the IC separation of organic compounds. This has a three-pronged effect. First, the organic modifier can reduce undesirable secondary interactions between the analyte and the SP, leading to improved peak shapes. Second, it is widely accepted that the ion-exchange separation of organic compounds is likely to proceed via both an electrostatic (ion-exchange) interaction as well as an adsorption (hydrophobic) interaction. The inclusion of an organic modifier in the eluent can minimize these adsorption interactions and in doing so reduce the retention for more hydrophobic compounds. The third effect that can arise from the addition of an organic modifier results if the change in retention caused by the addition of the modifier differs between the analytes being studied. If this is the case it is then possible to change the selectivity of the system by changing the concentration of organic modifier in the eluent.

Organic modifiers in IC have been mainly used to improve peak shape and/or reduce elution time as discussed above. Chen and co-workers used IC to separate artificial sweeteners [17] and methylxanthine derivatives [18,19]. In each of these reports it was concluded that the resultant separation mechanism included both an ion-exchange component as well as a hydrophobic adsorption component. The addition of methanol [17] and acetonitrile [18,19] led to reduced retention for all compounds studied, though the magnitude of the reduction differed depending on the charge and hydrophobicity of the particular analyte. Ding and Mou used IC for the separation of tetracyclines using a polymeric cation-exchange column [20]. The tetracyclines were separated chiefly on their hydrophobicity, with the addition of acetonitrile leading to large reductions in retention time, although this was not accompanied by any selectivity changes. While hydrophobicity was cited as the main mechanism of separation, increasing the concentration of HCl (H^+ being the competing ion) in the eluent also led to a decrease in retention time. This suggested the presence of an electrostatic interaction concurrent with the hydrophobic interaction mentioned above. Ouyang et al. separated ephedrine and related compounds in pharmaceutical preparations using cation-exchange IC [21]. The addition of acetonitrile was again found to improve peak shapes as well as to reduce elution times, although again no selectivity changes were observed.

The use of organic modifiers specifically as selectivity modifiers in IC has been reported by numerous researchers. In 1983 Issaq and Muschik [22] investigated the effect of different organic modifiers on the cation-exchange separation of methapyrilene from structurally similar compounds. Interestingly, it was observed that increasing the concentration of acetonitrile from 50 to 75% resulted in an increase in the separation time for the five analytes studied from 13 to 23 min. Furthermore, it was observed that while the separations obtained using methanol and tetrahydrofuran (THF) were inferior to those obtained using acetonitrile, they did show differing elution orders, and hence differing selectivity, for the five components. These results suggested that the organic modifier does not simply affect selectivity by changing the hydrophobicity of the eluent, but rather specifically modifies the nature of the analyte-SP interaction. Laurent et al. used alumina as an ion-exchange SP for the separation of basic drugs [23], with a particular focus on the use of organic modifiers rather than pH to influence retention and selectivity. In the absence of organic modifier it was found that the system was unable to separate a homologous series of primary amines. While the addition of methanol, acetonitrile, and THF led to significant increases in resolution, the elution order remained the same over the full range of organic modifier concentration (0–50%). It was also found for all compounds that as the organic modifier concentrations increased there was a corresponding increase in retention up to a maximum, irrespective of the particular organic modifier. The exact organic modifier concentration that corresponded to this maximum retention was dependent on the analyte and the particular organic modifier used. Methanol was found to show a smaller effect and therefore required higher concentrations than acetonitrile or THF. At higher levels of organic modifier, retention started to decrease, suggesting that the separation mechanism was made up of two competing interactions. The initial increase in retention with rising organic modifier concentrations contrasted with the findings outlined above where increasing levels of organic modifier led to decreases in retention. This finding presumably results from a specific interaction on the alumina SP. The first of the two interactions, namely the increase in retention as the organic modifier concentration rises from zero, was attributed to enhanced solvation of the buffer cation, while the subsequent decrease in retention as the modifier concentrations further increased was attributed to a decrease in the analyte ionization. This dual interaction mechanism was used to optimize the separation of a mixture of six alkaloids, all of which behaved in a similar fashion to the primary amines described above. It was found that the effect of modifier concentration differed for each of the alkaloids investigated, resulting in different selectivity changes depending on the particular modifier, as shown in Figure 5.6. The use of eluents consisting of a ternary mixture of water and two of the modifiers was also investigated. This approach allowed an element of "fine tuning" to be introduced into the system, although the retention changes observed on moving from one organic modifier to another were relatively small.

Work by Zakaria et al. on the IC separation of pharmaceutically related anions and cations also found that, depending on the analyte, the non ion-exchange component of the separation was likely to proceed via at least two analyte-SP interactions [24]. This produces a different relationship between retention and methanol concentration for each of the compounds investigated. Some compounds exhibited a decrease in retention over the full range of methanol concentration (0–75% (v/v/)), others an increase, and the rest an initial decrease followed by an increase as the methanol levels increased further. Examples of these relationships are shown in Figure 5.7. Interestingly, this initial decrease in retention followed by a subsequent increase is the opposite trend to that shown by Laurent et al. [23] on alumina columns and was attributed to the combination of a reversed-phase adsorption interaction (which decreases as the methanol concentration increases) and another unknown interaction, possibly hydrophilic in nature, which increases as the methanol concentration increases.

5.4 MODELLING THE EFFECT OF METHANOL AT ISOCRATIC ION-EXCHANGE CONDITIONS

From the above discussion it can be seen that the effect of organic modifier on the separation of organic ions is not straightforward. The effect on retention from increasing levels of organic modifier may be an increase, a decrease, or a combination of the two. Obviously, this variability makes the theoretical approach to modelling difficult to apply since knowledge of the exact interactions involved is required before any modelling can take place. For this reason, it is simpler to use the empirical approach to describe the way in which an organic modifier may potentially affect the retention of organic and pharmaceutically relevant ions.

Zakaria et al. used an empirical approach to describe the observed relationship between the percentage of methanol and retention for a series of pharmaceutically relevant anions and cations [24]. Under isocratic conditions the observed plots of $\log(k)$ versus %methanol could be fitted accurately to a quadratic curve, irrespective of whether the trend in retention was increasing, decreasing, or a combination of the two. This curve can be fitted using three experimentally derived points obtained at three different methanol concentrations, which then makes it possible to predict retention under isocratic conditions at any methanol concentration (preferably within the range of the three experimental data points).

The effect on peak width from the addition of methanol was also investigated with the aim of being able to generate a predicted chromatogram from three initial experiments. Peak width was found to also approximately follow a quadratic curve, so that the peak width data for the same three conditions used above could be utilized

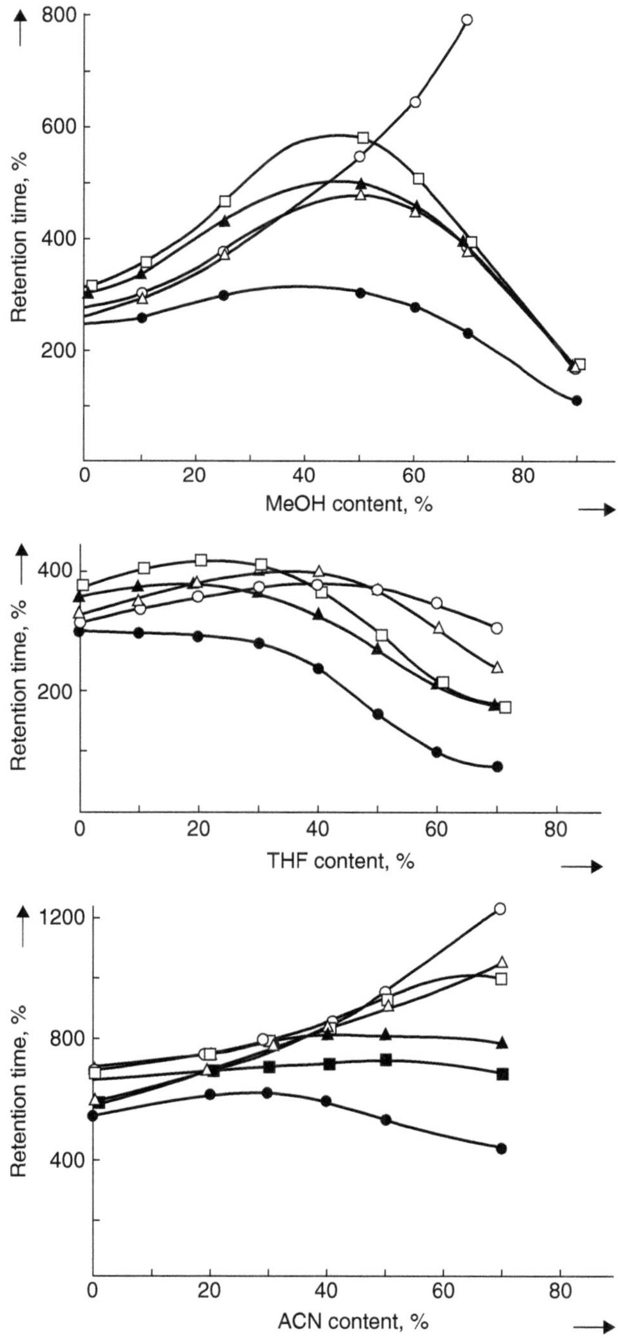

Figure 5.6. Retention behavior of alkaloids with increasing concentration of organic modifier. Cocaine (•), dihydromorphine (∗), morphine (□), dihydrocodeine (△), ephedrine (◦), brucine (×). (Taken from [23].)

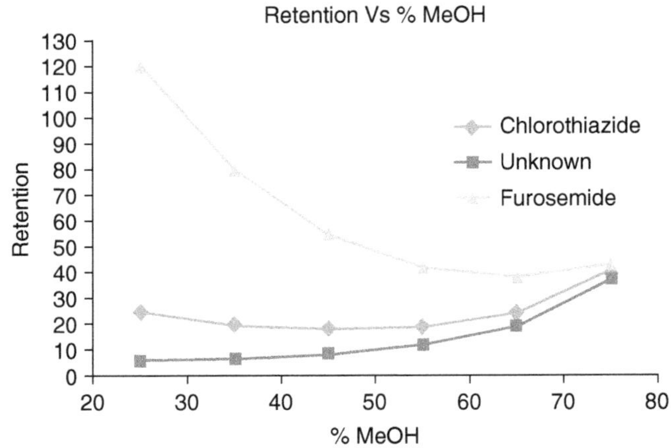

Figure 5.7. Retention vs. %MeOH for three of the test compounds investigated. (Taken from [24].)

to predict peak widths over the full range of methanol concentrations with acceptable accuracy.

5.5 TWO-DIMENSIONAL MODELLING OF ION-EXCHANGE AND METHANOL EFFECTS—ISOCRATIC

From Sections 5.2 and 5.4 above it can be seen that the use of an empirical approach to modelling for both ion-exchange and methanol effects under isocratic conditions can be successfully applied to the IC separation of pharmaceutically related compounds. The next logical step is to combine these two empirical models into a single two-dimensional model that can deal with both effects simultaneously and can accurately describe the entire two-dimensional experimental space comprising the concentration of competing ion on one axis and the concentration or organic modifier on the other.

From an examination of the empirical IC model of Madden and co-workers [9–11] and that of Zakaria et al. for the effect of methanol [24], it can be seen that the underlying process of each model is very similar, although the ion-exchange and methanol models are based on linear and quadratic retention relationships, respectively. Furthermore, Zakaria and co-workers found that the linear relationship between log(k) and log(*competing ion*) is valid over the full range of methanol concentration investigated (0–75%(v/v)), although the values of a_i and b_i in the resultant plots were not constant. Likewise, the quadratic relationship between concentration of methanol and log(k) is valid over the full range of competing ion concentrations investigated (generally 10–75 mM). This independence between the ion-exchange and methanol retention relationships means that the two empirical models can be combined to produce one two-dimensional model. This was demonstrated by Zakaria et al. who used an in-house developed Microsoft Excel spreadsheet to take six experimentally derived

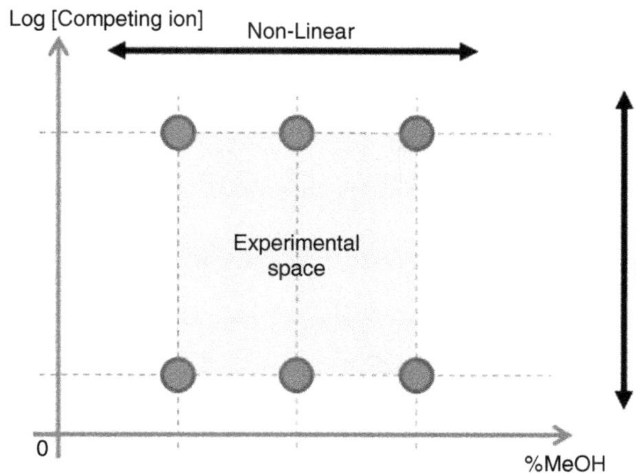

Figure 5.8. Experimental space used by Zakaria et al. for two-dimensional IC modeling. The two variables used are competing ion and methanol concentrations.

points within the two-dimensional experimental space and used the two relationships outlined above to predict the observed retention time for a series of pharmaceutically relevant anions and cations [24]. The six points used for modelling are shown in Figure 5.8 and comprised the four corner points of the experimental design space, as well as an additional two points at intermediate methanol concentrations. The two-dimensional retention model was used to predict retention, and therefore elution time, at any point within the experimental space with correlation at nine conditions for anions and 19 conditions for cations showing average errors less than 5% and 6% for anions and cations, respectively. Figure 5.9a shows the correlation between predicted and observed retention time for pharmaceutically relevant anions using the two-dimensional model. Four of the eluent conditions used to evaluate accuracy of the model were actually taken from outside the experimental design space shown in Figure 5.8, while for cations, Figure 5.9b, 13 of the 19 conditions were taken from outside the experimental space. This highlights the ruggedness of the model. Peak width prediction was also included in the spreadsheet and it was found that, unlike retention time, width prediction was more sensitive to the choice of initial retention data conditions. This resulted in width prediction at points outside the experimental design space giving large systematic errors, whereas results from within the experimental space gave good estimations and yielded accurate predicted chromatograms.

5.6 MODELLING METHANOL GRADIENT SEPARATIONS UNDER ISOCRATIC ION-EXCHANGE CONDITIONS

From the above discussions it can be seen that there are two major classes of interaction (ion-exchange and hydrophobic/hydrophilic adsorption) that affect the IC separation of organic and pharmaceutically relevant ions. For ion-exchange the easiest means to

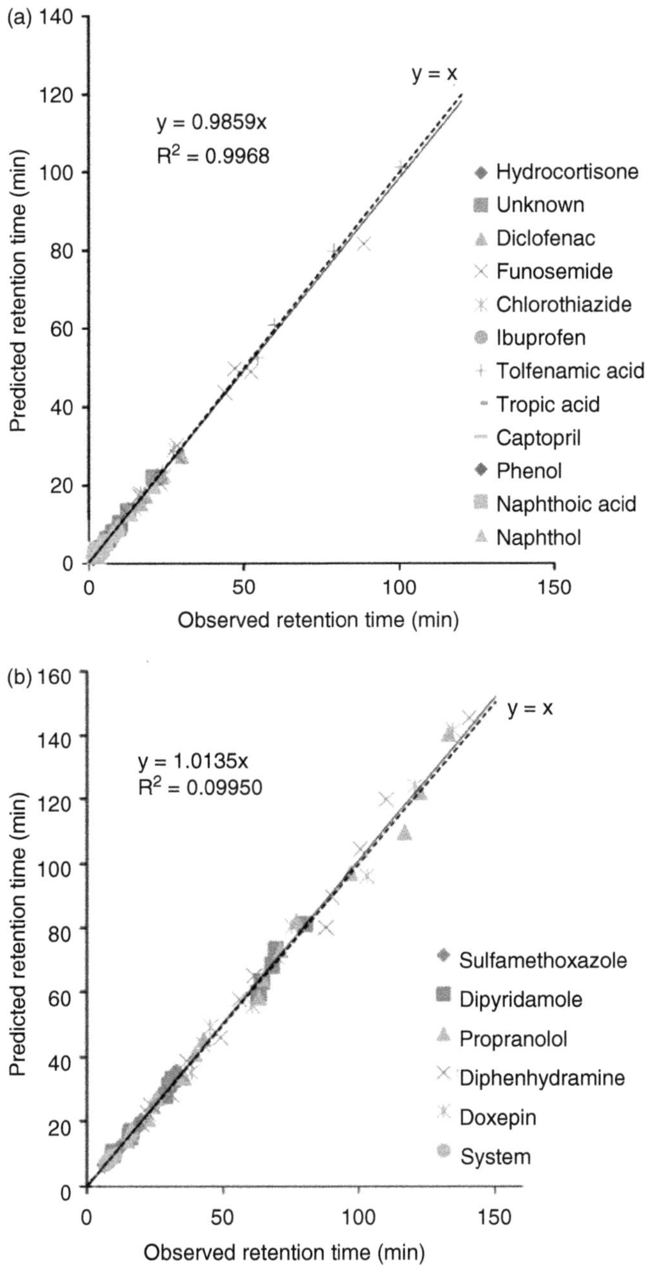

Figure 5.9. Observed and predicted retention times for anions (a) and cations (b) using the two-dimensional model of Zakaria et al. The solid line represents the line of best fit through the predicted vs. observed data while the dotted line represents $y = x$, i.e. perfect agreement between observed and predicted values. (Taken from [24].)

control the extent of the interaction is to change the concentration of competing ion in the eluent, thereby altering the amount of competition the analyte ions experience when competing for the ion-exchange sites on the SP. Modelling of this effect has been described in detail in Section 5.2, including the development of this model from isocratic to gradient and finally to multi step gradient profiles. Adsorption interactions have been shown to be controlled by the addition of organic modifier to the eluent, and in Section 5.4 the modelling of isocratic separations using different levels of methanol at constant competing ion concentration has been described. The logical next step is to investigate the effect of methanol gradients on the observed IC separation of pharmaceutically relevant anions and cations.

Modelling of retention using methanol gradients is common in RP-HPLC and has been included in the DryLab computer simulation software of Snyder and co-workers [25]. A limited amount of experimental gradient data on the particular system being investigated is required for the predictions to be generated. The effect of organic modifier is based on the empirical relationship:

$$\log k^l = \log k_w - S\varphi \tag{6}$$

where k_w and S are constants characteristic of the solvent and analyte while φ is the volume fraction of the organic solvent in the eluent. It can be seen from Eq. (6) that an increase in the organic modifier concentration should decrease the retention of all analytes over the full working range of the equation. Section 5.3 showed that when using IC columns this was not necessarily the case and that, in fact, the relationship between ϕ and retention time could be positive, negative or even vary over the range of organic modifier concentrations used. This fact makes the gradient modelling process more difficult than the equivalent ion-exchange gradient system reported by Shellie et al. [12] where the relatively simple expression of Jandera and Churacek [13] for gradient elution was used. An empirical integration approach can therefore be used to predict the elution times under various linear gradient conditions. This approach is based on the assumption that the velocity at which an analyte moves through the column is constant under isocratic conditions. The velocity, v(t), at a particular time is dependent in this case on the concentration of methanol in the region of the analyte band, i.e. $M(t)$. As the slope of the linear gradient is known, an equation for the percentage of methanol at time t can be written as,

$$M(t) = M_i + rt \tag{7}$$

where M_t is the initial methanol percentage, r is the ramp rate in %/min and t is the time. The distance, $S(t)$, that the analyte band has travelled at time t is given by:

$$S(t) = \int v(t)dt = \int \frac{L}{t_R(t)}dt \tag{8}$$

where L is the length of the column used and $t_R(t)$ is an expression for retention time in terms of time along the methanol gradient.

From Eq. (8) it is possible to calculate the retention time for an analyte by setting $S(t) = L$ and solving for t. For compounds that have not eluted by the end of the linear gradient step their displacement along the column can be calculated by solving for $S(t_s)$, where t_s is the length of the gradient.

The most important aspect of the above procedure is that an accurate approximation of retention time versus percentage methanol is obtained since all resultant calculations are derived from this relationship. These plots are generated from limited experimental data and then approximated using a suitable fitting curve. Zakaria et al. [24] have showed that for the pharmaceutically relevant anions and cations investigated the resultant t_R vs %MeOH plots could generally be described using either a quadratic $(ax^2 + bx + c)$ or power (ax^b) approximation (Figure 5.10). In the few cases where neither of these approximations accurately fitted the observed curves, the straight line segments between the experimental points could be considered individually using the integration method above and summed together to give a final total elution time (linear-segment method). However, this final method was more difficult computationally and had the limitation that it was only applicable to gradients that had starting and end points within the range of the original experimental data. Both equation approximations (quadratic and power) can be used outside of this range which was useful for instances where the particular gradient being used started at methanol levels at which it was impractical to obtain isocratic data due to large retention times and/or poor peak shapes. The number of points used to generate the experimental t_R versus %MeOH plot is also important since the more points used the more accurate will be the resultant curve approximation. Zakaria et al. found that at least four points were needed to ensure accurate results [24]. The above method of prediction was combined with a modified peak compression factor model for peak width which allowed for acceptable width prediction.

Application of the above model to a subset of seven of the pharmaceutically relevant anions and five of the cations was performed using linear and multi step gradient profiles. In the case of anions, the model was applied to 19 linear gradient separations as well as 29 multi step gradient separations. In all cases, correlation between observed and predicted retention time was found to be very good with correlation coefficients of 0.998 for the linear gradient separations and 0.998 and 0.995 for the multi step separations at 30 mM (Figure 5.11a) and 50 mM hydroxide, respectively. For cations, the described model was applied to separations at 14 different conditions consisting of 1, 2, and 3-step gradient profiles, with a resultant correlation coefficient of 0.997 between observed and predicted retention times (Figure 5.11b).

As mentioned previously, the integration approach of Zakaria et al. [24] requires a good approximation of the retention time versus percentage methanol plots. The use of Microsoft Excel to perform all the calculations and display the resultant chromatograms therefore represents a major limitation of the process in its current form because the selection of the most accurate curve estimation, $t_R(M)$, is achieved by visual inspection only. The introduction of a criterion-based selection process would alleviate this problem and aid in the automation of the model. A further practical problem associated with the use of Excel is that the constants associated with $t_R(M)$ cannot be automatically generated and entered into the appropriate cells,

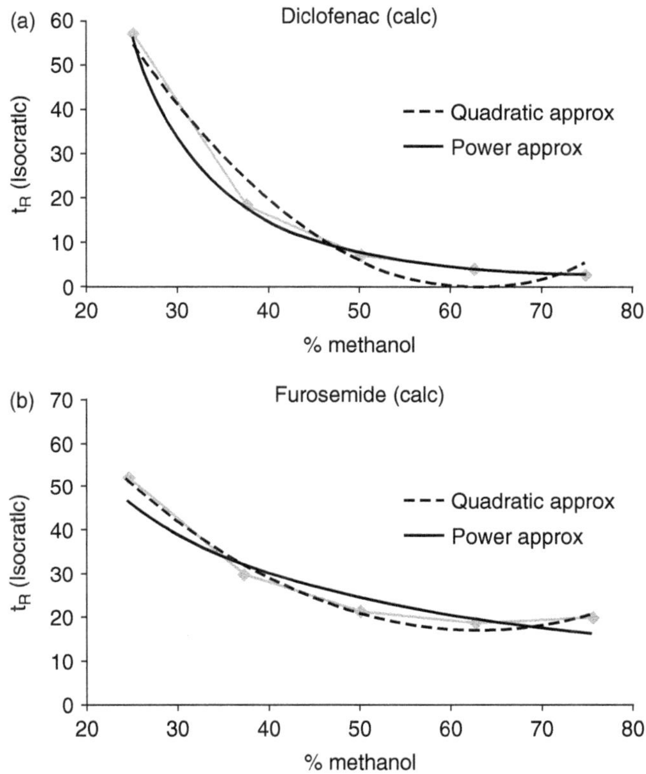

Figure 5.10. Comparison of curve approximations used to describe t_R vs. %MeOH plots. (Taken from [24].)

so that manual transcription is required each time a variation in conditions leads to a change in these constants.

5.7 COMBINING ION-EXCHANGE AND REVERSED-PHASE INTERACTIONS

Section 5.5 described the isocratic modelling of the two-dimensional space using competing ion and percentage methanol as variables, while Section 5.6 described the method used to model methanol gradient based separations under isocratic ion-exchange conditions. Zakaria et al. [24] showed that it is possible to combine the two models, both in Excel, into a single spreadsheet-based model. It is then possible for the isocratic t_R versus %MeOH plots to be generated at any competing ion concentration using only the two-dimensional model and the initial six data points required for its use. The resulting spreadsheet, see Figure 5.12, can therefore not only accurately describe isocratic separations at any point within the competing-ion/methanol two-dimensional space, but can also describe any methanol gradient or multi step gradient

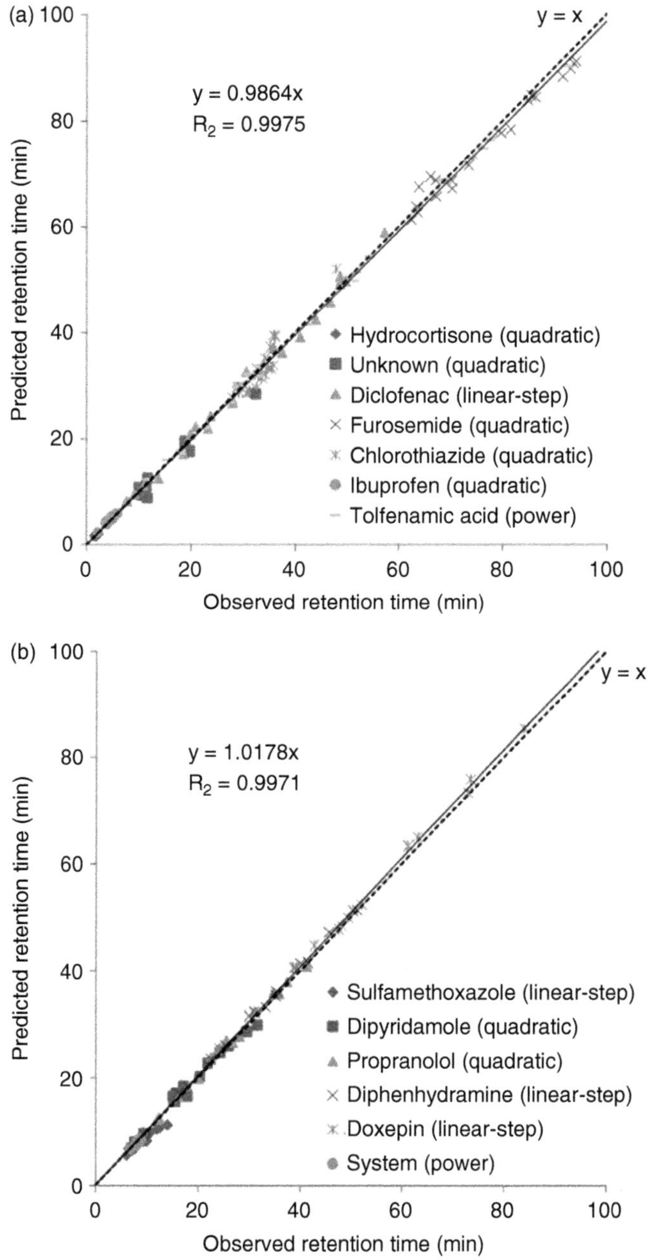

Figure 5.11. Correlation between observed and predicted retention times. (a) Pharmaceutical anions using 20 multi step conditions comprising 2 and 3-step methanol gradient profiles at 30 mM hydroxide. (b) Pharmaceutical cations using 14 multi step conditions comprising 2 and 3-step methanol gradient profiles at 20 mM MSA. The solid and dotted lines have the same meaning as in Figure 5.9. (Taken from [24].)

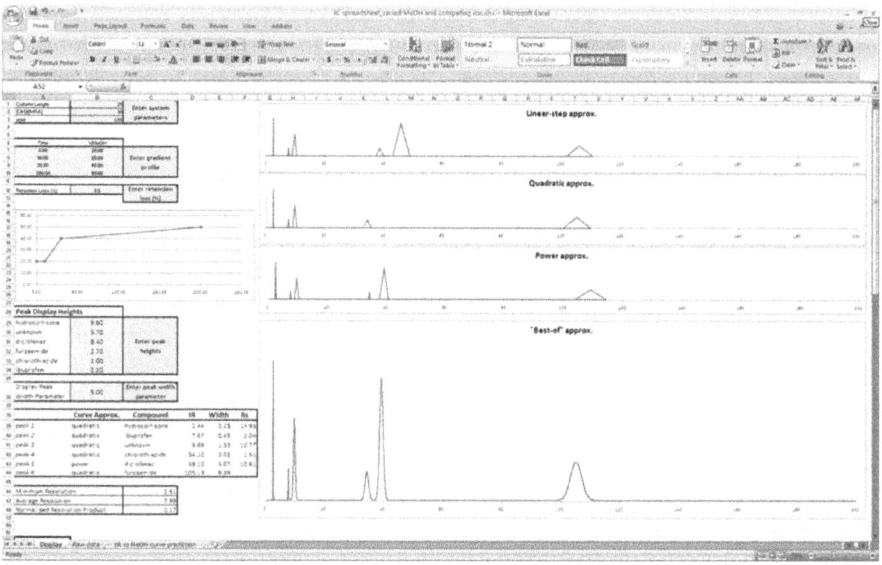

Figure 5.12. Methanol gradient and two-dimensional isocratic spreadsheet of Zakaria et al. [24].

separation, provided the concentration of competing ion is kept constant during the separation. This represents a simple to use yet powerful model that, in conjunction with the ion-exchange model described in Section 5.2, covers isocratic ion-exchange under any conditions within the appropriate two-dimensional experimental space, covers competing-ion gradient and multi step gradient separations at constant organic modifier concentration, and covers organic modifier gradients and multi step gradient separations at constant competing-ion concentrations.

The above modelling therefore covers all likely separation conditions apart from the use of two concurrent gradients (competing ion and methanol gradients) which with modern instrumentation can be easily generated. While practically such a system is easy to set up and use, the complexity introduced to the modelling is significant, such that it is no simple matter to combine the individual gradient models together. Furthermore, it is arguable whether the benefits introduced by the use of two concurrent gradients justify the additional computational complexity.

5.8 OPTIMIZATION

The ability to predict chromatograms introduces the possibility to optimize a particular separation using a pre-defined optimization criterion, such as maximizing the resolution of the least-resolved pair of analytes in the chromatogram. The software can generate all possible chromatograms attainable over the given experimental space and, on the basis of the selected optimization criterion, can identify the optimal separation conditions. This process of optimization is simple, robust and easily applied to a range of systems. Single variable systems are the simplest to optimize since a plot of resolution criteria versus the chosen system variable (e.g., competing-ion concentration) can be easily prepared and the optimum conditions identified. Such an optimization system has been demonstrated by Madden et al. [11] as Virtual Column software (Figure 5.2). Two-variable systems, such as those using carbonate-bicarbonate eluents in IC, or the competing ion versus %methanol system described above can also be relatively easily optimized using this procedure. In this case, the plot of resolution criterion versus system variables becomes a three-dimensional surface with the resolution criterion on the z-axis and the two experimental variables on the x- and y-axis. This has been demonstrated by Madden et al. using a carbonate-bicarbonate system [11]. Zakaria et al. have also used this approach to optimize the isocratic separation of pharmaceutically relevant anions using IC [14] and in the two-dimensional modelling of competing-ion and %methanol [24]. In both cases, the resolution of the least-resolved pair of analytes was calculated over the entire experimental space and the separations with the best resolution and with the best compromise of resolution and separation time were chosen.

The introduction of gradients dramatically increases the number of potential experimental variables in the system. A simple linear gradient, for example, already has three variables (starting concentration, ramp rate, and total time of ramp) with this number increasing as more steps are added to the gradient profile. It is therefore impractical to calculate the resolution criterion at every possible gradient profile in the same manner as used for isocratic separations. To overcome this, an iterative approach can be followed whereby a starting condition is initially chosen and the resolution criterion calculated. Each of the variables can then be varied slightly, either simultaneously or successively, with the resolution criterion calculated at each change. This process is repeated until optimum conditions are found that are satisfactory for the particular application. This method of optimization was performed by Zakaria et al. for the IC separation of pharmaceutically relevant anions using linear gradients and multi step elution profiles [14], see Figure 5.13. The multi step gradient optimization was limited to 2-step gradient profiles, due to the number of variables, and initially only focused on retention time with no consideration of peak width. To ensure that the optimum that was found was indeed global rather than local, the iterative process was started at several different initial conditions within the experimental space and the resultant optima then compared to ensure that the best result was chosen. It was found that the optimized linear gradient and the optimized 2-step gradient separations gave similar resolutions with both successfully separating all 9 compounds, although the 2-step separation took 160 min while the linear gradient separation was completed in 120 min. In all cases the differences between observed and predicted retention times were less than 5%.

A similar process of optimization was again undertaken by Zakaria et al. for the IC separation of pharmaceutically relevant anions and cations using methanol gradients [24]. Again, to limit the potential number of variables an iterative approach was adopted. This was set up by keeping the starting point of the methanol gradient constant at 25%MeOH and the end point constant at 100 min and 75%MeOH, but permitting two intermediate points to be varied, thereby producing a 3-step gradient separation with only four variables. Each was varied simultaneously and the optimization criterion calculated for each separation condition. The resultant optimized separation conditions were then run experimentally and Figure 5.14 shows the experimental and predicted separations. The average error in the predicted retention time was less than 2.4% over the seven compounds. Predicted peak areas were greater than those obtained experimentally due to the limitations of the simple peak width prediction model used.

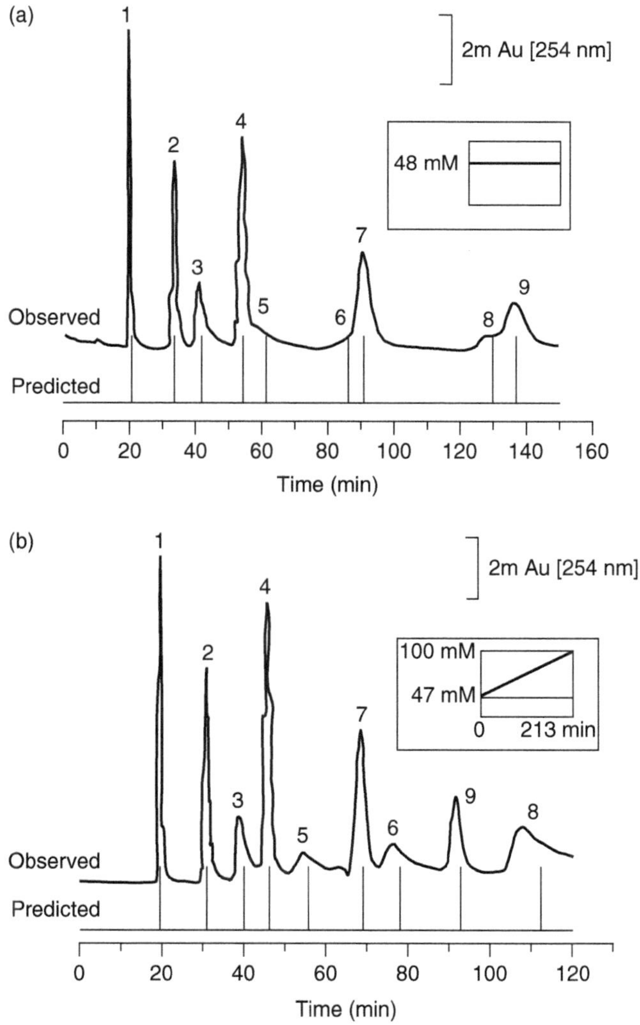

Figure 5.13. Optimized separation of nine pharmaceutically relevant anions using the minimum resolution criterion. (a) Optimized isocratic separation. (b) Optimized linear gradient separation. (c) Optimized 2-step gradient separation. (Taken from [14].)

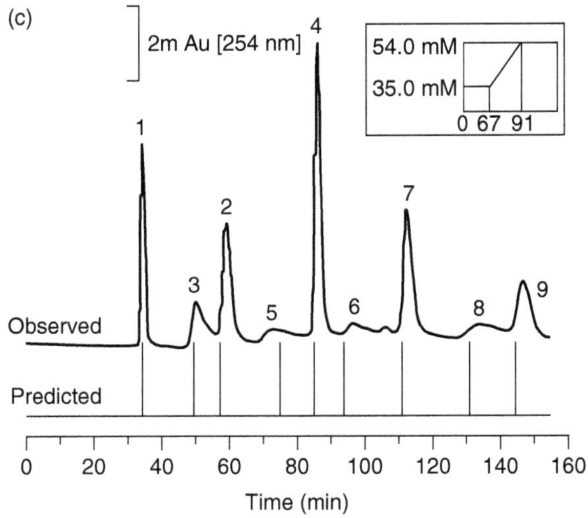

Figure 5.13. (*Continued*)

5.9 SUMMARY

The application of IC to the separation of pharmaceutically relevant compounds has progressed greatly over the last few decades to the point where it now has great potential as a complementary technique to RP-HPLC in the pharmaceutical industry. Modelling of these IC separations has also shown some success. This is because the presence of organic modifier does not appear to dramatically alter the ion-exchange process, so that in IC systems where an organic modifier is present traditional ion-exchange modelling is still applicable.

Although an organic modifier does not appear to greatly influence the ion-exchange process, it does play a major role in the overall separation and can have a large impact on observed retention times. Organic modifiers can therefore be used as an additional selectivity modifier, giving rise to different selectivity changes than those achievable by variation of the ion-exchange interaction. Therefore, in systems where the concentration of organic modifier is varied there needs to be a modification in the modelling procedure to take this effect into account. The effect of organic modifiers in an IC separation is relatively complex and it is likely that there are multiple analyte-stationary phase interactions that are affected directly. This complicates the modelling process since each analyte may exhibit unique selectivity effects. For this reason the simplest and most practical approach to the modelling of these effects is the use of an empirical model based on observed data, rather than attempting to apply a fully theoretical retention model. Such an empirical model has been demonstrated for isocratic (constant organic modifier concentration) and gradient (linear gradients of organic modifier) separations and has shown accurate

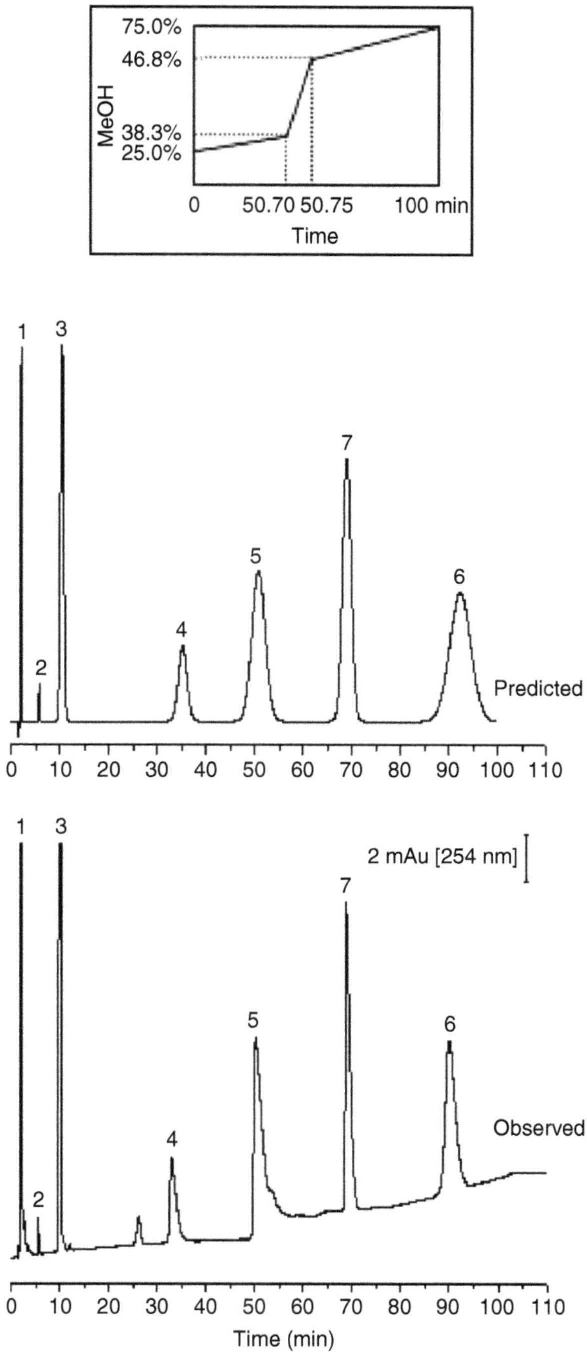

Figure 5.14. Optimized 3-step methanol gradient separation of pharmaceutically relevant anions using the normalized resolution product criterion. (Taken from [24].)

predictive ability for a variety of pharmaceutically relevant anions and cations and over a range of anion- and cation-exchange columns.

To date, the major limitation with the use of IC for the separation of relatively large organic anions and cations is the poor peak efficiencies which result on ion-exchange columns. Detection is also a significant limitation, with the vast majority of published work focusing only on the use of UV-Vis detection and UV-absorbing analytes. Conventional suppressed conductivity detection is generally inapplicable because the analytes are usually weakly conducting after suppression due to either their physical size or partial ionization. Since a large number of potential analytes of pharmaceutical significance do not contain UV-Vis chromophores, there is a need for alternative universal modes of detection. Nebulization detectors, such as evaporative light scattering detectors, are therefore likely to have major potential so that the use of IC in the pharmaceutical industry can be extended over a wide range of analytes.

REFERENCES

1. Horvath CG, Lipsky SR. Use of liquid ion exchange chromatography for the separation of organic compounds. Nature 1966;211:748–749.
2. Megoulas NC, Koupparis MA. Ion-chromatographic determination of carbocisteine in pharmaceuticals based on non-suppressed conductometric detection. J. Chromatogr. A 2004;1026:167–174.
3. Biemer TA. Analysis of saccharin, acesulfame-K and sodium cyclamate by high-performance ion chromatography. J. Chromatogr. 1989;463:463–468.
4. Qu F, Qi Z-H, Liu K-N, Mou S-F. Determination of aspartame by ion chromatography with electrochemical integrated amperometric detection. J. Chromatogr. A 1999;850:277–281.
5. Zhu Y, Guo Y, Ye M, Fritz JS. Separation and simultaneous determination of four artificial sweeteners in food and beverages by ion chromatography. J. Chromatogr. A 2005;1085: 143–146.
6. Knox JH, Jurand J. Separation of morphine alkaloids, heroin, methadone and other drugs by ion-exchange chromatography. J. Chromatogr. 1973;82:398–401.
7. Murgia E, Richards P, Walton HF. Liquid chromatography of xanthines, analgesic drugs and coffee. J. Chromatogr. 1973;87:523–533.
8. Perez JL, Bello MA. Determination of paracetamol in dosage forms by non-suppressed ion chromatography. Talanta 1999;48:1199–1202.
9. Madden JE, Haddad PR. Critical comparison of retention models for optimization of the separation of anions in ion chromatography I. Non-suppressed anion chromatography using phthalate eluents and three different stationary phases. J. Chromatogr. A 1998;829:65–80.
10. Madden JE, Haddad PR. Critical comparison of retention models for the optimization of the separation of anions in ion chromatography II. Suppressed anion chromatography using carbonate eluents. J. Chromatogr. A 1999;850:29–41.
11. Madden JE, Shaw MJ, Dicinoski GW, Avdalovic N, Haddad PR. Simulation and optimization of retention in ion chromatography using virtual column 2 software. Anal. Chem. 2002;74:6023–6030.

12. Shellie RA, Ng BK, Dicinoski GW, Poynter SDH, O'Reilly JW, Pohl CA, Haddad PR. Prediction of analyte retention for ion chromatography separations performed using elution profiles comprising multiple isocratic and gradient steps. Anal. Chem. 2008;80:2474–2482.
13. Jandera P, Churacek J. Gradient elution in liquid chromatography II. retention characteristics (retention volumn, band width, resolution, plate number) in solvent-programmed chromatography - theoretical considerations. J. Chromatogr. 1974;91:223–235.
14. Zakaria P, Dicinoski G, Ng BK, Shellie RA, Hanna-Brown M, Haddad PR. Application of retention modelling to the simulation of separation of organic anions in suppressed ion chromatography. J. Chromatogr. A 2009;1216:6600–6610.
15. Jin M, Zhu Y. Ion chromatography-atmospheric pressure chemical ionization mass spectrometry for the determination of trace chlorophenols in clam tissues. J. Chromatogr. A 2006;1118:111–117.
16. Zakaria P, Dicinoski G, Hanna-Brown M, Haddad PR. Application of the linear solvent strength model to the separation of pharmaceutical cations using ion exchange columns. 2009.
17. Chen Q-C, Mou S-F, Liu K-N, Yang Z-Y, Ni Z-M. Separation and determination of four artificial sweeteners and citric acid by high-performance anion-exchange chromatography. J. Chromatogr. A 1997;771:135–143.
18. Chen Q-C, Mou S-F, Hou X-P, Ni Z-M. Simultaneous determination of caffeine, theobromine and theophylline in foods and pharmaceutical preparations by using ion chromatography. Anal. Chim. Acta 1998;371:287–296.
19. Chen Q-C, Wang J. Simultaneous determination of artificial sweeteners, preservatives, caffeine, theobromine and theophylline in food and pharmaceutical preparations by ion chromatography. J. Chromatogr. A 2001;937:57–64.
20. Ding X, Mou S. Ion chromatographic analysis of tetracyclines using polymeric column and acidic eluent. J. Chromatogr. A 2000;897:205–214.
21. Ouyang J, Gao W, Baeyens WRG, Delanghe JR. Determination of ephedrine and related compounds in pharmaceutical preparations by ion chromatography with direct conductivity detection. Biomed. Chromatogr. 2005;19:266–271.
22. Issaq HJ, Muschik GM, Ion exchange high performance liquid chromatography separations using a high content organic modifier mobile phase. J. Liq. Chromatogr. 1983;6:825–831.
23. Laurent CJCM, Billiet HAA, deGalan L, The use of organic modifiers in ion exchange chromatography on alumina. The separation of basic drugs. Chromatographia 1983;17:394–399.
24. Zakaria P, Dicinoski G, Hanna-Brown M, Haddad PR. Prediction of the effects of methanol and competing ion concentration on retention in the ion chromatographic separation of anionic and cationic pharmaceutically related compounds. J. Chromatogr. A 2010;1217:6069–6076.
25. Dolan JW, Lommen DC, Snyder LR., Drylab computer simulation for high-performance liquid chromatographic method development II. gradient elution. J. Chromatogr. 1989;485:91–112.

PART II

PHARMACEUTICAL APPLICATIONS

6

ION CHROMATOGRAPHY IN PHARMACEUTICAL DRUG ANALYSIS[1]

Lokesh Bhattacharyya

Division of Biological Standard and Quality Control, Office of Compliance and Biologics Quality, Center for Biologics Evaluation and Research, Food and Drug Administration, Rockville, MD

6.1 INTRODUCTION

Ion chromatography (IC) received wide recognition from the analytical community in the biotechnology industry, as well as from academic researchers, after Lee and his associates introduced in the late 1980s the High Performance Anion Exchange Chromatography (HPAEC) with Pulsed Amperometry Detection (PAD) for the analysis of mono- and oligosaccharides derived from glycoproteins [1]. It provided the first general chromatographic method for the analysis of mono- and oligosaccharide (glycan) moieties of glycoproteins and glycolipids, which is robust enough to provide precision, accuracy and reproducibility consistent with the current industry practice. The interest of the biotechnology community in IC grew further when technology was

[1]The findings and conclusions in this chapter have not been formally disseminated by the Food and Drug Administration and should not be construed to represent any Agency determination or policy.

Applications of Ion Chromatography for Pharmaceutical and Biological Products, First Edition.
Edited by Lokesh Bhattacharyya and Jeffrey S. Rohrer.

introduced to combine IC systems on-line with more analytically powerful techniques, such as mass spectrometry [2,3], and a HPAEC-PAD method was introduced for the analysis of amino acids and peptides [4].

However, the interest of the pharmaceutical scientists remained luke-warm toward IC for several reasons. Many of the drug molecules are not stable enough in dilute acid or alkali solutions, used commonly in IC, to permit their analysis. Another problem is that the drug molecules typically have extended hydrophobic surfaces, which interact with the hydrophobic surfaces of silica-based and polymeric chromatographic support materials. The strong hydrophobic interaction between the hydrophobic surfaces of drug molecules and the polymer-based chromatographic supports in an ionic environment that is typical of IC was not favorable for the analysis of such molecules. Furthermore, such molecules often have low solubility in water and require organic solvents for good solubility. The suppressors and the electrochemical detection systems, traditionally used in conjunction with IC, have not operated well in the presence of organic solvents. These factors did not encourage pharmaceutical scientists to apply IC in the analysis of pharmaceutical drugs. On the other hand, the ability to use relatively high concentrations of organic liquids in normal- and reversed-phase chromatographies to help minimize or eliminate hydrophobic interactions of analytes with the chromatographic supports and permit good solubilities of analytes in mobile phases, and the ease of using mass spectrometers with reversed-phase chromatography, resulted in strong interest in applying these techniques in the pharmaceutical industry. In addition, the theory and the applications, particularly of normal- and reversed-phase chromatographies have been developed since the 1970s [*cf*. 5,6]. Thus, the applications of IC have been limited to the types of drug molecules that are ionic (or strongly polar) or have extended hydrophilic surfaces, most notably aminoglycoside antibiotics and bisphosphonates. Even as late as 2000, USP-NF had no general chapter on IC and only a few monographs that described IC-based test methods [7].

With the increasing demand of newer types of pharmaceutical drugs (new molecular entities), use of different counter ions to improve stabilities and solubility properties of pharmaceutically active drug molecules, availability of mixed mode columns, and use of detection systems that have higher tolerance for or are not affected by organic solvents, there has been an increase in the popularity of IC applications in the pharmaceutical industry. In addition, the recent increase in environmental consciousness together with cost considerations has contributed toward greater interest in IC. In most instances, IC uses dilute solutions of acids, alkalis or salts, with little or no organic solvents, as eluents. Such eluents can be conveniently disposed off after neutralization, sometimes with additional dilution with water, in compliance with the environmental requirements. Organic solvents, on the other hand, are costly to buy and their safe disposal is often complex, environmentally unfriendly, and costly. The recent shortage of acetonitrile, one of the common solvents used in reversed-phase chromatography, also raised the possibility of shortage of critical organic solvents in future, which is a point to keep in mind in the selection of the chromatographic method to be used in pharmaceutical drug analysis.

IC has been used widely in the quantitation and characterization of drug substances, counter ions, and excipients, as well as in the quantitation of active and inactive components of drug products. It has been used also in the identification and characterization of impurities as well as in other applications, including dissolution tests. USP33-NF28 has several monographs that use one or more IC-based test procedures and two general chapters on IC (<345> and <1065>) (see Chapter 16 for more details on IC in USP-NF).

This chapter will review briefly some of the applications of IC in the analysis of drugs with emphasis on the active drug molecules and excipients. The analyses of impurities and counter ions are not discussed in detail because they are presented elsewhere in this book (Chapters 7 and 13, respectively). The chapter also provides insight into the selection of eluents and columns.

6.2 MOBILE PHASE AND COLUMN SELECTION

The resolution of analytes by IC involves ionic mechanisms of separation. Thus, factors affecting ionic environment, such as pH of the mobile phase, play a critical role in the separation of analytes. Change of pH of the mobile phase will have little impact on the ionic state of the analyte if they are fully ionized (cations or anions). However, decrease in pH of the mobile phase will result in an increase in the analyte cationic species (number, charge or both) due to increased protonation, when the analytes are polar basic compounds (e.g., amines). The retention times of such analytes are expected to increase during cation-exchange chromatography. The effect of pH on the analytes depends upon the extent of protonation, which is determined by how far the eluent pH is from the analyte pK_b. Decrease in pH will also increase the concentration of cationic species in the mobile phase. Since the elution depends upon the concentration of mobile phase cations, the retention times of the analyte cations are expected to decrease. Finally, decrease in pH of a mobile phase will also increase the protonation of the anionic functional groups on the stationary support, at which the cation exchange takes place, and suppress their dissociation. Thus, their ability to bind the analyte cations will decrease resulting in a decrease in the retention time. The latter two factors act contrary to the first factor. Thus, decrease in the pH of a mobile phase decreases the retention time of fully ionized cations due to the latter two factors. On the other hand, for polar basic analytes, the effect of the first factor will be predominant when the mobile phase pH is close to the pK_b of the analyte and the retention time will increase with the decrease in pH. However, when the mobile phase pH is significantly lower than the pK_b of the analyte, the second and the third factor will be dominant and the retention time is expected to decrease. Similar considerations will also apply for anion-exchange chromatography, however, with the increase of pH of the mobile phase (increase in alkali concentration). Thus, it is critical to evaluate the effect of the pH of the mobile phase on the separation characteristics, including peak resolution and column capacity (theoretical plates), and choose a condition that permits optimum separation.

Dilute nitric acid has been used as the eluent in many cation-exchange and ion-exclusion separations. Nitric acid absorbs UV with absorption maximum in 220–245 nm range. Thus, the use of nitric acid as eluent permits indirect UV (negative peak) detection of analytes. However, nitric acid is a strong oxidizing agent. Thus, some drugs, e.g., amines and thiols, may be oxidized by nitric acid. Under such circumstances, other UV absorbing compounds may be included in the eluent (e.g., trimesic acid), if indirect UV detection is desired.

Most of the pharmaceutical drugs contain extended hydrophobic surfaces. These drugs often interact strongly with the polymeric chromatographic support material, resulting in longer retention times and poor peak shapes with considerable peak tailing. Thus, it is important to select a column that shows low hydrophobic interaction with the analyte molecules. Organic solvents are often included in the mobile phases, sometimes as high as 70%, to circumvent this problem. Inclusion of organic solvents reduces hydrophobic interaction with the stationary phase resulting in reduced retention times and improved (symmetrical or near symmetrical) peak shapes.

In a few instances, the use of columns having poor hydrophobic characters resulted in coelution of analytes, elution with inadequate resolution or elution within the void volume. Employing a column that has some hydrophobic character, results in adequate separation. For example, when caffeine was eluted from an IonPac AS4A-SC column (Dionex) at a high pH, it was eluted within the void volume [8]. But, the PAX-100 column (Dionex) which is somewhat hydrophobic in character increased the retention of caffeine. In this example, additional studies showed that the retention of caffeine involved a hydrophobic interaction with the column.

A review of the application of IC in the analyses of different types of drugs is presented below.

6.3 ANTIBIOTICS

USP-NF General Chapter <81>, Antibiotics—Microbial Assay, describes procedures for the determination of activity (potency) of antibiotics by their inhibitory effects on the growth of microorganisms under suitable conditions [9]. The procedures are tedious, labor intensive, require days to complete, and have poor precision, compared to chromatographic methods of analysis.

6.3.1 Aminoglycoside Antibiotics

Aminoglycoside antibiotics are one of the important classes of antibiotics used as tablets, capsules, solutions, suspensions, ointments, creams, injections and other dosage forms, either alone or in combination with other drugs for topical, oral, ophthalmic, injection, intravenous and other routes of administrations [9]. Such antibiotics include amikacin, gentamicin, kanamycin, neomycin, netilmicin, paromomycin, sisomicin, streptomycin, tobramycin, and others. As the name indicates, the aminoglycosides are carbohydrates or contain significant carbohydrate moieties. They lack suitable chromophores for detection by UV absorbance and cannot be detected by this means unless the concentration is high.

Like other carbohydrates, aminoglycosides are oxidizable and can be detected and quantitated by PAD. PAD is an electrochemical detection method and does not require the analyte molecules to have chromophores to absorb light. This technique oxidizes molecules at an applied voltage resulting in electricity. The charge (number of electrons) flowing through the circuit is proportional to the analyte concentration. The chromatogram is obtained by plotting charge against time. Therefore, the ingredients that are not oxidizable at the applied voltage are not seen in the chromatogram making it simpler to analyze and interpret. The design and the mechanism of action of PAD are discussed extensively in Chapter 3.

HPAEC with PAD (HPAEC-PAD) is an accurate and robust technique for the assay (quantitation) and the analyses of impurities of aminoglycosides; with broad linear ranges, high repeatability and reproducibility, and low detection limits. The columns that are typically suitable for the analyses of carbohydrates, e.g., hydroxyl-selective anion exchange columns, such as, CarboPac PA1 and CarboPac MA1 (Dionex), are also suitable for the analysis of aminoglycosides.

The procedures for the assay of the drug substance/active component in the USP33-NF28 [9] monographs of three aminoglycosides and their dosage forms involve HPAEC-PAD. The monographs of amikacin, kanamycin and their dosage forms employ CarboPac MA1 columns (Dionex), a high capacity anion-exchange column functionalized with trimethylamine (USP L47 column), eluted with 0.115 M sodium hydroxide at room temperature. The aminoglycosides are detected by PAD using a 3-potential waveform. The assay procedure for streptomycin sulfate and its dosage forms uses CarboPac PA1 columns (Dionex), containing quaternary amine functionalized latex beads coated on polystyrene-divinylbenzene copolymer support (USP L46 column). The mobile phase is 70 mM sodium hydroxide. The column is eluted under ambient conditions and monitored by PAD using a 4-potential waveform (see Chapter 3 for details of 3- and 4-potential waveforms).

More details of the applications of IC in the assay and impurity analyses of aminoglycoside antibiotics, particularly of tobramycin and neomycin, are discussed in Chapter 8 and will not be discussed here.

6.3.2 Other Antibiotics

Tetracyclines are a group of broad-spectrum antibiotics, which are so named for the presence of four hydrocarbon rings in their structures. One of the earliest examples of the application of IC in the analysis of pharmaceutical drugs dates back to 1974 when Butterfield et al. [10] reported simultaneous determination of tetracycline (TC) and rolitetracycline (RTC) in two intravenous and one intramuscular formulations, in which RTC is the active component. A pellicular cation-exchange column, CP-128 (Pellionex) is used. The column is eluted at 25°C with a mobile phase containing 3 mM EDTA, 0.1 M sodium hydroxide and 40% methanol, adjusted to pH 4.35 with acetic acid. The column was monitored by UV detection at 254 nm. The results show that the RTC peak is fully resolved from TC and 4-epitetracycline, a degradation product of RTC, as well as from the excipients, ascorbic acid and lidocaine, present in the formulations. TC was resolved from ascorbic acid at half-peak height but coeluted with lidocaine. However, the latter excipient was present in the intramuscular formulation only. The results show linearity in the range 8–44 mg/mL (2–11 μg/injection) with the precision of 1.4% (RSD) for both tetracyclines. The mean recoveries for RTC and TC are 98.9 ± 0.6% and 100.2 ± 2.8%, respectively. The compendial microbiological method could not quantitate RTC and TC separately because the test organisms are sensitive to both actives. However, the sum total of the contents of RTC and TC in the formulations obtained by the IC method was found to be close to the values obtained by the official microbiological method.

Another method was reported [11], which can resolve four tetracyclines, TC, oxytetracycline (OTC), chlortetracycline (CTC) and doxycycline (DC), on an Omni-Pac PCX-100 column (Dionex) eluted with 0.2 M hydrochloric acid containing 27.9% acetonitrile at 16 ± 2°C with detection at 350 nm. The results suggest that the separation is due to a combination of ion exchange and hydrophobic interaction mechanisms. The method was validated for linearity, precision, and accuracy. However, some of the impurities and degradation products were not resolved from the actives under the separation conditions. Thus, the method is not suitable for the assay and purity test of the drugs. However, the authors applied the method to monitor tetracyclines in the waste water from the reactors at the manufacturing plants. The detection limits were found to be 10 μg/L for OTC and TC, and 20 μg/L for CTC and DC.

Erythromycin is an antibiotic with antimicrobial activity similar to that of penicillins and is often used for people who are allergic to penicillins. USP-NF describes an IC-based procedure for the erythromycin assay in the monographs of Erythromycin Ointment and Erythromycin Ophthalmic Ointment [9]. The method uses a chromatographic column packing consisting of ethylvinylbenzene-divinylbenzene copolymer coated with quaternary ammonium functionalized latex particles consisting of styrene cross-linked with divinylbenzene (USP L50 column, OmniPac PAX 500). The column is eluted with a dilute sodium hydroxide solution containing 50% Acetonitrile. An electrochemical detector consisting of a glassy carbon electrode is used in integrated pulsed amperometric mode with a 3-potential waveform.

Amoxicillin and *flucloxacillin* are β-lactam antibiotics used to treat bacterial infections. Liu et al. [12] analyzed the drug substances, amoxicillin and flucloxacillin, and an injection formulation containing both actives on a strong cation exchange column, Zorbax 300-SCX (Agilent) at 25°C by isocratic elution with 25 mM ammonium dihydrogen phosphate-acetonitrile (95:5) at pH 2.6 with UV detection at 225 nm. The study indicated both ion-exchange and hydrophobic interaction mechanisms contributing to the separation (mixed-mode). Results showed that the degradation products were baseline resolved from the actives concluding that the method is stability indicating. The method was validated for linearity ($r^2 \sim 1.000$ for both actives in 50–150% range of the respective target analysis concentrations), repeatability (RSD $\leq$ 0.5%), intermediate precision (RSD $\leq$ 1.2%) and accuracy (spike recovery ~100% for both drug substances). A similar IC method was described for the analysis of amoxicillin and flucloxacillin in capsules [13]. The method uses a LUNA SCX (Phemomenex) strong cation exchange column. The mobile phase is a 40 mM tetrabutyl ammonium bromide containing 1% acetonitrile and adjusted to pH 7.0 with orthophosphoric acid. The column was eluted under ambient condition and monitored at 254 nm. The results show baseline resolution of the flucloxacillin and amoxicillin peaks with excellent linearity

($r^2 = 0.9997$) in the concentration range 0.26–0.76 mg/mL, accuracy (spike recovery 98.8–100.6%) and precision (RSD $< 0.7\%$) of measurements for both components.

6.4 BISPHOSPHONATES

Bisphosphonates (also known as diphosphonate) are a class of drugs that prevent the loss of bone mass and are used to treat hypercalcemia, osteoporosis, and similar diseases. The bisphosphonate drugs are characterized by the presence of a P-C-P "backbone" with two phosphonate (PO_3) groups covalently linked to the carbon—hence the name "bisphosphonate". Each phosphonate group contains two dissociable protons. They are water soluble compounds, anionic at moderate pH, with low solubility in organic solvents. Thus, bisphosphonates are ideally suited for analysis by IC. In addition, they lack suitable chromophores for UV detection and have insufficient volatility for gas chromatographic analysis. Bisphosphonates have a strong tendency to chelate metal ions that are often leached from the chromatographic systems and/or column resins. Thus, analysis of these drugs requires use of metal-free chromatographic systems. Typically, commercial IC systems are metal-free systems because the methods frequently use dilute acids, alkalis or salt solutions as eluents. Several IC methods have been reported, which use anion-exchange columns packed with styrene, methacrylate, hydroxyethyl methacrylate or ethylbenzene copolymerized with divinylbenzene to form the stationary phase, which is functionalized on the surface with quaternary ammonium ions. The elution has been performed frequently with dilute nitric acid, which absorbs UV and permits indirect UV detection. However, other methods of detection, including nonsuppressed negative conductivity detection and suppressed conductivity detection have been reported. Several other methods, which involve IC followed by either post-column reaction, where complicated and specialized equipment is necessary, or online detection by flame photometry or inductively coupled plasma, were published. As novel and elegant as they are, such methods are not suitable for routine analysis in quality control laboratories. Many of the methods are stability indicating and can resolve bisphosphonates from impurities.

A stability-indicating IC method for the determination of disodium clodronate in bulk drug materials and in pharmaceutical formulations, in which the analytes are detected by UV absorption at 300 nm after post-column derivatization with iron (III) in acidic solution, was reported [14]. The separation is achieved on a poly(styrene-divinylbenzene) copolymer column (IonPac AS7, Dionex) using 40 mM nitric acid as the mobile phase at a flow rate of 0.5 mL/min. The post-column reagent consists of 1.25 mM ferric nitrate in 1.4% perchloric acid. The clodronate peak is eluted at about 11 minutes, well separated from its potential degradants and impurities. The method has been validated for specificity, precision (0.7–1.3%), linearity ($r^2 = 0.9999$), range (50–175% of the target analysis concentration), and accuracy (99.5–100.7%). Analysis of capsules, tablets and an injection formulation show that they contain 100.7%, 101.5% and 95.7% of the labeled amounts of clodronate, respectively.

Tsai et al. reported an IC method with nonsuppressed conductivity detection for the determination of alendronate in pharmaceutical dosage formulations [15].

The method eliminated the need for pre- or post-column derivatization. An IC-Pak HR column (Waters), packed with trimethylammonium functionalized polymethacrylate resin, was eluted with 1.6 mM nitric acid (pH ~ 2.5). Because of the high conductance of nitric acid, eluting alendronate produces a negative signal due to the uptake of H^+ by the alendronate ion. The negative signal is detected as positive peak by switching polarity of the output signal. An OmniPac PAX-100 column (Dionex) containing a highly cross-linked styrene-divinylbenzene polymeric support functionalized with quaternary ammonium ion on the surface, also gave comparable results when eluted with 1.76 mM nitric acid and 20% acetonitrile, included as the organic modifier, except that the OmniPac PAX-100 column yielded a shorter retention time and better peak shape for alendronate in the presence of acetonitrile than the IC-Pak HR column (in the absence of any organic solvent). The OmniPac PAX-100 column has significant hydrophobic character. Therefore, inclusion of acetonitrile was necessary to minimize hydrophobic interaction.

The validation of the methods using both columns with respective run conditions yielded similar results. The results showed that the method demonstrated adequate precision (RSD ~ 2%) and accuracy (spike recovery ~100%) for the assay of alendronate in an intravenous solution and a tablet formulation, with linearity ($r^2 > 0.999$) in the range of 40–160% of the target assay concentration (0.05 mg/mL). Somewhat higher RSD values are due to variation in peak integration in the presence of the high background due to nitric acid. The specificity of the assay was demonstrated by baseline separation of the alendronate from its aminohydroxypropyl analog as well as its thermal degradation products. The results also demonstrate that both methods are stability-indicating. Furthermore, analysis of the tablets and the intravenous liquid formulation show that the alendronate peak is fully resolved from the excipient peaks. The alendronate content was ~99% of the label claim for the tablets and ~101% of the label claim for the intravenous formulation.

A similar method was reported for the analysis of pamidronate, an amino bisphosphonate, by Hartigh et al. [16] who used IC-Pak HR (Waters) anion-exchange column and 2 mM nitric acid (pH 2.7) as the eluent, and observed the analyte peaks (negative) by indirect conductivity detection. However, when the authors used 25 mM succinic acid (pH 3.0) the positive peaks were observed by the conductivity detection. This is attributable to the fact that succinic acid, being a weak acid, provides significantly lower background conductance compared to nitric acid, permitting positive peaks of pamidronate. The column was operated at $35 \pm 5°C$. The method was shown to be stability indicating, being able to separate pamidronate from its known degradation products. The results of method validation showed that the values of validation characteristics were comparable to other chromatographic methods: $r^2 = 0.9992$ in the range 200 ng to 10 μg per 20 μL injection, intra- and interday precisions ~2.1% and ~2.7%, respectively, and accuracy (spike recovery) ~102%. The results of the analysis of pamidronate tablets found the active drug content to be ~103%. The authors also showed that the same method can be used as a stability-indicating assay for foscarnet, an antiviral drug used in the treatment of HIV-infection. The impurities and degradation products are baseline resolved from the drug substance. However, the method has not been validated for its application to foscarnet.

Another closely-related chromatographic procedure, however, using indirect UV-detection was reported for the quantitation of alendronate, etidronate and clodronate in drug substances and dosage forms [17]. The report compared performance of four anion-exchange columns, IC-Pak HR (Waters), IonPac AS7 and AS7A (Dionex) and HEMA 1000Q (Meta-Chem), and found that the results are very similar when eluted with dilute nitric acid solution, which absorbs in UV in the 220–245 nm range, thereby permitting indirect UV detection of the analytes and, at the same time, maintaining the charge state of the bisphosphonates at −1. The detection maxima of the negative peaks differ slightly among the three bisphosphonates in the range 230–245 nm. The polarity of the detector was reversed to permit observing the analytes as positive peaks. Etidronate possesses a higher negative charge under acidic conditions than alendronate because the latter contains an amino group, remembering that bisphosphonates contain two phosphonate groups, each containing dissociable protons. Thus, etidronate requires a higher eluent concentration (7.2 mM nitric acid) than alendronate (1.6 mM nitric acid). However, the etidronate injectable contains chloride (saline), which coeluted with etidronate when eluted with 7.2 mM nitric acid. The peaks could be baseline resolved by including KNO_3 at 60:40 (HNO_3 to KNO_3) molar ratio, thereby keeping the nitrate concentration to 7.2 mM but increasing the pH from 2.2 to 2.7. The increase in pH increases the negative charge of etidronate and increases its retention time but not that of ionic chloride ion, thus resulting in their resolution. Clodronate contains two electronegative chloride substituents, which makes it more acidic than etidronate and requires 12 mM nitric acid for elution. The validation studies showed that the method could provide precise (RSD 0.36–1.28%) and accurate (spike recovery 100.1–100.6%) results with appropriate linearity ($r^2 = 0.999$) in the 40–160% range of the respective target analytical concentrations. The method specificity was demonstrated by showing that the bisphosphonates can be resolved from their respective process-related impurities and thermal degradation products. Analysis of alendronate tablets and etidronate and clodronate injectable solutions show contents 99–101% of the label claims for the actives.

The chromatography method was somewhat modified by the same laboratory for IC-ion spray mass spectrometric (IC-ISP-MS) analysis of alendronate in a tablet formulation [18]. Nitric acid was replaced by 0.1% formic acid—5% acetonitrile as the eluent (pH 3.0). Since the solvent contains only volatile components, it was not necessary to include a micromembrane suppressor between the column outlet and the mass spectrometer inlet, as is typically required when eluents contain nonvolatile components, such as sodium hydroxide [3]. The total ion current chromatogram showed the formation of a small quantity of excipient-alendronate adduct. This adduct was not seen in the chromatogram when nitric acid was used as the eluent [15]. The alendronate peak eluting at about 10 minutes was shown to contain no coeluting species by IC-ISP-MS-MS. IC-ISP-MS showed that the alendronate peak consists only of a monovalent ion of m/z 248, $(M\text{-}H)^-$ species (M = alendronic acid). The IC-ISP-MS results also show that alendronate can form cluster anions with acid molecules, including a dimer and a trimer of alendronic acid, presumably through the formation of hydrogen bonds. IC-ISP-MS-MS study showed evidence of the cleavage of the C-P bond(s) to be the dominant fragmentation pathway of alendronate.

Quattrocchi et al. reported a similar method for the assay and analysis of impurities of olpadronate using an IC-Pak HR column, dilute nitric acid solution at pH 2.9 as the eluent and indirect UV-detection at 235 nm [19]. The results show baseline resolution of olpadronate from the product- and process-related impurities, phosphite, phosphate, chloride and methanesulfonic acid. The method was validated for the application for the olpadronate assay and analysis of its impurities.

An IC method with indirect UV detection has been reported recently, which uses Phenosphere SAX column (Phenomenex) for the analysis of etidronate and clodronate, and Sphereclone SAX column (Phenomenex) for the analysis of alendronate and pamidronate [20]. The mobile phases are 20 mM sodium citrate buffers at pH 3.6 and 4.6, respectively. The columns are operated at 30°C. UV detection was employed at 226 nm for the former and at 222 nm for the latter. The citrate buffer permitted the use of silica-based columns and the peaks are eluted much earlier than reported with the previously described methods at a significantly lower flow rate; etidronate and clodronate were eluted at about 3.5 min at 0.30 mL/min flow rate, and alendronate and pamidronate were eluted at about 4.5 min at 0.25 mL/min flow rate. There is no need to resolve the bisphosphonate peaks because there is no dosage form that contains more than one active. Analysis of tablets showed that the bisphosphonates are well resolved from the respective excipients. The validation studies show that the method precision is <2.0% (RSD) with linearity ($r^2 > 0.998$) within the range 50–400 μg/mL for etidronate and clodronate, and 100–500 μg/mL for alendronate and pamidronate, and the accuracy is (spike recovery) 97–103% for all bisphosphonates.

An assay procedure for pamidronate in pharmaceutical dosage forms employs a column packed with hydroxyethyl-methylacrylate polymer with quaternary amine functionalities and phthalate counterion (Universal Anion column, Alltech), an eluent consisting of 5 mM potassium nitrate adjusted to pH 3.5 with nitric acid, and a refractive index detector [21]. The report showed resolution of pamidronate from its impurities, phosphite and phosphate, and the excipients present in different dosage forms, including sarine, β-alanine, pH adjusters, binders, and other inactive ingredients. The validation of the method studied linearity, precision and accuracy. The contents of pamidronate were reported to be 99.6%, 101%, and 100% of the label claims in ampoule solutions, coated capsules and tablets, respectively, with RSD < 1.0% (n = 6) in all cases.

Sparidans et al. reported an IC method using an IC-Pak HC column (Waters) in which the bisphosphonates were complexed in-line with copper (II) ion under acidic conditions to yield an UV-absorbing complex [22]. Pamidronate and olpadronate and their pharmaceutical dosage forms were analyzed at 30°C using 1.5 mM nitric acid containing 0.5 mM copper (II) nitrate as eluent and 254 nm detection. The interday precision was found to be around 1% for the drug substances, 1.8% for pamidronate in injection concentrate and 5.2% for olpadronate in tablets. The results showed that the dosage forms contain 99.8% of pamidronate and 105.5% of olpadronate of the respective label claims.

IC hyphenated with inductively coupled plasma-mass spectrometry (ICP-MS) has been reported recently for the analysis of alendronate and etidronate [23]. The method is based on ICP-MS detection at $m/z = 31$, corresponding to phosphorous present in bisphosphonates. The chromatography was performed using AS7 analytical column (Dionex). A metal trap column, MFC-1 (Dionex), was inserted between the pump and the injector to remove trace amounts of metal ions from the eluent as bisphosphonates have strong affinity to bind metal ions. The eluents were 1.5 mM and 15 mM solutions of nitric acid for alendronate and etidronate, respectively. Since $m/z = 31$ involves polyatomic interference due to $^{15}N^{16}O^{+}$, $^{14}N^{16}O^{1}H^{+}$, and $^{12}C^{1}H_3^{16}O^{+}$ ions, the report conducted extensive studies on the optimization of the ICP conditions.

The Pamidronate Disodium drug substance monograph in USP33-NF28 describes the use of Super-Sep IC Anion column (Metrohm), an anion-exchange column containing porous polymethacrylate, functionalized with quaternary ammonium groups, for the assay and the Related Compound test (Test 2) [9]. The column is eluted isocratically at 35°C with a dilute solution of formic acid adjusted to pH 3.5 with sodium hydroxide and the effluent is monitored by refractive index detection. The same method is applied also for the assay of pamidronate in Pamidronate Disodium for Injection monograph [9].

The monograph of Risedronate Sodium drug substance employs an IonPac AS7 anion exchange column (Dionex) for the assay and the Related Compound test (Test 1) [9]. A dilute solution of disodium edetate (1.8 g/L) adjusted to pH 9.5 with sodium hydroxide is used as the mobile phase. The column is eluted isocratically and detected at 263 nm. The same procedure is used also for the assay of the active and as part of the dissolution test in the monograph of Risedronate Sodium Tablet.

6.5 OTHER DRUGS

Ribavirin is an antiviral drug indicated for severe respiratory tract, hepatitis C and other viral infections. The USP-NF procedures for the assay and chromatographic purity of ribavirin in Ribavirin (drug substance) and Ribavirin for Inhalation Solution (product) monographs involve separation on a strong cation-exchange resin consisting of sulfonated cross-linked styrene-divinylbenzene copolymer (USP L17 column, Aminex Fast Acid Analysis, Bio-Rad) [9]. The column is eluted at 65 ± 5°C with water adjusted to pH 2.5 with sulfuric acid and monitored by UV-detection at 207 nm.

Hydroxyzine is a first-generation antihistamine drug used primarily for the treatment of itching, allergies, hyperalgesia, motion sickness-induced nausea, and insomnia, as well as for the treatment of mild anxiety. The chromatographic procedure used for assay of hydroxyzine in USP-NF monographs of Hydroxyzine Hydrochloride Tablets, Hydroxyzine Pamoate Capsules, and Hydroxyzine Pamoate Oral Suspension, and for the Dissolution test for Hydroxyzine Pamoate Capsules involves cation-exchange chromatography on a USP L9 silica gel column containing a chemically bonded, strongly acidic cation-exchange coating [9]. The dosage forms are dissolved and diluted in methanol, and injected to a Partisil 10 SCX column (Whatman), eluted isocratically with a 3.4 g/L potassium phosphate monobasic solution containing 50% methanol and monitored by UV detection at 232 nm.

The USP-NF method for the assay of *butoconazole* in Butoconazole Nitrate Vaginal Cream also involves a USP L9 column (Partisil SCX), converted to the potassium form [9]. The column is eluted with a dilute potassium acetate buffer, pH 4.3, containing 65% methanol and monitored by UV-detection at 225 nm. The sample is prepared by extracting the cream with methanol.

Methenamine and its salts are often used in the treatment of urinary tract infection. The USP-NF monographs of the formulations of methenamine and its salts, use three different nonselective, multistep procedures for the methenamine assay [9]. However, a stability-indicating IC method has been reported for the assay of methenamine in the drug substance and in the pharmaceutical formulations containing methenamine mandelate and methenamine hippurate. The method involves ion-exchange mechanism of separation with UV detection at 212 nm [24]. The procedure uses a Zorbax SCX-300 column (Agilent) and isocratic elution with 0.1 M sodium perchlorate-Acetonitrile (30:70) at pH 5.8. The method is optimized by the inclusion of acetonitrile and perchlorate, and by adjusting pH to reduce the elution time (5 min) and improve peak shape (tailing factor of methenamine peak: ~2.5). The samples were prepared by dissolving tablets in the eluent and clarifying by filtration. Forced degradation studies showed that the methenamine peak is well resolved from its degradation products. Additional studies have indicated that methenamine is well resolved from the excipients present in the tablets. The method is validated for linearity ($r^2 > 0.9999$ in the range 0.25–50 mM of methenamine), precision (RSD 0.2–1.8%), accuracy (spike recovery 99–101%) and specificity. The assay of methenamine in commercially available tablets show results in the range 98–102% of the label claim.

Paracetamol (acetaminophen) is a widely used over-the-counter analgesic (pain reliever) and antipyretic (fever reducer) medication. An IC procedure for the simultaneous determination of paracetamol and the related compound, acetylsalicylic acid, in five over-the-counter solid dosage forms was reported [25]. The method uses an IC-Pak A HR column (Waters) packed with polymethacrylate resin containing quaternary ammonium functional group, which is eluted isocratically with a 5 mM lithium hydroxide solution containing 5% Acetonitrile and UV-detection at 300 nm. The solid dosage forms were dissolved in the mobile phase and cleaned up by passing through a C18 SEP-Pak Plus cartridge (Waters) before injection. The results show baseline resolution between paracetamol and acetylsalicylic acid peaks. The linearity of the method was demonstrated for both components ($r = 0.9998$ and 0.9995) in the concentration ranges 0.5–7 mg/L and 0.5–12 mg/L, respectively. The results of the analyses of the dosage forms show that they contain 90–108% and 93–106% of the label claim of paracetamol and acetylsalicylic acid, respectively.

The *oxymorphone* assay in the USP-NF monoghraph of Oxymorphone Hydrochloride Suppositories involves Zipax SAX column (Dupont), a strong anion-exchange packing made by chemically bonding a quaternary amine to a solid silica spherical core (USP L12 column), which is eluted with 0.05 M sodium borate solution adjusted to pH 9.1. The column is monitored by UV-detection at 254 nm [9].

Beta blockers are a class of drugs used for various indications, particularly for the management of cardiac arrhythmias, cardiac protection after heart attack, and to treat hypertension. Propranolol is the first clinically important beta blocker introduced in the late 1950s and has been considered to be one of the most important medicines of the 20th century. A method for the determination of propranolol, atenolol, metoprolol, alprenolol, oxprenolol and acebutolol in seven tablet formulations and a capsule has been reported [26]. The method uses an IC-Pak CM/D column (Waters) with isocratic elution with 50 mM nitric acid containing 4% (v/v) acetonitrile and UV-detection at 270 nm. The results show that four of the six beta-blockers, propranolol, atenolol, metoprolol and acebutolol can be resolved on the column. This does not pose any disadvantage because beta-blocker dosage forms only contain one of the active ingredients in a particular formulation. The tablets were dissolved in 50 mM nitric acid and the solutions were passed through a C18 Sep-Pak Plus SPE cartridge (Waters) to remove neutral organics, which were found to bind to the column through hydrophobic interactions. The absence of significant excipient peaks in the chromatograms of tablet samples show that most of the excipients were removed by the extraction process. The method is validated for linearity (r^2 0.9938–0.9998), range (different for different actives), intra- and interday precisions (RSD 2.5–3.0% and ~4.2%, respectively), accuracy (spike recovery 98–101%) and detection limits (0.1–2.7 mg/L). The analysis of the dosage forms show that the contents of the beta-blockers are 94–99% of the label claims. The assay results were compared by the Student t test using a previously published orthogonal method (reversed phase chromatography) at 5% significance level and found to have no significant difference.

Beta-agonists are bronchodilators that open the airways by relaxing the muscles around the airways, which may tighten during an asthma attack or in chronic obstructive pulmonary disease. A method for the simultaneous determination of four beta2-agonists (salbutamol, fenoterol, clorprenaline, and clenbuterol) in pharmaceutical dosage forms and plasma using IC with nonsuppressed conductivity detection was reported [27]. A Metrosep Cation 1–2 (Metrohm) column is eluted isocratically with 1.8 mM nitric acid containing 2% acetonitrile under ambient conditions. The results show better resolution of the peaks when nitric acid is used compared to sulfuric, phosphoric and citric acids (all at pH 2.98). The retention times were reduced and the peak shapes improved significantly by the inclusion of acetonitrile in the eluent, suggesting hydrophobic interactions between the analytes and the chromatographic support materials. However, when the acetonitrile concentration is higher than 2%, the peaks begin to come too close and are not baseline resolved. The validation studies show that the method provides linear response (r^2 0.9935–0.9970) over three orders of magnitude of concentration (ng/mL to μg/mL) for the four drug substances. The detection limits (2–10 ng/mL) are 3–10-fold lower compared to the detection limits obtained by the UV detection indicating that the UV-detection is not sensitive enough for these actives. Analysis of tablets shows inter-day precisions to be 1.8–2.2% and 1.1–2.8% (RSD) for salbutamol and clorprenaline tablets with spike recoveries 97.2–100.4% and 101.0–104.1, respectively. No results were presented for the fenoterol and clenbuterol containing dosage forms. However, the low detection limits of the method permitted its use for the detection and quantitation of these drugs simultaneously in plasma.

The same laboratory reported that the above method can be used also for the simultaneous determination of *ephedrine, pseudoephedrine* and *norephedrine* [28]. Method validation studies showed linearity (r^2 0.9897–0.9989) over two orders of magnitude of concentration (μg/mL), with precision <2% (RSD) and the detection limits 0.02–0.03 μg/mL for the drug substances. The analysis of the dosage forms showed that the content of ephedrine is 98.2% of the label claim in an injection formulation, and that of all three actives are in the range of 99.8–101.1% of the respective label claims from three bulk drug formations (powder).

Bethanechol is quaternary ammonium compound, which is functionally classified as a cholinergic agonist. Bethanechol is used in the treatment of urinary retention. The USP-NF monograph of Bethanechol Chloride (drug substance) describes two different IC-based procedures for the assay and the Related Compounds test. The assay uses IC-Pak C M/D column (Waters), a strong cation exchange column containing porous silica coated with a copolymer of polybutadiene and maleic acid (USP L55 column). The column is eluted at 30–35°C isocratically with a solution containing edetic acid, nitric acid and 5% acetonitrile, and monitored by the conductivity detection. The related compound test also uses the same column and conductivity detector but the column is eluted isocratically with 20 mM methanesulfonic acid. The procedures for the Assay and the limit of 2-hydroxypropyltrimethyl ammonium chloride (2-HPTA), the primary degradation product of bethanechol, in the monograph of Bethanechol Chloride Injection employ IonPac CS14 columns (Dionex), a weak cation-exchange column containing ethylvinylbenzene cross-linked with divinylbenzene with carboxylate functional groups on the surface (USP L53 column). The columns are eluted with 20 mM methanesulfonic acid at 30°C and monitored by suppressed conductivity detection.

The monograph of Bethanechol Chloride Tablets uses IC-based procedures for the assay, and related compounds and dissolution tests. The assay and the dissolution test use IC-Pak C M/D column, which is eluted at 30–35°C isocratically with a solution containing edetic acid, nitric acid and 5% acetonitrile, and monitored by the conductivity detection. However, the procedure for the related compounds test involves an IonPac CS14 column eluted with 20 mM methanesulfonic acid at 30°C and monitored by suppressed conductivity detection.

Chapter 11 of this book presents an extensive discussion on the development and validation of a simple procedure for the analysis of bethanechol and another cholinergic agonist, *carbachol*, as well as their degradation products.

Chen et al. reported two methods for the simultaneous determinations of three *methylxanthine* alkaloids, caffeine, theobromine and theophylline, in food and pharmaceutical formulations [8]. One of the methods involves cation exchange chromatography using HPIC-CS3 column (Dionex) and elution with 100 mM hydrochloric acid with UV detection at 274 nm. Nitric acid results in oxidation of certain organic compounds, particularly when they contain aromatic amines. Hence, hydrochloric acid is chosen as the eluent. The second, an anion exchange method, employs an OmniPac PAX-100 column (Dionex), which is eluted with 15 mM potassium hydroxide containing 1% acetonitrile. The elution behavior showed that the retention time of the caffeine peak is independent of the potassium hydroxide concentration in the range 10–300 mM while those of the theobromine and theophylline peaks decrease with increasing potassium hydroxide concentration. The retention time of caffeine decreases and the peak shapes of all three analytes improved when acetonitrile is included in the eluent. However, the separation performance did not change significantly when the acetonitrile concentration was increased from 1% to 2.5% but the column back-pressure increases. The authors concluded that an adsorption-partition mechanism was responsible for the separation of caffeine but the separation of theobromine and theophylline involved ion exchange. The validation results of the cation exchange method showed linearity ($r^2 \geq 0.9998$) in the range 0.1–20 μg/mL, with precision 0.34–0.73% (RSD) and detection limits 0.03–0.1 μg/mL for all actives. The anion exchange method also showed comparable results for each validation characteristic. Three pharmaceutical dosage forms, aminophylline tablet, aminophylline injection and caffeine-sodium benzoate tablet, were analyzed by both methods. The results of the recovery studies from the aminophylline injection by the cation exchange method showed 98.7%, 99.7% and 102.3% recovery for caffeine, theobromine and theophylline, respectively. The corresponding values obtained by the anion exchange method are 99.6%, 99.1% and 101.7%.

Carbocisteine (S-carboxymethylcysteine) is a mucolytic and expectorant drug administered as syrups and oral granules under several commercial names. An IC method was developed and validated using a cation exchange column, IonPac CS14 (Dionex) [29]. The column is eluted isocratically with 0.25 mM solution of trifluoroacetic acid and monitored by nonsuppressed conductivity detection. The validation results show a linear calibration curve ($r \geq 0.9999$) in the range 17–400 μg/mL, with a detection limit of 5.6 μg/mL and a precision of 1.5% (RSD). Four syrups containing Carbocisteine were analyzed after dilution with the mobile phase. The results show Carbocisteine content to be 96.8–102.8% of the labeled content with no interference from the excipients.

Citric acid and inorganic *citrates* are active ingredients in many dosage forms. They are used in the treatment of manic-depressive illness, as anticoagulants to preserve blood and plasma for transfusion, as laxatives to treat constipation and to empty and prepare colon for rectal and colon examinations and surgery, as systemic alkalizers to help contain certain kidney stones, to provide rehydration treatment, and as ingredients in rectal enemas. An IC method for the assay of citrate in nine dosage formulations was reported [30], which involved elution of PRP-X100 anion-exchange column (Hamilton) by 0.875 mM trimesic acid at pH 10.0 and indirect UV detection at 280 nm. Trimesic acid contains UV-absorbing chromophores. The negative citrate peak was recorded as a positive peak by reversing input polarity of the integrator. The results showed linear detector response ($r \geq 0.9996$) over the range 1–12 μg citrate injected with 98–100% spike recovery and 0.25–2.10% (RSD) precision. The formulations were found to contain 94–104% of the label claim of citrate. The run condition of the method was optimized specifically for tricarboxylic acids, eliminating interference with other ingredients in the formulations.

An ion-exclusion chromatographic method for the determination of citrate and acetate in pharmaceutical formulations was reported [31], in which Aminex HPX-87H column (Bio-Rad) was eluted with 25 mM sulfuric acid at 60°C and monitored by UV detection at 280 nm. Samples were cleaned up using an SPE cartridge (Waters) before chromatography to remove trace amounts of hydrophobic components, thus reducing the run time from 180 minutes to 35 minutes. Validation studies show that the citrate and acetate responses are linear ($r > 0.998$) in 80–120% of the target concentration range with precision ≤0.27% RSD and spike recovery (accuracy) 99.9–100.3%.

An IC procedure for the simultaneous determination of total citrate (citric acid/citrate) and phosphate in nine pharmaceutical formulations was reported [32]. These formulations were selected to represent 18 dosage forms for which the respective monographs in USP27-NF22 include assay for citric acid and/or citrate. These monographs described seven different techniques for the assay, including wet chemistry analysis, titration, colorimetry, ion-exchange chromatography, ion-exclusion chromatography, and reversed-phase chromatography. In addition, there were several monograph-specific variations related to sample preparation and analysis procedures. Furthermore, monographs of some of the anticoagulants also include phosphate assays that use wet-chemistry procedures. These procedures often involve several steps and are tedious, labor intensive and time consuming. In the reported method citrate and phosphate were separated in less than 10 minutes by anion-exchange chromatography using the hydroxide-selective AS11 column (USP L61, Dionex), 20 mM potassium hydroxide (electrolytically generated) or sodium hydroxide (manually prepared) as the eluent, and suppressed conductivity detection. The method showed linear responses over the concentration ranges 0.2–100 μg/mL for citrate ($r^2 > 0.9990$) and 0.2–60 μg/mL ($r^2 = 0.9999$) for phosphate, with limits of quantitation (signal-to-noise ratio = 10) of 0.2 μg/mL for both analytes. The accuracy of the procedure, determined by spiked recovery measurements, was 95–105% for all formulations. The intra- and the interday precisions were reported to be <1 and <2% (RSD), respectively, for both analytes. The method was optimized specifically for trivalent anions, eliminating interference with other ingredients in the formulations. This procedure was later included in USP-NF as the General Chapter <345>, Assay for Citric acid/citrate and Phosphate, and in the monographs of the dosage forms in 2009 [9].

6.6 DISSOLUTION TESTS

IC has been used in a few instances for the determination of analytes as part of dissolution tests. Figure 6.1 shows an example of application of IC in the dissolution test of calcium citrate tablets in Intestinal Fluid, USP using IonPac CS14 column (Dionex). The column is eluted with 20 mM methanesulfonic acid and detected by suppressed conductivity detection. Dissolution of calcium was monitored over time. The results also show that other ions present in the dissolution medium were well resolved among themselves and also from the calcium ion.

USP33-NF28 has six monographs in which IC is employed in the dissolution tests of tablets and capsules [9]. They are monographs of Bethanechol Chloride Tablet, Etidronate Disodium Tablet, Risedronate Sodium Tablet, Fexofenadine Hydrochloride and Pseudoephedrine Hydrochloride Extended-Release Tablets (Tests 1 and 4), Hydroxyzine Pamoate Capsules, and Meclizine Hydrochloride Tablets. In all cases, the active drugs are assayed in the dissolution media to determine the amount of drug dissolved with time. Some of these procedures are described above.

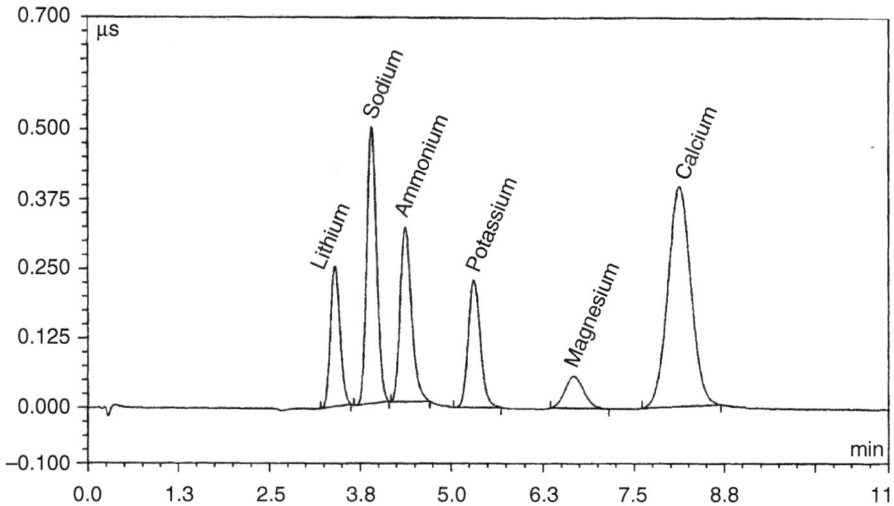

Figure 6.1. Analysis of the dissolution medium in the dissolution test of calcium citrate tablets in Intestinal Fluid, USP. Column: IonPac® CS14 (Dionex); guard column: IonPac® CG14; eluent: 20 mM methanesulfonic acid; flow rate: 1 mL/min; temperature: ambient; injection volume: 25 µL; detection: suppressed conductivity.

6.7 IMPURITIES

Several applications of IC in the analysis of impurities have been mentioned briefly earlier in this chapter as part of the discussion of the analysis of the drugs. An extensive discussion on the application of IC in the analysis of impurities is presented in another chapter in this book (see Chapter 7). However, it may be of interest to discuss one study that described an IC method for the analysis of impurities of clodronic acid/disodium clodronate [33] to illustrate how the selection of method is influenced by the intended application. Of the potential impurities arising from the synthesis of clodronic acid, monochloromethylenebisphosphonic acid and methylenebisphosphonic acid are tetraprotic acids. Other probable impurities and degradation products may have charges varying from −3 to −1 (e.g., intermediates of synthesis, carbonylbisphosphonate, phosphate, chloride, nitrate). As discussed above, the IC methods for the assay of bisphosphonates, including clodronate, have typically employed acidic eluents (around pH 3) [8,9,14–23]. At these low pH values of the eluents clodronate and the bisphosphonate impurities are predominantly monovalent, as are the inorganic impurities. Thus, it is difficult to separate the impurities using a mobile phase at a low pH. An alternate approach is to use an alkaline eluent at a pH at which all analytes, bisphosphonates and inorganic impurities, will be fully ionized and the separation can be effected by utilizing the difference in their charge densities. The study reported baseline separation of the active, the impurities and the degradation products using an anion exchange column (IonPac AS5, Dionex), a 20–100 mM sodium hydroxide gradient at 1 mL/min flow rate, and suppressed conductivity detection using an Anion Self-regenerating Suppressor (ASRS-1, Dionex). The optimum separation was obtained when the column was operated at 45°C. The validation results show that the method is linear ($r^2 > 0.997$) over two orders of magnitude of concentrations for the impurities/degradation products with the quantitation limits varying between 0.12 and 0.49 μg/mL.

6.8 EXCIPIENTS

Excipients are inactive ingredients used in pharmaceutical formulations for several purposes, including as stabilizers, antimicrobial preservatives, antioxidants, tablet binders, tablet disintegrants, solubilizing agents, coloring agents, antifoaming agents, aerosol propellants, and many others [*cf*. 9]. IC has been widely used for the detection and quantitation of sugar and sugar alcohol excipients as well as for the analysis of their impurities using both ion-exclusion and ion-exchange chromatographies. These compounds are polar with many hydroxyl groups which form an extended hydrophilic surface, easily oxidizable at the electrode surface and lack UV-absorbing chromophores. The IC methods are typically isocratic, use a simple mobile phase (water, salt solutions or alkalis) and can resolve sugar and sugar alcohol isomers. Ion-exclusion chromatography has the further advantage of achieving significantly different selectivities using different cationic forms of the stationary phase to achieve desired separation.

USP monographs use ion exclusion methods utilizing L17 (for propylene glycol, alcohol, glucose, erythritol), L19 (for inositol, isomalt, tagatose), L34 (for lactitol, maltitol, and sorbitol) and L58 (for maltose, trehalose) columns for assay and in some cases for the Related Compound test of the excipients as well as for the assay of excipients in formulations [9]. These columns consist of sulfonated cross-linked styrene-divinylbenzene copolymers, however, differ in their cationic counter ion. L17 is the hydrogen form, L19 is the calcium form, L34 is the lead form and L58 is the sodium form of the column. The eluents are dilute sulfuric acid for L17 (10 mN or 0.1%), water or calcium acetate solution for L19, and water for L34 and L58 columns. The columns are operated at elevated temperatures, between 50°C and 85°C and the effluents are monitored using refractive index detectors at or near ambient temperatures.

Rezex™ brand of strong cation-exchange columns have been used in ion-exclusion chromatography of sugar and sugar alcohol excipients [34]. Separation of eleven sugars and sugar alcohols has been reported using a 300 mm Rezex RPM Monosaccharide (Pb^{2+} form) column (Phenomenex) operated at 75°C using water as the mobile phase and refractive index detection. However, the total run time is about 40 minutes. Where application permits, shorter run time can be obtained by using a shorter column. Different selectivities may be achieved using Rezex RCM Monosaccharide or Rezex RCM Sugar Alcohol cation exchange columns (both Ca^{2+} form) with refractive index or evaporative light scattering detection. These columns use water as the mobile phase and require operation at 80–85°C for optimum separation.

An extensive study comparing the performances of the IonPac ICE-AS1 ion-exclusion column, and the CarboPac PA 10 and CarboPac MA1 ion-exchange columns with pulsed amperometry detection for the quantitation of sugar, sugar alcohol and glycol excipients in solid and liquid pharmaceutical formulations has been reported [35]. Propylene glycol, glycerol, sorbitol, mannitol, maltitol, and sucrose were separated baseline resolved on the CarboPac MA1 column operated under ambient conditions and isocratic elution with 480 mM sodium hydroxide. The same excipients and other sugars were separated by isocratic elution of the CarboPac PA10 column with 18 mM sodium hydroxide. However, propylene glycol and glycerol were eluted near the void volume and were not baseline resolved. Propylene glycol, glycerol, and other sugar alcohols are separated on the IonPac ICE-AS1 column using 100 mM sodium perchlorate as the mobile phase at ambient temperature. The run times for IonPac ICE-AS1 and CarboPac PA10 were 10–15 min but those for CarboPac MA1 was much longer. The report concluded that the selection of method would depend on the intended application. If the excipients consist of propylene glycol, glycerol, and sugar alcohols but no sugar, IonPac ICE-AS1 could be the choice. If the excipients consist of sugars and sugar alcohols but propylene glycol and glycerol are absent, CarboPac PA10 may be used. CarboPac MA1 could resolve all of these excipients simultaneously but needs considerably more run time. In all cases, the actives and other excipients of the dosage forms are not oxidizable at the electrode under the applied voltage conditions and were not seen in the chromatogram (see Chapte 3 for more details on the mechanism of action of PAD). The method employing CarboPac MA1 was validated for specificity, linearity, range, detection limit, repeatability and accuracy.

REFERENCES

1. Lee YC. High performance anion-exchange chromatography for carbohydrate analysis. Anal. Biochem. 1990;189:151–162.
2. Conboy JJ, Henion JD, Martin MW, Zweigenbaum JA. Ion chromatography/mass spectrometry for the determination of organic ammonium and sulfate compounds. Anal. Chem. 1990;62:800–807.
3. Niessen WMA, van der Hoeven RAM, van der Greef J, Schols HA, Voragen AGJ, Bruggnik C. Recent progress in high-performance anion-exchange chromatography-thermospray mass spectrometry of oligosaccharides. J. Chromatogr. 1993;647:319–327.
4. Clarke AP, Jandik P, Rocklin RD, Liu Y, Avdalovic N. An integrated amperometry waveform for the direct, sensitive detection of amino acids and amino sugars following anion-exchange chromatography. Anal. Chem. 1999;71:2774–2781.
5. Melander WR, Horvath C. Reversed-phase Chromatography. In: Horvath C, editor. High-performance Liquid Chromatography: Advances and Perspectives, Volume 2, New York: Academic Press, 1980; p. 113–319.
6. Grushka, E, Grinburg N, editors. Advances in chromatography, Volume 48, Boca Raton, Florida: CRC Press; 2010.
7. Bhattacharyya L. Ion chromatography in biological and pharmaceutical drug analysis: USP perspectives, presented at the Intl. IC Symp. Baltimore: October 2002.

8. Chen Q, Mou S, Hou X, Ni Z. Simultaneous determination of caffeine, theobromine and theophylline in foods and pharmaceutical preparations by using ion chromatography. Anal. Chim. Acta 1998;371:287–296.
9. USP33-NF28, Rockville: US Pharmacopeial Convention;2010.
10. Butterfield AG, Hughes DW, Wilson WL, Pound NJ. Simultaneous high-speed liquid chromatographic determination of tetracycline and rolitetracycline in rolitetracycline formulations J. Pharm. Sci. 1974;64:316–320.
11. Ding X, Mou S. Ion chromatographic analysis of tetracycles using polymeric column and acidic eluent. J. Chromatogr. A 2000;897:205–214.
12. Liu H, Wang H, Sunderland VB. An isocratic ion exchange HPLC method for the simultaneous determination of flucloxacillin and amoxicillin in a pharmaceutical formulation for injection. J. Pharm. Biomed. Anal. 2005;37:395–398.
13. Shanmugasundaram P, Raj RK, Mohanrangan J, Devdaas G, Arunadevi M, Maheswari R, Aanandhi MV. Simultaneous estimation of amoxicillin and flucloxacillin in its combined capsule dosage form by HPLC. Rasayan J. Chem. 2009;2:57–60.
14. Kosonen JP. Determination of disodium clodronate in bulk material and pharmaceuticals by ion chromatography with post-column derivatization. J. Pharm. Biomed Anal. 1992;10:881–887.
15. Tsai EW, Ip DP, Brooks MA. Determination of alendronate in pharmaceutical dosage formulations by ion chromatography with conductivity detection. J. Chromatogr. 1992;596:217–224.
16. Hartigh JD, Langebroek R, Vermeij, P. Ion-exchange liquid chromatographic analysis of bisphosphonates in pharmaceutical preparations. J. Pharm. Biomed. Anal. 1993;10:977–983.
17. Tsai EW, Chamberlin SD, Forsyth RJ, Bell C, Ip DP, Brooks, MA. Determination of biphosphonate drugs in pharmaceutical dosage formulations by ion chromatography with indirect UV detection. J. Pharm. Biomed. Anal. 1994;8:983–991.
18. Qin XZ, Tsai EW, Sakuma T, Ip DP. Pharmaceutical application of liquid chromatography-mass spectrometry. II. Ion chromatography-ion spray mass spectrometric characterization of alendronate. J. Chromatogr. A 1994;686:205–212.
19. Quattrocchi O, Frisardi L, Iglesias M, Noya M, Caputto M, Ferraris D, Siliprandi D, Piccinni E. Ion-exchange chromatographic determination of olpadronate, phosphite, phosphate, chloride and methanesulfonic acid. J. Pharm. Biomed. Anal. 2001;24:1011–1018.
20. Fernandes C, Leite RS, Lancas FM. Rapid determination of bisphosphonates by ion chromatography with indirect UV detection. J. Chromatogr. Sci. 2007;45:236–241.
21. Quitasol J, Krastins L. Analysis of pamidronate disodium in pharmaceutical dosage forms by ion chromatography. J. Chromatogr. A. 1994;671:273–279.
22. Spardans RW, Hartigh JD, Vermeij P. High-performance ion-exchange chromatography with in-line complexation of bisphosphonates and their quality control pharmaceutical preparations. J. Pharm. Biomed. Anal. 1995;13:1545–1550.
23. Kovacevic M, Gartner A, Novic M. Determination of biphosphonates by ion chromatography-inductively couples plasma mass spectrometry. J. Chromatogr. A 2004; 1039:77–82.
24. Pavitrapok C, Williams, DA. Determination of methenamine, methenamine mandelate and methenamine hippurate in pharmaceutical preparations using ion-exchange HPLC. J. Pharm. Biomed. Anal. 2006;40:1243–1248.

25. Perez JL, Bello MA. Determination of paracetamol in dosage forms by non-suppressed ion chromatography. Talanta. 1999;48:1199–1202.
26. Ghanem R, Bello MA, Callejón M, Guiraúm A. Determination of beta-blocker drugs in pharmaceutical preparations by non-suppressed ion chromatography. J Pharm. Biomed. Anal. 1996;15:383–388.
27. Shen S, Quyang J, Baeyens WRG, Delanghe JR, Yang Y. Determination of β2-agionists by ion chromatography with direct conductivity detection. J. Pharm. Biomed. Anal. 2005;38:166–172.
28. Quyang J, Gao X, Baeyens WRG, Delanghe JR. Determination of ephedrine and related compounds in pharmaceutical preparations by ion chromatography with direct conductivity detection. Biomed. Chromatogr. 2005;19:266–271.
29. Megoulas NC, Koupparis MA. Ion-chromatographic determination of carbocisteine in pharmaceuticals based on non-suppressed conductometric detection. J. Chromatogr. A 2004;1026:167–174.
30. Chalgeri A, Tan HIS. Indirect photometric detection for the determination of citrate in pharmaceutical matrices by ion chromatography. J. Pharm. Biomed. Anal. 1996;14:835–844.
31. Karmarkar S, Koberda M, Monami J, Kotecki D, Garber R. Validated ion-exclusion chromatographic method for citrate and acetate in medical fluids. J. Chromatogr. A 2004;1039:147–153.
32. DeBorba BM, Rohrer JS, Bhattacharyya L. Development and validation of an assay for citric acid/citrate and phosphate in pharmaceutical dosage forms using ion chromatography with suppressed conductivity detection. J. Pharm. Biomed. Anal. 2004;36:517–524.
33. Taylor, GE. Determination of impurities in clodronic acid by anion-exchange chromatography. J. Chromatogr. 1997;770:261–271.
34. Phenomenex Inc. TN-1054 Applications: Optimizing the analysis of sugar alcohol excipients in pharmaceutical tablet formulations using Rezex™ ion exclusion HPLC columns.
35. Dionex Corporation. Application Note 117: Quantification of carbohydrates and glycols in pharmaceuticals.

7

ANALYSIS FOR IMPURITIES BY ION CHROMATOGRAPHY

Shreekant V. Karmarkar

Baxter Healthcare, Round Lake, IL

7.1 INTRODUCTION

Ion Chromatographic (IC) methods, from inception of the technology in 1975 [1], have been popularly used for the determination of impurities in drinking and waste water in the environmental industry, such as determination of common anions in waters per US EPA method 300.0 [2] and determination of perchlorate in water per US EPA method 314.0 [3], as well as trace levels of anions and cations in high purity water [4]. This chapter will focus on IC analysis for impurities in pharmaceutical samples. Specifically, it will cover the ICH and compendial guidelines on impurities and discuss the IC niche in the analysis for impurities. Finally, a summary of IC applications in the USP-NF for impurities and degradation products analysis in drug substances and products will be presented.

Applications of Ion Chromatography for Pharmaceutical and Biological Products, First Edition.
Edited by Lokesh Bhattacharyya and Jeffrey S. Rohrer.

7.2 REGULATORY GUIDELINES ON IMPURITIES

Per the ICH Q3A(R2) guidance document [5], an impurity is any component of a drug substance that is not the chemical entity defined as the drug substance. Likewise, per ICH Q3B(R2) guidance document [6], an impurity is any component of a drug product that is not the drug substance or an excipient in the drug product. It should be noted that these guidance documents apply to impurities in chemically synthesized drug substances and the corresponding drug products but not to biological and biotechnological products, as well as pharmaceuticals produced by a biological processes such as fermentation (i.e., antibiotics).

In the ICH Q3A guidance document impurities are classified into three categories; organic impurities (process and drug related), inorganic impurities, and residual solvents. The organic impurities in the drug substance can arise from the manufacturing process or storage of the drug substance. These impurities include starting materials, by-products, intermediates, degradation products, reagents, ligands, catalysts, and others. Reporting thresholds for the identification and qualification of these impurities depend upon the maximum daily dose, as summarized in Table 7.1a. As described in the ICH documents, reporting threshold is the limit above which an impurity in a drug substance or a degradation product in a drug product should be reported to the regulatory authorities. For example, reporting thresholds are 0.05% and 0.03% for drug substance with maximum daily dose (MDD) of ≤2 g/day and ≥2 g/day, respectively (Table 7.1). ICH documents define identification and qualification thresholds as the limits above which an impurity in the drug substance or a degradation product in the drug product should be identified or qualified. An impurity or a degradation product is considered as identified when a structural characterization has been achieved. Qualification of an impurity in the drug substance is the process of acquiring and evaluating data that establishes the safety of an individual impurity or a given impurity profile at the level(s) specified. Likewise, qualification of a degradation product in the drug product is the process of acquiring and evaluating data that establishes the safety of a degradation product or a given degradation profile at the level(s) specified. It is worthwhile noting that identification and qualification thresholds for a MDD of ≤2 g/day (Tables 7.1a and 7.1b) are calculated as a smaller value among the percentage of the MDD (e.g.,, 0.2% of MDD of >100 mg–2 g for a drug product, Table 7.1b) and a

TABLE 7.1a. Impurities in New Drug Substance per ICH Q3A (R2)

Maximum Daily Dose[1]	Reporting Threshold[2,3]	Identification Threshold[3]	Qualification Threshold[2,3]
≤2 g/day	0.05%	0.10% or 1.0 mg per day intake (whichever is lower)	0.15% or 1.0 mg per day intake (whichever is lower)
>2 g/day	0.03%	0.05%	0.05%

[1]The amount of drug substance administered per day.
[2]Higher reporting thresholds should be scientifically justified.
[3]Lower thresholds can be appropriate if the impurity is usually toxic.

TABLE 7.1b. Thresholds for Degradation Products in New Drug Products per ICH Q3B (R2)

Maximum Daily Dose[1]	Reporting Threshold[2,3]
≤1 g	0.1%
>1 g	0.05%

Maximum Daily Dose[1]	Identification Threshold[2,3]
<1 mg	1.0% or 5 μg TDI, whichever is lower
1 mg–10 mg	0.05% or 20 μg TDI, whichever is lower
>10 mg–2 g	0.2% or 2 mg TDI, whichever is lower
>2 g	0.10%

Maximum Daily Dose[1]	Qualification Threshold[2,3]
<10 mg	1.0% or 50 μg TDI, whichever is lower
10 mg–100 mg	0.5% or 200 μg TDI, whichever is lower
>100 mg–2 g	0.2% or 3 mg TDI, whichever is lower
>2 g	0.15%

[1]The amount of drug substance administered per day.
[2]Thresholds for degradation products are expressed either as a percentage of the drug substance or as total daily intake (TDI) of the degradation product. Lower thresholds can be appropriate if the degradation product is unusually toxic.
[3]Higher reporting thresholds should be scientifically justified.

quantity in mg (e.g., 3 mg total daily intake for a MDD of >100 mg −2 g for a drug product, Table 7.1b). For example, for a MDD of 1.9 g of a drug product, the total daily intake would be 3.8 mg (0.2% of 1.9 g) if % value were to be used. Therefore, for this drug product, the identification threshold would be 3 mg and not 3.8 mg.

The degradation products in drug products are a subject of the ICH Q3B(R2) guidance document. Similar to the drug substance, the reporting, identification, and qualification thresholds for the degradation products depend upon the maximum daily dose as summarized in Table 7.1b. Further guidance is provided in the ICH Q6A [7] guidance, which states that organic impurities arising from degradation of the drug substance and impurities that arise from the manufacturing process should be monitored in the drug product. Quantitation limit expectations for an impurities method are dictated by these thresholds. As such, the impurities method is considered satisfactory when it can furnish a quantitation limit that is minimally the same, and preferably less than, the required reporting threshold.

For biotechnological/biological products, the guidance on test procedures and acceptance criteria is published in the ICH Q6B document [8]. The document provides guidance on tests to be used for structural characterization and confirmation, and for determining physicochemical properties of biotechnological/biological products, including that of the impurities present in them. The document also describes process-related impurities, contaminants, and product related impurities, including degradation products. The readers are encouraged to read these guidance documents before determining the specific needs of the products they are interested in and for selection of methods to be used in impurity analysis. The selected method must be such as to conform to the requirements described in these guidance documents.

7.3 IC METHODS AS COMPLEMENTARY TO OTHER HPLC METHODS

HPLC methods have become the workhorse in the pharmaceutical industry for assay and determination of impurities in drug substances and products. The IC methods provide unique separation and detection capabilities, i.e., separation on ion-exchange or ion-exclusion columns followed by conductivity, pulsed amperometric, or other detection. Hence these methods are able to provide avenues that are complementary (or orthogonal) to common HPLC methods (i.e. reversed-phase separation followed by absorbance detection). These IC methods are often capable of measuring compounds at concentrations that are not possibly using common HPLC methods. A summary of published literature on IC methods for impurities as ions amenable to conductivity or UV detection and for molecules that possess oxidizable groups detected using pulsed amperometry is provided in Table 7.2.

7.4 IC IN THE ANALYSIS OF IMPURITIES

The importance of IC in analyzing impurities is illustrated below by describing selected IC methods for the analysis of (a) impurity in a cephalosporin drug (cefepime) determined as an organic amine; (b) impurities of an anticonvulsant (topiramate) and of an AIDS treatment drug (indinavir sulfate) analyzed as organic and inorganic anions; (c) impurities in heparin separated on an anion-exchange column, and (d) contents of pharmacopeial-grade cleaning detergents measured as organic anions. Lastly, IC methods with pulsed amperometric detection have been described for impurities in tobramycin and heparin, which are separated by anion-exchange chromatography.

7.4.1 Cationic Impurity in Cefepime

Cefepime is a fourth generation cephalosporin drug. Degradation of cefepime includes cleavage of the side chain and opening of the β-lactam ring to yield 2-[(2-amino-4-thiazolyl)((Z)-methoxyimino)acetyl]aminoacetoaldehyde and N-methylpyrrolidine (NMP) (Figure 7.1). In the IC methods, NMP is separated on a cation-exchange column and detected by suppressed or nonsuppressed conductivity detection [9,10]. The current USP monograph method [20] uses a silica-based strong cation-exchange column (L52 phase) with which the cefepime peak in the samples elutes as a broad peak at about 55 min. Also, it is necessary to flush the column with a rinse solution

TABLE 7.2. Summary of IC Applications for Impurities Published in the Literature

Application	LOQ	Comments	Ref.
Suppressed or non-suppressed conductivity detection			
N-methylpyrrolidine impurity in cefepime hydrochloride by cation-exchange IC and nonsuppressed conductivity detection	0.4 μg/mL, 0.005%	NMP peak was well resolved from common inorganic cations, cefepime, and Arginine present in Cefepime for Injection with a runtime of about 45 min. Column rinse step to remove cefepime from the column, per the USP method, is not necessary.	9
N-methylpyrrolidine impurity in cefepime hydrochloride by cation-exchange IC and suppressed conductivity detection	0.10 μg/mL, 0.001%	NMP peak was well resolved from common inorganic cations, cefepime, and Arginine present in Cefepime for Injection with a runtime of about 35 min. Column rinse step to remove cefepime from the column, per the USP method, is not necessary.	10
Clodronic acid impurities by anion-exchange IC and suppressed conductivity detection	The lowest concentration of spiked solution: 0.047% for methylenebisphosphonate to 0.299% for sulfate	The direct analysis method eliminates the need for pre- or post-column derivatization. Metal trap column was necessary to prevent the column from fouling. Total runtime: about 20 min	11
Monoethylsulfate in Indinavir sulfate by anion-exchange IC and suppressed conductivity detection	74 ng/mL	The IC method was advantageous over the less accurate and time consuming titrimetric method and also advantageous over the less sensitive capillary electrophoresis method. Total runtime of about 30 min.	12
Sulfate and sulfamate in topiramate drug substance and finished product by anion-exchange IC and suppressed conductivity detection	0.05% for sulfamate and 0.1% for sulfate	The IC method can also be used for topiramate assay in drug substance and drug product, and, therefore, it provides total molar accountability.	13

(continued)

TABLE 7.2. (*Continued*)

Application	LOQ	Comments	Ref.
Traces (EDTA) of CIP-100 (Clean-in place, industrial detergent) detergent by anion-exchange IC and suppressed conductivity detection	0.39 ppm	Method for cleaning validation using EDTA as the measure of CIP-100 present in cleaning validation samples.	14
Human salivary anionic analysis by anion-exchange and ion-exclusion IC and suppressed conductivity detection	Not established	Inorganic (chloride, phosphate, nitrate, sulfate, and thiosulfate) and organic (lactate and acetate) anions determined using IC and IEC methods, respectively.	15
UV/VIS detection			
Heparin: DS and OSCS by anion-exchange IC	Not established, the standard is prepared at 1%	Total runtime of about 75 min	16
Integrated/Pulsed amperometric detection			
Tobramycin and its impurities by anion-exchange IC	0.016 to 0.027% kanamycin B using AAA-Direct waveforms	Aminoglycosides contain oxidizable amine and hydroxyl groups that can be detected electrochemically with a couple of advantages: a) at the selected oxidation potential for the analyte, the other compounds remain undetected and b) simplified analysis since there is no derivatization step. Total run time of about 20 and 30 min for tobramycin and paromomycin, respectively.	17, 18, 19
Paromomycin and its impurities by HPAEC with IPAD	0.10 μM		
Heparin: Galactosamine	0.29%	Total runtime of about 35 min	16

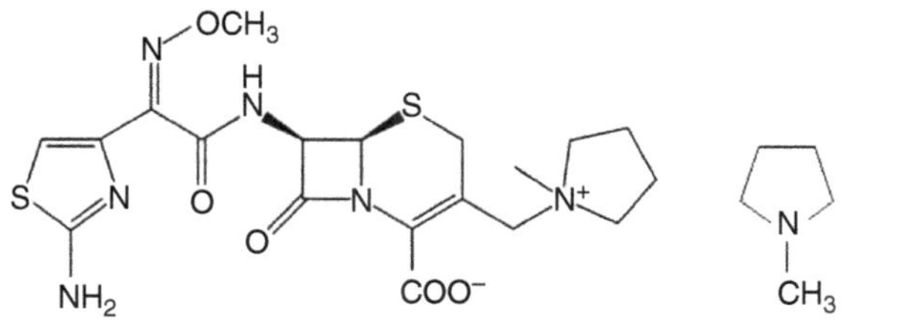

Figure 7.1. Structures of Cefepime and its degradants, *N*-methylpyrrolidine.

having higher acetonitrile content and re-equilibrate the column with the mobile phase prior to the next injection. The whole process could take about 2–3 h for each run. This limitation has been overcome in a couple of IC procedures published in the literature. A silica-based, weak cation-exchange column, packed with a resin containing polybutadiene-maleic acid functional group (Metrosep C2-150 column from Metrohm), was used with nonsuppressed conductivity detection [9]. With this method, column rinse step was not necessary and the cefepime peak is eluted with a total run time of about 45 min. Also, arginine (present as a buffering agent in the cefepime injection product) and NMP peaks are fully resolved. In another IC method with suppressed conductivity detection [10] employing a hydrophilic, carboxylate-functionalized cation exchanger was found to be advantageous since total run time was about 35 min using only aqueous mobile phase without having to use acetonitrile (Figure 7.2). Both methods are capable of furnishing quantitation limits (Table 7.2) that are far less than the USP limits for NMP in the drug substance and drug product of NMT (Not more than) 0.3% and NMT 1.0%, respectively.

7.4.2 Anionic Impurities in Topiramate and Indinavir Sulfate

Topiramate is an anticonvulsant drug that is stable under ambient storage conditions. However, it can degrade at elevated temperatures and humidity into organic degradation products, as well as sulfamate and sulfate, as illustrated in Figure 7.3 [13]. Sulfamate and sulfate are produced stoichiometrically during topiramate degradation and can be determined using suppressed conductivity detection. The separation was accomplished using Dionex IonPac AG5A (4.0 mm × 50 mm guard) and IonPac AS5A (4.0 mm x 150 mm analytical) columns at ambient conditions with a gradient of deionized water (eluent A) and 50 m*M* NaOH (eluent B). The initial composition of 4% B was linearly increased to 50% B from 2 to 15 min, and changed back to 4% B at 16th min with a run time of 30 min. The injection volume was 20 μL. The method could also be used for the topiramate assay and, therefore, it provided total molar accountability.

Compared with the originally used indirect UV detection method [24], the suppressed conductivity method also provided better precision (1.5% RSD for the suppressed conductivity method versus the 5.5% RSD for the earlier indirect UV detection method) and three-fold decrease in the limit of quantitation.

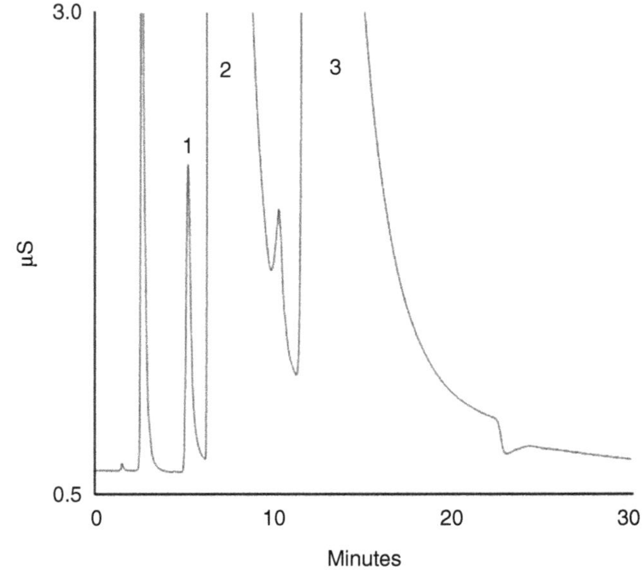

Figure 7.2. Determination of NMP in a simulated Cefepime for Injection. Ion chromatographic conditions: column, IonPac® CG17 guard and CS17 analytical; eluent, 6 mM MSA from 0 to 7.5 min, 85 mM from 7.5 to 20 min, and 6 mM from 20 to 30 min, 0.4 mL/min; column temperature, 40° C; injection volume, 5 μL; suppressed conductivity detection. Sample: 10 mg/mL cefepime hydrochloride + 7.25 mg/mL arginine. Peaks: 1, NMP (0.25%); 2, arginine; and 3, cefepime. [Chromatogram obtained from Dionex Corporation with permission].

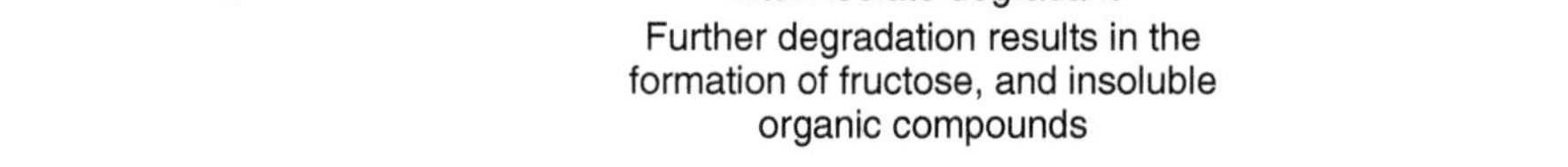

Figure 7.3. Topiramate and its degradation products.

Similarly monoethylsulfate, an anionic impurity of indinavir sulfate, has been determined using suppressed conductivity detection [12]. Indinavir sulfate, the active ingredient of Crixivan®, is a specific and potential inhibitor of HIV-1 protease. It is widely used in the treatment of AIDS. Monoethylsulfate is a potential degradation product of indinavir sulfate and therefore monitored in stability studies. Using the IC method with an anion-exchange separation, the peak was well separated from other peaks in the sample matrix with a runtime of about 25 min. The IC method had an LOQ of 24 ng/mL, much better than the LOQ of 880 ng/mL by the capillary electrophoresis method.

7.4.3 Trace Level Ions in Pharmacopeial-grade Cleaning Detergents

Detergents are often used to clean the manufacturing equipment. In a guidance on the validation of cleaning processes, the Food and Drug Administration (FDA) stated that any residues from the cleaning process itself (detergents, solvents, etc.) should be removed from the equipment [25]. According to the material safety data sheet of a commercially available industrial detergent, Clean-in-place (CIP-100), is composed of water, KOH, EDTA, and surfactants [14]. An anion-exchange IC method with suppressed conductivity detection has been reported for the determination of EDTA, in which EDTA concentration was used as a measure of trace amount of CIP-100 present in cleaning validation samples [14]. The EDTA peak was well separated from other components of CIP-100 with a total runtime of about 12 min. The direct detection allowed sample injection without prior derivatization. The method gave an established LOQ of 0.39 ppm EDTA.

Likewise, an anion-exchange separation followed by direct UV detection at 220 nm has been reported for dimethylbenzenesulfonate (DBS) in another industrial detergent, MICRO® [26]. With a total runtime of about 10 min, the DBS peak was well separated from other peaks in MICRO®, and the method furnished an LOQ of about 0.1 ppm.

7.4.4 Absence of Oversulfated Chondroitin Sulfate in Heparin

Because of the suspected adverse events associated with the contamination of heparin with oversulfated chondroitin sulfate (OSCS), the USP revised the Heparin Sodium monograph and proposed introduction of a strong anion-exchange method for identification of dermatan sulfate, heparin and OSCS. Using a relatively high-capacity anion-exchange column (IonPac AS11), the IC method separates OSCS, Heparin, and dermatan sulfate peaks with a total runtime of about 75 min (Figure 7.4).

7.4.5 Methods with Pulsed Amperometric Detection

Aminoglycosides (e.g., amikacin, gentamicin, and paromomycin) are a group of structurally similar antibiotics used to treat infections caused by gram-negative bacteria. Like other aminoglycosides, tobramycin lacks a suitable UV chromophore, but tobramycin and its impurities (Figure 7.5) can be detected using high-performance anion-exchange chromatography (HPAEC) with pulsed amperometric detection (PAD) [17,18]. The method was applied for analysis of tobramycin at different stages of production, starting from fermentation step with *Streptomyces tenebrarius*,

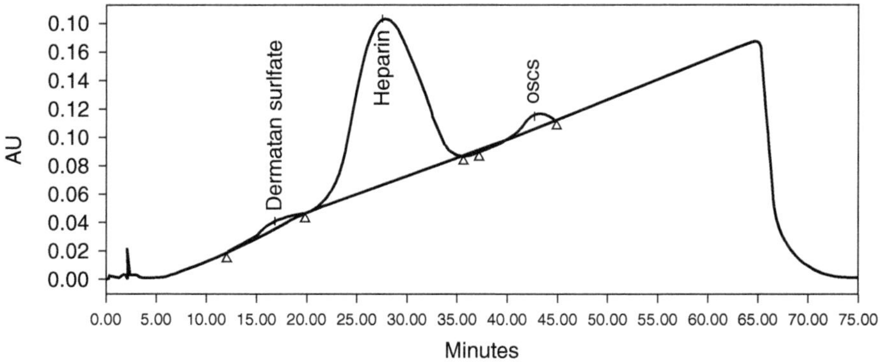

Figure 7.4. Chromatogram showing separation of dermatan sulfate and over-sulfated chondroitin sulfate in heparin. Ion chromatographic conditions: columns, IonPac AG11 guard and AS11 analytical; eluent A, 0.4 g $NaH_2PO_4 \cdot H_2O$/L pH 3.0 with H_3PO_4; eluent B, (0.4 g $NaH_2PO_4 \cdot H_2O$ + 140 g $NaClO_4$)/L pH to 3.0 with H_3PO_4; gradient, 20% B at 0 min to 90% B at 60 min as a linear gradient, 20% B at 61 min to 75 min at 0.22 mL/min; column temperature of 40° C; 10 μL injection volume; refrigerated autosampler; and UV detection at 202 nm. Heparin is prepared at 20 mg/mL, and DS and OSCS are each at 1% of heparin.

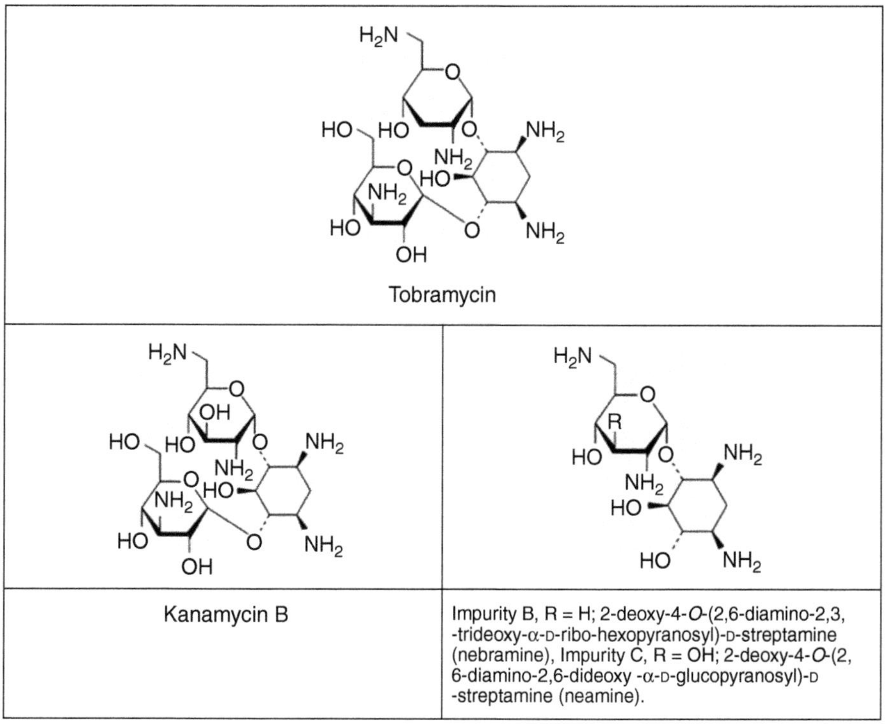

Figure 7.5. Tobramycin and its impurities.

TABLE 7.3. Ion chromatographic Methods for Impurities in USP-NF and Proposed Methods in USP-PF

Monograph	Test	Column	Source
	Conductivity detection		
Fenoldopam Mesylate	Limit of Iodide: NMT 0.2%	Anion-Exchange	USP
Calcium Gluconate	Limit of oxalate: NMT 0.01%	L12	USP
Fosinopril Sodium	Related compounds (Test 2): NMT 0.3%	L12	USP
Ammonia N 13 Injection	Radiochemical purity: Equal peak areas for radioactive and non-radioactive peaks.	L17	USP
Cefepime Hydrochloride	Limit of *N*-methylpyrrolidine: NMT 0.3%	L52 with L17 guard placed in between pump and injector	USP
		Cation-exchange column[1]	USP-PF 36(1)[1]
Sodium Fluoride F 18 Injection	Radiochemical purity: Radiochemical activity of the major peak NLT 95% of the total radioactivity.	L17	USP
Irbesartan	Limit of azide: NMT 10 ppm (0.001%)	L31	USP
Topiramate	Limit of sulfamate and sulfate: NMT 0.10% for each.	L46	USP
Bethanechol Chloride Injection	Limit of 2-hydroxypropyl-trimethyl ammonium chloride: NMT 4.0%	L53	USP
Bethanechol Chloride	Related Compounds: NMT 1.0% for hydroxypropyltrimethyl ammonium and NMT 0.1% for any other impurity. Conductimetric detection	L55	USP
Bethanechol Chloride Tablets	Related Compounds: NMT 1.0% for hydroxypropyltrimethyl ammonium and NMT 0.2% for any other impurity	L55	USP

(continued)

TABLE 7.3. (*Continued*)

Monograph	Test	Column	Source
	Conductivity detection		
Enoxaparin Sodium Injection	Free sulfate content: NMT 0.12%	L61	USP
Betadex Sulfobutyl Ether Sodium	Limit of sodium chloride: NMT 0.2%	L50	USP-PF 36(2)[2]
	Limit of 4-hydroxybutane-1-sulfonic acid: NMT 0.09%	L61	USP-PF 36(2)[2]
	Limit of Bis(4-sulfobutyl) ether disodium: NMT 0.05%	L61	USP-PF 36(2)[2]
	UV detection		
Gonadorelin Hydrochloride	Limit of acetate: NMT 1.0%	L17	USP
Ribavirin for Inhalation Solution	Chromatographic purity: Any individual impurity NMT 0.25%.	L17	USP
Foscarnet Sodium	Limit of phosphate and phosphite: NMT 0.3%	L23	USP
Calcitonin Salmon	Related peptides and other related substances (Test 2): NMT 0.1% (any individual impurity) to 0.6% of Calcitonin Salmon related compound B.	L9	USP
Raclopride C 11 Injection	Chemical purity: Purity NLT 98%	L9	USP
Cisplatin	Limit of transplatin: NMT 2.0%	L9	USP
Metformin Hydrochloride	Related compounds: NMT 0.02% of related compound A	L9	USP
Thiothixene	Limit of (E)-thiothixene: NMT 1.0%	L9	USP
	UV detection		
Risedronate Sodium	Related Compounds (Test 1): NMT 0.10% of any individual impurity	L48	USP
Aprotinin	Limit of *N*-pyroglutamyl-aprotinin: NMT 1.0%.	L52	USP

	Related Compounds: NMT 0.5%	L52	
Fumaric Acid	Limit of maleic acid: NMT 0.1%	L17	NF
Malic Acid	Fumaric and maleic acids (content): NMT 1.0% fumaric acid and NMT 0.05% maleic acid	L17	NF
Heparin Sodium	Absence of oversulfated chondroitin sulfate: No peaks corresponding to oversulfated chondroitin sulfate should be detected eluting after the heparin peak	L61	USP-PF35(5)[3]
	RI detection		
Pamidronate Disodium	Related compounds (Test 2): NMT 0.5% phosphate determined as ortho-phosphoric acid and NMT 0.5% phosphate determined as phosphorous acid and NMT 0.5% of total phosphate and phosphite combined.	L23	USP
Isomalt	Related compounds: NMT 0.5%	L19	NF
	Pulsed amperometric detection		
Fludeoxyglucose F 18 Injection	Limit of 2-chloro-2-deoxy-d-glucose: NMT 1 mg per dose.	L46	USP
Betadex Sulfobutyl Ether Sodium	Limit of beta cyclodextrin: NMT 0.1%	L61	USP-PF 36(2)[2]
Heparin Sodium	Limit of galactosamine in total hexosamine, NMT 1%	L69	USP-PF 35(5)[3]

[1]L designation not assigned in the USP-PF [21].
[2]Reference 22.
[3]Reference 23.

through crude and final products (see Chapter 8 of this book for more details), and also in forced degraded solutions [17,18]. Using the AAA-Direct waveform, the HPAEC-PAD method can detect 0.016 to 0.027 mole percent of 750 pmol tobramycin [17].

Determination of galactosamine in total hexosamine in Heparin has been discussed in Chapter 15 of this book. The method had an LOQ of 0.29% galactosamine in total hexosamine with a total run time of about 35 min.

7.5 IC METHODS FOR IMPURITIES IN THE USP-NF AND USP-PF

A summary of IC methods for various organic and inorganic impurities in the USP-NF and USP-PF is presented in Table 7.3. The detection techniques used are suppressed or nonsuppressed conductivity, UV, RI, and pulsed amperometry. Several different L-series packings for ion-exchange and ion-exclusion separation have been identified in the USP. The methods included in Table 7.3 show that the IC methodology is capable of furnishing low quantitation limits, thereby allowing tighter test limits, e.g., NMT 0.01% oxalate in calcium gluconate to NMT 0.02% related compound A in Metformin HCl.

7.6 CONCLUSIONS

In the past twenty years, IC methodology has established a niche in the analysis of impurities in pharmaceutical drug substances and drug products with applications for organic and inorganic ions using conductivity, UV, amperometric, or RI detection. Additionally, the IC method with PAD has become an important technique for oxidizable organic impurities such as galactosamine in heparin and impurities of aminoglycosides. Like other HPLC methods, the ICH guidelines are applicable in establishing detection limit and quantitation limit expectations of the IC methods for impurities in drug substance and drug products. Over 25 IC methods for impurity analysis have been accepted in USP with a newer IC methods appearing in USP-PF as proposed methods (see Chapter 16 of this book for more details). Many of these applications have very low quantitation limits thereby allowing tighter limits of impurities, e.g., NMT 0.01% oxalate in calcium gluconate to NMT 0.02% related compound A in Metformin HCl.

REFERENCES

1. Small H, Stevens T, Bauman W. Novel ion exchange chromatographic method using conductimetric detection, Anal. Chem. 1975; 47: 1801–1809.
2. USEPA method 300.0, Rev. 2.1. Determination of inorganic anions by ion chromatography; 1993.
3. USEPA method 314.0., Rev. 1.0. Determination of perchlorate in drinking water using ion chromatography; 1999.

4. Dionex Corporation. Application Update 142: Improved determination of trace anions in high purity waters by high-volume direct injection with the EG-40; 2001.
5. ICH Harmonised Tripartite Guideline ICH Q3A(R2). Impurities in new drug substances; 2006.
6. ICH Harmonised Tripartite Guideline ICH Q3B(R2). Impurities in new drug products; 2006.
7. ICH Harmonised Tripartite Guideline ICH Q6A. Specifications: Test procedures and acceptance criteria for new drug substances and new drug products: Chemical substances; 1999.
8. ICH Harmonised Tripartite Guideline ICH Q6B. Specifications: Test procedures and acceptance criteria for biotechnological/biological products. 1999.
9. Subramanian NH, Thygarajan S, Manigandan P, Jeevan RG, Radhakrishnan G. An improved ion chromatographic method for fast and sensitive determination of *N*-methylpyrrolidine in cefepime hydrochloride. J. Chromatogr. Sci., 2009; 47: 549–552.
10. Dionex Corporation. Application Note 199: Determination of *N*-methylpyrrolidine in cefepime using a reagent-free ion chromatography system; 2008.
11. Taylor GE. Determination of impurities in clodronic acid by anion-exchange chromatography. J. Chromatogr. 1997; 770: 261–271.
12. Prasanna SJ, Sharma HK, Mukkanti K, Sivakumaran M, Pavan Kumar KSR, Kumar VJ. Validation of a sensitive ion chromatography method for determination of monoethylsulfate in Indinavir sulfate drug substance. J. Pharmaceutical Biomed. Anal. 2009; 50: 1065–1069.
13. Micheel AP, Ko CY, Guh HY. Ion chromatography method and validation for the determination of sulfate and sulfamate ions in topiramate drug substance and finished product. J. Chromatogr. 1998; 709: 166–172.
14. Resto W, Hernández D, Rey R, Colón, Zayas J. Cleaning validation 2: Development and validation of an ion chromatographic method for the detection of traces of CIP-100 detergent. J. Pharm. Biomed. Anal. 2007; 44: 265–269.
15. Chen Z, Darvell B, Leung V. Validation of ion chromatography for human salivary anionic analysis. Archives of Oral Biology 2004; 49: 855–862
16. Moore E. Heparin impurities testing, 3rd workshop on the characterization of Heparin products, July 27-28, 2009, United States Pharmacopial Convention Rockville, Maryland
17. Dionex Corporation. Application Note 61: Determination of Tobramycin and impurities using HPAE-PAD; 2004.
18. Hanko VP, Rohrer JS, Liu HH, Zheng C, Zhang S, Liu X, Tang X. Identification of tobramycin impurities for quality control process monitoring using high-performance anion-exchange chromatography with integrated pulsed amperometric detection. J. Pharm. Biomed. Anal. 2008; 47: 828–833.
19. Dionex Corporation. Application Note 186: Analysis of paromomycin by HPAE-IPAD, 2004.
20. USP 32-NF 27, Reagents. Rockville: United States Pharmacopeial Convention; 2009.
21. In-Process Revision: Cefepime Hydrochloride. USP-PF 36(1), Rockville: United States Pharmacopeial Convention; 2010.
22. In-Process Revision: Betadex Sulfobutyl Ether Sodium. USP-PF 36(2), Rockville: United States Pharmacopeial Convention; 2010.
23. In-Process Revision: Heparin Sodium. USP-PF 35(5), Rockville: United States Pharmacopeial Convention; 2009.

24. Li W, Rossi TM. Determination of Sulfamate and Sulfate as Degradation Products in an Antiepileptic Drug Using Ion Chromatography and Indirect UV Detection. 1995; 18: 917–923.
25. FDA Inspections, Compliance, Enforcement, and Criminal Investigations. Validation of cleaning process; 1993.
26. Dionex Corporation. Application Note 106: Ion Chromatography in the Pharmaceutical Industry; 1996.

8

ION CHROMATOGRAPHY ANALYSIS OF AMINOGLYCOSIDE ANTIBIOTICS

Valoran P. Hanko and Jeffrey S. Rohrer

Thermo Fisher Scientific, Sunnyvale, CA

8.1 INTRODUCTION

Aminoglycoside antibiotics are one of the older classes of antibiotics and are isolated from certain bacteria. Those with the suffix–mycin are produced from fermentation of different strains of the *Streptomyces* genus while those with the suffix–micin originate from *Micromonospora*. For example, tobramycin is purified from fermentation of *Streptomyces tenebrarius*. Examples of aminoglycoside antibiotics include tobramycin, streptomycin, neomycin, gentamicin, and netilmicin. Figure 8.1 shows the structures of tobramycin, neomycin B, kanamycin B, and neamine (neomycin A). In some cases the purified antibiotic is a complex of more than one compound and more than one of the compounds have antibiotic activity. Some aminoglycoside antibiotics (e.g., neomycin) are assayed by their inhibitory effects on the growth of microorganisms and the results are obtained as activities (or potencies) rather than antibiotic content. While most classes of pharmaceutical active ingredients are assayed by reversed-phase liquid chromatography with UV absorbance detection (HPLC), this has not been true for aminoglycoside antibiotics. Only a few aminoglycosides are assayed by liquid chromatography. This is due mainly to the fact that most aminoglycosides lack a good

Applications of Ion Chromatography for Pharmaceutical and Biological Products, First Edition.
Edited by Lokesh Bhattacharyya and Jeffrey S. Rohrer.

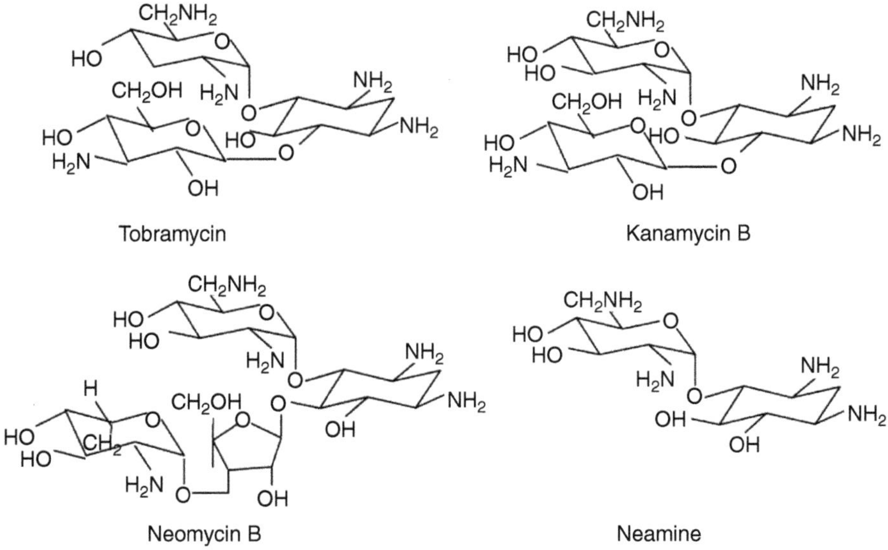

Figure 8.1. Structural diagrams of tobramycin, kanamycin B, neamine, and neomycin B.

chromophore and are therefore poorly detected by UV absorbance. Consequently, to assay these antibiotics by HPLC, the analyst performs either a precolumn derivatization to improve analyte retention and detection sensitivity, or installs a postcolumn derivatization setup for detection sensitivity. The inherent disadvantages of these approaches (i.e. complexity, labor, cost, etc.) may be related to the adoption of only one such test method as a United States Pharmacopeia-National Formulary (USP-NF) assay method for an aminoglycoside (gentamicin) [1]. The growth-inhibition assay procedures are still being used to assay many aminoglycoside antibiotics.

The difficulty of using conventional HPLC for aminoglycoside antibiotic assays led Statler to show that a type of ion chromatography (IC), high-performance anion-exchange chromatography with pulsed amperometric detection (HPAEC-PAD), could be used to assay tobramycin [2]. Seven years earlier Rocklin and Pohl showed that carbohydrates could be separated and detected by HPAEC-PAD [3]. Though most aminoglycosides and carbohydrates are not charged at neutral pH they can be ionized and separated at high pH. The high pH used for separation allowed these carbohydrates to be easily detected by PAD using a gold electrode. This detection was sensitive and required no sample derivatization. Two years after Rocklin and Pohl's publication, Polta, et al. showed that aminoglycosides could be detected by PAD after separation on a polymeric reversed-phase column [4]. While aminoglycosides are not as easily ionized as most monosaccharides, Statler demonstrated that aminoglycosides could be ionized and retained on the anion-exchange column used by Rocklin and Pohl, the CarboPac PA1 (named IonPac AS6 in 1983), but a weaker hydroxide eluent than used for monosaccharides (<5 mM vs. 16 mM NaOH) was required. The method also required the postcolumn addition of 300 mM NaOH to achieve the necessary detection sensitivity. Statler used a gradient of 1 to 5 mM NaOH to separate tobramycin and assay the tobramycin content of the drug product Tobrex. Kaine et al. developed a similar method to assay gentamicin sulfate [5]. Their method used 3 mM NaOH for 5 min followed by a gradient of 3–5 mM NaOH from 5 to 20 min and postcolumn addition of NaOH. The method resolved the four major compounds of the gentamicin C complex, C_1, C_{1a}, C_2, and C_{2A}. Their HPAEC-PAD assay yielded results similar to those in the manufacturer's label. This suggests that the method is comparable to the USP-NF assay for the drug substance, which uses precolumn derivatization with *o*-phthalaldehyde and subsequent detection at 330 nm [1]. Dionex Corporation also published an application note that showed the separation of neomycin B and neomycin C in both a neomycin standard and a topical lotion using a CarboPac PA1 column. This separation also used a hydroxide eluent gradient (0–4 mM over 10 min) and postcolumn addition of 300 mM sodium hydroxide to aid PAD detection [6].

These IC methods had limitations. The postcolumn addition of NaOH was inconvenient and increased the method complexity. Reproducible preparation of dilute NaOH eluents (< 5 mM) was difficult due to the inherent solubility of carbon dioxide in basic solutions. This results in carbonate in the eluent, which is a stronger eluent than the NaOH, and ultimately causes retention time changes and impacts method reproducibility. Not all aminoglycoside antibiotics are as weakly retained as gentamicin, tobramycin, and others. Those that are more strongly bound to the anion-exchange column have assays that use more concentrated eluents that are easier to prepare. For example, 70 mM NaOH is required to elute streptomycin in less than 15 min.

The methods discussed above also had poor long-term peak area reproducibility due to the electrochemical waveform used for detection. This waveform, a 3-potential waveform, causes gradual loss of gold from the working electrode surface resulting in a gradual loss of analyte peak area. Though this does not impact analyte quantification during any two-day period of analysis, it does make comparing peak area responses from month to month difficult. In addition, the detection of impurity peaks is compromised with an older working electrode.

In the mid to late 1990s the lab of Hoogmartens published a series of papers that showed another liquid chromatography approach for determining neomycin sulfate, kanamycin sulfate, netilmicin sulfate, amikacin, gentamicin sulfate, and tobramycin, which lack good chromophores and are not strongly bound to the CarboPac PA1 column [7–12]. These assays used either a three-component or four-component mobile phase (tetrahydrofuran (THF), sodium octane sulfonate, potassium dihydrogen phosphate, and sodium sulfate) and a 100 angstrom pore polymeric reversed-phase column controlled at a temperature higher than ambient, and subsequent PAD detection on a gold working electrode after postcolumn addition of NaOH. The assay procedures from these publications for tobramycin, neomycin sulfate, gentamicin sulfate, and netilmicin sulfate were adopted as European Pharmacopeia (EP) methods. The EP assay for kanamycin sulfate is a microbial assay and the assay for amikacin uses a precolumn derivatization with 2,4-dinitrobenzene followed by a reversed-phase separation and detection at 340 nm.

Difficulties in reproducing the EP methods for tobramycin, neomycin sulfate, gentamicin sulfate, and netilmicin sulfate have been reported. Reproducible preparation of the mobile phase that also requires pH adjustment is difficult, and purity of the THF and sodium octane sulfonate can impact method performance. There were also reports of column reproducibility problems. These methods have some of the same limitations of the early IC methods. The addition of postcolumn NaOH can be problematic and the use of the 3-potential waveform causes destruction of the working electrode. Cao et al. studied the method for netilmicin and published a method that addressed some of the weaknesses of EP method [13]. They reported that a column with a silica-based polar-embedded stationary phase reduced separation time from 45 to 15 min. Cao et al. altered the composition of the 4-component mobile phase to achieve the faster analysis time. They also explored the electrochemical conditions and were able to replace the destructive 3-potential waveform with a non-destructive 4-potential waveform. The combination of these two major changes also improved the assay sensitivity. The authors also explored other electrochemical conditions (working electrode size, reference electrode, and gasket thickness) and noted the importance of reporting these parameters when publishing an electrochemical assay method. Two of the authors of the publications that were the basis of the EP monograph methods published a paper that provided additional guidelines for successful execution of these methods [14]. A more recent publication suggested that these additional details did not resolve the problems with the EP monograph methods [15]. This publication reported improvements for electrochemical detection of tobramycin and gentamicin. The authors found the originally prescribed electrochemical conditions caused working electrode fouling and they modified the waveform to prevent such fouling. They also found that the prescribed polymeric reversed-phase column had poor efficiency under the prescribed eluent conditions, compromising detection of impurities. To remedy this situation they used the same column as in reference 13.

Since the development of the original HPAEC-PAD methods for aminoglycoside antibiotics, there have been a number of improvements to HPAEC-PAD technology. There are now five additional columns in the CarboPac family with differences in capacity and selectivity. The series of three potentials applied to the gold working electrode, the PAD waveform, was changed to a waveform that delivered more stable performance and eliminated the need for post-column addition of NaOH in all but a few applications. In 1997 there was a further improvement of the waveform for carbohydrate analysis that significantly improved long-term peak area reproducibility [16]. The improved waveform uses four potentials and is commonly referred to as the "quad potential" or "4-potential" waveform. At approximately the same time an integrated pulsed amperometric detection (IPAD) waveform was developed that could detect both amino acids and carbohydrates on a gold electrode in highly basic eluents [17]. The same time period also saw the introduction of an eluent generation device [18]. This device was capable of preparing high purity hydroxide eluents at low concentrations from deionized water and a cartridge installed in the eluent generator (EG). This eluent is essentially carbonate-free. Martens and Loeffelmann showed that the EG could be successfully applied to HPAEC-PAD carbohydrate analysis with better retention time reproducibility [19]. Finally, to minimize working electrode to working electrode differences in analyte response, disposable gold working electrodes, produced by sputtering gold on a polymeric surface that was first sputtered with titanium to promote gold adhesion, were introduced for HPAEC-PAD [20].

As a consequence of these improvements, some aminoglycosides now have HPAEC-PAD methods as USP monograph assay methods. The USP monograph assay method for streptomycin uses a CarboPac PA1 (USP L46) column, a 70 mM sodium hydroxide eluent, and the 4-potential PAD waveform. USP-NF included new IC-based assay methods in the monographs of amikacin, amikacin sulfate, amikacin sulfate injection, kanamycin sulfate, kanamycin sulfate capsules, and kanamycin injection. These methods use a CarboPac MA1 (USP L47) column, 115 sodium hydroxide as eluent, and the 3-potential PAD waveform.

Given the problems of the original HPAEC-PAD methods for aminoglycosides weakly retained on the CarboPac PA1 column and the difficulties with the EP monograph methods for these aminoglycosides, we decided in 2003 to attempt to improve the HPAEC-PAD methods for weakly retained aminoglycosides. We believed that the improvements in HPAEC-PAD technology since the publication by Statler [2] would allow us to design more rugged and reproducible methods.

8.2 TOBRAMYCIN

Starting with tobramycin, the aminoglycoside analyzed by Statler, we evaluated newer CarboPac columns, including the CarboPac MA1 (USP L47) used in the USP monograph assay method for kanamycin sulfate drug substance. We found that the PA1 remained the best column to retain tobramycin and separate it from its major impurities kanamycin B and neamine [21]. Figure 8.2 shows separation of tobramycin from kanamycin B (peak 4) and neamine (peak 5) using the PA1 column. Nebramine, another major impurity of tobramycin, should also be present, but absence of a commercially available standard precluded unambiguous identification of the nebramine peak. However, discussions with a tobramycin manufacturer that uses a HPAEC-PAD method led us to believe the nebramine is peak 7, eluting after tobramycin, peak 6. We found that the best resolution of tobramycin and its impurities was with 2 mM potassium hydroxide produced by an electrolytic eluent generator. Multiple efforts to reproduce the separation in Figure 8.2, and the separations discussed later in this chapter, with manually prepared hydroxide eluents were unsuccessful. The major issue was the poorer peak shapes observed with the manually prepared eluents. The eluent generator produces essentially carbonate-free hydroxide eluents, while even the most careful manual preparation of hydroxide eluents contains carbonate. We speculate that the aminoglycosides are more efficiently eluted from the stationary phase by hydroxide compared to carbonate.

Because aminoglycosides have hydroxyl groups and amine groups we evaluated both the PAD waveform designed for carbohydrates and a waveform designed for amino acid determinations [17]. The waveform designed for amino acid determinations detects amines and hydroxyls. While the carbohydrate waveform detects using a gold surface, the amino acid waveform detects using a gold oxide surface. The amino acid waveform produces the gold oxide surface in situ on the gold working electrode and

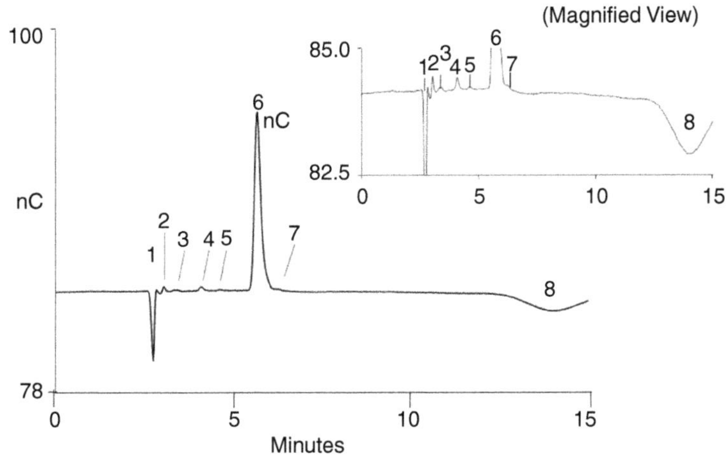

Figure 8.2. Analysis of Tobramycin by IC. The sample (20 μL of 1.07 μM tobramycin) was separated on a CarboPac PA1 column set (4 × 50 mm and 4 × 250 mm) at 30°C with 2.0 mM KOH produced by an eluent generator flowing a 0.5 mL/min. The detection was with PAD using a disposable gold working electrode. Peaks: 1-void volume, 2, 3, and 7—unidentified impurities, 4—kanamycin B, 5—neamine, 6—Tobramycin, 8—oxygen dip. The inset magnifies the chromatogram for an improved view of the impurities.

then removes oxide. In this manner the current produced from forming the necessary oxide surface is accounted for in its removal by the later steps of the integration portion of the waveform (i.e. the steps between 0.46 s and 0.56 s). Table 8.1 displays the carbohydrate and the amino acid waveforms. Both waveforms could directly detect aminoglycosides with high sensitivity, but we chose the amino acid waveform for its better tobramycin sensitivity ([2–4] times greater signal-to-noise ratio), concluding the latter waveform to be more suitable for determining low level impurities (Figure 8.3). The magnitude of the difference in sensitivity is dependent on the noise level. The carbohydrate waveform has better long term peak area reproducibility; therefore if tobramycin assay is the reason for analysis, this is the preferred waveform.

During our initial work with tobramycin we did not have a drug product as all tobramycin drug products in the U.S.A. require a prescription. Therefore we used tobramycin sold by the Sigma Aldrich Chemical Company as tobramycin drug substance. This sample was used to evaluate the method's ability to assay tobramycin and to accurately determine its impurities. For assay we determined the range, linearity, detection and quantitation limits for tobramycin and kanamycin B impurity, intermediate precision of tobramycin peak area and retention time, and evaluated robustness. To evaluate robustness we intentionally varied column temperature, flow rate, eluent concentration, water source, eluent generator cartridge, disposable working electrode, and column (four columns). The method proved to be robust and reproducible for tobramycin and its impurities. The tobramycin sample was found to have 0.24% kanamycin B and 0.73% neamine. We concluded that the IC assay could be used for both assay and related substances analyses of tobramycin and could replace the TLC

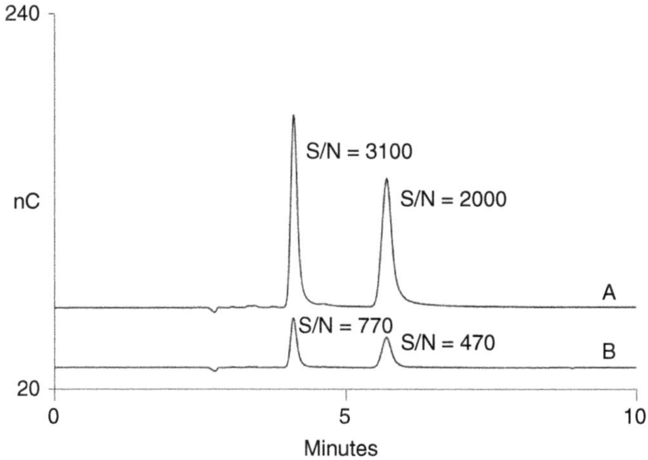

Figure 8.3. Comparison of Tobramycin and Kanamycin B sensitivities using (A) the IPAD Waveform and (B) the PAD Waveform. The 20 μL sample had concentrations of 10 μM each tobramycin and kanamycin B and was separated by the conditions in Figure 8.2.

TABLE 8.1. Comparison of the Carbohydrate and Amino Acid Waveforms.

Carbohydrate Waveform			Amino Acid Waveform		
Time (s)	Potential (V) vs Ag/AgCl	Integration	Time (s)	Potential vs pH	Integration
0.0	0.1		0.0	0.13	
0.2	0.1	Begin	0.04	0.13	
0.4	0.1	End	0.05	0.33	
0.41	−2.0		0.21	0.33	Begin
0.42	−2.0		0.22	0.55	
0.43	0.6		0.46	0.55	
0.44	−0.1		0.47	0.33	
0.50	−0.1		0.56	0.33	End
			0.57	−1.67	
			0.58	−1.67	
			0.59	0.93	
			0.60	0.13	

assay in the USP monograph for tobramycin. Table 8.2 presents a brief summary of much of the data from our study.

Given the low ionic strength eluent and weak retention of tobramycin relative to other carbohydrates, we were concerned that the separation could be easily altered by salt in the sample. A systematic study showed that the separation would be altered with sample salt concentrations of 5 mM and greater (20 μL injection). A known amount of sodium chloride was added to tobramycin to yield varying sodium

TABLE 8.2. Summary of the Studies of the Analytical Parameters of the Tobramycin IC Assay.

Parameter	Test(s) or Criteria	Result(s)
Detection Limit of Tobramycin (20 μL injection)	3× the average peak to peak noise	0.24 pmol
Detection Limit of Kanamycin B (Impurity) (20 μL injection)	3× the average peak to peak noise	0.15 pmol
Quantitation Limit of Tobramycin (20 μL injection)	10× the average peak to peak noise	0.79 pmol
Quantitation Limit of Kanamycin B (Impurity) (20 μL injection)	10× the average peak to peak noise	0.49 pmol
Identification	Separation of a reference standard and separation of tobramycin from its known major impurity Kanamycin B	The method passed both tests.
Range (Tobramycin)	Plot of peak area versus response factor (peak area/amount injected) for varying concentrations of a standard. A change >10% from the average is considered out of range.	4-300 pmol 4-200 pmol
Range (Kanamycin B)	Same as for Tobramycin	
Linearity (Tobramycin)	Linear regression of triplicate injections of 2, 20, 100, 200, and 300 pmol	$r^2 = 0.9994$ y-intercept (nC·min) = 0.336 Slope (nC·min/pmol) = 0.0917
Linearity (Kanamycin B)	Linear regression of triplicate injections of 2, 20, 100, and 200, pmol	$r^2 = 0.9958$ y-intercept (nC·min) = 0.380 Slope (nC·min/pmol) = 0.0929
Intermediate Precision—Tobramycin and Kanamycin B Retention Times	Measure retention time with continuous analysis of 200 pmol each over 7 days	Tobramycin-5.74 ± 0.02 min Kanamycin B- 4.12 ± 0.01 min (572 injections)
Intermediate Precision—Tobramycin and Kanamycin B Peak Areas	Measure peak area with continuous analysis of 200 pmol each over 7 days	Tobramycin-18.52 ± 0.42 nC·min Kanamycin B- 17.81 ± 0.33 nC·min

Accuracy for Kanamycin B Impurity Measurement	Measure recovery of varying amounts of kanamycin B added to varying amounts of tobramycin	High recovery for all concentrations.
Robustness Test #1—Eluent Concentration	Change concentration ±10% and evaluate chromatography.	Resolution of tobramycin and kanamycin B > 4 for both changes
Robustness Test #2—Change of Eluent Generator Cartridge or Source Water	Change cartridge and evaluate chromatography. Change source deionized water and evaluate chromatography.	No statistically significant change in tobramycin and kanamycin B retention times.
Robustness Test #3—Change of Disposable Gold Working Electrode	Compare the ratio of the mean peak areas from four 200 pmol injections of tobramycin and kanamycin B on 3 electrodes from the same lot and 4 electrodes from different lots.	Tobramycin to kanamycin B ratio for 3 electrodes from the same lot = 1.02 ± 0.01
Robustness Test #4—Column Temperature	Change temperature ±10% and evaluate chromatography.	Changes in retention time, peak area, background, and noise did significantly alter the chromatography
Robustness Test #5—Flow Rate	Change flow rate ±10% and evaluate chromatography.	Resolution of tobramycin and kanamycin B >4 for both changes
Robustness Test #6—Column change	Evaluate the chromatography using 4 columns manufactured over a 4 year period.	Good resolution on each column with tobramycin retention time RSD of 7.0%

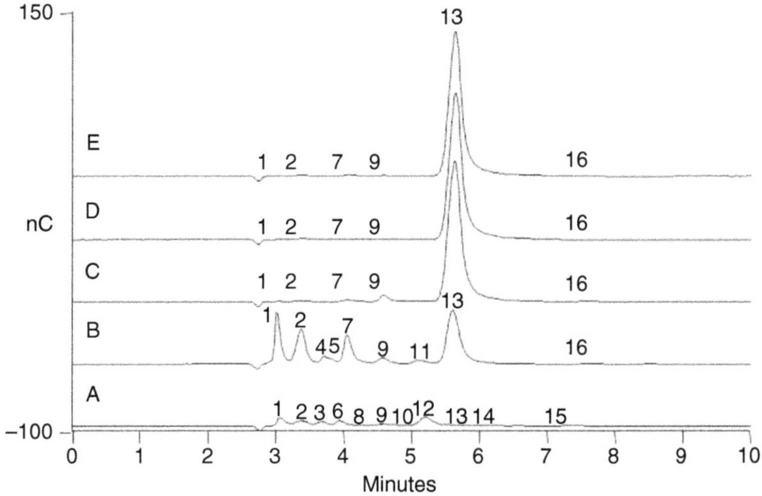

Figure 8.4. Analysis of tobramycin samples from different stages of production. The chromatograms are: (A) fermentation broth, (B) ammonium hydroxide hydrolyzed fermentation broth, (C) crude tobramycin, (D) finished product batch #1, (E) finished product batch #2. The chromatographic conditions are those in Figure 8.2. The identities of peaks 1, 3, 4, 5, 8, 14, and 15 are unknown. Peaks 6, 11, and 12 are identified tentatively as carbomyl-kanamycin B, deoxystreptamine-kanosamide, and carbomyl-tobramycin, respectively. Peaks 2, 7, 9, 10, 13, and 16 are 2-deoxystreptamine, kanamycin B, neamine, kanamycin A, tobramycin, and nebramine, respectively. Note: some peaks are too small to be seen at the scale chosen to best display the major peaks.

chloride concentrations in the injected sample having a constant tobramycin concentration. Tobramycin peak shape and retention time were evaluated in all samples and peak shape and retention time were impacted at sodium chloride concentrations >5 mM. A survey of tobramycin products revealed that the dilution necessary to place the tobramycin concentration in the linear range of the assay would produce salt concentrations well below 5 mM. While this method can successfully assay tobramycin in drug products and determine tobramycin impurities, we concluded that it would probably not be able to assay low concentrations of tobramycin in high ionic strength samples (e.g., serum).

The publication of our tobramycin work led to a collaboration from which we were provided more complex samples containing tobramycin. Crick Pharma and Liuzon New North River Pharmaceutical Company provided us with samples from different stages of commercial scale manufacture of tobramycin where it is produced by fermentation of *Streptromyces tenebrarius* [22]. Although some of these samples were unpurified fermentation samples, they were simply diluted with deionized water prior to analysis. To produce tobramycin, the fermentation products must be hydrolyzed with ammonium hydroxide and then subjected to additional purification steps. Figure 8.4 shows that the IC method could successfully monitor the production and purification of tobramycin. During this study we were able to either identify or tentatively identify a number of degradation products of tobramycin using controlled acid and base hydrolysis of tobramycin and the standards from our earlier tobramycin studies. This was possible due to a publication by Bandl and Gu [23]. In their paper they studied the acid and base degradation pathways of tobramycin and reported the products of both. Both degradations yield some of the same products. Using their findings, an understanding of the retention behavior of some of the aminoglycosides related to tobramycin on the CarboPac PA1 column, and predictions of which of the degradation products would be detected by IPAD, we were able identify kanosamine, nebramine, and 2-deoxystreptamine, three compounds for which there are no commercially available standards. For example, Brandl and Gu reported that the acid degradation of tobramycin yields nebramine and kanosamine while the acid hydrolysis of kanamycin B yield neamine and kanosamine. After performing the acid hydrolysis of both, each yields two major peaks. In the separation of the acid hydrolysate of kanamycin B we are able to identify the neamine peak with a standard. The other peak has the same retention time as one of the two peaks in the tobramycin acid hydrolysate. We believe this peak is kanosamine as both degradations are reported to yield it. By process of elimination the second peak in the tobramycin hydrolysate is nebramine. It elutes after tobramycin which is where we earlier speculated it would elute. We were also able to tentatively identify other peaks, including the precursors to tobramycin produced during fermentation. These precursors and degradation products do not have adequate UV absorbance to allow them to be monitored by reversed-phase HPLC. The high resolution and high detection sensitivity of the IC method facilitates in-process impurity profiling and product stability monitoring.

8.3 NEOMYCIN

After successfully using IC to assay tobramycin we evaluated whether the same general method using the CarboPac PA1 column, a pure hydroxide eluent produced by an eluent generator, and detection on a gold electrode with the same waveform used for tobramycin analysis could be applied to neomycin analysis [24]. Unlike tobramycin, there are several over-the-counter products that contain neomycin, which enabled us to evaluate the method's applicability to drug product analysis. Neomycin consists of a number of aminoglycosides with antibiotic activity, but the most predominant compound with the highest antibiotic activity is neomycin B. With the goals of maximizing neomycin B separation from the other aminoglycosides and having a short separation time for assay, we empirically determined that a 2.4 mM potassium hydroxide eluent met both goals (Figure 8.5, chromatogram B). These conditions provided good separation of neomycin from neomycin A (neamine) and other impurities and were used to evaluate the purity of two neomycin sulfates; the reference standard from the USP (USP RS) and the commercial grade product from Sigma Aldrich. The USP RS had 0.2% neamine and 0.67% of another major compound presumed to be neomycin C, while the other preparation had 1.16% neamine and 5.1% of the other major compound. As the USP monograph assay method for neomycin sulfate is a microbiological assay, there are no impurity limits. The EP does have impurity limits for neamine, neomycin C, other individual impurities, and total impurities. Both preparations of neomycin sulfate meet the EP impurity requirements. Prior to measuring impurities we determined the range, linearity, detection and quantitation limits of neomycin B, intermediate precision of the neomycin B peak area and the retention time, and evaluated robustness by the same criteria as used for tobramycin. Like the method for tobramycin, the method for neomycin sulfate proved to be robust and reproducible for impurity analysis. Table 8.3 presents a brief summary of much of the data from our study. We also found that the method was unaffected by sample salt concentration less than 10 mM, and as with tobramycin, the dilutions necessary to assay neomycin products place their salt concentrations well under 10 mM.

To evaluate the method for suitability to assay neomycin, we measured neomycin in three topical first aid creams. Two of these products also contained other active ingredients (e.g., bacitracin). After extracting neomycin from each of the products we assayed neomycin content by making the assumption that the neomycin B peak represented the amount of neomycin as determined by the microbiological assay. Figure 8.5 chromatogram A shows the neomycin B determination of one of the three products which contained bacitracin and polymyxin B. For each product the amount of neomycin determined using the IC assay was between 90 and 130% of the label value (determined by the USP microbiological assay procedure) and within the limits of the USP monographs for these products. This suggested that the IC method could accurately assay these products by making the assumption that the neomycin B peak represents the neomycin activity determined by the USP assay procedure.

The success of our IC assay for the three topical products led to increased interest into whether the assay could be applied to the other neomycin-containing drug products in the USP-NF [25]. There are 65 drug product monographs in the USP-NF that use the aforementioned microbial method to assay neomycin. These products can be grouped into four classes; creams, solids, liquids, and suspensions. The USP kindly provided us with one product from each class. In short, we found that the assay was precise and accurate for all four classes of products, as judged by comparing the values we determined to the labeled values.

To assay the cream we used the same sample preparation developed for our earlier studies, while the sample preparation for the other three products was straight forward. A solution was prepared from the solid, and that solution, the liquid product, and the suspension (previously centrifuged) were simply diluted to put the neomycin concentration in the calibration range of the IC assay. Samples were assayed with two chromatography systems and on three separate days for each system. To compare to the label values, we again assumed that the amount of neomycin B represented the neomycin activity in the product. For the four products the % of label values (from the microbiological assay) on each system over three days were 100 ± 3 and 103 ± 4, 104 ± 5 and 107 ± 12, 109 ± 3 and 113 ± 4, and 109 ± 6 and 108 ± 6, for the solid, liquid, suspension, and cream, respectively. In addition to that favorable comparison, the assay results of drug product samples spiked with a known amount of neomycin also suggested that the method was accurate. This study was also conducted on two systems over three separate days for each

TABLE 8.3. Summary of the Studies of the Analytical Parameters of the Neomycin Sulfate IC Assay.

Parameter	Test(s) or Criteria	Result(s)
Detection Limit of Neomycin B (20 μL injection)	3× the average peak to peak noise	0.21 pmol
Quantitation Limit of Neomycin B (20 μL injection)	10× the average peak to peak noise	0.72 pmol
Identification	Separation of a reference standard and separation of neomycin from a known major impurity neamine	The method passed both tests.
Range	Plot of peak area versus response factor (peak area/amount injected) for varying concentrations of a standard. A change >10% from the average is considered out of range.	6-200 pmol
Linearity	Linear regression of triplicate injections of 4, 8, 12, 16, 20, 40, 80, 120, 160, and 200 pmol	$r^2 = 0.9987$ y-intercept (nC·min) = 0.3519 Slope (nC·min/pmol) = 0.1394
Intermediate Precision–Retention Time	Measure retention time with continuous analysis of 200 pmol each over 10 days	7.45 ± 0.05 min (822 injections)
Intermediate Precision–Peak Area	Measure peak area with continuous analysis of 200 pmol each over 10 days	29.92 ± 0.40 nC·min (822 injections)
Accuracy for Neamine Impurity Measurement	Measure recovery of varying amounts of neamine added to varying amounts of neomycin	High recovery for all concentrations.
Robustness Test #1—Eluent Concentration	Change concentration ±10% and evaluate chromatography.	Resolution of neomycin B and nearest impurity >5 for both changes

(*continued*)

TABLE 8.3. (*Continued*)

Parameter	Test(s) or Criteria	Result(s)
Robustness Test #2—Column Temperature	Change temperature ±10% and evaluate chromatography.	Changes in retention time, peak area, background, and noise did significantly alter the chromatography
Robustness Test #3–Flow Rate	Change flow rate ±10% and evaluate chromatography.	Resolution of neomycin B and all impurities >2 for both changes
Robustness Test #4—Column change	Evaluate the chromatography using 4 columns manufactured over a 4 year period.	Good resolution on each column with neomycin B retention time RSD of 5.6%
Accuracy for Neomycin B Measurement in Drug Products	Spike known amounts of neomycin into three neomycin-containing creams and measure recovery.	Recoveries were 99.6 ± 2.5%, 95.2 ± 3.9%, and 96.3 ± 1.2%.
Accuracy for Neomycin B Measurement in Drug Products	Determine the amount of neomycin in three neomycin-containing creams and compare to the labeled value.	Found 118.6, 143.4, and 154.3% of labeled value (the product with 118.6% was measured on three days and the other two products measured on a single day.

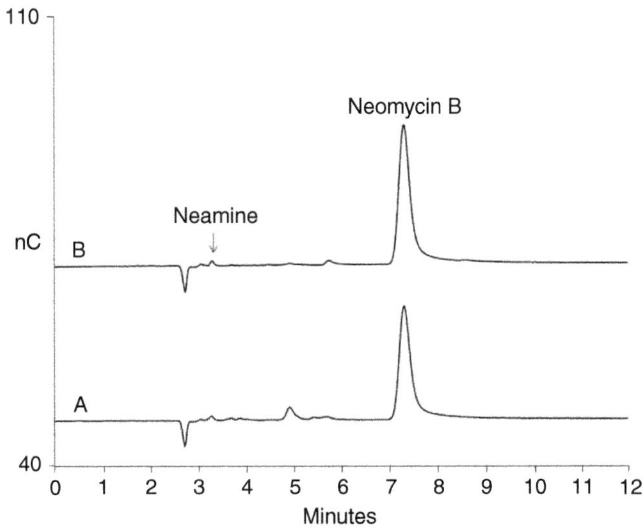

Figure 8.5. Analysis of a neomycin standard and a neomycin-containing first aid cream. Each sample was separated on a CarboPac PA1 column set (4 × 50 mm and 4 × 250 mm) at 30°C with 2.4 mM KOH produced by an eluent generator flowing a 0.5 mL/min. The detection was with IPAD using a disposable gold working electrode. Chromatogram A shows the prepared sample for the first aid cream (20 μL) and Chromatogram B shows the standard (2.0 μM, 20 μL).

system. Samples were prepared at 70, 100, and 150% of the targeted concentration for the assay (3.07 μg/mL, 20 μL injection) and then spiked at 20% of the target concentration. The average percent recoveries of the four products assayed on two systems were 99.4 ± 2 and $98.6 \pm 4, 98.2 \pm 4$ and $94.3 \pm 10, 93.7 \pm 3$ and 85.3 ± 7, and 93.7 ± 4 and 93.3 ± 4, for the solid, liquid, suspension, and cream, respectively.

Chromatography of each of the four samples revealed no peaks from the drug products that interfered with measurement of neomycin B peak area. In fact very little else was detected other than neomycin B and its related compounds. The exception was the suspension product which includes propylene glycol (Figure 8.6). It is detected by IPAD on the gold electrode, but elutes at about 3 min, near the column's void volume, and well-resolved from neomycin B. We concluded that it was probable that the IC assay could replace the USP microbiological assay for neomycin drug products but that a side by side controlled comparison of the two assays was required to reach a definitive conclusion.

8.4 OTHER AMINOGLYCOSIDES

It should be possible to develop IC-based assays similar to the assays presented in this chapter for other aminoglycosides, especially those that are poorly retained on the CarboPac column with stronger eluent conditions. There is already a published

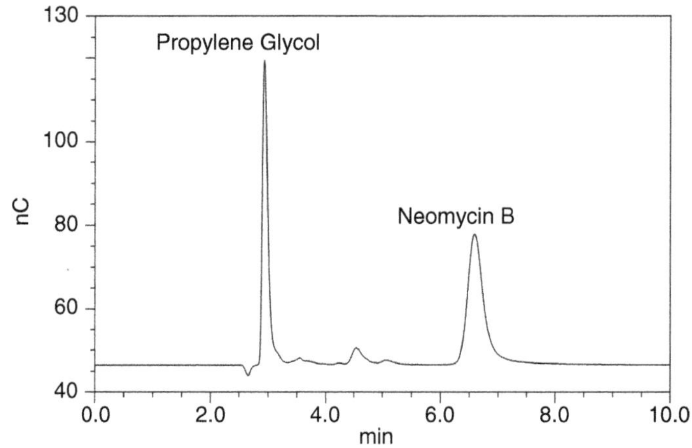

Figure 8.6. Determination of Neomycin in Cortisporin® Suspension. The 20 μL sample was analyzed using the same conditions as in Figure 8.5.

assay for paromomycin in both the drug substance and in the drug product, Humatin [26]. This assay uses a CarboPac PA1 column, 1.8 mM potassium hydroxide eluent produced by an eluent generator and flowing at 0.5 mL/min, and the amino acid waveform. Gentamicin should have a similar separation to that published by Kaine, et al. [5] though with eluent generator produced KOH sharper peaks might be expected, and to assay a product multiple peak areas would need to be combined. Keeping all conditions the same as the tobramycin and neomycin separations except the KOH concentration; we found that netilmicin was retained for 8.74 min with 25 mM KOH and amikacin for 6.08 min at the same concentration.

8.5 CONCLUSION

This chapter showed that IC can be used to successfully assay tobramycin drug substance, neomycin drug substance, and neomycin in four types of drug products. The same IC methods for tobramycin and neomycin can also be used to determine impurities (related substances), and to follow the progress of tobramycin production. IC has also been applied successfully to the analysis of amikacin, streptomycin, kanamycin B, gentamicin, and paromomycin.

REFERENCES

1. USP 32-NF 27. Rockville: United States Pharmacopeial Convention;2009: 2489.
2. Statler J. Determination of tobramycin using high-performance liquid chromatography with pulsed amperometric detection. J. Chromatogr. 1990;527:244–246.

3. Rocklin RD, Pohl CA. Determination of carbohydrates by anion exchange chromatography with pulsed amperometric detection. J. Liq. Chromatogr. 1983;6:1577–1590.
4. Polta JA, Johnson DC, Merkel KE. Liquid chromatographic separation of aminoglycosides with pulsed electrochemical detection. J. Chromatogr. 1985;324:406–414.
5. Kaine LA, Wolnik KA. Forensic investigation of gentamicin sulfates by anion-exchange ion chromatography with pulsed electrochemical detection. J. Chromatogr. A 1994;674:255–261.
6. Dionex Corporation. Application Note 66: Neomycin in Topical Solutions.
7. Adams E, Schepers R, Roets E, Hoogmartens J. Determination of neomycin sulfate by liquid chromatography with pulsed electrochemical detection. J. Chromatogr. A 1996;741:233–240.
8. Adams E, Dalle J, De Bie E, De Smedt I, Roets E, Hoogmartens J. Determination of kanomycin sulfate by liquid chromatography with pulsed electrochemical detection. J. Chromatogr. A 1997;766:133–139.
9. Adams E, Puelings D, Rafiee M, Roets E, Hoogmartens J. Determination of netilmicin sulfate by liquid chromatography with pulsed electrochemical detection. J. Chromatogr. A 1998;812:151–157.
10. Adams E, Van Vaerenbergh G, Roets E, Hoogmartens, J. Determination of amikacin by liquid chromatography with pulsed electrochemical detection. J. Chromatogr. A 1998;819:93–97.
11. Adams E, Roelants W, De Paeper R, Roets E, Hoogmartens, J. Analysis of gentamicin by liquid chromatography with pulsed electrochemical detection. J. Pharm. Biomed. Anal. 1998;18:689–698.
12. Szunyog J, Adams E, Roets E, Hoogmartens, J. Analysis of tobramycin by liquid chromatography with pulsed electrochemical detection. J. Pharm. Biomed. Anal. 2000;23: 891–896.
13. Cao Y, Cheng, J, Mou S, Jandik, P. Improving sensitivity and robustness of chromatographic assays for netilmicin. Pharmacopeial Forum 2004;30(4):2–8.
14. Adams E, Hoogmartens, J. The application of pulsed electrochemistry to the detection of aminoglycoside antibiotics in liquid chromatography. Pharmeuropa 2004;16:222–227.
15. Ghinami C, Giuliani V, Menarini A, Abballe F, Travaini S, Ladisa T. Electrochemical detection of tobramycin or gentamicin according to the European Pharmacopoeia analytical method. J. Chromatogr. A 2007;1139:53–56.
16. Rocklin RD, Clarke AP, Weitzhandler, M. Improved long-term reproducibility for pulsed amperometric detection of carbohydrates via a new quadruple-potential waveform. Anal. Chem. 1998;70:1496–1501.
17. Clarke AP, Jandik P, Rocklin RD, Liu Y, Avdalovic N. An integrated amperometry waveform for the direct, sensitive detection of amino acids and amino sugars following anion-exchange chromatography. Anal. Chem. 1999;71:2774–2781.
18. Liu Y, Avdalovic N, Pohl C, Matt R, Dhillon K, Kaiser R. An online, high-purity acid and base eluent generation system for ion chromatography. Am. Lab. 1998;30(23):162–173.
19. Martens DA, Loeffelmann KL. Automatic generation of ultra-pure hydroxide eluent for carbohydrate analysis of environmental samples. J. Chromatogr. A 2004;1039:33–37.
20. Cheng, J, Jandik, P, Avdalovic, N. Development and characterization of microfabricated gold working electrodes for high-performance anion-exchange chromatography with integrated pulsed amperometric detection. Anal. Chem. 2003;75:572–579.

21. Hanko, VP, Rohrer, JS. Determination of tobramycin and impurities using high-performance anion-exchange chromatography with integrated pulsed amperometric detection. J. Pharm. Biomed. Anal. 2006;40:1006–1012.
22. Hanko, VP, Rohrer, JS, Liu, HH, Zheng, C, Zhang, S, Liu, X, and Tang, X. Identification of tobramycin impurities for quality control process monitoring using high-performance anion-exchange chromatography with integrated pulsed amperometric detection. J. Pharm. Biomed. Anal. 2008;47:828–833.
23. Brandl, M, Gu, L. Degradation of tobramycin in aqueous solution. Drug Dev. Indust. Pharm. 1992;18:1423–1436.
24. Hanko, VP, Rohrer, JS. Determination of neomycin sulfate and impurities using high-performance anion-exchange chromatography with integrated pulsed amperometric detection. J. Pharm. Biomed. Anal. 2007;43:131–141.
25. Hanko, VP, Rohrer, JS. Suitability of a liquid chromatography assay of neomycin sulfate to replace the microbiological assay for neomycin in USP monographs. J. Pharm. Biomed. Anal. 2010;51:96–102.
26. Dionex Corporation. Application Note 186: Analysis of Paromomycin by HPAE-IPAD.

9

USE OF CATION-EXCHANGE ION CHROMATOGRAPHY IN THE ANALYSIS OF PHARMACEUTICALS

Christopher Pohl, Maria Rey, and Rosanne Slingsby

Thermo Fisher Scientific, Sunnyvale, CA

9.1 INTRODUCTION

Many active pharmaceutical ingredients (APIs) in drug substances are amines, which can be protonated to form cations with both ionic and lipophilic characters. Most drug substances are utilized in the salt form for improvement of solubility, physicochemical stability, processing properties, and biopharmaceutical characteristics without altering their basic chemical structures [1]. The contents of the salt counter-ions and excipients of the drug products need to be determined during pharmaceutical development [2]. In the case of acidic drug substances, the counter-ions are cations. It is important to identify and quantify impurities in drug substances, as they can cause undesired side effects and their amounts need to be limited (see Chapter 7 for more details).

Genotoxic impurities (GTIs) in pharmaceuticals at trace levels are an increasing concern to both the pharmaceutical industry and regulatory agencies due to their potential for human carcinogenesis [3]. These include basic molecules that can be protonated to form cations, such as hydroxylamines and hydrazines.

Ion-exchange stationary phases provide good retention and selectivity for charged analytes that are poorly retained by reversed-phase HPLC. In this chapter, cationic

Applications of Ion Chromatography for Pharmaceutical and Biological Products, First Edition.
Edited by Lokesh Bhattacharyya and Jeffrey S. Rohrer.

analytes that are UV/VIS detection-challenged (i.e. with low-to-nonexistent extinction coefficients) and that are of pharmaceutical interest will be discussed in terms of their chromatographic separation and detection.

9.2 DETECTORS IN CHROMATOGRAPHIC SEPARATIONS

UV/VIS detection, using either a variable wavelength or a diode array detector, is the most widely used type of detection in pharmaceutical HPLC analyses. However, there are many compounds of significant biological, pharmaceutical, or drug development interest that either lack a chromophore or do not provide the sensitivity to achieve the minimum detection levels (MDLs) required.

Detection techniques based on fluorescence have provided some of the highest sensitivities available in LC, but it is a specific detector that senses only those substances that fluoresce. Unfortunately, relatively few compounds fluoresce in a practical range of wavelengths, though some can be made to fluoresce by forming appropriate derivatives.

A refractive index detector is a more universal detector but is not practical when gradient chromatography is used. Unlike refractometry, evaporative light scattering detectors (ELSD) and charged aerosol detectors (CAD) can be used with gradient chromatography. These too can detect compounds that lack UV chromophores, but their response is highly dependent on eluent composition and analyte volatility, as volatile analytes may be lost in the interface during solvent evaporation. They are also limited in sensitivity and dynamic range, with the CAD detector providing greater sensitivity, wider linear response range, and better precision than the ELSD [4–6]. For drug substances that have low solubility in water, the high percentage of organic solvent required with HILIC/CAD mobile phases make this method more appropriate than ion chromatography (IC) [2]. Organic solvent-rich mobile phases tend to enhance signal in ELSD, CAD, and electrospray-mass spectrometric (ESI-MS) detectors [7] (see Chapter 10 for more on the application of CAD in IC).

Electrochemical detectors do not rely on analytes having a chromophore, but the analytes must be electroactive. These detectors are therefore more selective. Typically, analytes having hydroxyl, amino, or not fully oxidized sulfur-containing groups in their structure are electrochemically active and therefore detectable. In general, with cation-exchange separations there is the need for postcolumn addition of base to neutralize the acidic eluent, which also makes available the lone pair electrons of the analyte to react with the electrode.

Mass Spectrometry is the most useful detector and most specific as well, but is expensive and requires highly trained personnel.

IC with suppressed conductivity detection is a well established and popular technique for the analysis and quantification of charged species. It is often used as an alternative method to reversed-phase HPLC in the pharmaceutical industry, and also to provide confirmation of the ionic species. Furthermore, ion-exchange separations coupled with suppressed conductivity detection are in many cases the analytical tool of choice for analytes that do not contain a chromophore. In this chapter we will emphasize the utility of cation-exchange chromatography with suppressed conductivity detection for analytes that are difficult to detect by UV/VIS.

9.3 CLASSES OF CATIONS OF INTEREST IN THE PHARMACEUTICAL INDUSTRY THAT DO NOT HAVE CHROMOPHORES

9.3.1 Alkanolamines

The main function of the alkyl alkanolamines is to increase the solubility of other components and enhance solution stability. For example, *N*,*N*-dimethylethanolamine, monoisopropanolamine, and *N*-methylethanolamine are used in the synthesis of antihistamines among other pharmaceutical products, and *N*-methyldiethanolamine is an intermediate in the production of analgesics. In pharmaceutical formulations, monoethanolamine is primarily used for buffering or in preparation of emulsions.

Tris Buffer, tris(hydroxymethyl)aminomethane, can serve as a buffer, neutralizer, solubilizer, and stabilizer for the active drug components in ophthalmic and other formulations. It is also used as an enhancer of antimicrobial effects in contact lens cleaning and in disinfecting products. Hydroxylamine and *N*-alkylhydroxylamines are used as reducing agents in pharmaceutical processes, and can be separated as cations after protonation in acidic eluents. Generally, hydroxylamines are mutagenic in nature, so their quantitative determination is very important. As Figure 9.1 shows, electrochemical detection of the above (with postcolumn addition of base), in addition to being much more selective, provides at least a factor of 10 lower MDLs than indirect conductivity detection [8]. The weak base nature of these analytes makes suppressed conductivity detection unsuitable for these analytes.

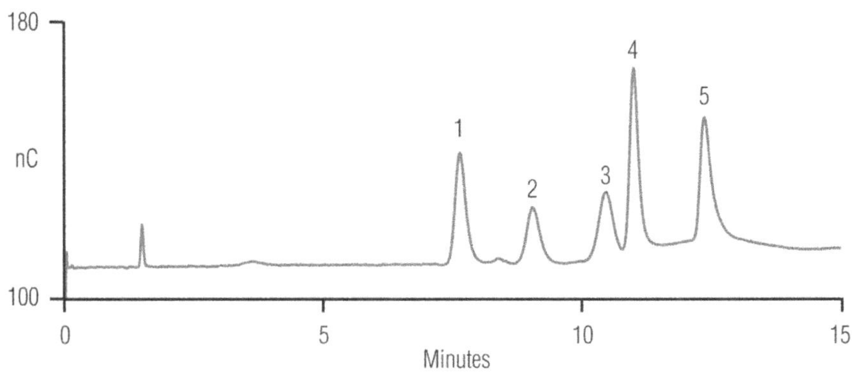

Figure 9.1. Electrochemical detection of hydroxylamines. Peaks: (1) hydroxylamine (1 mg/L), (2) methylhydroxylamine (5 mg/L), (3) ethylhydroxylamine (40 mg/L), (4) dimethylhydroxylamine (5 mg/L), (5) diethylhydroxylamine (8.5 mg/L). Separation column: IonPac CS17 (4 × 250 mm); Injection: 25 μL; Gradient elution: 1.8 mM MSA isocratic to 8 min, gradient to 16 mM MSA at 11 min, isocratic to 15 min, 1.8 mM MSA at 15.1 min; Flow rate: 1.0 mL/min; Temperature: 30°C; Detection: Integrated Pulsed Amperometry with Au electrode, AAA waveform (37), 1 M NaOH post column addition at 1 mL/min.

9.3.2 Linear Aliphatic and Alicyclic Amines

Hydrazine is another highly reactive base and reducing agent that can be carcinogenic and mutagenic and is used in the synthesis of some pharmaceutical intermediates and drug substances. Alicyclic structures containing aliphatic amine functional groups have been the subject of a number of pharmacodynamic studies, especially for their effect on blood pressure [9]. Several pharmaceutical products use aliphatic amines as raw materials in their production, primarily ethylenediamine and piperazine. Cyclohexylamine is used in the production of cyclamate sweeteners, and cyclohexylamine derivatives have been found to be potent cardioplegic agents. Cyclohexylamines and aminoadamantanes have also been found to have antimicrobial properties. The morpholines are also used in the preparation of pharmaceuticals. Ethylamines, such as 2-ethylhexylamine, are used as pharmaceutical intermediates. Amylamine (pentylamine) and tert-butylamine are intermediates in the synthesis of a thromboxane antagonist [10]. *N*-butylamine, *N*-octylamine and 3-dimethylamino propylamine are utilized in the manufacture of pharmaceuticals.

9.3.3 Biogenic Amines

The biogenic amines, such as putrescine, spermidine, and spermine, are found in nearly every cell type, and are required to support a wide variety of cellular functions including cell proliferation and growth. Several studies have revealed higher concentrations of biogenic amines in cancer patients compared to healthy individuals. Many substituted alkylpolyamines have been found to possess potent anti-tumor or anti-parasitic activity, supporting the hypothesis that polyamine-based small molecules can potentially be developed for use as biochemical probes [11]. Many unsubstituted monoalkylamines, amino alcohols, and some diamines have been found to inhibit protein degradation (lysosomotropic), making these compounds useful tools in the study of intracellular protein metabolism [12]. There is a need to determine amine cross-linking residues in pharmaceutical polymers and their decomposition product [13]. Ethylenediamine and polyethylene amines are used as precursors to pharmaceutical products [14]. Ethylenediamine, because of its bifunctional nature, can form heterocycles, such as imidazolidines, which are pharmaceutically active compounds for treatment of inflammatory diseases [15]. Triethylenetetramine has been found to provide topical anti-inflammatory activity for skin disorders [16]. Diamine derivatives have been used to enhance the radiosensitization of hypoxic (poorly oxygenated) tumors with reduced toxic side effects [17].

9.3.4 Inorganic Cations

The determination of alkali and alkaline earth metals is also of significant importance in clinical chemistry. Cations routinely monitored include lithium, whose salts are well established for treatment of psychiatric disorders such as bipolar manic depression, and to minimize the neutropaenic effects of anti-cancer chemotherapy [18]. Ionized calcium measurement is frequently performed during major surgical procedures in which large volumes of blood anti-coagulated with citrate are given. Magnesium is frequently monitored in physiological fluids for hypo- and hypermagnesemia. Rubidium has been found to take the place of potassium in the sodium-potassium pump [19]. Very high rubidium levels combined with low potassium can put muscles into a state of semi-paralysis. Depression has also been linked to rubidium. Ammonia levels are useful in the diagnosis of Rye's syndrome because blood ammonia is often elevated before liver enzymes. Its sensitive detection by methods other than IC is very difficult. For urine samples, no sample preparation is normally required. Plasma and serum samples have to be deproteinized with acetonitrile before injection. Even with this preparation, some physiological fluid samples may contain organic material capable of fouling an ion-exchange column; therefore the use of a guard column is strongly recommended. Other inorganic cations of interest include manganese, whose complexes are used for their paramagnetism and are present in some MRI contrast agents. The use of rubidium 87 in imaging heart disease by MRI has been reported. Strontium 89 has been used as a palliative treatment for bone metastases in cancer patients. The use of strontium salts and pharmaceutical compositions thereof for the treatment of arthrosis has been reported [20]. Barium has been used in the opacification of the digestive tract.

9.3.5 Cationic Counter-ions

The separation and quantification of counter-ions in the pharmaceutical industry is an important determination. The selection of the correct salt form can prevent repeating toxicology, biological, and stability studies. Solubility, dissolution rate, stability, and drug hygroscopic properties can be influenced by the counter-ions. Formation of salts is probably the only chemical technique available to change aqueous solubility and dissolution rate without changing the API molecule. The counter-ions most widely used to form API salts of acidic molecules have been sodium, calcium, and potassium [21]. There is however a recent trend toward using a wider variety of counterions, such as magnesium, zinc, lysine, tromethamine, diethanolamine, diethylamine, piperazine, and meglumine, depending upon whether the drug is given to the patient topically, orally, or by injection. Sodium as a counterion was found to be the most effective at reducing the bitterness of selected oral pharmaceuticals [22].

9.4 CATION-EXCHANGE STATIONARY PHASES

9.4.1 History

Ion-exchange stationary phases provide good retention and selectivity for charged analytes that are poorly retained by reversed-phase HPLC. IC was introduced in 1975 by Small, Stevens, and Baumann [23]. This new analytical method allowed the sensitive detection of ions using suppressed conductivity detection. The original components included a low capacity, surface-sulfonated polystyrene/divinylbenzene (PS/DVB) stationary phase followed by a packed-bed suppressor in the hydroxide form and a conductivity detector. These surface-sulfonated PS/DVB copolymers typically had exchange capacities between 0.005 and 0.1 meq/g. The polymer beads were originally packed in glass column hardware but more recently PEEK tubing has been used because it is inert to acids, bases, and most organic solvents, and has much higher tensile strength than glass.

In 1979, Fritz et al. [24] described an alternative separation and detection scheme for inorganic anions, utilizing non-suppressed conductivity detection.

In 1985, latex based columns delivered significant improvement in peak efficiencies due to their high mass transfer rate. These columns consist of a lightly sulfonated PS/DVB substrate to which fully aminated latex beads with a much smaller diameter are attached by electrostatic interactions. The positively charged substrate, thus produced, is electrostatically covered by a second layer of latex beads which carry the actual exchange function in the form of sulfonate groups [25]. Divalent cations such as alkaline-earth metals are very strongly retained by the sulfonic acid functional groups in the column, and cannot be eluted with dilute mineral acids. To the acidic eluent, a divalent eluent cation such as 2,3-diaminopropionic acid (DAP) must be added to elute these cations in a reasonable time. DAP typically exists in the mobile phase as a mixture of both monovalent and divalent cations but in the suppressor it is in its zwitterionic form and thus has very low intrinsic conductance.

In 1987 Schomburg et al. developed a silica-based polymer-coated carboxylic acid cation-exchanger [26]. It consists of a copolymer that is derived from a mixture of butadiene and maleic acid in equal parts and has low cation-exchange capacity so that it is used with non-suppressed conductivity detection. The high selectivity for hydronium ions of these weak acid functional groups allowed the separation of alkali and alkaline earth metals, and ammonium in less than 20 min using only tartaric acid, a mild complexing agent, as the eluent. The complexation of the divalent cations reduces their effective positive charge, thus decreasing their retention on the stationary phase.

The combined limitations of the silica substrate and the relatively high pKa of the stationary phase functional groups only allows for a relatively narrow sample and eluent pH operating range of pH 2–8. At pH <2, the ion exchange sites in the stationary phase are nearly fully protonated and thus provide minimal retention, while at pH >8 the silica substrate will be corroded due to alkaline hydrolysis.

In 1992, Dionex Corporation introduced the IonPac CS12, a polymer-based cation-exchange column with grafted carboxylate functional groups for IC with suppressed conductivity detection. The column utilizes a grafted macroporous, high surface area polymeric substrate that provides a much higher cation-exchange capacity than the Schomburg column. In addition, use of reduced pKa carboxylic acid functional groups also enhances retention. Because weak acid cation-exchange materials have an unusually high affinity for hydronium ion, the stationary phase charge density is greatly reduced when used under low pH conditions due to protonation of many of the weak acid cation exchange sites in the stationary phase. Accordingly DAP is no longer required in the eluent. Furthermore, the suppressed conductivity background is significantly lower with a simple acid eluent. Newer hydronium-selective carboxylate-functionalized polymeric stationary phases used with suppressed conductivity detection will be discussed in this chapter.

9.4.2 Cation-exchange Chromatography with Suppressed Conductivity Detection

The separation mechanism of IC is based on an ion-exchange equilibrium process occurring between the sample ions and eluent ions with the ion-exchange functional groups bonded to the stationary phase. A typical solid phase consists of ethylvinylbenzene cross-linked with divinylbenzene (EVB/DVB) and functionalized with either sulfonic acid groups or carboxylic acid groups as exchange sites for the separation of cations. The anionic counter-ions of both the acidic eluent and the analytes are exchanged for hydroxide ions in the suppressor. Thus, when suppressed conductivity is used, the hydronium ions in the eluent combine with the hydroxide ions from the suppressor forming water, dramatically reducing the conductivity of the eluent. At the same time, the analyte response is maximized by the suppressor supplying analyte cations with the more conductive hydroxide ion as the analyte counter-ion, resulting in significant improvement in the signal-to-noise ratio compared to non-suppressed conductivity detection.

9.4.3 Suppressed versus Nonsuppressed Cation-exchange Chromatography

The technological difference between suppressed and nonsuppressed IC is that nonsuppressed IC does not use a suppressor; hence it is also known as indirect conductivity and Single-Column IC (SCIC). SCIC is limited to low-capacity stationary phases and dilute acidic eluents, as the background will be proportional to the eluent concentration and its specific conductance. Associated detector noise is related to eluent conductivity when operating in SCIC mode via a square root relationship. In order to avoid excessively high noise associated with high background conductivity when operating in the SCIC mode, it is necessary to avoid the use of concentrated eluents. This precludes the use of high capacity stationary phases, as such columns require high acid concentrations to elute the analytes within a reasonable time. Suppressed and nonsuppressed conductivity may be differentiated in terms of detection limits, linear range, column capacity, and the ability to perform gradient separations. References [27] and [28] discuss this subject in more detail.

9.4.3.1 Detection Limits. With a given column, using the same chromatographic system and a modest 3 mM MSA eluent concentration, the signal in the nonsuppressed conductivity mode is between 1 and 1.6 times higher than in the suppressed mode, depending on the analyte. The noise, however, is about 25 times higher in a nonsuppressed system, and changes in temperature can affect the baseline drift considerably. Thus, for a signal-to-3x noise ratio for MDL calculation, suppressed conductivity offers 10 to 25 times better (i.e. lower) MDL for the common six cations (lithium, sodium, ammonium, potassium, magnesium, calcium). Using higher eluent concentrations would result in even poorer MDL values for nonsuppressed IC.

9.4.3.2 Linear Range. With suppressed conductivity IC much higher eluent concentrations can be used without adversely affecting system noise, thus enabling the use of relatively high capacity columns. For this reason, columns used with suppressed conductivity typically provide a linear range of more than three orders of magnitude in terms of analyte concentration. However, for analytes that are weakly basic, such as ammonium ion and most aliphatic and aromatic amines, a nonlinear calibration curve is observed and a quadratic fit is typically required for such analytes. Weak bases are converted by the suppression reaction to the corresponding cationic hydroxides which are in equilibrium with the neutral amine form of the analyte. Because the equilibrium is concentration dependent, this results in ionization suppression at high analyte concentrations. On the other hand, in nonsuppressed IC, a linear calibration is observed for the weak bases as well, but the calibration can extend only up to three orders of magnitude due to the lower signal-to-noise ratio values of SCIC.

9.4.3.3 Column Capacity. The requirement of a low capacity column for nonsuppressed detection restricts its ability to analyze high ionic strength matrices. The dynamic range is also lowered to avoid overloading the column. See Figure 9.2 for a comparison of sample overloading in a low capacity silica-based column (Dionex, IonPac SCS1, 4 × 250 mm, 318 μeq/column) and in a high capacity polymeric column (Dionex, IonPac CS16, 5 × 250 mm, 8400 μeq/column). Even though the SCS1 has about twice the chromatographic efficiency of the CS16, under overload conditions the much lower chromatographic efficiency and poor peak symmetry with the SCS1 considerably lowers resolution of the first three monovalent cations. The high capacity of the CS16 (with a 25 μL injection it can tolerate up to 130 mM acid concentration in the sample (pH 0.9)) permits it to analyze the sample that overloads the SCS1. The SCS1 in the nonsuppressed mode already shows peak splitting for lithium, magnesium, and calcium when the sample contains as little as 25 mM acid (pH 1.6). See Figure 9.3 for a comparison of sample pH effect on a high and a low capacity column.

9.4.3.4 Gradient Separations. For suppressed systems, high eluent concentrations may be used without any significant change in the background conductance, as long as the suppressor capacity is not exceeded, with the help of a continuously regenerated cation trap column (CR-CTC) to remove any cationic impurities from the eluent. Thus, it is possible to perform gradient elution with minor changes in baseline, improving MDLs due to the focusing effect of the gradient, and decreasing the total analysis time required for samples which contain polyvalent analytes. See Figure 9.4 as an example of isocratic versus gradient elution.

9.4.4 IC-MS

When using electrospray ionization-mass spectrometric detection (ESI-MS), analyte signal suppression can occur if the analyte co-elutes with other ions. In addition mixed forms of the analyte, i.e. analyte plus various counterions, will effectively lower the signal for one form, thereby lowering sensitivity. Thus, having a separation column and suppressor as the front-end to the ESI-MS helps avoid these problems [29]. The results are less ion suppression, predictable ion and cluster formation, and overall better sensitivity compared to flow injection.

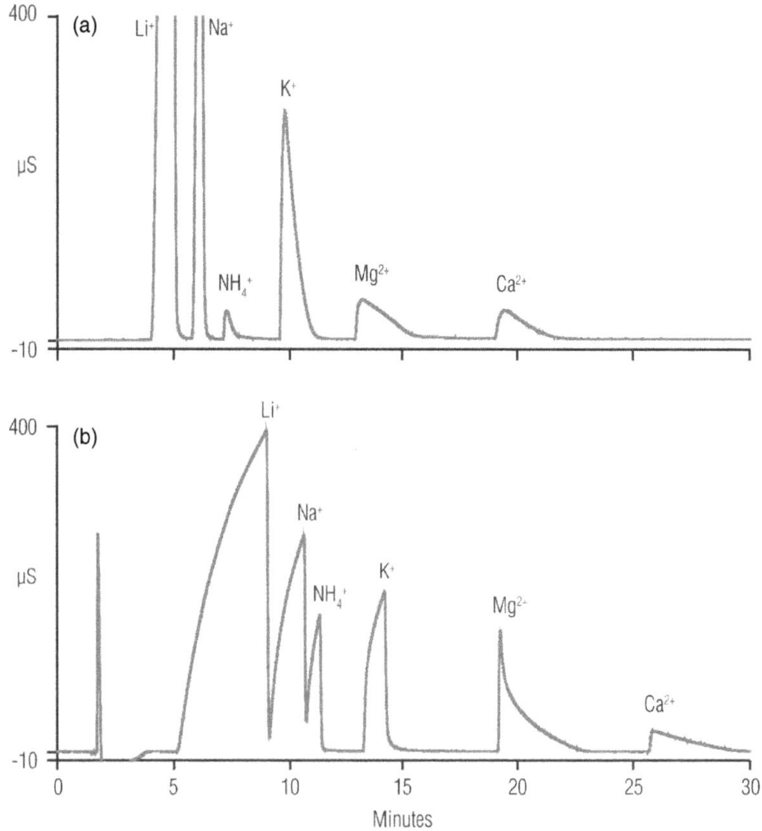

Figure 9.2. Column overloading comparison in a suppressed and nonsuppressed cation-exchange chromatography system. (a) Suppressed conductivity detection. Separation column: IonPac CS16 (5 × 250 mm); Injection: 25 μL; Eluent: 30 mM MSA; Flow rate: 1.0 mL/min; Temperature: 40°C; Detection: suppressed conductivity. (b) Nonsuppressed conductivity detection. Separation column: IonPac SCS1 (4 × 250 mm); Injection: 25 μL; Eluent: 3 mM MSA; Flow rate: 1.0 mL/min; Temperature: 30°C; Detection: nonsuppressed conductivity Samples: lithium (1000 mg/L), sodium (1000 mg/L), ammonium (250 mg/L), potassium (1000 mg/L), magnesium (250 mg/L), calcium (250 mg/L).

Some amines and ions have a very low mass-to-charge ratio (m/z) such that choice of the mass spectrometer is important. Most quadrupole-type mass spectrometers lose sensitivity below about 70 m/z and therefore show poor detection of low mass ions. There is a single quadrupole ESI-MS instrument that was specifically designed for low mass detection, the MSQ-Plus.

Both single quadrupole MS and triple quadrupole MS/MS detection can be useful in IC-MS. If the analyte produces a stable relatively high abundance fragment, then MS/MS detection can provide better specificity and sensitivity than a single quadrupole

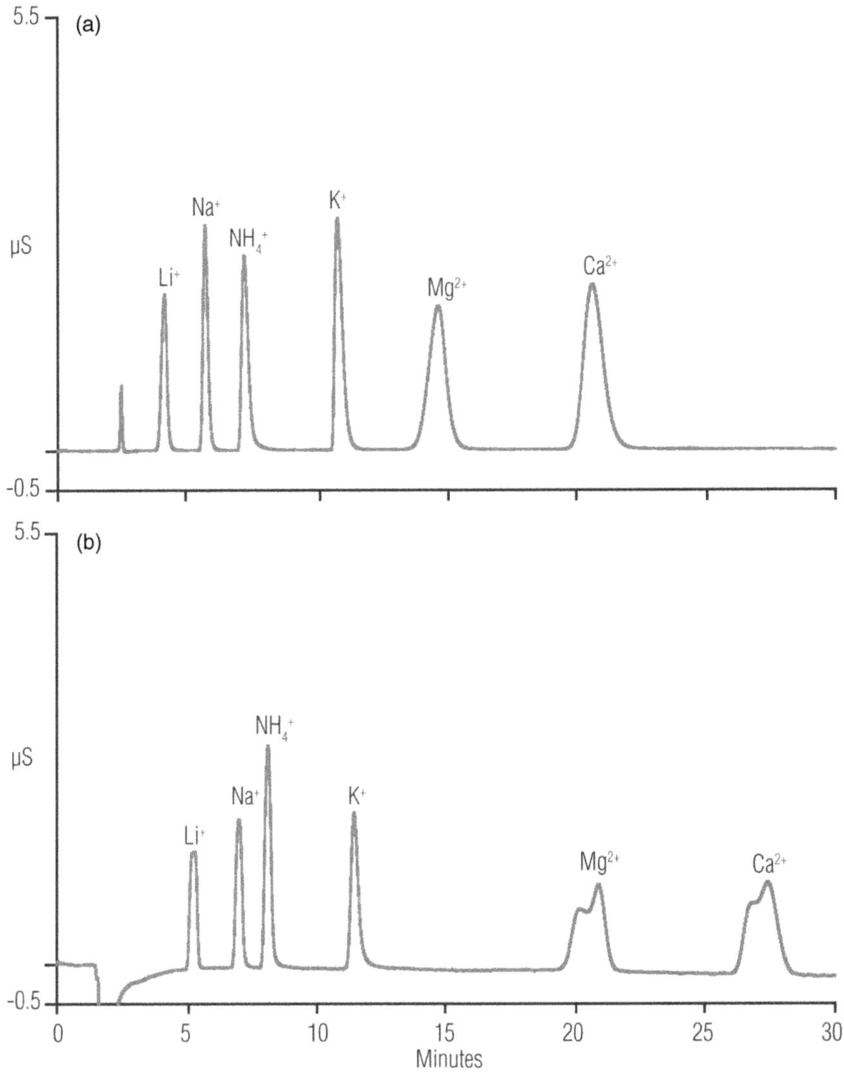

Figure 9.3. Sample pH effect comparison in a suppressed and nonsuppressed cation-exchange chromatography system. (a) Suppressed conductivity detection. Separation column: IonPac CS16 (5 × 250 mm); Injection: 25 µL; Eluent: 30 mM MSA; Flow rate: 1.0 mL/min; Temperature: 40°C; Detection: suppressed conductivity. Sample (in 130 mM H^+ or pH 0.89): lithium (0.2 mg/L), sodium (0.8 mg/L), ammonium (1.0 mg/L), potassium (2.0 mg/L), magnesium (1.0 mg/L), calcium (2.0 mg/L). (b) Nonsuppressed conductivity detection. Separation column: IonPac SCS1 (4 × 250 mm); Injection: 25 µL; Eluent: 3 mM MSA; Flow rate: 1.0 mL/min; Temperature: 30°C; Detection: non-suppressed conductivity. Sample (in 25 mM H^+ or pH 1.6): lithium (0.2 mg/L), sodium (0.8 mg/L), ammonium (1.0 mg/L), potassium (2.0 mg/L), magnesium (1.0 mg/L), calcium (2.0 mg/L).

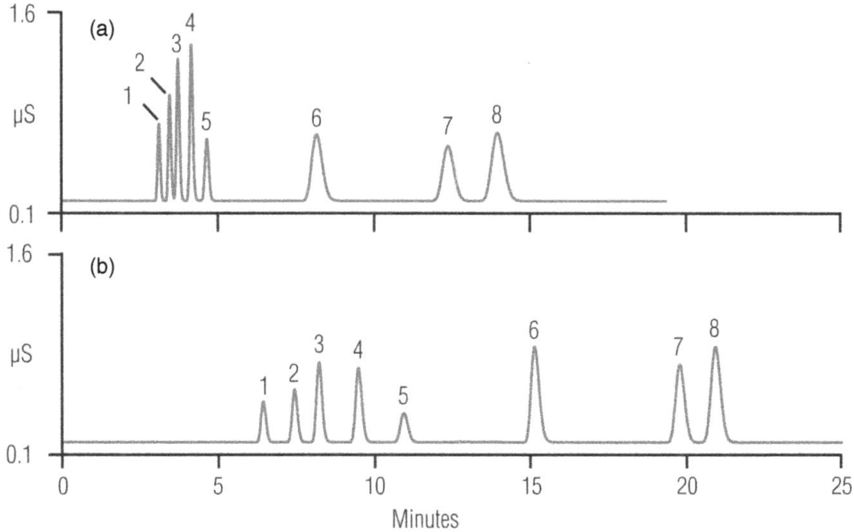

Figure 9.4. Comparison of isocratic and gradient elution on the IonPac CS17 column (4 × 250 mm). (a) Isocratic elution. Injection: 25 µL; Eluent: 6 mM MSA; Flow rate: 1.0 mL/min; Temperature: 30°C; Detection: suppressed conductivity. (b) Gradient elution. Injection: 25 µL; Eluent: 2 mM MSA isocratic to 11 min, 7 mM MSA at 11.1 min, isocratic to 22 min, 2 mM MSA at 22.1 min; Flow rate: 1.0 mL/min; Temperature: 30°C; Detection: suppressed conductivity. Sample: (1) lithium (0.1 mg/L), (2) sodium (0.4 mg/L), (3) ammonium (0.5 mg/L), (4) potassium (1.0 mg/L), (5) dimethylamine (1.0 mg/L), (6) triethylamine (7.0 mg/L), (7) magnesium (0.5 mg/L), (8) calcium (1.0 mg/L).

instrument. The single quadrupole instruments are generally easier to use but can show more interference in complex matrices.

With the electrospray interface, sensitivity and instrument performance are better at lower flow rates, so the separator columns used with IC-MS should be narrow bore i.d. (ideally 2 mm i.d. or less) so as not to require flow rates above 0.30 mL/min.

In cation IC coupled with mass spectrometry (IC-MS), the suppressor removes eluent and analyte counter-ions and replaces them with hydroxide ions, so that the effluent from the suppressor consists of only water and the analytes of interest. This relatively "clean" stream reaching the mass spectrometer makes the system more rugged by reducing maintenance. MDLs are improved as the electrospray signal is considerably more stable and ionization suppression is minimized.

Baseline resolution of the analytes when using MS is generally not required due to the selectivity of MS, although mixtures containing extreme ratios of analyte concentrations may require baseline resolution to avoid co-ion suppression of ionization. See Figure 9.5 for an example of an IC-MS screening gradient of small cations.

The cation-exchange selectivity of the IonPac CS18 column coupled with the mass selective detection of ESI-MS can provide fast peak identification and quantification of small alkanolamines (Figure 9.6). Cation analyses can be accelerated by using

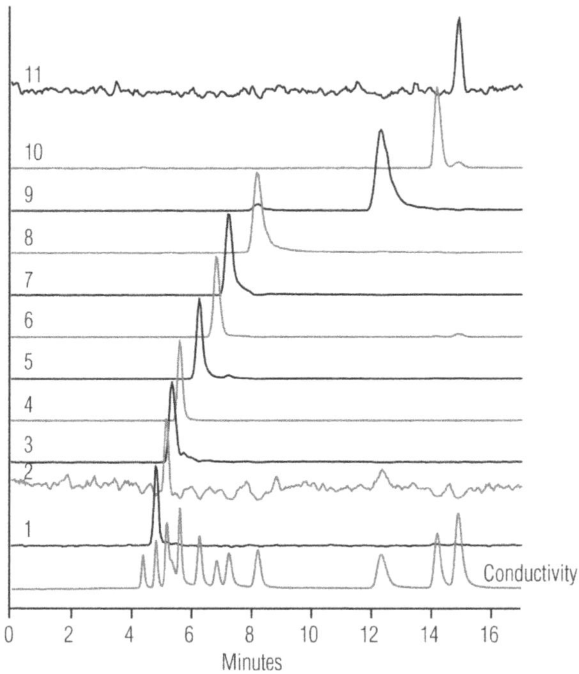

Figure 9.5. An IC-MS screening gradient for small amines and other cations. Separation column: IonPac® CS17 (2 × 250 mm); Injection: 25 μL; Eluent: 1 to 10 mM MSA in 10 min; Flow rate: 0.3 mL/min; Temperature: 30°C; Detector #1: suppressed conductivity (bottom trace) with CSRS 2 mm suppressor, external water regenerant, 20 mA applied current; Post column solvent: 90/10 v/v acetonitrile/deionized water at 0.2 mL/min; Detector #2: ESI-MS; Cone voltage: All peaks, 50V; ESI-MS source temperature: 400°C. Peaks: (1) Sodium, *m/z* 23; (2) Ammonium, *m/z* 18; (3) Hydrazine, *m/z* 33; (4) Potassium, *m/z* 39; (5) 2-(2-Aminoethoxy)ethanol, *m/z* 106; (6) 5-Amino-1-pentanol, *m/z* 104; (7) Morpholine, *m/z* 88; (8) 2-Diethylaminoethanol, *m/z* 118; (9) Cyclohexylamine, *m/z* 100; (10) Magnesium, 24; (11) Calcium, *m/z* 20. *Note*: lithium, the first peak in the conductivity trace, is not detected in the ESI-MS.

steeper gradients or higher eluent concentrations, and partial coelution due to this can be resolved in the MS detector. In this case a single quadrupole mass spectrometer with a highly sensitive low-mass filter was used so that cations with mass-to-charge ratios as low as 23 could be detected with good sensitivity. The alkanolamines elute in about 5 min and are resolved from sodium, which can cause signal suppression if coeluting. A 10 μL injection volume was used in this application, but even lower detection limits can be achieved by injecting as much as 100 μL of sample. The use of postcolumn solvent aids the electrospray process and increases sensitivity for these analytes 5 to 10-fold, with acetonitrile better than methanol.

A conductivity detector and an MS detector can be run in series, with the conductivity detector placed after the suppressor and ahead of the MS detector. The use of cation-exchange coupled with UV, suppressed conductivity, and MS detection in

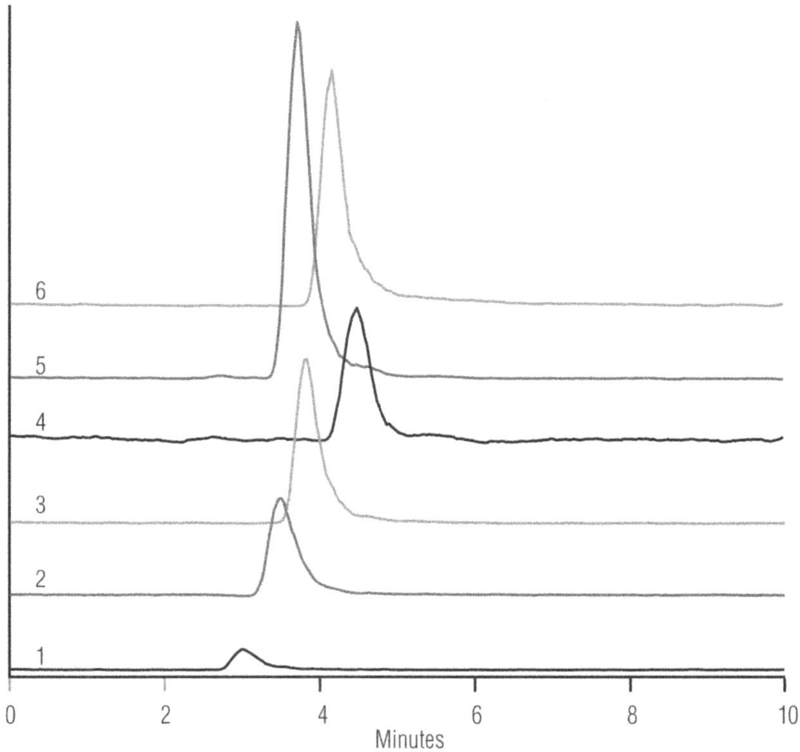

Figure 9.6. Fast peak identification and quantitation of small alkanolamines with IC-MS Separation column: IonPac® CS18 (2 × 250 mm); Injection: 10 μL; Eluent: 5 mM MSA; Flow rate: 0.3 mL/min; Temperature: 30°C; Suppressor: CSRS 300, 2 mm, external water regenerant, 20 mA applied current; Post column solvent: 90/10 v/v acetonitrile/deionized water at 0.2 mL/min; Detection: ESI-MS; Cone voltage: All peaks, 50V; ESI-MS source temperature: 400°C. Peaks: (1) sodium, *m/z* 23 (1 mg/L); (2) ethanolamine, *m/z* 62 (100 μg/L); (3) methylaminoethanol (100 μg/L); (4) diethylaminoethanol (100 μg/L); (5) diethanolamine (100 μg/L); (6) triethanolamine (100 μg/L).

series has been reported for the determination of biogenic amines profiles [30]. In this application, long term stability of the MS signal was attributed to the use of a suppressor device before the MS ionization.

9.4.5 Newer Hydronium-selective Stationary Phases

9.4.5.1 Fast Analysis. By modifying the formulation, the grafting process and reducing the raw EVB/DVB macroporous resin particle size, chromatographic efficiencies were much improved with the IonPac CS12A column. In addition to hydronium-selective carboxylate functional groups, this column also contains phosphonic acid groups. It is compatible with 100% organic solvents, such as acetonitrile, as well as 100% aqueous eluents. The CS12A is a rugged column which can be

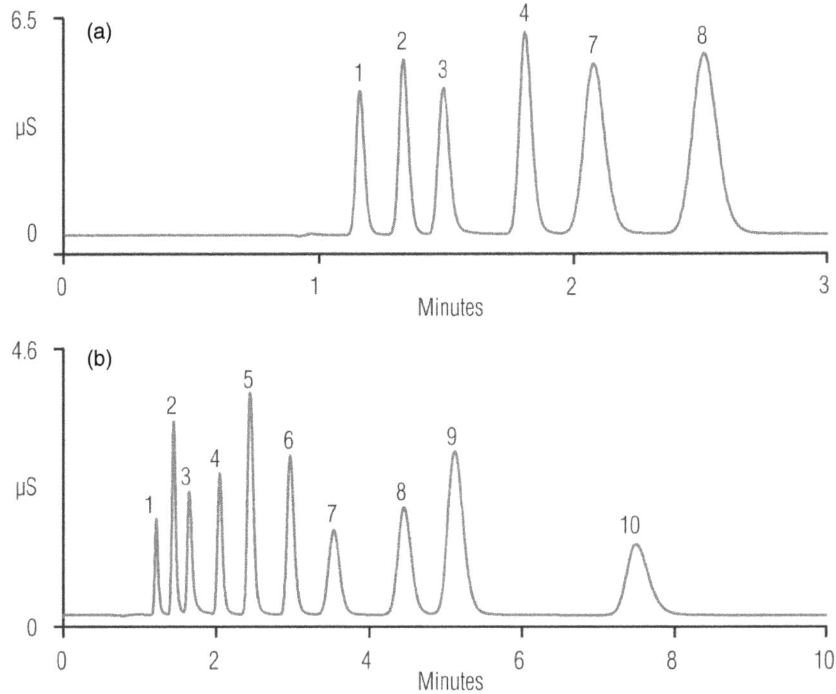

Figure 9.7. Isocratic fast cation separations with the IonPac CS12A column (3 × 150 mm). (a) Injection: 25 µL; Eluent A: 33 mM MSA; Flow rate A: 0.8 mL/min; Temperature: 30°C; Detection: suppressed conductivity. Sample: (1) lithium, (2) sodium, (3) ammonium, (4) potassium, (7) magnesium, (8) calcium. (b) Injection: 25 µL; Eluent B: 20 mM MSA; Flow rate B: 1.0 mL/min; Temperature: 30°C; Detection: suppressed conductivity. Sample: (1) lithium, (2) sodium, (3) ammonium, (4) potassium, (5) rubidium, (6) cesium, (7) magnesium, (8) calcium, (9) strontium, (10) barium.

used up to 60°C and is available in a number of different analytical column formats: 4 × 250 mm, 2 × 250 mm, 3 × 150 mm and 2 × 100 mm. The latter is ideally suited for IC-MS applications. The IonPac CS12A 3 × 150 mm was developed for enhanced speed of analysis. It is based on smaller particle size EVB/DVB 55% crosslinked substrate which provides higher chromatographic efficiencies, allowing the use of a short column format for faster analysis times. As Figure 9.7 illustrates, the six common cations can be determined isocratically within 3 min, and the Group I and Group II cations plus ammonium ion within 8 min.

9.4.5.2 High Loading Capacity. The grafting process was modified with the IonPac CS16 column to allow for a much higher quantity of cation-exchange sites, using the same carboxylic acid functional monomer as the CS12A. The column i.d. was also increased to 5 mm from the typical 4 mm. The resulting column has a cation-exchange capacity of 8400 µeq/column, or about three times

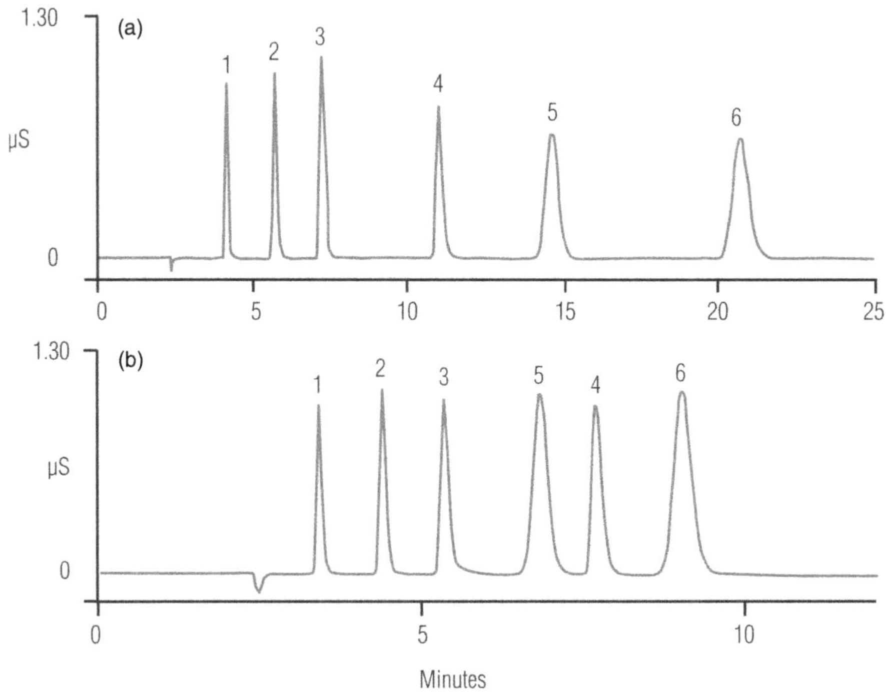

Figure 9.8. High resolving power of the high capacity IonPac CS16 column (5 × 250 mm). (a) Eluent A: 30 mM MSA. (b) Eluent B: 48 mM MSA. The higher eluent concentration allows for much faster elution with still excellent peak resolution. Notice the elution order change for magnesium. Injection: 25 µL; Flow rate: 1.0 mL/min; Temperature: 40°C; Detection: suppressed conductivity. Peaks: (1) lithium (0.1 mg/L), (2) sodium (0.4 mg/L), (3) ammonium (0.5 mg/L), (4) potassium (1.0 mg/L), (5) magnesium (0.5 mg/L), (6) calcium (1.0 mg/L).

the cation-exchange capacity of a 4 × 250 mm CS12A column. The CS16 column has the highest cation-exchange capacity of the columns mentioned in this chapter. Due to its high capacity, it is the best choice for samples with very acidic pH values (as low as pH 0.9) and can tolerate higher analyte concentrations than other columns. It was specifically designed for samples with high ionic matrices and for samples with extreme concentration ratios of cations eluting adjacent to one another. Thus, it can be used for the determination of samples with 10,000:1 ratios of sodium ion concentration to ammonium ion concentration. This column avoids sample preparation that is needed for lower capacity columns. The high resolving power of this column can be seen in the common cation separation shown in Figure 9.8. The IonPac CS16 column can be used with up to 100% organic solvents and at temperatures up to 60°C, which allows the analyst to use temperature and solvent to affect separation.

9.4.5.3 Polyvalent Cations and Hydrophobic Amines. Use of an increased pKa carboxylic acid functional group, a hydrophilic co-monomer, and a primary layer coating with a neutral monomer on a macroporous resin results in

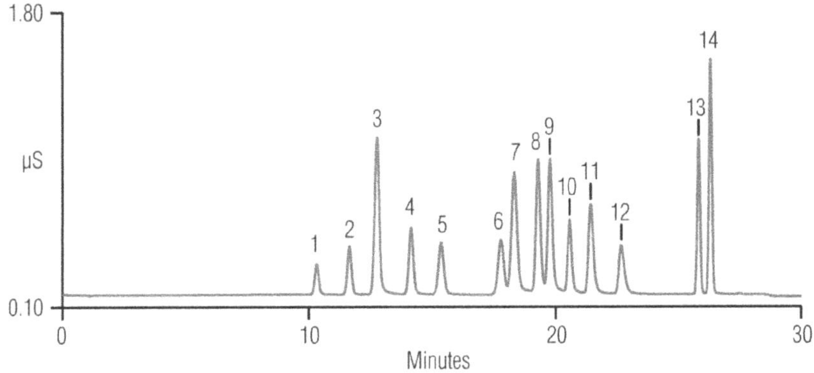

Figure 9.9. Gradient elution (without organic solvent) of linear aliphatic amines and the Group I and Group II cations with the IonPac CS17 column (2 × 250 mm). Injection: 25 µL; Eluent: gradient from 0.5 to 1.1 mM MSA in 12 min, to 1.3 mM MSA at 16 min, to 10 mM MSA at 25 min, isocratic to 27 min, to 0.5 mM MSA at 27.1 min.; Flow rate: 0.40 mL/min; Temperature: 40°C; Detection: suppressed conductivity. Peaks: (1) lithium (0.03 mg/L), (2) sodium (0.10 mg/L), (3) ammonium (0.34 mg/L), (4) potassium (0.25 mg/L), (5) ethylamine (0.50 mg/L), (6) propylamine (0.75 mg/L), (7) tert-butylamine (1.25 mg/L), (8) sec-butylamine (1.25 mg/L), (9) iso-butylamine (1.25 mg/L), (10) n-butylamine (3.75 mg/L), (11) 1,2-dimethylpropylamine (1.00 mg/L), (12) di-n-propylamine (1.00 mg/L), (13) magnesium (0.14 mg/L), (14) calcium (0.34 mg/L). *Note*: the MSA eluent generator with a CR-CTC (continuously recycled cation trap column) was used to provide the almost flat baseline while performing the gradient.

considerably more hydrophilic character for the IonPac CS17 column than with other columns already described. In addition, the CS17 has significantly lower cation-exchange capacity, making it better suited to gradient applications [31]. Relatively hydrophobic analytes can be eluted without the necessity of resorting to the addition of organic solvent to the eluent, making this a more "green" column and less expensive to use (reducing the expense of purchasing and disposal of the organic solvent). In addition, it is always more desirable to use eluents that do not contain organic solvents, as the electrolytic suppressor can then be used in its most convenient "eluent recycle" or AutoSuppression mode [32]. The CS17 is however compatible with 100% of the typical organic solvents such as acetonitrile, acetone, and alcohols without loss of performance. The relatively flat baselines observed with gradient elution (Figures 9.9 and 9.10) are possible due to the very high purity eluent provided by the eluent generator [33], and the continuously regenerated cation trap column (CR-CTC), which removes impurities from the deionized water used to supply the eluent generator, and a clean analytical column.

9.4.5.4 Early-eluting Hydrophilic Amines and Polyvalent Cations. Compared to the Ion Pac CS17 column, the IonPac CS18 column utilizes an alternative primary graft layer and a more hydrophobic co-monomer in the functional layer. The selectivity differences between the two columns are subtle, with the CS18

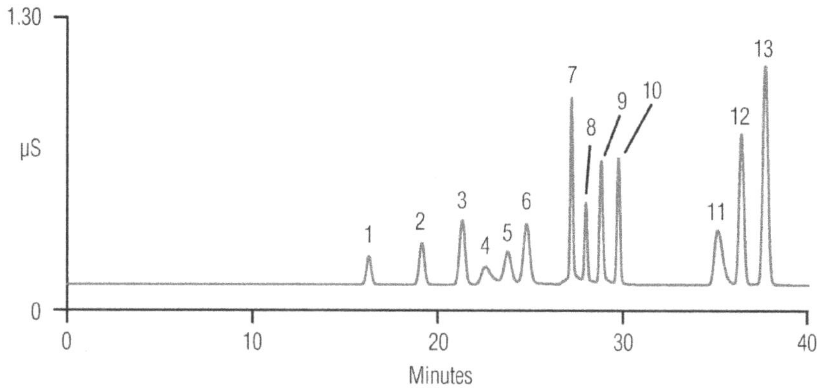

Figure 9.10. Gradient elution of alkanolamines, alicyclic amines and other cations with the IonPac CS17 column (2 × 250 mm). Injection: 25 μL; Eluent: gradient from 0.5 to 0.7 mM MSA in 25 min, to 4 mM MSA at 25.1 min, isocratic to 27 min, to 6.5 mM MSA at 27.1 min, isocratic to 37 min, to 0.5 mM MSA at 37.1 min; Flow rate: 0.35 mL/min; Temperature: 30°C; Detection: suppressed conductivity. Peaks: (1) lithium (0.025 mg/L), (2) sodium (0.10 mg/L), (3) ammonium (0.125 mg/L), (4) hydrazine (1.00 mg/L), (5) ethanolamine (0.250 mg/L), (6) potassium (0.250 mg/L), (7) 2-(2-aminoethoxy)ethanol (1.00 mg/L), (8) 5-amino-1-pentanol (0.500 mg/L), (9) morpholine (1.00 mg/L), (10) 2-diethylamino ethanol (1.00 mg/L), (11) cyclohexylamine (1.00 mg/L), (12) magnesium (0.125 mg/L), (13) calcium (0.250 mg/L). *Note*: the MSA eluent generator with a CR-CTC (continuously regenerated cation trap column) was used to provide the almost flat baseline while performing the gradient.

being a bit more hydrophobic and slightly lower capacity than the CS17. The CS18 is compatible with up to 20% organic solvent (acetonitrile and acetone) and can be used at temperatures up to 50°C. The CS18 offers better resolution among early eluting peaks as is shown in Figure 9.11 (peaks 1–3). This figure also exhibits the greater retention and inferior peak shape of triethylamine, the most hydrophobic of the analytes in the injected sample, with CS18.

Figure 9.12 shows the effect of temperature on the retention and elution of the more hydrophobic cations. Notice that at the lower temperature potassium elutes after dimethylamine, and how triethylamine's chromatographic efficiency and asymmetry are improved by use of elevated temperature. The CS18's superior resolution of early eluting cations is demonstrated in Figure 9.13. Notice also the flat baseline even with gradient elution.

Because most biogenic amines lack a suitable chromophoric or fluorophoric group, methods for their determination required pre- or postcolumn chemical derivatization in the past, leading to labor-intensive methods, long analysis times, and poor reproducibility. Since the introduction of the IonPac CS17, due to its hydrophilicity, selectivity, and relatively low capacity, its use in methods for the separation and detection of biogenic amines with suppressed conductivity detection have been described as "simple and most complete" [34]. The newer IonPac CS18, with hydrophobicity slightly higher than the CS17, provides even better resolution

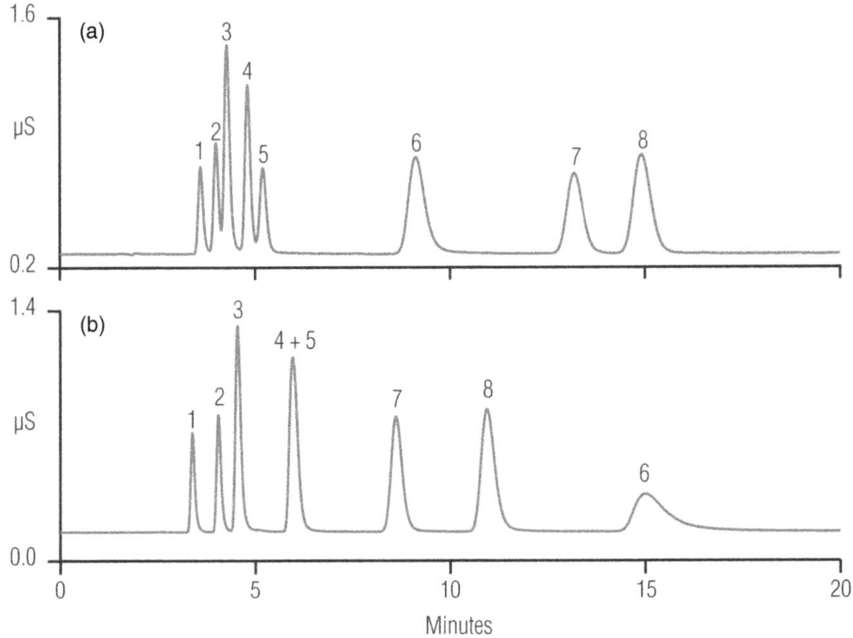

Figure 9.11. Comparison of the IonPac CS17 (2 × 250 mm) with the IonPac CS18 (2 × 250 mm) column. (a) Separator column: IonPac CS17 (2 × 250 mm). (b) Separator column: IonPac CS18 (2 × 250 mm). Injection: 5 μL; Eluent: 6 mM MSA; Flow rate: 0.25 mL/min; Temperature: 30°C; Detection: suppressed conductivity. Peaks: (1) lithium (0.1 mg/L), (2) sodium (0.4 mg/L), (3) ammonium (0.5 mg/L), (4) potassium (1.0 mg/L), (5) dimethylamine (1.0 mg/L), (6) triethylamine (7.0 mg/L), (7) magnesium (0.5 mg/L), (8) calcium (1.0 mg/L).

of the pair: putrescine/cadaverine [35]. Most amino acids are weakly retained on the CS18 column and therefore do not interfere with most of the biogenic amines [36]. This application is shown in Figure 9.14. A simple methanesulfonic acid gradient condition is suitable for use with suppressed conductivity detection, integrated pulsed amperometric detection (IPAD), UV detection, and MS detection, for further examination and characterization of biogenic amines in complex matrices.

The CS18 has the lowest capacity of the polymeric carboxylated columns, and as such can be used in either the suppressed or the nonsuppressed conductivity mode with a 4 mM MSA eluent. In Figure 9.15 we can see a comparison of the polymeric IonPac CS18 and the silica-based IonPac SCS1 both used with nonsuppressed conductivity detection. Due to its low capacity and modest hydrophobicity, polyvalent amines can be separated using a relatively low acid concentration and therefore suppressible eluent (Figure 9.16). As Figure 9.17 illustrates, the slight hydrophobic character of this column allows baseline resolution of alkyl diamines differing in only one carbon chain length. Figure 9.18 shows the versatility of the CS18 column through the separation of Group I and Group II cations, small alkyl amines, alkanolamines, and alicyclic amines with a simple MSA gradient.

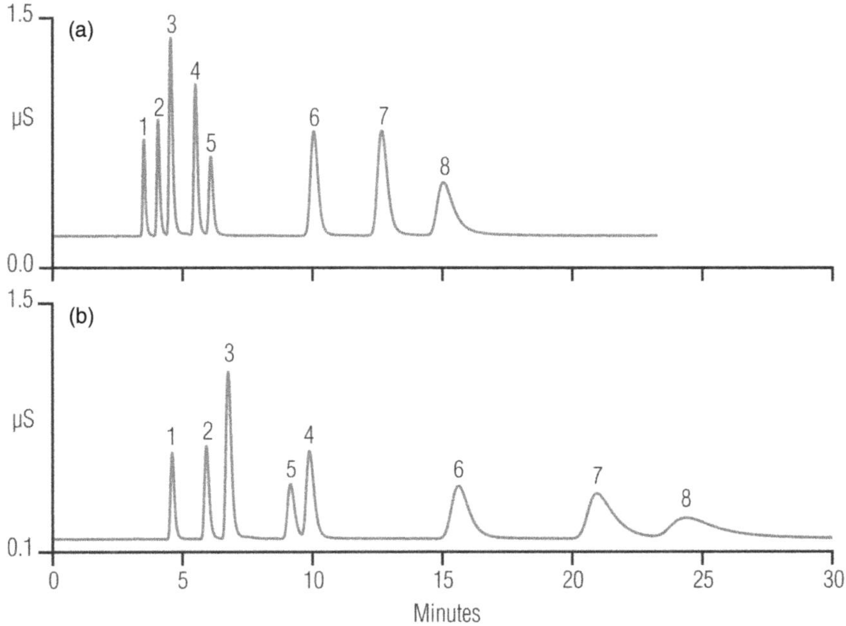

Figure 9.12. Temperature effect on elution of cations with the IonPac CS18 (2 × 250 mm) column. (a) A: Temperature: 50°C; (b) Temperature: 15°C Injection: 5 μL; Eluent: 5 mM MSA; Flow rate A: 0.25 mL/min; Detection: suppressed conductivity. Peaks: (1) lithium (0.1 mg/L), (2) sodium (0.4 mg/L), (3) ammonium (0.5 mg/L), (4) potassium (1.0 mg/L), (5) dimethylamine (1.0 mg/L), (6) triethylamine (7.0 mg/L), (7) magnesium (0.5 mg/L), (8) calcium (1.0 mg/L).

9.4.5.5 *High Chromatographic Performance using Nonsuppressed IC.* The IonPac SCS1 column, similar to the Schomburg column but with superior chromatographic efficiency, consists of a weak carboxylic acid preformed functional polymer that coats a 4.3 μm silica substrate. It offers higher chromatographic efficiency than polymeric-based substrates (better by about a factor of two for monovalents), but as with other silica-based columns, it has the typical pH limitations discussed earlier (should be used between pH 2–7). It was developed for nonsuppressed IC use, therefore has the lowest cation-exchange capacity of the columns mentioned in this chapter. Even though the column is typically used with a 3 mM MSA eluent, it is not recommended to be used with a MSA eluent generator, due to increased noise in the nonsuppressed conductivity mode. The SCS1 is compatible with 100% of typical HPLC organic solvents (acetonitrile, acetone), but alcohols should be avoided as well as prolonged use at temperatures above 35°C. For lowest background noise and drift performance, the column should be thermostated while in use. The SCS1 is recommended when extended calibration linearity for ammonium or weak basic amines is required. An advantage of nonsuppressed IC is that it can be used to analyze certain transition metals (Figure 9.19). However, in most real samples their detection limits can be too high with nonsuppressed IC. Most transition

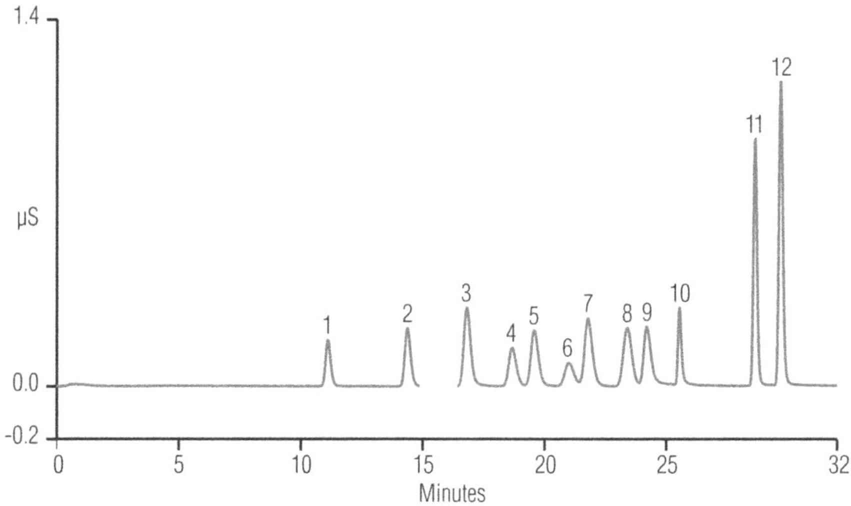

Figure 9.13. Elution of ethanolamines, methylamines and the common six cations on the IonPac CS18 column (2 × 250 mm) Injection: 5 μL; Eluent: gradient from 0.5 to 1.0 mM MSA in 20 min, to 6 mM MSA at 22.4 min, isocratic to 27 min, to 8.0 mM MSA at 28 min, to 0.5 mM MSA at 28.1 min; Flow rate: 0.30 mL/min; Temperature: 35°C; Detection: suppressed conductivity. Peaks: (1) lithium (0.1 mg/L), (2) sodium (0.4 mg/L), (3) ammonium (0.5 mg/L), (4) ethanolamine (1.0 mg/L), (5) methylamine (1.0 mg/L), (6) diethanolamine (2.0 mg/L), (7) potassium (1.0 mg/L), (8) dimethylamine (0.8 mg/L), (9) triethanolamine (9.0 mg/L), (10) trimethylamine (3.0 mg/L), (11) magnesium (0.5 mg/L), (12) calcium (1.0 mg/L). *Note*: the MSA eluent generator with a CR-CTC (continuously regenerated cation trap column) was used to provide the almost flat baseline while performing the gradient.

and lanthanide metals are incompatible with suppressed conductivity detection. Postcolumn reagent absorbance detection is typically used. The Dionex transition metal column IonPac CS5A is optimized for this mode of operation. Applications that require eluent gradients or proportioning from two or more eluent bottles are also not recommended for the SCS1. Without the use of a suppressor, the background shift and the noise will be quite significant, making quantification of the analytes difficult.

9.5 CONCLUSION

In this chapter an overview has been given on the application of IC for the determination of cations of interest in the pharmaceutical industry. The discussion has been focused on analytes which are not well suited for UV/Vis detection, but for which conductivity detection works well. The merits and shortcomings of suppressed and nonsuppressed conductivity have been briefly discussed, as well as the importance of IC-MS. A review of the latest packed-bed polymeric cation-exchange stationary phases for small cation analysis, developed at our company, has been given emphasizing their differences and utilities in the pharmaceutical industry.

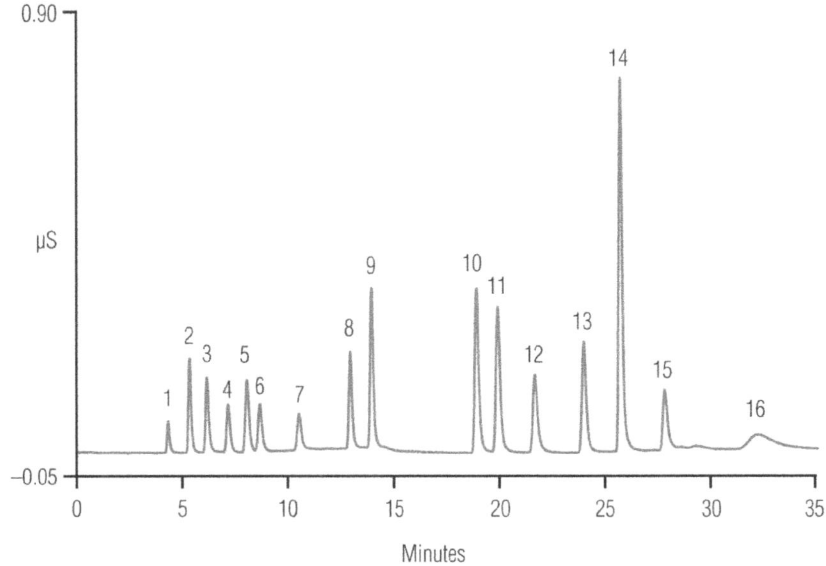

Figure 9.14. Elution of biogenic amines, methylamines and the common six cations on the IonPac CS18 column (2 × 250 mm). Injection: 5 µL; Eluent: 3 mM MSA, isocratic to 5 min, gradient to 18 mM MSA at 20 min, to 45 mM MSA at 25 min, isocratic to 35 min, to 3 mM MSA at 35.1 min; Flow rate: 0.25 mL/min; Temperature: 40°C; Detection: suppressed conductivity Peaks: (1) lithium (0.02 mg/L), (2) sodium (0.08 mg/L), (3) ammonium (0.10 mg/L), (4) methylamine (0.25 mg/L), (5) potassium (0.20 mg/L), (6) dimethylamine (0.20 mg/L), (7) trimethylamine (0.75 mg/L), (8) magnesium (0.10 mg/L), (9) calcium (0.20 mg/L), (10) putrescine (1.10 mg/L), (11) cadaverine (1.10 mg/L), (12) histamine (1.60 mg/L), (13) agmatine (1.40 mg/L), (14) spermidine (4.00 mg/L), (15) spermine (0.40 mg/L), (16) phenethylamine (1.50 mg/L). *Note*: the MSA eluent generator with a CR-CTC (continuously regenerated cation trap column) was used to provide the almost flat baseline while performing the gradient.

9.6 APPENDIX: OTHER PUBLISHED APPLICATIONS WITH THE CATION-EXCHANGE COLUMNS DISCUSSED ABOVE

9.6.1 IonPac CS12A

Ions in Physiological Fluids. Dionex Application Note 107.

One of the most important applications of IC in clinical chemistry is the analysis of physiological fluids, such as urine, plasma, and serum, for inorganic and organic cations. For plasma and serum, acetonitrile deproteinization procedure is used. Some physiological fluid samples may contain organic materials capable of fouling an ion-exchange column, therefore the use of a guard column is strongly recommended for the analysis.

Determination of Choline in Dry Milk and Infant Formula. Dionex Application Note 124.

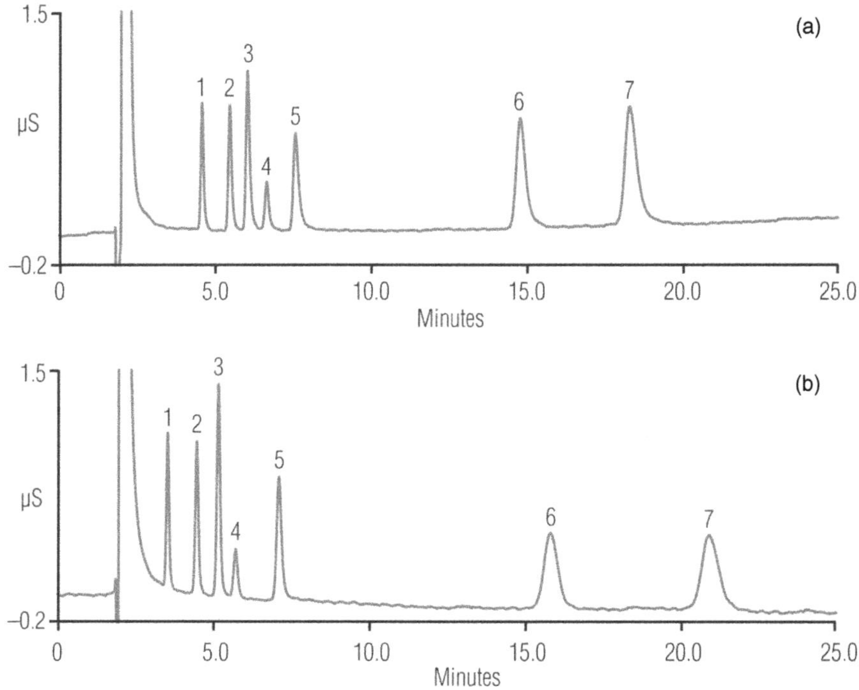

Figure 9.15. Comparison of silica-based IonPac SCS1 column (2 × 250 mm) and polymer-based IonPac CS18 column (2 × 250 mm), both used with nonsuppressed conductivity detection. (a) Separator column: IonPac SCS1 (2 × 250 mm). (b) Separator column: IonPac CS18 (2 × 250 mm). Injection: 5 μL; Eluent: 4 mM MSA; Flow rate: 0.25 mL/min; Temperature: 30°C; Detection: Nonsuppressed conductivity; Chromatography system: Dionex ICS 2000. Peaks: (1) lithium (0.1 mg/L), (2) sodium (0.4 mg/L), (3) ammonium (0.5 mg/L), (4) ethanolamine (0.5 mg/L), (5) potassium (1.0 mg/L), (6) magnesium (0.5 mg/L), (7) calcium (1.0 mg/L).

Choline plays an important role in cardiovascular and liver health and in reproduction and development. It may even help improve memory and physical performance. Choline is essential to proper metabolism, and is therefore added to many food and nutrition sources. The IonPac CS12A column was used for this application which provides accurate results for ppm amounts of choline in milk samples.

9.6.2 IonPac CS16

Determination of Inorganic Cations and Ammonium in Environmental Waters by Ion Chromatography using the IonPac CS16 Column. Dionex Application Note 141.

The high capacity of the IonPac CS16 column enables the analysis of a wide range of environmental waters, and resolves trace ammonium in the presence of a 10,000-fold higher concentration of sodium. The retention time and the peak area precisions are each 1% (RSD) or less for standards in the low mg/L range.

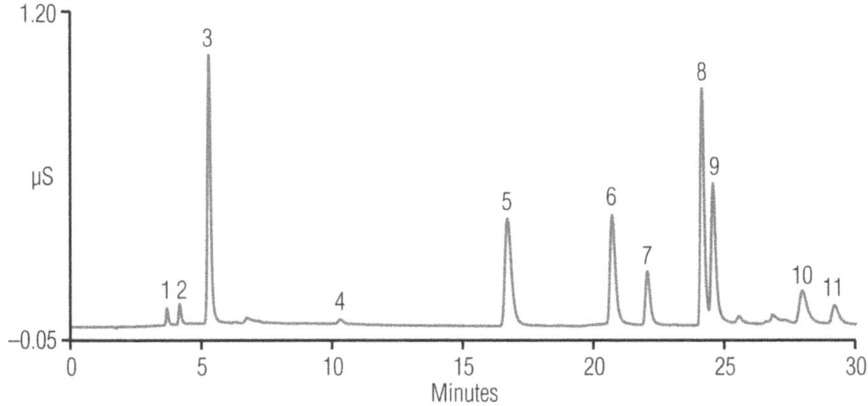

Figure 9.16. Polyamines and the six common cations with the IonPac CS18 column (2 × 250 mm). Injection: 5 μL; Eluent: 3–10 mM MSA gradient in 20 min, gradient to 40 mM MSA at 30 min, gradient to 45 mM MSA at 45 min, to 3 mM MSA at 45.1 min; Flow rate: 0.30 mL/min; Temperature: 40°C; Detection: suppressed conductivity. Peaks: (1) sodium, (2) ammonium, (3) ethylamine (2.0 mg/L), (4) calcium, (5) ethylenediamine (2.0 mg/L), (6) 3-dimethylaminopropylamine (2.4 mg/L), (7) diethylethylenediamine (2.4 mg/L), (8) Bis(2-aminopropyl)amine (2.4 mg/L), (9) diethylenetriamine (2.0 mg/L), (10) triethylenetetramine (2.0 mg/L), (11) paraquat (2.0 mg/L) *Note*: the MSA eluent generator with a CR-CTC (continuously regenerated cation trap column) was used to provide the almost flat baseline while performing the gradient.

9.6.3 IonPac CS17

Li F, Liu HY, et al. Simultaneous determination of dimethylamine, trimethylamine and trimethylamine-N-oxide in aquatic products extracts by ion chromatography with non-suppressed conductivity detection. J. Chromatogr. A 2009; 1216:5924–5926.

IC with suppressed conductivity detection can not detect trimethylamine oxide (TMAO) because it is a predominantly zwitterionic species and thus TMAO will exhibit no conductivity under these conditions.

Closmann F, Nguyen T, Rochelle GT. Methyldiethanolamine/Piperazine as a solvent for CO_2 capture. Energy Procedia 2009;1:1351–1357.

The above solvent blend was investigated as an alternative for CO_2 capture from coal-fired power plants. The IonPac CS17 with suppressed conductivity detection and MSA eluent was used to determine the amounts of methyldiethanolamine and piperazine remaining in each amine sample.

Slingsby R, Rey M. IC-MS for Pharmaceuticals. Paper presented at IC Symposium 2001 [29].

The CS17 was used with MSA eluent and suppressed conductivity detection followed by MS for the detection and confirmation of the presence of cystamine, a small, hydrophobic, disulfide-containing alkyl diamine. Cystamine has been studied for its antioxidant effects and its potential for the treatment of Huntington's disease. It can also be used in the synthesis of a heparin antagonist and fructose bisphosphatase.

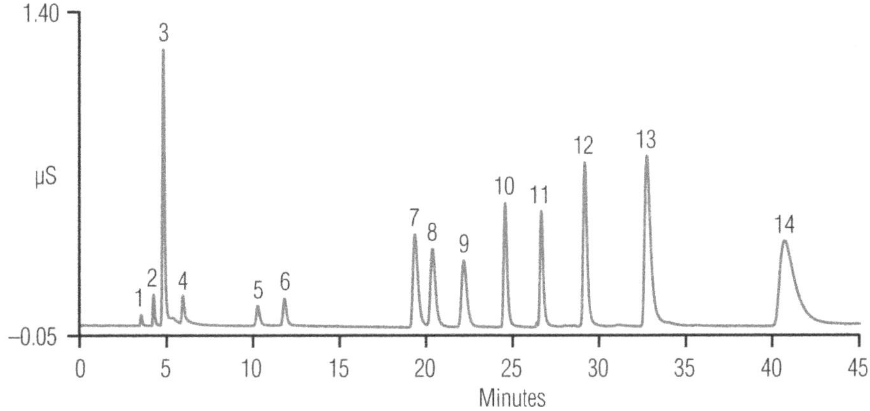

Figure 9.17. Alkyl diamines and the common six cations on the IonPac CS18 column (2 × 250 mm). Injection: 5 μL; Eluent: 3 to 10 mM MSA gradient in 20 min, gradient to 40 mM MSA at 30 min, gradient to 45 mM MSA at 45 min, to 3 mM MSA at 45.1 min; Flow rate: 0.30 mL/min; Temperature: 40°C; Detection: suppressed conductivity. Peaks: (1) lithium (0.02 mg/L), (2) sodium (0.08 mg/L), (3) ammonium, (4) potassium (0.20 mg/L), (5) magnesium (0.10 mg/L), (6) calcium (0.20 mg/L), (7) ethylenediamine (2.0 mg/L), (8) putrescine (2.0 mg/L), (9) cadaverine (2.0 mg/L), (10) 1,6-hexanediamine (2.0 mg/L), (11) 1,7-heptanediamine (2.0 mg/L), (12) 1,8-octanediamine (2.0 mg/L), (13) 1,9-nonanediamine (6.0 mg/L), (14) 1,10-decanediamine (10.0 mg/L). *Note*: the MSA eluent generator with a CR-CTC (continuously regenerated cation trap column) was used to provide the almost flat baseline while performing the gradient.

Determination of Melamine in Milk by Ion Chromatography with UV Detection. Dionex Application Note 231.

While melamine is a cation at neutral pH, it is not fully ionized at pH 7 and therefore, suppressed conductivity does not provide sensitive detection for this compound. In this case, more sensitivity and selectivity was achieved using the CS17 with a 10–30 mM MSA gradient and UV detection at 240 nm. Melamine was found as a contaminant in milk, milk-containing products and pet food, deliberately adulterated in China to increase the apparent protein content. Infant and pet deaths resulted as a consequence of this adulteration. Note that no organic solvent was needed for elution of the melamine from the CS17 column.

Determination of *N*-Methylpyrrolidine in Cefepime using a Reagent-Free Ion Chromatography System. Dionex Application Note 199.

Cephalosporins are curently the most prescribed class of antibiotics for the treatment of bacterial infections. Cefepime is a semi-synthetic, fourth generation cephalosporin used in the treatment of pneumonia and other infections. Cefepime is unstable, degrading to *N*-methylpyrrolidine (NMP), which is potentially toxic to patients. Thus, the determination of NMP in cefepime is critical to assess the purity of the product. The hydrophilicity of the IonPac CS17 column with an MSA gradient and suppressed conductivity detection enabled the fast elution of a previously highly-retained cefepime. The common cations do not interfere with

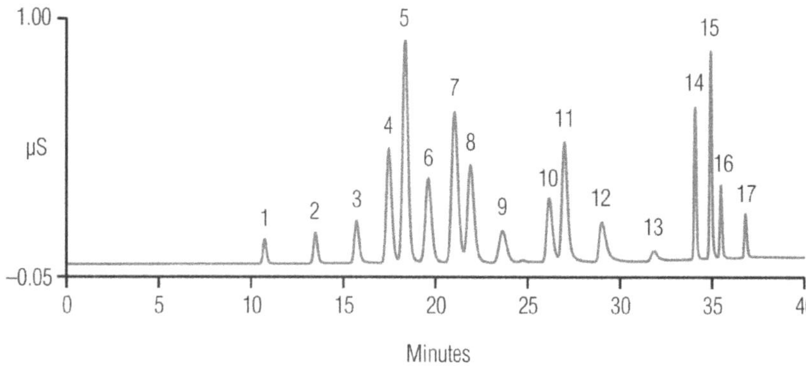

Figure 9.18. Gradient elution of various types of amines and the common six cations on the IonPac CS18 column (2 × 250 mm). Injection: 5 µL; Eluent: gradient from 0.5 to 1.0 mM MSA in 20 min, to 4 mM MSA at 28 min, to 11 mM MSA at 34 min, isocratic to 40 min, to 0.5 mM MSA at 40.1 min; Flow rate: 0.30 mL/min; Temperature: 50°C; Detection: suppressed conductivity Peaks: (1) lithium (0.05 mg/L), (2) sodium (0.2 mg/L), (3) ammonium (0.25 mg/L), (4) ethanolamine (3.0 mg/L), (5) methylamine (3.6 mg/L), (6) potassium (0.5 mg/L), (7) ethylamine (3.0 mg/L), (8) dimethylamine (1.4 mg/L), (9) N-methyldiethanolamine (3.0 mg/L), (10) morpholine (3.2 mg/L), (11) 1-dimethylamino-2-propanol (3.7 mg/L), (12) N-methylmorpholine (7.5 mg/L), (13) butylamine (1.5 mg/L), (14) magnesium (0.25 mg/L), (15) calcium (0.5 mg/L), (16) strontium (0.5 mg/L), (17) barium (0.5 mg/L). *Note*: the MSA eluent generator with a CR-CTC (continuously regenerated cation trap column) was used to provide the almost flat baseline while performing the gradient.

the determination of NMP. The low baseline background and noise enabled the quantification of 0.001% NMP in cefepime hydrochloride.

Determination of Carbachol in Ophthalmic Solutions using a Reagent-Free Ion Chromatography System. Dionex Application Note 194.

Carbachol is a choline ester and a positively charged quaternary ammonium compound used for ophthalmic applications such as glaucoma treatment. Diminished concentrations of carbachol in an ophthalmic formulation may prevent effective reduction of intraocular pressure. In alkaline solutions, carbachol degrades to choline, a member of the B vitamin group. Bethanechol chloride is a quaternary ammonium compound that is structurally and pharmacologically related to carbachol. With the IonPac CS17 and suppressed conductivity detection, carbachol, bethanechol, and choline can be resolved from each other and from the common cations isocratically in 25 min. The hydrolysis product of bethanechol under basic conditions is 2-hydroxypropyltrimethyl ammonium chloride, and this was found to be also resolved from the other analytes of interest and did not interfere.

9.6.4 IonPac SCS1

Johns C, Shellie RA, Potter OG, O'Reilly JW, Hutchinson JP, Guijt RM, Breadmore MC, Hilder EF, Dicinoski GW, Haddad PR. Identification of homemade inorganic

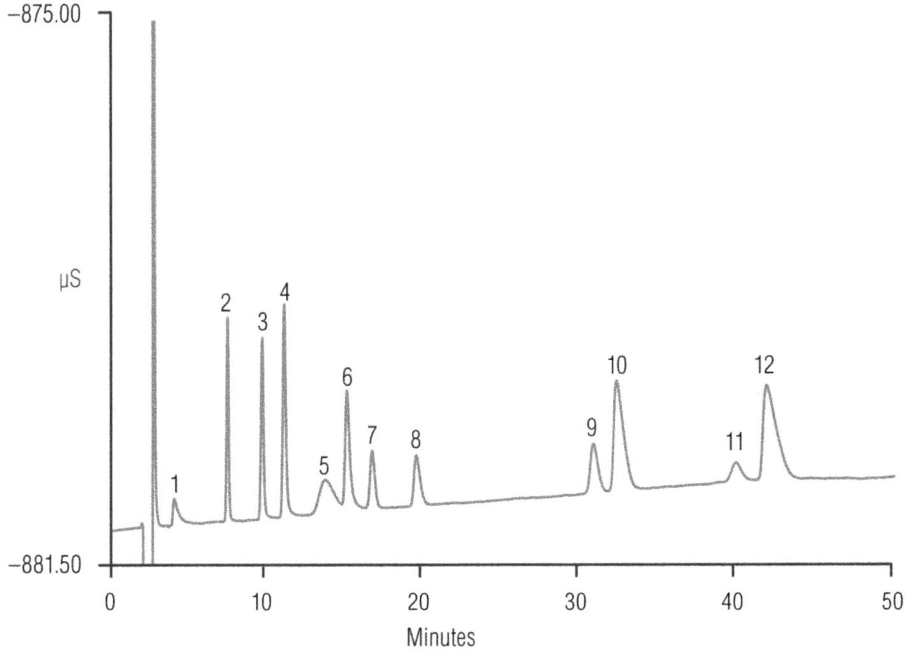

Figure 9.19. Separation of transition metals and the common six cations on the IonPac SCS1 column with nonsuppressed conductivity detection. Columns: SCG 1 (4 × 50 mm) guard and SCS 1 (4 × 250 mm) separator; Injection: 25 µL; Eluent: 4 mM tartaric acid/2 mM oxalic acid; Flow rate: 1.0 mL/min; Temperature: 30°C; Detection: nonsuppressed conductivity. Peaks: (1) copper (10 mg/L), (2) lithium (0.5 mg/L), (3) sodium (2.0 mg/L), (4) ammonium (2.5 mg/L), (5) nickel (10 mg/L), (6) potassium (5.0 mg/L), (7) zinc (10 mg/L), (8) cobalt (10 mg/L), (9) manganese (10 mg/L), (10) magnesium (2.5 mg/L), (11) cadmium (10 mg/L), (12) calcium (5.0 mg/L).

explosives by ion chromatographic analysis of post-blast residues. J. Chromatogr. A 2008;1182:205–214.

Target cations comprised ammonium, methylammonium, ethylammonium, potassium, sodium, barium, strontium, magnesium, manganese, calcium, zinc, chromium (III) and lead (II). Suppressed conductivity detection is not applicable for all analyte cations, as lead (II) and zinc (II) are prone to precipitation in the suppressor. Therefore, the IonPac SCS1 column was used for this application with nonsuppressed conductivity detection.

REFERENCES

1. Brittain HG. Developing an appropriate salt form for an active pharmaceutical ingredient. APR 2009; November/December: 62–65.
2. Huang Z, Richards MA, Zha Y, Francis R, Lozano R, Ruan J. Determination of inorganic pharmaceutical counterions using hydrophilic interaction chromatography coupled with a Corona® CAD detector. J. Pharm. Biomed. Anal. 2009;50:809–814.

3. Liu DQ, Sun M, Kord AS. Recent advances in trace analysis of pharmaceutical genotoxic impurities. J. Pharm. Biomed. Anal. 2010;51(5):999–1014.
4. Gamache PH, McCarthy RS, Freeto SM, Asa DJ, Woodcock MJ, Laws K, Cole RO. HPLC analysis of nonvolatile analytes using charged aerosol detection. LCGC North America 2005;23(2):151–161.
5. Vervoort N, Daemen D, Torok G. Performance evaluation of evaporative light scattering detection and charged aerosol detection in reversed phase liquid chromatography. J. Chromatogr. A 2008;1189:92–100.
6. Hazotte A, Libong D, Matoga M, Chaminade P. Comparison of universal detectors for high-temperature micro liquid chromatography. J. Chromatogr. A 2007;1170:52–61.
7. Mitchell CR, Bao Y, Benz NJ, Zhang S. Comparison of the sensitivity of evaporative universal detectors and LC/MS in the HILIC and the reversed-phase HPLC modes. J. Chromatogr. B 2009;877:4133–4139.
8. Prokai AM, Ravichandran RK. Simultaneous analysis of hydroxylamine, N-methylhydroxylamine and N,N-dimethylhydroxylamine by ion chromatography. J. Chromatogr. A 1994;667:298–303.
9. Burger A, Standridge RT, Stjernstrom NE, Marchini P. Alicyclic alkylamines and alkanolamines. J. Med.Chem. 1961;4(3):517–562.
10. Jagota NK, Chetram AJ, Nair JB. Ion chromatography of amylamine and tert-butylamine in pharmaceuticals. J. Chromatogr. A 1996;739:343–349.
11. Boncher T, Bi X, Varghese S, Casero Jr RA, Woster PM. Polyamine-based analogues as biochemical probes and potential therapeutics. Biochemical Society Transactions 2007;35, part 2: 356–363.
12. Dale MM, Ladd R. A comparison of dimaprit, nordimaprit, methylamine and chloroquine as inhibitors of mitogen-induced lymphocyte activation. Br J Pharmacol. 1984;83(1):293–298.
13. Zhang T, Watson DG, Lu D, Carr D, Trager L. Determination of cross-linking residues in a pharmaceutical polymer by liquid chromatography-high resolution full scan mass spectrometry. Talanta 2008;76(3):509–512.
14. Kotti SRSS, Timmons C, Li G. Vicinal diamino functionalities as privileged structural elements in biologically active compounds and exploitation of their synthetic chemistry. Chemical Biology & Drug Design 2006;67:101–114.
15. Neises B, Wehner V, Stilz HU. inventors; Aventis Pharma Deutschland GmbHUS, assignee. Imidazolidine derivatives, the production thereof, their use and pharmaceutical preparations containing the same. US Patent 6667334. 2003 Dec 23.
16. Nelson EL. inventor; Nelson Research & Development Company (United States of America), assignee. Topical triethylenetetramine-containing pharmaceutical compositions and methods of use. Canadian Patent Number CA 1146070. 1983 May 10.
17. Hofer KG, Yang L. inventors; Florida State University, assignee. Radiosensitizing diamines and their pharmaceutical preparations, US Patent 5700825. 1997 Dec 23.
18. Blower PJ. Inorganic pharmaceuticals. Annu. Rep. Prog. Chem., Sect A 2001;97:587–603.
19. Glynn IM, Richards DE. Occlusion of rubidium ions by the sodium-potassium pump: its implications for the mechanism of potassium transport. J. Physiology 1982;330:17–43.
20. Tsouderos Y, Deloffre P, Wierzbicki M. inventors; Adir et Compagnie, assignee. Use of strontium salts and pharmaceutical compositions thereof for the treatment of arthrosis. US Patent 5856356. 1999 Jan 5.

21. Paulekuhn GS, Dressman JB, Saal C Trends in active pharmaceutical ingredient salt selection based on analysis of the orange book database. J. Med. Chem. 2007;50(26): 6665–6672.
22. Keast RSJ, Breslin PAS. Modifying the bitterness of selected oral pharmaceuticals with cation and anion series of salts. Pharmaceutical Research 2002;19(7):1019–1026.
23. Small H, Stevens TS, Baumann WC. Novel ion exchange chromatographic method using conductimetric detection. Anal. Chem. 1975;47:1801–1809.
24. Gjerde DT, Fritz JS, Schmuckler G. Anion Chromatography with Low-Conductivity Eluents. J. Chromatogr. 1979;186:509–519.
25. Weiss J. Ion Chromatography. Second Edition; ISBN 3-527-28698-5.
26. Kolla P, Kohler J, Schomburg G. Polymer-coated cation-exchange stationary phases on the basis of silica. Chromatographia 1987;23:465–472.
27. Rey M, Bordunov A, De Borba B, Slingsby R, Pohl C. Cation Exchange Chromatography with Conductivity Detection: Suppressed or Nonsuppressed ? Poster presented at IC Symposium 2003;San Diego, CA.
28. Comparison of Suppressed to Nonsuppressed Conductivity Detection for the Determination of Common Inorganic Cations. Dionex Application Note 157.
29. Slingsby R, Rey M. IC-MS for Pharmaceuticals. Paper presented at IC Symposium 2001; Chicago, IL.
30. Saccani G, Tanzi E, Pastore P, Cavalli S, Rey M. Determination of biogenic amines in fresh and processed meat by suppressed ion chromatography-mass spectrometry using a cation-exchange column. J. Chromatogr. A 2005;1082:43–50.
31. Rey M, Pohl C. Novel cation-exchange column for the separation of hydrophobic and/or polyvalent amines. J. Chromatogr. A 2003;997:199–206.
32. Rabin S, Stillian J, Barreto V, Friedman K, Toofan M. New membrane-based electrolytic suppressor device for suppressed conductivity detection in ion chromatography. J. Chromatogr. 1993;640:97–109.
33. Liu Y, Avdalovic N, Pohl C, Matt R, Dhillon H, Kiser R. An on-line, high purity acid and base eluent generation system for ion chromatography. Am. Lab. 1998; November: 48C–54C.
34. Cinquina AL, Cali A, Longo F, De Santis L, Severoni A, Abballe F. J. Determination of biogenic amines in fish tissues. Chromatogr. A 2004;1032:73–77.
35. De Borba BM, Rohrer JS. Determination of Biogenic Amines in Alcoholic Beverages by Ion Chromatography with Suppressed Conductivity Detection and Integrated Pulsed Amperometric Detection. J. Chromatogr. A 2007;1155:22–30.
36. Determination of Biogenic Amines in Fermented and Non-Fermented Foods using Ion Chromatography with Suppressed Conductivity and Integrated Pulsed Amperometric Detections. Dionex Application Note 183.
37. Clarke AP, Jandik P, Rocklin RD, Liu Y, Avdalovic N. An integrated amperometry waveform for the direct, sensitive detection of amino acids and amino sugars following anion-exchange chromatography. Anal. Chem. 1999;71:2774–2781.

10

COMPREHENSIVE APPROACHES FOR MEASUREMENT OF ACTIVE PHARMACEUTICAL INGREDIENTS, COUNTER-IONS, AND EXCIPIENTS USING HPLC WITH CHARGED AEROSOL DETECTION

Christopher Crafts, Bruce Bailey, Paul Gamache, Xiaodong Liu, and Ian Acworth

Thermo Fisher Scientific, Chelmsford, MA

10.1 INTRODUCTION

Optimizing a pharmaceutical formulation is a significant aspect of the drug development process. For an active pharmaceutical ingredient (API) with ionizable functional groups, appropriate choice of salt forms is critical to improve the physicochemical properties including aqueous solubility, hygroscopicity, solution pH, melting point, dissolution rate, chemical stability, crystal form, and mechanical properties. Judicious salt selection provides the pharmaceutical chemists and formulation scientists an opportunity to overcome undesirable characteristics without changing its chemical structure. Proper salt and excipient selection can thus permit the development of dosage forms with the desired bioavailability, stability, manufacturability, and patient acceptance. A sound appreciation of the influence of excipient and salt forms on the properties of active ingredients must be obtained early in the process to help ensure selection of the best formulation and to prevent costly reformulation efforts. Chromatographic methods are widely used in the salt and excipient selection process and in monitoring the quality of the formulation throughout the development process. Given

Applications of Ion Chromatography for Pharmaceutical and Biological Products, First Edition.
Edited by Lokesh Bhattacharyya and Jeffrey S. Rohrer.

the wide diversity of substances that must be measured and, often, the limited quantity of available material, there exists a need for methods that allow both comprehensive measurement and acceptable analytical figures of merit.

Today, approximately 50% of all drug molecules used in medicinal therapy are administered as salts. Although a number of counter-ions can be used, the top ten salt forms in the USP29-NF24 (2006) were hydrochloride, sulfate, acetate, phosphate, chloride, citrate, maleate, sodium, potassium, and calcium [1]. The most common salt forms in the Cambridge Structural Database (CSD) were chloride, bromide, nitrate, ammonium, sulfate, tosylate, phosphate, L-tartrate, ethylenediamine, and maleate [2]. Due to their wide diversity in chemical properties (e.g., charge, hydrophobicity, presence of a chromophore), APIs and counter-ions are usually analyzed by various chromatographic methods that require different separation columns and/or different detection techniques. In this chapter we (a) explore how new column technologies, including the Acclaim® Trinity P1 column, can be used in conjunction with charged aerosol detection (CAD) to obtain a more comprehensive analysis of APIs, and their corresponding counter-ions (nonvolatile); (b) show that the sensitivity (subnanogram) and wide dynamic range (four+ orders of magnitude) of CAD can make the measurement of trace impurities to $<0.1\%$ possible; (c) illustrate that as these methods use volatile mobile phases, a mass spectrometer can be used in parallel with CAD for more in-depth characterization of the sample; and finally (d) discuss how columns can be combined serially to extend the range of analytes measured. Before proceeding, however, we will first briefly review some of the new columns that make simultaneous analysis of APIs and their counter-ions possible and then discuss the advantages of CAD over other "universal" detectors.

10.2 SIMULTANEOUS MEASUREMENT OF APIS AND COUNTER-IONS

10.2.1 Novel Columns

Due to the complex nature of many drug formulations (e.g., they may contain anions, cations, and neutral species) complete separation and characterization of components in a single analysis can be challenging. However, the simultaneous measurement of anions, cations, and neutral species is possible when using a trimodal stationary phase that can provide cation-exchange, anion-exchange, and reversed-phase retention mechanisms. Recently, a trimodal column based on Nanopolymer Silica Hybrid (NSH) technology was commercialized by Dionex, Inc. under the trade name of Acclaim Trinity™ P1 [3]. The column packing material consists of high-purity porous spherical silica gel coated with charged nanopolymer particles. The inner-pore area of the silica gel is modified with an organic layer that provides both reversed-phase and anion-exchange properties. The outer surface is modified with cation-exchange functionality. This chemistry arrangement ensures distinctive spatial separation of the anion-exchange and cation-exchange regions, which allows both retention mechanisms to function simultaneously and be controlled independently. Other stationary phases with both anion-exchange and cation-exchange functional groups have been used for simultaneous separation of anions and cations. This includes a zwitterionic column (SeQuant ZIC-HILIC by Merck) [4] and two amphoteric columns (Obelisc R or Obelisc N by SIELC) [5]. A disadvantage of these columns is that the functional groups may neutralize one another, due to their proximity, if both functional groups are ionized. Thus, they behave as a salt exchange material rather than an anion-exchanger and a cation-exchanger when both functional groups are ionized. As a result, method development is not straightforward. In addition, the SeQuant HILIC and Obelisc N can only be used in HILIC mode which requires high organic solvent, and the resultant lack of reversed-phase retention makes them unsuitable for analyzing hydrophobic APIs.

Another commercially available column with anion-exchange, cation-exchange, and reversed-phase properties (Scherzo SM-C18 by Imtakt) [6] is based on blending two different packing materials—one part RP/anion-exchange mixed-mode particles and one part RP/cation-exchange mixed-mode particles. Each of these two packings consists of covalently bonded silica gel, modified in the bonding step by a mixture of C18 and corresponding ion-exchange bearing silyl ligands. Because it is difficult to control the ratio of all three functional groups during the bonding process and because there are differences in the hydrolytic stability of these ligands, severe selectivity drifting and manufacturing reproducibility are major concerns.

10.2.2 "Universal" Detection

While the most commonly used detector for HPLC is based on UV absorbance, analytes must possess a sufficiently strong chromophore to be measured. For substances that lack or have only a weak chromophore (e.g., some APIs, many excipients, and most counter-ions), other complementary detectors must be used. Currently, two aerosol-based mass-dependent detection techniques are typically used to measure such analytes: evaporative light scattering detection (ELSD) and CAD.

ELSD is considered a mature technique and has been commercially available from several vendors for many years. The principles of ELSD are well documented and have been covered extensively elsewhere ([7] and references therein). Although ELSD enables the measurement of most nonvolatile analytes, it suffers from a number of limitations including: poor sensitivity (Figure 10.1; Table 10.1), a complex non-linear response, narrow dynamic range, and wide response variability between analytes (Figure 10.2) [8–13]. As a result, method development and calibration can be complex. ELSD has been used to measure anions [14,15], and for the simultaneous determination of inorganic anions and cations [16]. ELSD has also been used to detect and quantify counter-ions from pharmaceutical salt forms [17–19]. More recently, Risley and Pack used HILIC-ELSD for the simultaneous determination of cations and anions, as well as APIs and counter-ions [10].

CAD was introduced in 2005 and the principle of operation has been described in the literature [20–23]. CAD has been shown to overcome many of the issues found with ELSD. CAD is more sensitive (subnanogram levels), has a wider dynamic range (over four orders of magnitude), better precision and, although response is non-linear, calibration is less complex than with ELSD. Furthermore, CAD response among

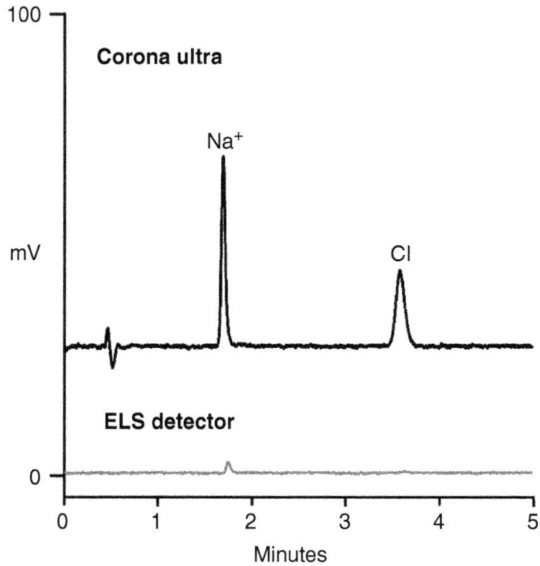

Figure 10.1. Separation of sodium and chloride (5 ng on column) using an Acclaim Trinity P1 3 μm 3.0 × 50 mm column with flow at 0.6 mL/min. Analytes were measured using either a charged aerosol detector (Corona *ultra*) or ELSD (Sedex-85). Mobile phase: 60/40 v/v acetonitrile /15 mM ammonium acetate buffer, pH = 5.2. Corona *ultra* settings: Nebulizer T−35°C, Range −100 pA full scale, Filter - Off. Sedex-85 settings: Evaporation T − 45°C; Gain - 10; Filter - Off.

TABLE 10.1. Signal to Noise Ratio and Calculated Limits of Detection for Sodium and Chloride Using the Conditions Described in Figure 10.1 (based on a signal/noise of 3:1).

Detector	Na^+ (S/N)	Na^+ (LOD)	Cl^- (S/N)	Cl^- (LOD)
Corona *ultra*	71	0.2 ng	27	0.9 ng
Sedex-85	13	1.2 ng	2	>11 ng

analytes has been found to be more uniform than with ELSD (i.e. less dependence on analyte nature, Figure 10.2) and is far simpler than ELSD to operate [11,12,24–29]. For these reasons, CAD is being implemented throughout the pharmaceutical and biopharmaceutical industry from discovery, through research and development to manufacturing quality control (QC). Some examples include the use of CAD for compound library management and in support of compound synthesis [26,30–32], detection of impurities, including potential genotoxins [22,29,33,34], QC of surfactants and excipients [1,24,35,36], detection of extractables and leachables [37], drug stability and mass balance studies [38,39], measurement of PEGylated biomolecules [40], and manufacturing cleaning validation and QC [41]. CAD in combination with HILIC has been used for the determination of anions [21], inorganic pharmaceutical counter-ions [24,42], and the simultaneous analysis of cations and anions [24,42–44].

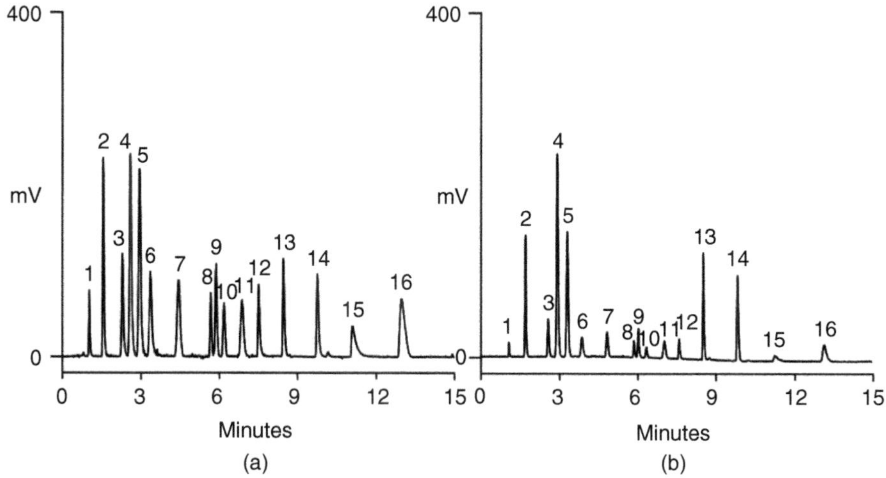

Figure 10.2. Separation of 15 pharmaceutical counter-ions with detection by Corona *ultra* (b) or ELS (a). Analytes were eluted using a ternary gradient with an Acclaim Trinity P1 3 μm 3.0 × 50 mm with flow at 0.6 mL/min. Column temperature: 30°C. Mobile Phase A: acetonitrile; Mobile Phase B: deionized water; Mobile Phase C: 200 mM ammonium acetate pH = 4.0. Gradient (time mins, %B, %C): 0, 35, 5; 2, 35, 5; 7, 0, 90; 15, 0, 90 with a total flow of 0.5 mL/min 10 minute pre-injection equilibration. Detector settings are same as in Figure 1. Peak Identification: 1. procaine, 2. choline, 3. tris, 4. sodium, 5. potassium, 6. meglumine, 7. mesylate, 8. maleate, 9. chloride, 10. bromide, 11. iodide, 12. phosphate, 13. malate, 14. tartrate, 15. citrate, 16. sulfate.

The simultaneous measurement of APIs, counter-ions and impurities has been demonstrated using either HILIC [24,42,44] or mixed-mode [39] separation with CAD.

10.2.3 Analysis of APIs and Counter-ions

The introduction of the Acclaim Trinity P1 column provides the analyst with the capability of resolving a broad range of hydrophobic and hydrophilic APIs that possess either acidic or basic functional groups along with their respective positive or negative counter-ions. This NSH column technology includes reversed-phase, anion-exchange, and cation-exchange functionalities that enable great chromatographic flexibility and selectivity by simply controlling the mobile phase composition. For example, the separation of the hydrophilic and basic API, 1,1-dimethylbiguanide, from its chloride counter-ion can be achieved using a mobile phase with a low (20%) organic content (Figure 10.3). Increasing the organic content to 75% enables the separation of more hydrophobic and acidic APIs, naproxen and diclofenac, along with their sodium counter-ion (Figures 10.4 and 10.5). The impact of changing the pH of the mobile on analyte retention is illustrated in Figure 10.4, whereas the effect of buffer strength is shown in Figure 10.5.

To illustrate a practical use of our approach, the Acclaim Trinity P1 column with CAD was used to determine the sodium content of two common pharmaceutical salts.

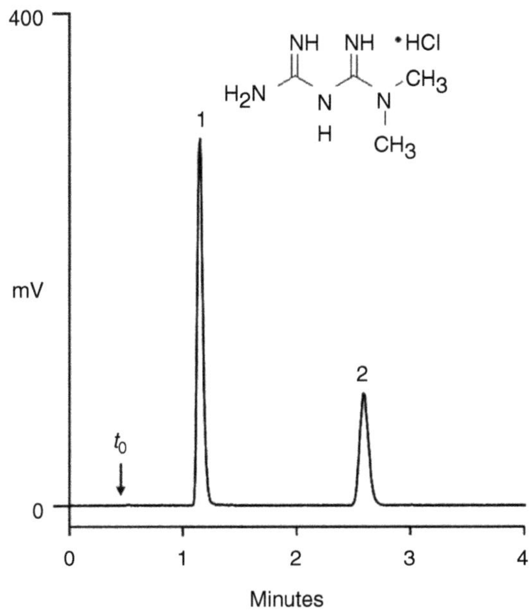

Figure 10.3. Analysis of 1,1-dimethylbiguanide hydrochloride (400 ng on column) separated using an Acclaim Trinity P1, 3 μm 3.0 × 50 mm with flow at 0.6 mL/min. Mobile phase 20/80 v/v acetonitrile/40 mM ammonium acetate, pH5.2. Detector settings: Corona *ultra* Gain = 100 pA; Filter = med; Neb Temp = 30°C. Peak identification: 1. 1,1-dimethylbiguanide, 2. chloride.

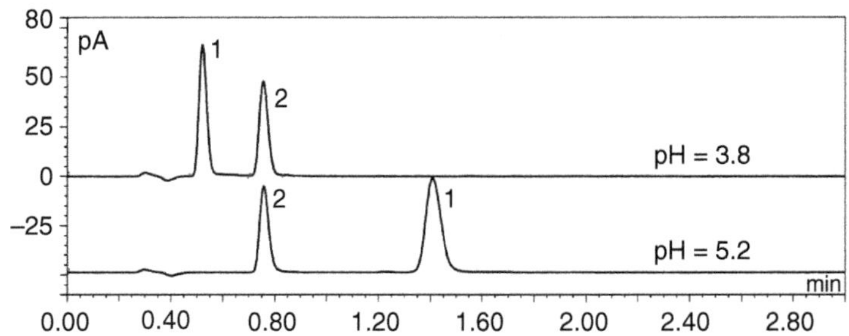

Figure 10.4. Overlay of chromatograms of naproxen sodium salt (~500 ng on column) separated using an Acclaim Trinity P1, 3 μm 3.0 × 50 mm flow at 1.0 mL/min. Mobile phase 75/25 v/v acetonitrile/120 mM ammonium acetate (30 mM total buffer strength) pH adjusted to 3.8 and 5.2. Detector settings: Corona *ultra*; Filter = med; Neb Temp = 25°C. Peak Identification: 1. naproxen, 2. sodium.

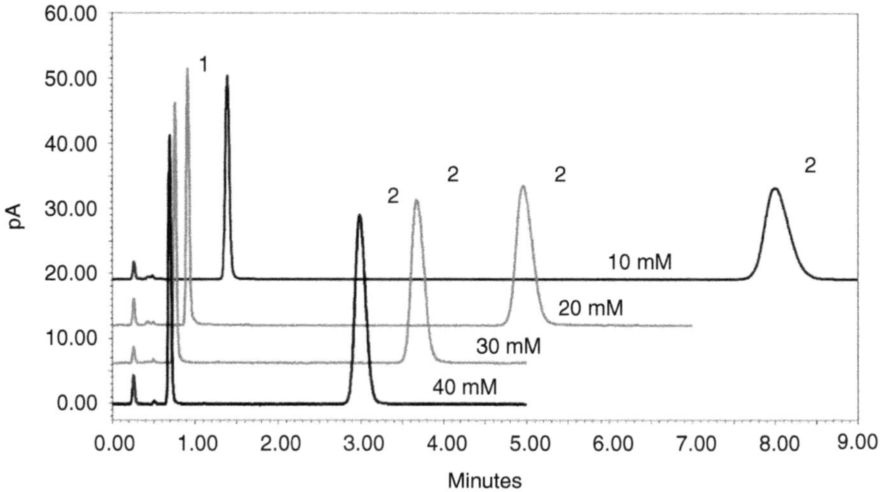

Figure 10.5. Overlay of chromatograms of diclofenac sodium salt (~500 ng on column) separated using an Acclaim Trinity P1, 3 μm 3.0 × 50 mm at 1.0 mL/min. Mobile phase 75/25 v/v acetonitrile/ ammonium acetate pH = 4.8, buffer strength adjusted to 10, 20, 30, and 40 mM total buffer strength. Detector settings Corona *ultra*; Filter = med; Neb Temp = 25°C. Peak Identification; 1. sodium, 2. diclofenac.

TABLE 10.2. Calculated % Sodium using Either a Seven Point Linear Curve or a 12 Point Polynomial Curve (Described in Figures 10.6 and 10.7, Respectively). Theoretical Values and Calculated Weight/Weight Percent are also Presented.

	% Sodium Content		
	Theoretical	Linear Calculation	Full Curve Calculation
Naproxen Sodium Salt	9.12%	9.14%	8.89%
Diclofenac Sodium Salt	7.23%	7.34%	7.02%

Naproxen sodium (Figure 10.4) and diclofenac sodium (Figure 10.5) were analyzed using the same chromatographic conditions (see figure legends for details). Depending on the required accuracy and concentration range, analyte quantification was evaluated using two approaches: linear regression and a non-linear fit. Although the CAD is non-linear over its full range, linear calibration can typically be employed for quantitation over a targeted range. For example, a linear fit for seven levels of sodium acetate is presented in Figure 10.6. Standard levels were chosen to comprise a relatively narrow range encompassing the expected sodium levels in both pharmaceutical samples. As shown in Table 10.2, the sodium content determined by linear calibration was in good agreement with the theoretical values. In cases where quantitation is required over a wider concentration range, a non-linear calibration fit can be used. Figure 10.7 shows an example of a non-linear calibration using 12 levels of sodium acetate spanning

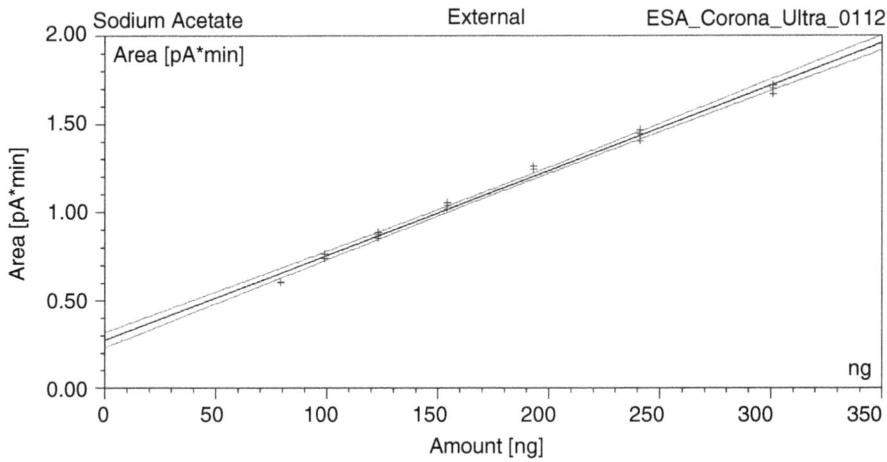

Figure 10.6. Sodium acetate from 80 to 300 ng on column (7 points with 3 injections at each point), fit with linear regression ($R^2 = 0.994$). Upper and lower 95% confidence limits are displayed.

four orders of magnitude in concentration. Over this range, a best fit was found using a quadratic calibration with mass-on-column on the y-axis and response (area) on the X-axis [39]. In this case, the full range experimental values were within 3% of the theoretical values while the linear fit (Figure 10.6) yielded results closer to the theoretical values (Table 10.2).

10.2.4 Determination of Trace Impurities

Limit testing of impurities requires accurate quantitation at low levels to ensure product safety and efficacy. Determination of impurities present at the 0.1% level or lower is often required by the pharmaceutical industry. The dynamic range and sensitivity of the CAD detector enables the simultaneous analysis of an API, its counter-ion, and trace impurities. To evaluate the ability of CAD to distinguish impurity levels around the 0.1% limit, a sample solution of diclofenac sodium was mixed with dilute solutions of ammonium chloride to achieve 0.1% and 0.15% (w/w) chloride 'spiked' levels. The chromatograms presented in Figure 10.8 clearly illustrate that chloride is well resolved from both sodium and diclofenac and that impurity levels below 0.1% can readily be determined. This approach was sufficiently sensitive to even measure contaminant chloride levels in unspiked samples. Table 10.3 shows the mean and relative standard deviation of the peak areas obtained with CAD for five analyses each of the spiked and unspiked diclofenac sodium. The precision obtained illustrates the ability of this analytical method to distinguish between 0.1% and 0.15% chloride levels. The table also shows the weight percent of chloride in each sample calculated with a four point linear regression curve from 3 to 25 ng chloride on column.

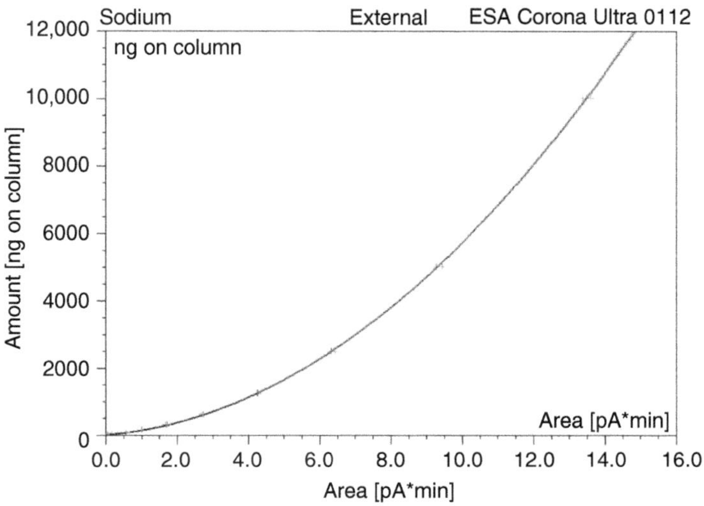

Figure 10.7. Sodium acetate from 5 ng to 10 μg on column (12 points with three injections at each point) fit with a polynomial and inversed X and Y axis ($R^2 = 0.9997$). Upper and lower 95% confidence limits are displayed.

TABLE 10.3. Mean and Relative Standard Deviation of the Areas of Chloride Peaks. The Calculated Weight Percent of Chloride in Diclofenac Sodium Samples is also Presented.

Sample	Raw Area (pA* min)	RSD ($n = 5$)	Calculated Chloride (w/w)
Diclofenac Standard	0.0231	13.01%	0.03%
Diclofenac + 0.1% Cl−	0.1069	2.39%	0.14%
Diclofenac + 0.15% Cl−	0.1318	1.49%	0.17%

10.3 ORTHOGONAL MASS SPECTROMETRY DETECTION

The mobile phases used with CAD are volatile and are therefore compatible with mass spectrometry (MS) detection. The incorporation of charged aerosol, UV and MS detectors into a single platform (Figure 10.9) enables more complete characterization of a sample [45]. For example, Figure 10.10 shows the analysis of 5,7-dihydroxytryptamine (DHT) creatinine sulfate (see ESA Application Note 70-8376 for other examples [46]). Using this approach all nonvolatile compounds can be measured by CAD; both volatile and nonvolatile compounds will be determined by UV detection as long as the compound contains a suitable chromophore; MS can be used to obtain mass/charge information for species that ionize.

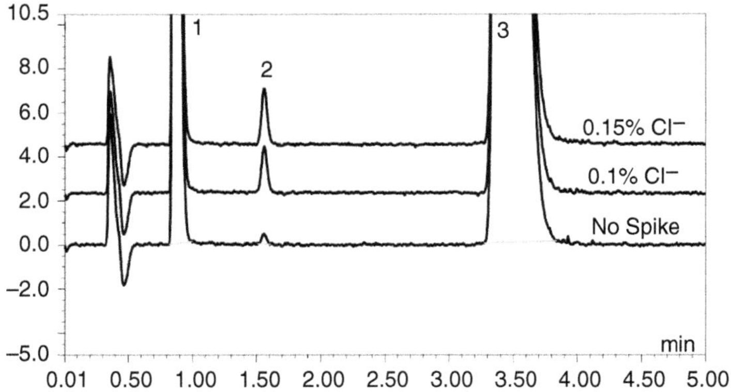

Figure 10.8. Overlay of chromatograms of diclofenac sodium salt samples (5.7 μg on column) with 0, 0.1% and 0.15% (w/w) chloride spikes. Conditions: Dionex Acclaim Trinity P1, 3 μm 3.0 × 50 mm flow at 0.8 mL/min. Mobile phase 75/25 v/v acetonitrile/120 mM ammonium acetate pH = 4.8. Detector settings: Corona *ultra*; Filter = med; Neb Temp = 25°C. Peak Identification; 1. sodium, 2. chloride, 3. diclofenac.

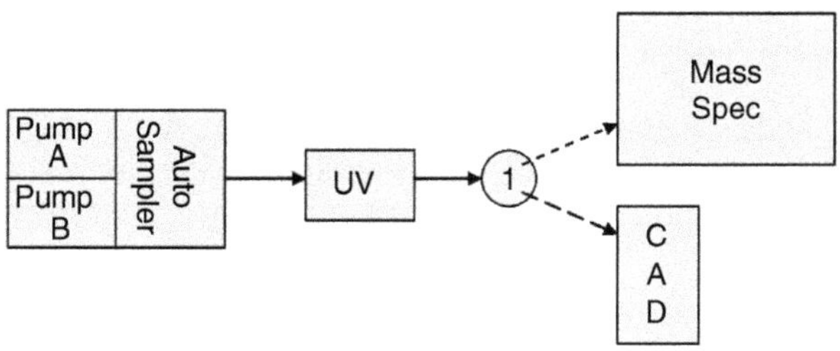

Figure 10.9. A schematic of the multi-detector platform. An adjustable flow splitter (1) was set to deliver 80 μL/min to the single-quad MS and 0.42 mL/min to the Corona CAD when using a flow-rate of 0.5 mL/min total flow.

10.4 EXCIPIENT CHARACTERIZATION

Excipients are inactive substances widely used in pharmaceutical formulations where they act as binders, fillers, lubricants, disintegrants, and preservatives or have other roles. The FDA mandates that all ingredients in a drug formulation must be guaranteed to be safe for human use. The characterization of an excipient, including its purity, the identities and quantities of its impurities, and the chemical degradation products, is critical.

One excipient, magnesium stearate, is commonly used as filler and as an anti-adherent/lubricant during the manufacture of tablets and capsules. The simultaneous measurement of both magnesium and stearate ions by HILIC-CAD was described by Huang et al. [24]. Although magnesium was well retained (~3.5 minutes), stearate

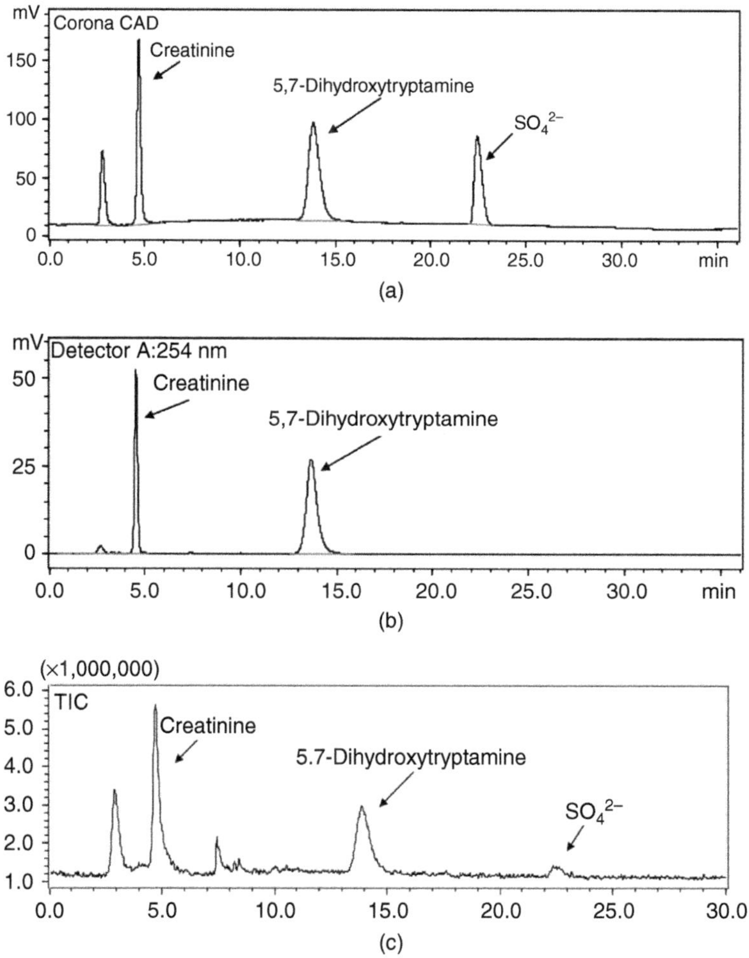

Figure 10.10. Separation of DHT creatinine sulfate (~2 μg on column) on a Sequant ZIC®-pHILIC, 5-μm 4.6 × 150 mm column at 30°C using a binary gradient flow at 0.5 mL/min. Mobile Phase A: 15/5/20/60 (v/v), 100 mM ammonium acetate (pH 4.6)/methanol/isopropyl alcohol/acetonitrile; Mobile Phase B: 50/5/20/25 (v/v) 30 mM ammonium acetate pH 4.6/methanol/isopropyl alcohol/acetonitrile. Gradient (time min., %B): 0, 20; 3, 20; 24, 70; 26, 70; 32, 15; 33, 20; 36, 20. (a) Corona CAD detection: 100 pA full scale; filter—none. (b) UV/Vis detection at 254 nm. (c) Total Ion Current mass spectral scan from 100 to 500 amu in positive ion mode. (d) Mass spectrum for creatinine (MW = 113.1) at retention time 5.7 minutes. (e) Mass spectrum for DHT (MW = 192.2) retention time 5.7 minutes. LC-ESI-MS: Acquisition Mode, Scan; Polarity, Positive; Event Time, 1.00 sec; Detector Voltage, 1.5 kV; Start m/z = 100; End m/z = 500; Scan speed, 500.

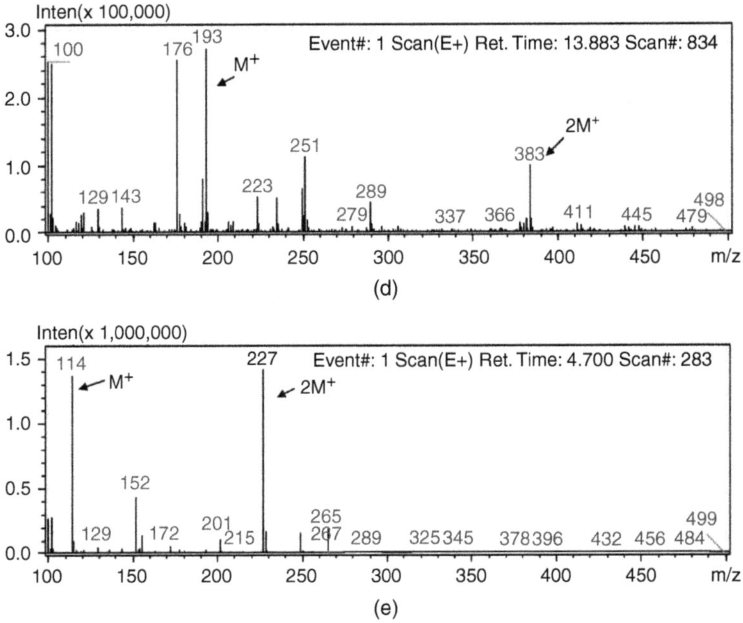

Figure 10.10. (*Continued*)

eluted in the solvent front. In order to retain hydrophobic ions under HILIC conditions, Louw et al, [45] coupled RP and HILIC columns in series and showed that this approach could be used to simultaneously measure both hydrophilic and hydrophobic pharmaceutical compounds [45]. Using a similar approach, we evaluated a gradient RP-HILIC-CAD method for the characterization of magnesium stearate (Figure 10.11). Using this method it was possible to measure the purity of the excipient and characterize trace impurities including fatty acids (oleic and palmitic acids) and cations at levels that would remain undetected using the ELSD.

10.5 CONCLUSIONS

The use of novel HPLC columns in combination with CAD enables the simultaneous analysis of APIs and counter-ions. The Acclaim Trinity™ P1 column is chromatographically flexible—analyte retention can readily and predictably be manipulated by changes in buffer strength, % organic, and pH. CAD enables the sensitive measurement of nonvolatile and many less volatile species. Together, they offer the analyst a rapid way to characterize a sample. For example, 15 of the 17 most common salt forms listed between USP and CSD can be measured in less than 14 min using this approach. This technique also enables the measurement of impurities, both inorganic and organic, to <0.1%. The analytical platform can be extended to include both UV and MS detectors enabling more in-depth characterization of the sample. Finally, the inability for

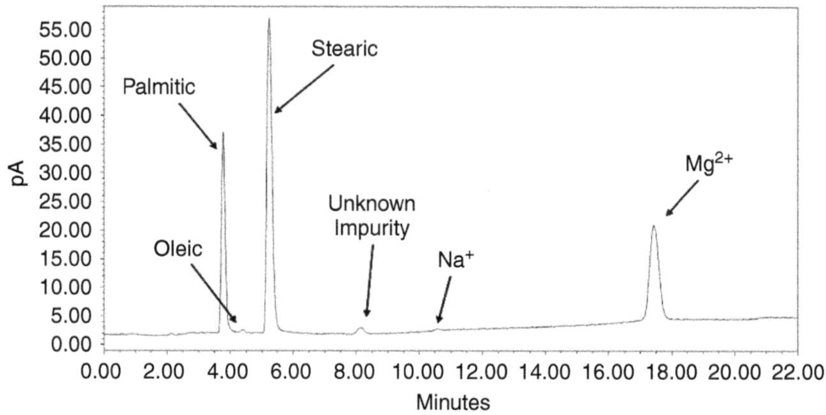

Figure 10.11. Analysis of magnesium stearate (Aldrich, Technical Grade - 65% stearate, 25% palmitate) (11 μg on column) separated using serially connected reversed phase and HILIC-mode chromatography columns. Column 1: Acquity UPLC BEH C181.7 μm 2.1 × 50 mm; column 2: SeQuant ZIC® -pHILIC 5 μm 50 × 4.6 at 40°C and a flow rate of 0.8 mL/min. Analytes were eluted using a binary gradient. Gradient (time mins, %B): 0, 10; 5, 10; 20, 65; 21, 65; 22, 10. Mobile Phase A: 15/75/5/5 200 mM ammonium formate pH = 3.2/acetonitrile/methanol/isopropyl alcohol; Mobile Phase B: 50/25/20/5 200 mM ammonium formate pH = 2.75/acetonitrile/methanol/isopropyl alcohol. Magnesium stearate was dissolved in hot ethanol prior to dilution in mobile phase A. Corona *ultra* settings: Filter = high; Neb Temp = 25°C.

HILIC-mode columns to retain hydrophobic analytes found in formulations (e.g., the excipient magnesium stearate) while measuring the API and counter-ion, can be overcome by using a reversed-phase column in series with the HILIC-mode column or by replacing the HILIC-mode column with the Acclaim Trinity™ P1 column.

REFERENCES

1. Kou D, Manius G, Zhan S, Chokshi HP. Size exclusion chromatography with Corona charged aerosol detector for the analysis of polyethylene glycol polymer. J. Chromatogr. A 2009;1216:5424–5428.
2. Haynes DA, Jones W, Samuel Motherwell, WD. Occurrence of pharmaceutically acceptable anions and cations in the Cambridge Structural Database. J. Pharm. Sci. 2005;94:2111–2120.
3. Dionex Acclaim Trinity P1 LC Column available at http://www.dionex.com/en-us/products/columns/lc/specialty/acclaim-trinity/lp-81754.html. Accessed 2010 Jul 8.
4. SeQuant ZIC-HILIC available at http://www.sequant.com/default.asp?ml=11528. Accessed 2010 Jul 8.
5. Sielc Products available at http://www.sielc.com/Products_Obelisc.html. Accessed 2010 Jul 8.

6. Imtakt HPLC Column Products available at http://www.imtakt.com. Accessed 2010 Jul 8.
7. Young CS, Dolan JW. Success with evaporative light-scattering detection. LCGC 2003;21:120–128.
8. Fries HE, Evans CA, Ward KW. Evaluation of evaporative light-scattering detection for metabolite quantification without authentic analytical standards or radiolabel. J. Chromatogr. B 2005;819:339–344.
9. Lane S, Boughtflower B, Mutton I, Paterson C, Farrant D, Taylor N, Blaxill Z, Carmody C, Borman P. Toward single-calibrant quantification in HPLC. A comparison of three detection strategies: evaporative light scattering, chemiluminescent nitrogen, and proton NMR. Anal. Chem. 2005;77:4354–4365.
10. Risley DS, Pack, BW. Simultaneous determination of positive and negative counter-ions using a HILIC method. LCGC 2006;24:776–785.
11. Vervoort N, Daemen D, Török G. Performance evaluation of evaporative light scattering detection and charged aerosol detection in reversed phase liquid chromatography. J Chromatogr. A. 2008;1189:92–100.
12. Zhang B, Li X, Yan B. Advances in HPLC detection – towards universal detection. Anal. Bioanal. Chem. 2008;390:299–301.
13. Webster GK, Jensen JS, Diaz AR. An investigation into detector limitations using evaporative light-scattering detectors for pharmaceutical applications. J. Chromatrogr. Sci. 2004;42:484–490.
14. El Haddad M, Lazar S, Akssira M, Dreux M. Isocratic conditions for analysis of inorganic anions by ion exchange chromatography with evaporative light scattering detection. J. Sep. Sci. 2002;25:23–28.
15. Mercier J-P, Elfakir C, Lafosse M, Dreux M, Tambute A. Hydrophilic anions analysis using reversed phase chromatography coupled with evaporative light-scattering detection. J. Liquid Chromatogr. Rel. Technol. 1999;22:267–279.
16. Mouchère F, El Haddad M, Elfakir C, Dreux M. Determination of inorganic cations and anions by ion-exchange chromatography with evaporative light-scattering detection. References and further reading may be available for this article. To view references and further reading you must purchase this article. J. Chromatogr. A 2001;14:167–173.
17. Lantz M, Risley DS, Peterson JA. Simultaneous resolution and detection of a drug substance, impurities and counter-ion using a mixed-mode HPLC column with ELSD. J. Liquid Chromatogr. Rel. Technol. 1997;20:1409–1422.
18. Peterson JA, Risley DS. Validation of an HPLC method for the determination of sodium in LY293111 sodium, a novel LTB_4 receptor antagonist, using evaporate light scattering detection. J. Liquid Chromatogr. 1995;18:331–338.
19. Risley, DS, Peterson, JA, Griffiths KL, McCarthy S. An alternative method for the determination of chloride in pharmaceutical drug substances using HPLC and evaporative light-scattering detection. LCGC 1996;14:1040–1047.
20. Dixon RW, Peterson DS. Development and testing of a detection method for liquid chromatography based on aerosol charging. Anal. Chem. 2002;74:2930–2937.
21. Gamache P, McCarthy R, Freeto S, Woodcock M, Laws K, Cole R. HPLC analysis of non-volatile analytes using charged aerosol detection. LCGC 2005; 23.
22. Reilly J, Everatt B, Aldcroft C. Implementation of charged aerosol detection in routine reversed phase liquid chromatography methods. J. Liq. Chromatogr. Rel. Technol. 2008;31:3132–3142.

23. Vehovec T, Obreza A. Review of operating principle and applications of the charged aerosol detector. J Chromatogr. A 2010;1217:1549–1556.
24. Huang Z, Richards MA, Zha Y, Francis R, Lozano R, Ruan J. Determination of inorganic pharmaceutical counter-ions using hydrophilic interaction chromatography coupled with a Corona CAD detector. J. Pharm. Biomed. Anal. 2009;50:809–814.
25. Lísa M, Lynen F, Holcapek M, Sandra P. Quantitation of triacylglycerols from plant oils using charged aerosol detection with gradient compensation. J Chromatogr. A, 2007;1176:135–142.
26. Loughlin J, Phan H, Wan M, Guo S, May K, Lin B. Evaluation of charged aerosol detection (CAD) as a complementary technique for high-throughput LC-MS-UV-ELSD analysis of drug discovery screening libraries. Am. Lab. September 2007.
27. McCarthy R, Gamache P, Asa D. Development and evaluation of Corona charged aerosol detection (CAD): a new universal detector for HPLC. LabPlus International June 2005.
28. Ramos RG, Libong D, Rakotomanga M, Gaudin K, Loiseau PM, Chaminade P. Comparison between charged aerosol detection and light scattering detection for the analysis of Leishmania membrane phospholipids. J Chromatogr, A, 2008;1209, 88–94.
29. Sun P, Wang X, Alquier L, Maryanoff C. Determination of relative response factors of impurities in paclitaxel with high performance liquid chromatography equipped with ultraviolet and charged aerosol detectors. J. Chromatogr. A 2008;1177:87–91.
30. Sinclair I, Charles I. Applications of the Charged Aerosol Detector in Compound Management J. Biomol. Screen 2009;14:531–537.
31. Brunelli C, G6recki T, Zhao Y, Sandra P. Corona-charged aerosol detection in supercritical fluid chromatography for pharmaceutical analysis. Anal. Chem. 2007;79:2472–2482.
32. Soman A, Jerfy M, Swanek F. Validated HPLC method for the quantitative analysis of a 4-methanesulfonyl-piperidine hydrochloride salt. J. Liq. Chromatogr. Rel. Technol. 2009;37:1000–1009.
33. Holzgrabe U, Nap CJ, Almeling S. Control of impurities in L-aspartic acid and L-alanine by high-performance liquid chromatography coupled with a corona charged aerosol detector. J. Chromatogr. A 2010;1217:294–301.
34. Yuabova ZR, Holschlag DR, Rodriguez CQ, Papov V, Qui F, McCaffery JF, Norwood D. Genotoxic Impurities: A Quantitative Approach. J. Liq. Chromatogr. Rel. Technol. 2008; 2318–2330.
35. Lobback C, Backensfeld T, Funke A, Weitschies W. Quantitative determination of nonionic surfactants with CAD. Chromatogr. Techniques. November 2007: 18–20.
36. Takahashi K, Kinugasa S, Senda M, Kimizuka K, Fukushima K, Matsumoto T, Shibata Y, Christensen J. Quantitative comparison of a corona-charged aerosol detector and an evaporative light-scattering detector for the analysis of a synthetic polymer by supercritical fluid chromatography. J. Chromatogr. A 2008;1193:151–155.
37. Yu X, Zdravkovic S, Wood D, Li C, Cheng Y, Ding X. A new approach to threshold evaluation and quantitation of unknown extractables and leachables using HPLC/CAD Drug Del. Tech. 2009;9:50–55.
38. Go'recki T, Lynen F, Szucs R, Sandra P. Universal response in liquid chromatography using charged aerosol detection. Anal. Chem. 2006;78:3186–3192.
39. Liu XK, Fang JB, Cauchon N, Zhou P. Direct stability-indicating method development and validation for analysis of etidronate disodium using a mixed-mode column and charged aerosol detector. J. Pharm. Biomed. Anal. 2008;46:639–644.

40. Gaberc-Porekar V, Zore I, Podobnik B, Menart V. Obstacles and pitfalls in the PEGylation of therapeutic proteins. Curr. Opin. Drug Disc. Devel. 2008;11:242–250.
41. Forsatz B, Snow NH. HPLC with charged aerosol detection for pharmaceutical cleaning validation. LCGC. September 2007.
42. Swartz M, Emanuele M, Awad A, Hartley D. Charged aerosol detection in pharmaceutical analysis: An overview. Chromatography Online. April 2009.
43. Crafts C, Acworth IN. Single run ion analysis using charged aerosol detection: Ion content of commercially available mineral waters. LCGC Application Notebook – June 2009:40.
44. Crafts C, Bailey B, Plante M, Acworth IN. Evaluation of methods for the simultaneous analysis of cations and anions using HPLC with charged aerosol detection and a zwitterionic stationary phase. J. Chromatogr. Sci. 2009b;47:534–539.
45. Louw S, Pereira AS., Lynen F, Hanna-Brown M, Sandra P. Serial coupling of reversed-phase and hydrophilic interaction liquid chromatography to broaden the elution window for the analysis of pharmaceutical compounds. J. Chromatogr. A. 2008;1208:90–94.
46. ESA Application Note 70-8376 available at http://www.esainc.com/docs/spool/70-8376-Simultaneous-Measurement-of-APIs-and-Counter-ions-by-UV-CAD-and-MS.pdf. Accessed 2010 Jul 8.

11

HIGH PERFORMANCE ION CHROMATOGRAPHIC ANALYSIS OF CHOLINERGIC COMPOUNDS: CARBACHOL AND BETHANECHOL, AND ASSOCIATED DEGRADATION PRODUCTS

David Trowbridge

Alcon Laboratories, Fort Worth, TX

11.1 INTRODUCTION

Carbachol (Ethanaminium, 2-[(aminocarbonyl)oxy]-*N*,*N*,*N*-trimethyl-chloride) (Figure 11.1) and bethanechol (*N*,*N*,*N*-trimethylpropan-1-aminium) (Figure 11.2) are both parasympathomimetic compounds, which mimic the action of acetylcholine to stimulate muscarinic receptors found in the body. Essentially, these compounds bind and activate the acetylcholine receptors. Thus this type of compound is classified as a cholinergic agonist. Carbachol and bethanechol are positively charged ammonium compounds. Carbachol is primarily used for various ophthalmic purposes in the treatment of glaucoma or for use during ophthalmic surgery. Bethanechol is primarily used for the treatment of urinary retention problems.

In 1969, Poe and Puckett described a method for the analysis of carbachol in aqueous solutions that utilized the chlorination of the free N-H moiety which in turn was utilized to oxidize iodide to iodine. Quantification was then accomplished by reacting the latter product with starch and measuring the resulting blue color [1]. This is also the method that is described in USP 32-NF 27. The major problems encountered with the method were:

Applications of Ion Chromatography for Pharmaceutical and Biological Products, First Edition.
Edited by Lokesh Bhattacharyya and Jeffrey S. Rohrer.

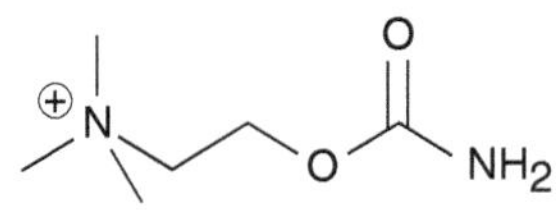

Figure 11.1. Structure of carbachol, systematic (IUPAC) Name: Ethanaminium, 2-[(aminocarbonyl)oxy]-*N, N, N*-trimethyl-chloride.

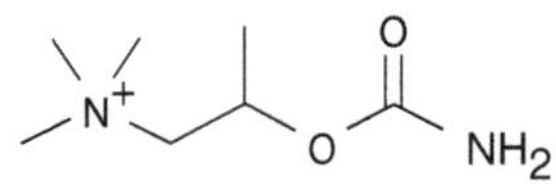

Figure 11.2. Structure of bethanechol, systematic (IUPAC) Name: *N, N, N*-trimethylpropan-1-aminium.

1. Measurement errors.
2. It was time and labor intensive.
3. The inability to discern the major degradation product of carbachol, choline.

Other methods reported in the literature utilize a post column enzyme reactor with an immobilized enzyme and electrochemical detection [2]. The method is based on a separation of choline and acetylcholine by cation-exchange HPLC followed by passage of the effluent thorough a post column reactor that is a mixture of acetyl cholinesterase and choline oxidase. The choline oxidase converts choline to betaine and hydrogen peroxide, and the acetyl cholinesterase converts acetylcholine to acetate and choline, which is converted to betaine and hydrogen peroxide by choline oxidase. The hydrogen peroxide produced is then detected electrochemically with a glassy carbon working electrode.

The disadvantages of this method are:

1. Requires specialized reactor equipment in the form of the post column enzyme reactor.
2. System preparation is labor intensive and time consuming.
3. Stability of the glassy carbon electrode.

It is of prime importance that an assay method not only has the ability to detect and quantify cholinergic compounds used within pharmaceutical preparations, but must also be able to quantify their primary degradation products in compliance with current ICH guidelines.

This chapter will explore a different methodology for the determination cholinergic compounds; ion chromatography (IC) with suppressed conductivity detection.

11.2 DESIGN OF THE IC METHOD

In designing an IC method one must consider both the separation (i.e. the column chemistry) and detection methods. This application uses an IonPac CS 17 column. The CS17 is a moderate capacity, hydrophilic, carboxylate-functionalized cation-exchange column. The substrate is composed of a macroporous resin bead of ethylvinylbenzene and the cation-exchange polymer consists of carboxylic acids grafted to the resin surface and in the pores. The CS17 was specifically designed for the separation of amines with a strong acid mobile phase in the absence of organic solvent. Low millimolar concentrations of methanesulfonic acid (MSA), an ideal strong acid for suppressed conductivity detection, are used to separate a wide variety of amines, including quaternary amines, with good peak shapes. The absence of organic solvents in the eluent allows for more sensitive detection. The CS17 with MSA eluent and suppressed conductivity detection was found to be ideal for determination of cholinergic compounds and their breakdown products.

11.3 EXPERIMENTAL DESIGN

This study was performed on two different integrated ion chromatography systems. A Dionex ICS 2000 system with the ability to use eluent generation for the MSA, and a Dionex DX-600 system that was not equipped with eluent generation capability. Both systems were equipped with the same type of columns and suppressors, and were run under similar conditions.

The following items were utilized during this study:

Equipment: ICS 2000 and DX-600 IC systems (Dionex)
IonPac CS17 column, 4 × 250 mm (Dionex)
IonPac CG17 guard, 4 × 50 mm (Dionex)

Reagents: Deionized water (18 mega-ohm or greater in resistivity)
MSA, Fluka Brand (98% or higher purity) or MSA from an eluent generator (Dionex) −5 mM

Conditions: Flow rate: 1.0 mL/min
Column temperature: 30°C
Injection volume: 25–30 μL
Detection: Suppressed conductivity, CSRS ULTRA 4 mm cation ion suppressor, recycle mode, 20 mA

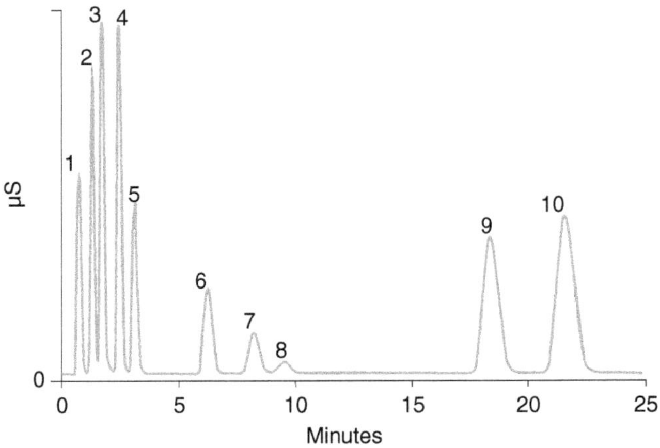

Figure 11.3. Separation of Group I and Group II cations from carbachol and choline. The analytes are separated on an IonPac CS17 (4 × 250 mm) column with IonPac CG17 guard (4 × 50 mm) with a 5 mM methanesulfonic acid eluent produced by an eluent generator and flowing at 1.0 mL/min. The column temperature is 30°C and the injection volume 25 µL. The analytes are detected by suppressed conductivity detection using a CSRS-ULTRA II suppressor. The peak identities with concentrations in mg/L are: 1-lithium, 0.1; 2-sodium, 0.4; 3-ammonium, 0.5; 4-potassium, 1.0; 5-dimethylamine, 1.0; 6-choline, 1.0; 7-carbachol, 1.0; 8-bethanechol, 1.0; 9-magnesium, 0.5; 10-calcium, 1.0.

Standard Preparations: USP standards for carbachol and bethanechol
Choline and 2-HPTA (2-hydroxypropyltrimethyl ammonium chloride) standards were prepared by hydrolyzing carbachol and bethanechol in 0.1N NaOH for five days.

Sample: Spiked over-the-counter (OTC) pharmaceutical preparations and a pharmaceutical preparation containing carbachol that is used as a surgical eye irritant.

11.4 CHROMATOGRAPHY AND SAMPLE ANALYSIS

Determinations of carbachol in commercially prepared pharmaceutical preparations could be compromised by various group I and II cations. Figure 11.3 shows a chromatogram of a 1 m carbachol standard with several commonly occurring cations that may be contained in different saline irrigating solutions [3]. All components were resolved from carbachol and choline, with the group I cations eluting well before, and the group II cations well after, the analytes of interest. The ten compounds in Figure 11.3 including carbachol, bethanechol, and choline are resolved on the CS17 cation-exchange column with a 5 mM MSA eluent. While suppressed conductivity has

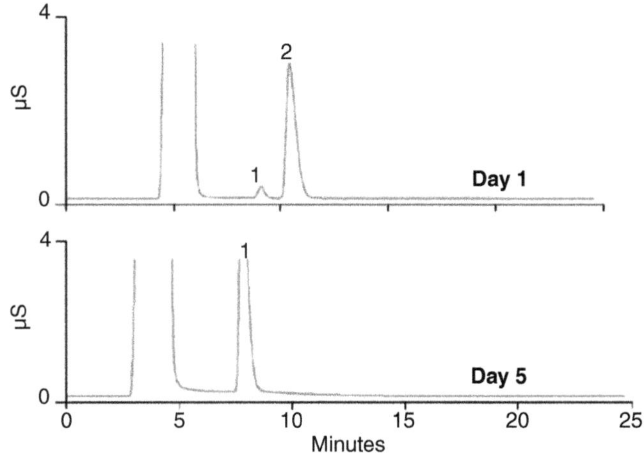

Figure 11.4. Base-catalyzed decomposition of carbachol to choline. Peak 1 is choline and peak 2 is carbachol. The concentrations of choline on days 1 and 5 are 42.2 and 502 mg/L, respectively and the concentration of carbachol on day one is 502 mg/L. The chromatography conditions are the same as Figure 11.3.

good sensitivity for all 10 compounds, the sensitivity for the quaternary ammonium compounds is less than that for the group I and II cations. This is expected as the charge to mass ratio will be less for quaternary ammonium compounds compared to the inorganic cations resulting in a lower equivalent conductance. The authors of reference 3 demonstrated the possibility that the same isocratic IC method could be used to determine the carbachol and bethanechol in pharmaceutical samples. For both compounds the authors conducted a forced degradation in alkaline solution (Figures 11.4 and 11.5). This yields choline from carbachol and 2-HPTA from bethanechol. Both studies indicated that degradation could be followed, if necessary, and that the resulting compounds would not interfere with assay of the analytes of interest. Lacking ophthalmic solutions containing carbachol, these authors added it to an OTC saline solution and an OTC contact lens solution. The solutions have high concentrations of group I cations and must be first diluted so as to not overload the column. As shown in Figure 11.6, 1 mg/L carbachol, bethanechol, or choline can be easily determined in both samples.

In this chapter the author applied the same method to a commercial product containing carbachol. This formulation is used during eye surgery as an irrigation solution. As a consequence of the intended application, the formulation has a very high salt content. The components of the formulation are as follows:

USP Sodium Chloride: 0.64% W/V
USP Potassium Chloride: 0.75% W/V
USP Calcium Chloride Dihydrate: 0.48% W/V

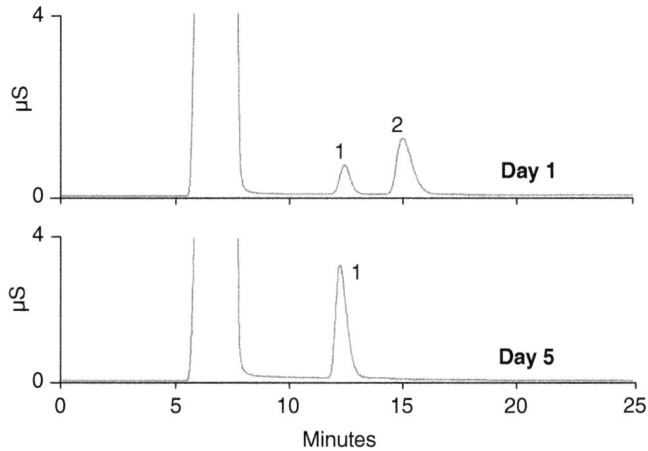

Figure 11.5. Conversion of bethenechol to 2-HPTA in alkaline solution. Peak 1 is 2-HPTA and peak 2 is bethanechol. The concentrations of 2-HPTA on days 1 and 5 are 101 and 501 mg/L, respectively, and the concentration of bethanechol on day one is 401 mg/L. The chromatography conditions are the same as Figure 11.3.

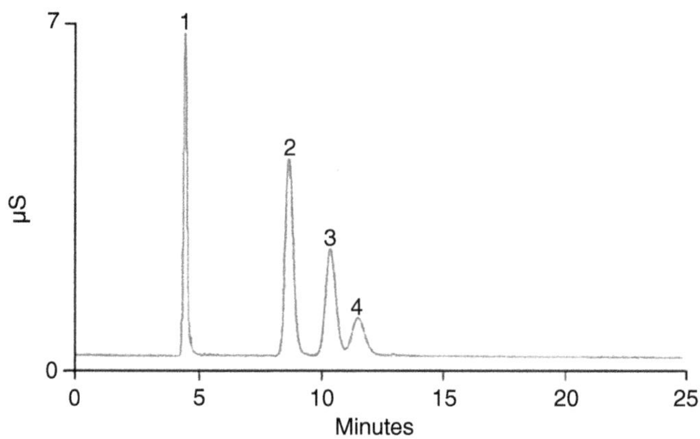

Figure 11.6. Determination of choline, carbachol, and bethanechol in lens and saline solutions. The peak identities with concentrations in mg/L are: 1- sodium, 0.031; 2-choline, 0.987; 3-carbachol, 0.967; 4- bethanechol, 0.985. The chromatography conditions are the same as Figure 11.3.

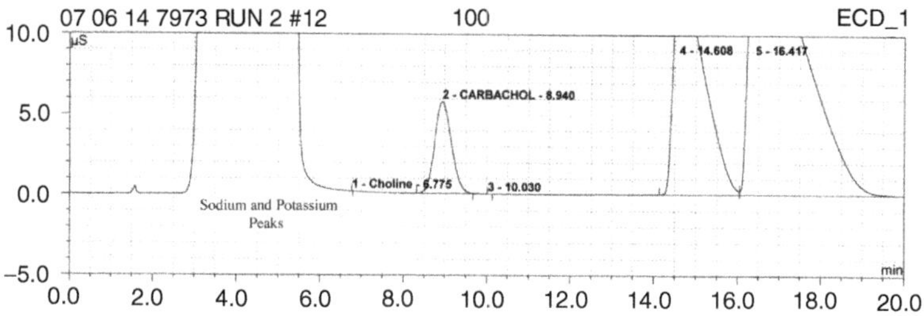

Figure 11.7. Separation of carbachol and choline from Group I and II cations in an eye irrigating formulation.

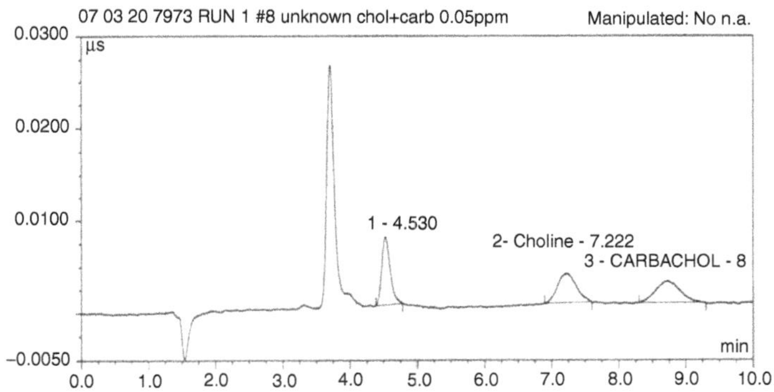

Figure 11.8. Determination of choline and carbachol at 0.05 mg/L.

USP Magnesium Chloride Hexahydrate: 0.03% W/V
USP Sodium Acetate Trihydrate: 0.39%
USP Sodium Citrate Dihydrate: 0.17% W/V
USP Carbachol: 0.01% W/V

As can be seen from the preceding list, there are high concentrations of group I and II cations in this formulation that could make the determination of carbachol problematic. However, Figure 11.7 shows that a simple 1:1 dilution of the formulation with purified water yields a chromatogram from which carbachol can be easily quantified. The resulting concentration of the carbachol was 50 mg/L. Quantification of the carbachol peak with an external standard showed that carbachol was fully recovered in the high ionic strength samples. In order to meet ICH guidelines for impurities in drug products, it is necessary that the method be able to detect 0.1% of carbachol concentration, or 0.05 mg/L [4]. This capability is demonstrated in Figure 11.8. That figure also shows choline, a known impurity of carbachol, which results from its degradation, at the same concentration. Table 11.1 presents the data for the aqueous standard curve of choline. The data, presented as a percentage of the diluted standard

TABLE 11.1. Aqueous Standard Curve Data for Choline.

Part per Million	Percentage of Diluted Standard	Area Response (μS)
0.05495	0.1	0.002
0.05495	0.1	0.002
0.05495	0.1	0.002
0.10990	0.2	0.0041
0.10990	0.2	0.0042
0.10990	0.2	0.0041
0.16490	0.3	0.0066
0.16490	0.3	0.0066
0.16490	0.3	0.0067
0.21980	0.4	0.0089
0.21980	0.4	0.0090
0.21980	0.4	0.0090
0.27480	0.5	0.0114
0.27480	0.5	0.0112
0.27480	0.5	0.0116
0.54950	1.0	0.0260
0.54950	1.0	0.0263
0.54950	1.0	0.0264
1.09890	2.0	0.0579
1.09890	2.0	0.0579
1.09890	2.0	0.0588
1.6490	3.0	0.0907
1.6490	3.0	0.0905
1.6490	3.0	0.0902
2.7473	5.0	0.1580
2.7473	5.0	0.1588
2.7473	5.0	0.1578
$r^2 = 0.99888$		

Note: The above data presents the concentration as a percentage of the diluted standard used for the choline analysis and illustrates that it is possible to meet the ICH guidelines for achieving reproducible responses at 0.1% of the diluted run standard.

used for the choline analysis, illustrates the excellent sensitivity of the method, and that ICH guidelines that recommend achieving reproducible responses for 0.1% of the diluted run standard can be met. Demonstrating this capability is essential for successful degradation product analysis.

11.5 CONCLUSION

Since the introduction of IC in the late 1970s many unique applications have been developed for a variety of ionic species. The ability to discern an ever increasing variety of species by IC has been a result of both improvements in the technology, and unique ways of applying these advancements. From the simple analysis of group I and II cations to the subject of this particular article, substituted amines, IC with suppressed conductivity detection delivers direct sensitive determination of a wide range of cationic analytes.

The successful application of IC with suppressed conductivity detection to the analysis of cholinergic compounds has proven to be a valuable asset in the study of the stability profile of these drug components.

REFERENCES

1. Puckett R, Poe RD. Determination of carbachol in aqueous solutions. J. Pharm Sci. 1969;58:602–604.
2. Damsama G, Westerink BH. A simple, sensitive, and economic assay for choline and acetylcholine using HPLC, an enzyme reactor, and an electrochemical detector. Neurochem. 1985;45:1649–1652.
3. Dionex Corporation, Application Note 194: Determination of carbachol in ophthalmic solutions using a reagent-free ion chromatography system.
4. ICH Harmonised Tripartite Guideline ICH Q3B(R2). Impurities in new drug products; 2006.

12

ION CHROMATOGRAPHIC ANALYSIS OF PHARMACEUTICALS FOR AUTHENTICITY AND ADULTERATION

David S. Jackson

US Food and Drug Administration, Forensic Chemistry Center, Cincinnati, Ohio

12.1 INTRODUCTION

Ion chromatography (IC) has become a universal technique for the detection of anionic and cationic species with applications that have been developed in numerous fields of interest. Its use in pharmaceutical applications has increased in recent years as well. It was not until 2006 that a chapter on *Ion Chromatography* was added to the United States Pharmacopeia-National Formulary (USP-NF) [1]. Prior to this issue, the pharmacopeia had listed several chromatographic packings that were used for ion chromatographic analysis. In the 2009 USP-NF, a number of ion chromatographic analyses using conductivity detection and pulsed amperometric detection were made official. The majority of these applications are for residual limits in pharmaceuticals: limit of oxalate, iodide, sulfate, sulfamate, phosphate, potassium, and sodium. Twenty-eight official methods were found that were specific to the analysis of either the active ingredient or a component of the active ingredient using conductivity or pulsed amperometric detection. Those specifically related to drug components included potassium perchlorate [2], bethanechol chloride [3], erythromycin [4], fludeoxy glucose injection [5], streptomycin sulfate [6], kanamycin sulfate [7], amikacin [8], and sodium

Applications of Ion Chromatography for Pharmaceutical and Biological Products, First Edition.
Edited by Lokesh Bhattacharyya and Jeffrey S. Rohrer.

fluoride [9]. Additionally, the USP-NF has a general chapter describing an IC procedure for assaying citrate or citric acid and phosphate content in drug products [10]. However, only 16 of over 60 monographs containing citric acid or citrate use IC for its determination (see Chapter 16 of this book for more details).

In addition to the compendial applications described above, the analysis of anions, cations, organic acids, and carbohydrates in pharmaceutical formulations can be extremely useful in determining drug authenticity and adulteration. Many pharmaceutical products contain excipients such as sugars, sugar alcohols, amino acids, salts, and organic acids which are added as fillers, binders, sweeteners, or preservatives. Saline and buffered solutions are often used as vehicles for intravenous and injectable drugs. Many of these excipients are amenable to IC analysis, and differences in the type or amount of an excipient present may be indicative of the use of unapproved processes, unapproved formulations, or authenticity issues. Active drug ingredients are often supplied as hydrochloride, phosphate, citrate or other salt forms which can be determined by IC as well. This information often proves to be useful in the investigation of drug authenticity and adulteration.

The Forensic Chemistry Center has been involved in numerous pharmaceutical authenticity and adulteration cases. There are many reasons why a pharmaceutical drug may become adulterated. Unscrupulous persons may commit fraud by selling expired, diluted, or unapproved products as authentic, good quality products to a consumer, a doctor's office, or an institution. A caregiver may steal a patient's pain medications for personal use to meet an illicit drug habit or to sell while giving the patient in need of medication a dosage of water, saline, or something inferior. Pharmaceuticals may become adulterated by the manufacturer in the production process; beyond the manufacturer in a doctor's office or storage facility; or at the point of use, i.e. where the medication is being administered.

The use of IC to verify drug authenticity has been described [11]. In one case, the determination of a lubricating agent, sodium laurel sulfate, was used to show that a generic drug manufacturer was not following the approved manufacturing process. In another instance, a veterinary drug product was labeled to contain one active ingredient while a different, less potent and less expensive active ingredient was used in the formulation. The substitute drug was available as the hydrochloride or phosphate salt. IC was used to determine the amounts of chloride and phosphate in the suspect products, and the concentrations of substitute drug and anions present were used to investigate the sources of the suspect samples.

High pH anion-exchange chromatography with pulsed amperometric detection (HPAEC-PAD) was used to separate and detect the primary gentamicin C components, C_1, C_{1a}, C_2, and C_{2a} in suspected counterfeit gentamicin samples [12]. Component ratios along with chromatographic profiles were used to compare authentic and suspect samples and to try to provide evidence of common sources for suspect samples.

IC is an important analytical tool for the characterization of various drug products. The analysis of anions, cations, organic acids, and carbohydrates in pharmaceutical formulations can be useful in determining drug authenticity and/or adulteration. Three cases will be discussed in this chapter in which IC was useful in examining the active pharmaceutical ingredient, counter-ion, or an excipient as part of forensic investigations of suspect pharmaceutical products.

12.2 RESULTS AND DISCUSSION

12.2.1 Case 1: Fentanyl Citrate Syringes Analyzed for Citrate

The Forensic Chemistry Center has provided forensic analysis in a number of cases involving tampering of injectable pain medications and flu vaccines at hospitals and clinics. Typically, pain medications have been stolen to support drug addictions. Often water or saline is added back into the vials or syringes in an attempt to cover up the tampering. In cases involving flu vaccines, the suspect samples are often found to be diluted for monetary gain. This has become more of an issue during periods of flu vaccine shortages.

In this particular case, the suspect, a certified registered nurse assistant, was allegedly tampering with injection syringes that were pre-filled in the hospital pharmacy with 50 mg/mL fentanyl. Fentanyl (*N*-phenyl-*N*-(1-phenethyl-4-piperidinyl) propanamide) is a strong analgesic often used to manage pain after surgery, in cancer therapy, and in other chronic pain management situations. Because of its potential for abuse, it is classified as a DEA Schedule II drug in the United States [13]. The suspect was allegedly stealing the 50 mg/mL fentanyl from the syringes and replacing all or part of it with water. It was also alleged that the suspect was knowingly administering water or diluted fentanyl citrate to hospital patients while using the full strength drug himself in order to feed his drug addiction.

Twenty-three syringes of suspect fentanyl citrate were analyzed and compared to authentic fentanyl citrate using IC for the analysis of citrate, and liquid chromatography/ion trap mass spectrometry (LC-MS^n) for the analysis of fentanyl. The molar ratio of fentanyl to citrate is 1:1. Anion-exchange chromatography with suppressed conductivity detection was used to quantitate the citrate in the suspect and control samples and to show the degree of dilution of the pain killer. The concentration of citrate found was used to calculate the concentration of fentanyl in the suspect syringes, and these results were compared with the results obtained using LC-MS^n.

Table 12.1 shows the chromatographic conditions used for analysis (Case 1). The USP-NF citrate procedure calls for an L61 column packing; a low capacity anion-exchange column [10]. However, in this case, a Dionex IonPac AS11-HC high capacity anion-exchange column with 30 mM hydroxide eluent was used. It has the same selectivity as the L61 packing, but with a higher capacity for separation. Figure 12.1 shows chromatograms of a 20 mg/L citrate standard and a deionized water method blank. A calibration curve was established using 4, 10, 20, and 40 μg/mL citrate standards. Suspect samples were analyzed in triplicate without dilution. The RSD for individual sample quantities ranged from 1.2% to 4.4% with an average of 2.5%. Spike recoveries for citrate in the suspect samples ranged between 95% and 107% for two spike levels, 4 μg/mL and 20 μg/mL.

TABLE 12.1. Chromatographic Conditions Used in the Three Cases Discussed.

Case 1: Fentanyl Citrate Injection Analysis for Citrate

Column:	Dionex AS11-HC analytical and guard column (2 mm)
Eluent:	30 mM KOH
Flow rate:	0.4 mL/min
Injection volume:	10 μL
Detection mode:	Suppressed conductivity
Suppressor setting:	Dionex ASRS (2 mm) in recycle mode, 100 mA
Calibration range:	4 μg/mL—40 μg/mL citrate
Dilution:	Sample analyzed neat

Case 2: Botulinum Neurotoxin Type A Vials Analyzed for Sucrose

Column:	Dionex Carbopac PA-1 analytical and guard column (4 mm)
Eluent:	160 mM NaOH
Flow rate:	1.0 mL/min
Injection volume:	20 uL
Detection mode:	Pulsed amperometric detection with gold working electrode and pH/Ag/AgCl reference electrode
Calibration range:	1.0 μg/mL—100 μg/mL sucrose
Dilution:	Sample diluted 0.01 mL → 0.11 mL distilled deionized water

Waveform setting:	Step	Potential	Time
	E1	0.1V	400msec
	E2	−2.0V	10msec
	E3	0.6V	0msec
	E4	−0.1V	60msec
		Integrate 200 to 400 msec	

Case 3: Suspect Oral Suspension Analyzed for Bethanechol

Column:	Dionex IonPac CS14 analytical and guard column (4 mm)
Eluant:	20 mM methanesulfonic acid
Flow rate:	1.0 mL/min
Injection volume:	50 uL
Detection mode:	Suppressed conductivity
Suppressor setting:	Dionex CSRS-ULTRA in recycle mode with 80 mA current
Calibration range:	0.011 mg/mL—1.1 mg/mL bethanechol chloride
Dilution:	Sample diluted 1.0 mL → 2.0 mL distilled deionized water

A Dionex ICS-3000 system consisting of a DP-1 pump, DC-1 conductivity detector, AS-1 auto sampler, and EG-2 eluent generator and a Dionex DX-600 IC system consisting of a GP50 gradient pump, a ED50 with conductivity and pulsed amperometric detectors, EG40 eluent generator and AS50 autosampler were used to perform analyses.

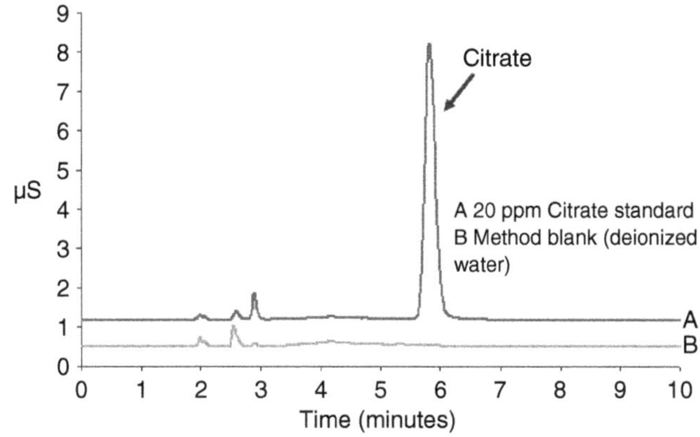

Figure 12.1. Chromatogram of 20 mg/L citrate standard and deionized water blank.

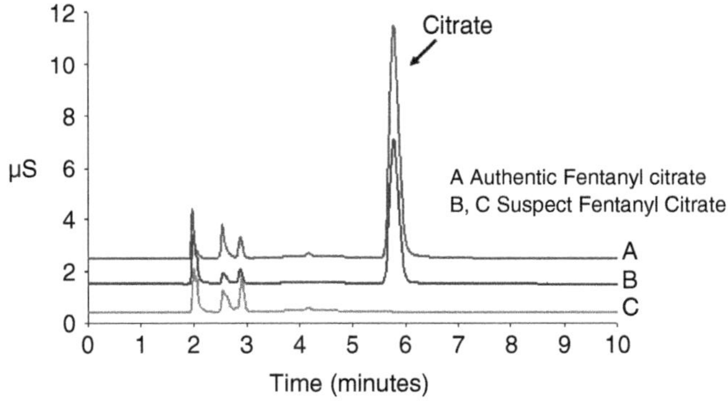

Figure 12.2. Chromatogram of authentic and suspect fentanyl citrate injectables.

Figure 12.2 shows a chromatogram (A) of the authentic 50 µg/mL fentanyl injectable compared to two of the suspect samples. Citrate was found in the suspect sample associated with chromatogram (B) at approximately 60% of the level found in the authentic product. Whereas no citrate was detected in the suspect sample associated with chromatogram (C). Table 12.2 shows the citrate concentrations and the calculated fentanyl concentrations (based on the citrate concentration) for the authentic and suspect samples. Citrate was not detected in ten of the suspect syringes indicative of virtually a 100% substitution of pain killer. In thirteen syringes, citrate was detected at a level between 52% and 67% of the citrate level found in authentic product.

The LC-MS^n results for fentanyl in the samples compared well with the calculated fentanyl values from the citrate determination (Table 12.2). It was determined that suspect samples in which no citrate or fentanyl was detected by IC or LC-MS^n contained less than 0.2 µg/mL fentanyl. Suspect samples in which both citrate and

TABLE 12.2. Sample Results Summary for Suspect Fentanyl Citrate Syringes.

Suspect Fentanyl Citrate samples	Citrate Conc. (μg/mL)	Calculated Fentanyl Citrate Conc. based on Citrate Conc. (μg/mL)	Fentanyl Citrate Conc. based on LC-MS^n analysis (μg/mL)
Authentic	26.5 ± 0.5	46	46
1–4	nd	nd	nd
5	15.2 ± 0.3	27	30
6	nd	nd	nd
7	13.7 ± 0.3	24	23
8–12	nd	nd	nd
13	16.0 ± 0.7	28	26
14	17.2 ± 0.2	30	26
15	15.4 ± 0.5	27	24
16	16.9 ± 0.6	30	23
17	17.4 ± 0.5	31	30
18	14.9 ± 0.3	26	21
19	16.8 ± 0.7	29	23
20	16.3 ± 0.3	29	24
21	16.5 ± 0.3	29	22
22	17.7 ± 0.3	31	27
23	14.9 ± 0.4	26	21

nd = not detected.

fentanyl were detected, contained approximately 42%–62% of the labeled amount of fentanyl. Using IC to detect the citrate counter ion, it was possible to show the degree of adulteration of the suspect fentanyl citrate injectables and to confirm LC-MS^n results obtained for fentanyl. This information was useful to investigators and helped to secure a guilty plea from the suspect in Federal Court on two counts of obtaining a controlled substance by misrepresentation.

12.2.2 Case 2: Botulinum Neurotoxin Type A Vials Analyzed for Sucrose

Unapproved drug products may not be safe and are not permitted to be sold in the United States. These products may contain the wrong active ingredient, the wrong amount of active ingredient, or even dangerous ingredients. An undercover purchase was made of two vials suspected of being an unapproved version of injectable botulinum toxin type A. An approved form of botulinum toxin type A is used in cosmetic procedures to temporarily diminish face wrinkles. The toxin's effect is to paralyze muscles and has thus also been used in clinical therapeutic applications on a variety of pathological conditions including dystonia [14].

The suspect samples were labeled in Spanish and declared to contain botulinum toxin type A and 25 mg/vial saccharose (sucrose), as well as dextran and gelatin. Sucrose and other ingredients like lactose, albumin and benzyl alcohol are added to

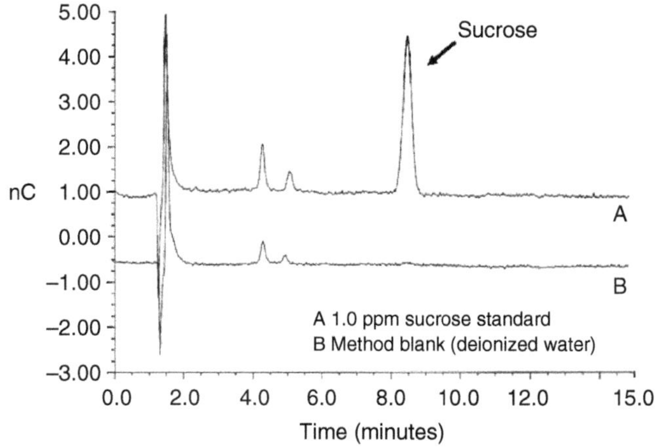

Figure 12.3. Chromatogram of 1 mg/L sucrose and a method blank (deionized water).

the neurotoxin protein for stabilization [14]. The product was tested for botulinum toxin type A using Enzyme-Linked Immunosorbant Assay (ELISA). HPAEC-PAD was used to analyze the vials for sucrose in an effort to verify the labeling. HPAEC-PAD has excellent sensitivity so that analysis can be performed on small sample sizes. Its selectivity is very good which makes it less prone to interferences (see Chapter 17 of this book).

The samples were analyzed for sucrose using a Dionex CarboPac PA1 column under conditions listed in Table 12.1 (Case 2). Figure 12.3 shows chromatograms obtained for a 1 mg/L sucrose standard with a retention time of 8.5 min and for a deionized water method blank. A calibration curve was prepared using 1, 2, 10, 50, and 100 μg/mL sucrose standards. Sucrose was not detected in the suspect product as shown in Figure 12.4, chromatogram B. Sucrose was detected when spiked into the suspect product at the labeled amount (25 mg/vial) as seen in Figure 12.4, chromatogram A. Spike recoveries of sucrose-fortified samples at the 25 mg/vial and 0.2 mg/vial levels were 91% and 102%, respectively. Both vials were determined to contain botulinum toxin type A at the level expected using ELISA. The product's lack of sucrose serves to demonstrate some of the potential problems with unapproved drugs including false labeling.

12.2.3 Case 3: Suspect Oral Suspension Analyzed for Bethanechol

The Forensic Chemistry Center often receives samples for forensic investigation as a follow-up to consumer complaints or adverse events reported through MedWatch, FDA's adverse event reporting system for human medical products. In this particular case, twin 7 month old infants were admitted to the hospital following what was believed to be an adverse reaction to a pharmacy compounded oral suspension of bethanechol. The suspect oral suspension was administered to the infants for gastro-esophageal reflux problems. After ingesting the product, both infants exhibited

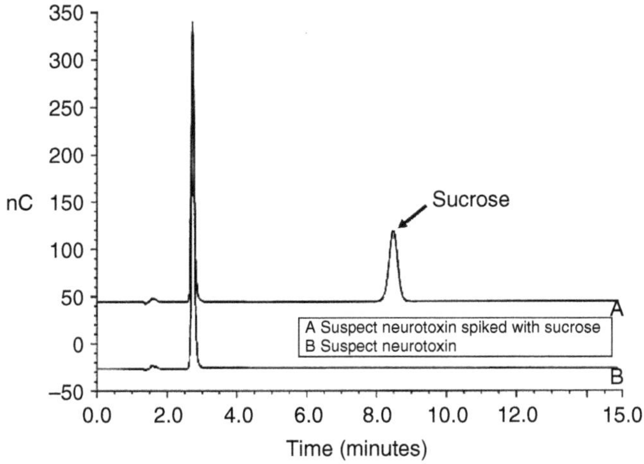

Figure 12.4. Chromatograms of suspect product vial and suspect vial spiked with sucrose.

respiratory distress requiring emergency care. Bethanechol is a cholinergic agent which is used to aid patients in urinary retention and in the reduction of gastroesophageal reflux [15,16].

The oral suspension prepared by the pharmacy was made up of ground bethanechol tablets, a sweet syrup vehicle, and an oral suspending vehicle. The suspension was labeled to contain 1 mg/mL bethanechol (2-[(Aminocarbonyl)oxy]-N,N,N-trimethyl-1-propanaminium chloride). The suspect oral suspension was received by the laboratory for forensic analysis. Early in the investigation, the pharmacy that prepared the oral suspension claimed that the suspension had been mistakenly prepared at a bethanechol concentration ten times stronger than its intended dosage. The work order indicated this 10-fold increase in bethanechol concentration even though the work instructions for preparation were at the correct level. Control components of bethanechol tablets, sweet syrup vehicle, and the oral suspending vehicle were obtained for comparison purposes. In addition, the laboratory prepared a control oral suspension from the components received using the work order instructions (1 mg/mL bethanechol).

A cation-exchange IC method with suppressed conductivity detection was used to assay the product for bethanechol [14]. The official USP assay for bethanechol calls for an L55 column packing; a strong cation-exchange column. The Dionex IonPac CS14 cation- exchange column is a weak cation-exchange column; however, it provided ample separation and selectivity for the bethanechol determination. Table 12.1 lists the separation conditions used. The laboratory initially analyzed the suspect product to verify labeled contents and to try to verify the mistake in the bethanechol concentration reported by the pharmacy. Figure 12.5 shows a chromatogram of a 0.11 mg/mL bethanechol chloride standard. When the suspect sample was analyzed, however, a peak corresponding to bethanechol was not detected. Figure 12.6 shows a chromatogram obtained for the suspect sample (chromatogram B)

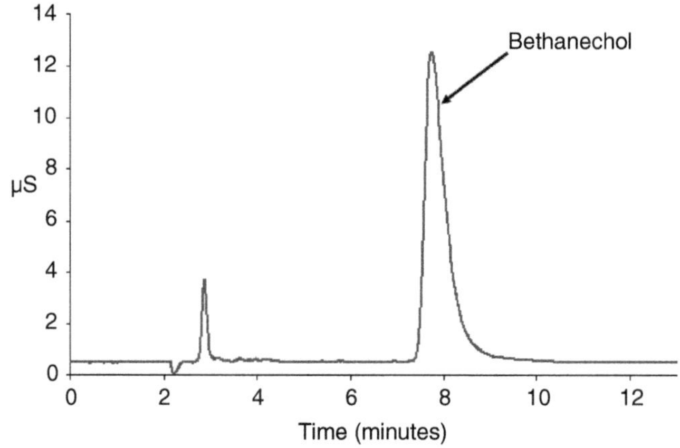

Figure 12.5. Bethanechol chloride standard in distilled deionized water (0.11 mg/mL).

and the laboratory prepared control bethanechol oral suspension (chromatogram A) which contains 1 mg/mL bethanechol. The sodium, magnesium, and calcium peaks seen in Figure 12.6 are attributed to the sweet syrup and the oral suspending vehicles used in making the suspension. Their concentrations are at a similar level in the suspect and control oral suspensions indicating that the two preparations were made in a similar manner. It appeared as if the active ingredient, bethanechol, was either not present in the suspect sample or could have potentially been degraded. The latter case seemed very unlikely as no evidence of a bethanechol peak or degradation peak was noted in the chromatogram. Spike recoveries of bethanechol to the suspect sample ranged from 88%–102% for spike levels of 0.09, 0.25, and 1.9 mg/mL.

Evidence that bethanechol was not present in the suspect sample was obtained by analyzing for the chloride counter ion using anion-exchange IC. Even if the bethanechol degraded, the amount of chloride from bethanechol chloride would not be expected to change. The sample was analyzed using a Dionex IonPac AS16 column with 23 mM KOH produced using an EG40 eluent generator. The flow rate was 1.3 mL/min. The ASRS was used in recycle mode and set at 100 mA. After accounting for the contribution of chloride in the sample from the flavoring syrup and the oral suspending vehicle, only 1.6% of the expected concentration of chloride for a 1 mg/mL bethanechol suspension (as bethanechol chloride) was found.

Further analysis using gas chromatography-mass spectrometry (GC-MS) of the suspect oral suspension indicated the presence of the drug baclofen, a muscle relaxer and antispastic agent, in the suspension. This was confirmed using LC-MS^n. Quantitative analysis using HPLC with UV detection showed a level of approximately 10 mg/mL baclofen in the suspect sample. Baclofen tablets are supplied in a 10 or 20 mg/tablet dose. It is surmised that the compounder accidentally picked up the baclofen bottle next to the bethanechol bottle for preparation and used it to prepare the bethanechol oral suspension.

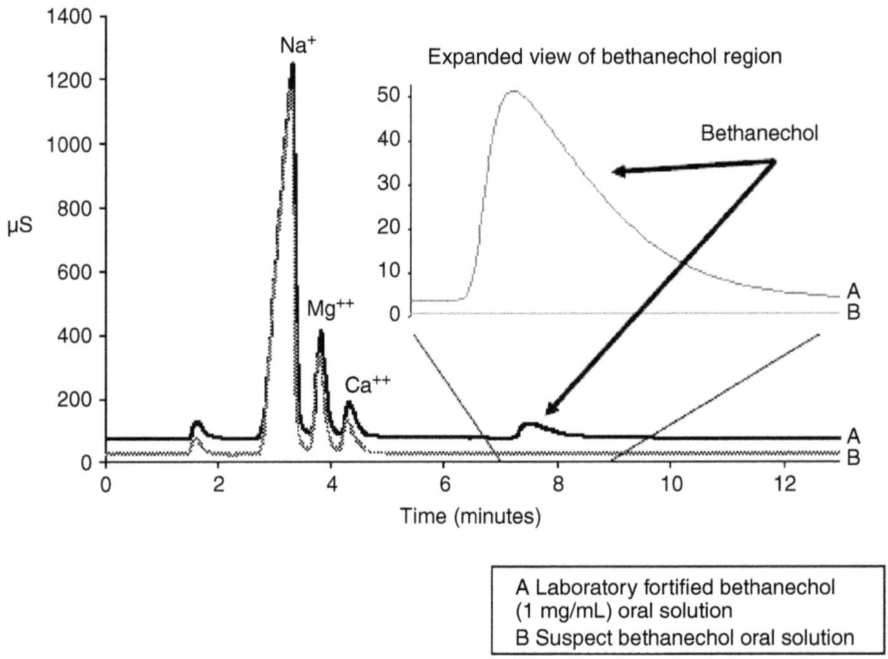

Figure 12.6. Chromatogram of suspect and laboratory fortified bethanechol oral solutions.

12.3 CONCLUSION

IC is an important tool for analytical problem solving. Anion-exchange chromatography with suppressed conductivity detection was used to determine citrate levels in fentanyl citrate syringes suspected of being tampered. The results provided evidence that the pain medication had been diluted or in some cases completely removed and substituted with water thus supporting the tampering allegation. An unapproved botulinum toxin type A product was analyzed using HPAEC-PAD. The results demonstrated that sucrose was not present in the product despite being on the label. The absence of the sucrose stabilizer could have had an effect on the stability of the neurotoxin protein which potentially could result in a sub-potent product or the presence of potentially harmful degradation products. A pharmacy compounded bethanechol oral suspension was analyzed using cation-exchange chromatography with suppressed conductivity detection after an apparent mistake in compounding that resulted in the hospitalization of two infants. IC showed that the compounder used the wrong drug in the making of the oral solution and that bethanechol was not present in the oral suspension. The three cases presented demonstrate the utility of ion chromatographic analysis in forensic investigations of pharmaceutical authenticity and adulteration.

12.4 ACKNOWLEDGEMENTS

The author would like to thank his co-workers Mantai Mesmer, Lorrie Lin, Elisa Nickum, Frank Platek, Timothy Yi, Cathy Dasenbrock and Doug Heitkemper for providing data and information related to the cases discussed and for technical review of the manuscript.

REFERENCES

1. USP 29-NF 24, General Chapter <1065> Ion Chromatography. Rockville: United States Pharmacopeial Convention; 2006:2898–2900.
2. USP 32-NF 27, Monograph: Potassium Perchorate, Assay. Rockville: United States Pharmacopeial Convention;2009:3351.
3. USP 32-NF 27, Monograph Bethanechol Chloride, Assay. ibid.:1669.
4. USP 32-NF 27, Monograph Erythromycin Ointment, Assay. ibid.:2285.
5. USP 32-NF 27, Monograph Fludeoxyglucose F18 Injection, Chemical Purity. ibid.:2406.
6. USP 32-NF 27, Monograph Streptomycin Sulfate, Assay. ibid.:3601.
7. USP 32-NF 27, Monograph Kanamycin Sulfate, Assay. ibid.:2734.
8. USP 32-NF 27, Monograph Amikacin, Assay. ibid.:1508.
9. USP 32-NF 27, Monograph Sodium Fluoride and Acidulated Phosphate Topical Solution. ibid.:3575.
10. USP 32-NF 27, General Chapter <345> Assay for Citric Acid / Citrate and Phosphate. ibid.: 143.
11. Kaine LA, Heitkemper DT, Jackson DS, Wolnik KA. Use of Ion Chromatography for the Verification of Drug Authenticity. J. Chromatogr. 1994;671:303–308.
12. Kaine LA, Wolnik KA. Forensic Investigation of Gentamicin Sulfates by Anion-exchange Ion Chromatography with Pulsed Electrochemical Detection. J. Chromatogr. 1994;674:255–261.
13. US Drug Enforcement Agency, Drug Scheduling. Available at http://www.justice.gov/dea/pubs/scheduling.html. 29 April 2010.
14. Stauros K, Botulinum Neurotoxin: The Ugly Duckling. European Neurology 2009;61(6):331–342
15. Dionex Corporation. Application Note 148: Determination of Bethanechol by Ion Chromatography.
16. Sondheimer JM, Arnold GL. Early Effects of Bethanechol on the Esophageal Motor Function of Infants with Gastroesophageal Reflux. J Pediatr Gastroenterol Nutr. 1986;5(1):47–51.

13

ION CHROMATOGRAPHY OF DRUG PRODUCT COUNTER-IONS AND OTHER IONS IN DRUG PRODUCTS

Brian M. De Borba and Jeffrey S. Rohrer

Thermo Fisher Scientific, Sunnyvale, CA

13.1 INTRODUCTION

Salt formation is a critical step during the drug development process to help promote solubility, improve stability, decrease toxicity, and reduce hygroscopicity [1]. Salt selection previously occurred during drug development, but most companies now perform this selection during the research stage to avoid repeating toxicological, formulation, and stability studies. Currently, most pharmaceuticals are administered in the salt form (~56%) compared to the free form (~44%) (2). Chloride and sodium counter-ions remain the dominant salt forms for the manufacture of basic and acidic drugs, respectively. These counter-ions have advantages over other salt forms due to their low molecular weight, higher aqueous solubility, and low toxicity [2]. Other common anionic salt forms include sulfate, bromide, mesylate, maleate, citrate, tartrate, phosphate, and acetate.

It is critical to quantify the concentration of the counter-ions in the drug formulations to establish stoichiometry, completeness of salt formation, and mass balance. Many of these ions are non-chromophoric, precluding their analysis by traditional

Applications of Ion Chromatography for Pharmaceutical and Biological Products, First Edition.
Edited by Lokesh Bhattacharyya and Jeffrey S. Rohrer.

reversed-phase liquid chromatography (HPLC) with UV detection. Ion chromatography (IC) is able to determine a wide range of chromophoric and non-chromophoric anionic and cationic pharmaceutical counter-ions and therefore is the preferred technique for pharmaceutical counter-ion determinations. Anions are separated on a high-performance anion-exchange column using either a carbonate/bicarbonate or hydroxide eluent and detected by suppressed conductivity detection. Cations are separated on a high-performance cation-exchange column using a dilute strong acid and also detected by suppressed conductivity detection. This chapter describes recent changes in the separation and detection components of the IC methods used to determine pharmaceutical counter-ions and recommends starting points for these analyses. The remainder of the chapter is devoted to examples of IC methods for counter-ion analysis.

13.2 IC METHODS FOR PHARMACEUTICAL COUNTER-ION ANALYSES

Traditionally, carbonate/bicarbonate eluents have been widely used for the determination of common inorganic anions in a variety of sample matrices, including those of pharmaceutical origin. Carbonate eluents have relatively low background conductivities due to the weakly dissociated carbonic acid produced in the suppressor. In addition, the elution strength can be varied by changing the carbonate/bicarbonate ratio (i.e. pH change) to further optimize the separation [3]. These eluents also offer advantages of being easier to prepare, and because they are used isocratically, additional equilibration time prior to the next injection is not required. However, carbonate eluents are not suitable for gradient separations due to the significant increase in background signal with increasing eluent concentration, even with suppressed conductivity detection, which precludes the determination of many analytes. This limits the number of anions that can be separated in a single injection and, therefore, carbonate eluents are commonly used only in the routine analysis of common monovalent and divalent counter-ions, such as chloride and sulfate. Deviations from a linear calibration curve can also be expected due to the formation of carbonic acid, which is a weak acid with a $pK_{a,1}$ of 6.35.

Hydroxide is the ideal eluent choice for the separation of anions by IC because it provides several advantages over traditional carbonate eluents. The primary advantage of a hydroxide eluent originates from the exceptionally low background conductivity of its suppression product, water. The lower background produces lower baseline noise and, therefore, lower detection limits relative to carbonate eluents. Another advantage is that increasing the hydroxide concentration during a gradient elution does not significantly influence the background signal. This assumes that the suppressor has enough capacity to continuously suppress an eluent of increasing concentration. Due to the capacity limitations of the early suppressors, gradient elution was impractical. Current suppressor designs enable continuous eluent suppression and, contrary to early suppressors, do not require the preparation of acid or base solutions for suppression [4]. These suppressors allow the use of high capacity columns that are ideal for separating a wide range of analytes in a single injection.

Hydroxide gradients are particularly advantageous for screening pharmaceuticals for the counter-anions and unintended anionic impurities that are present at significantly lower concentrations. An example chromatogram using gradient hydroxide elution is shown in Figure 13.1, which demonstrates the separation of a wide range of possible counter-ions used in drug substances. Such separation is impossible using a carbonate eluent [5]. Conversion of hydroxide to water by the suppressor results in calibration curves that are linear over a greater range than those obtained when carbonate eluents are used. However, the manual preparation of hydroxide eluents has been particularly challenging because carbonate can be introduced as an impurity from the reagents or through adsorption of carbon dioxide from the air. The presence of carbonate in hydroxide eluents can result in undesirable baseline shifts during the gradient, irreproducible retention times, and increased detection limits. Therefore, the introduction of an electrolytic eluent generator to produce high purity hydroxide eluent is a significant advancement in IC [6]. This device eliminates the challenges previously observed with manual eluents and improves system performance by reducing errors associated with carbonate contamination from manually prepared hydroxide eluents. Perhaps the greatest advantages of electrolytically generating the hydroxide eluent are that only deionized water is required as a carrier solution for the eluent and the hydroxide generated is carbonate-free. This system configuration minimizes

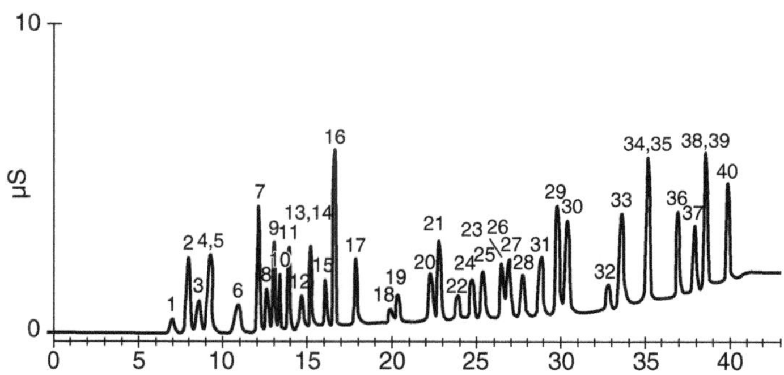

Figure 13.1. Gradient separation of various inorganic and organic anions on the IonPac AS11-HC. Eluent: (A) Deionized Water, (B) 5 mM NaOH, (C) 100 mM NaOH, (D) 100% Methanol; gradient: 80% A/20% B for 8 min, then to 65% A/15% C/20% D for 10 min, then to 50% A/30% C/20% D for 10 min, then to 40% A/60% C for 10 min; flow rate: 1.5 mL/min; column temperature: 30°C; detection: suppressed conductivity; injection volume: 10 μL; analyte concentrations: (1) 10 ppm quinate, (2) 3 ppm fluoride, (3) 10 ppm lactate, (4) acetate, (5) glycolate, (6) propionate, (7) formate, (8) butyrate, (9) methylsulfonate, (10) pyruvate, (11) chlorite, (12) valerate, (13) galacturonate, (14) monochloroacetate, (15) bromate, (16) 5 ppm chloride, (17) 10 ppm nitrite, (18) sorbate, (19) trifluoroacetate, (20) bromide, (21) nitrate, (22) glutarate, (23) 20 ppm carbonate, (24) 15 ppm succinate, (25) malate, (26) malonate, (27) tartrate, (28) maleate, (29) sulfate, (30) oxalate, (31) fumarate, (32) 20 ppm ketomalonate, (33) tungstate, (34) phosphate, (35) phthalate, (36) citrate, (37) isocitrate, (38) chromate, (39) *cis*-aconitate, (40) *trans*-aconitate.

the cost associated with pump maintenance and enables easy method transfer between laboratories, which is particularly important in regulated environments. A significant number of hydroxide-selective anion-exchange columns have been developed after the introduction of electrolytic eluent generation. Many of these columns offer moderate to high ion-exchange capacities for a high resolution separation of monovalent to polyvalent anions using a gradient elution with minimal shift in the baseline. Many of the columns developed in recent years also provide solvent compatibility, which can alter the selectivity of the separation. Although most IC pharmaceutical separations do not require solvent in the eluent, this characteristic has an important role in analyzing pharmaceuticals that are less water soluble.

The analysis of cations and selected amines can also be facilitated by using an electrolytically generated methanesulfonic acid (MSA) eluent to determine the counter-cations or cationic impurities in acidic pharmaceutical drugs. Although carbonate contamination is not an issue for acidic eluents, a high-purity and contaminant-free MSA eluent minimizes the increase in the background conductance during a gradient, and therefore greatly enhances the quality of the chromatogram [6]. Electrolytic generation of either an acid or base eluent produces highly reproducible chromatograms and, thus, improves the analytical results by removing eluent concentration and purity as variables.

13.3 APPLICATION EXAMPLES

Sulfate is the second most common anionic salt form used in the pharmaceutical industry [2]. Aminoglycosides, which are a well-known class of antibiotics, are an example of pharmaceuticals commonly prepared in this form. It is not uncommon for these compounds to contain up to 30% sulfate by weight. IC can be used to rapidly determine the sulfate concentration with good precision and accuracy. Paromomycin is an aminoglycoside antibiotic that is widely used in the treatment of various bacterial infections, such as leishmaniasis, cryptosporidiosis, and amebiasis, and is prepared in the sulfate form [7–10]. The sulfate concentration determined by IC in a USP paromomycin sulfate RS showed 23.6% (w/w), which is within 0.1% of the theoretical concentration with a sulfate recovery of 97.1% (Figure 13.2) [11]. In this example, a hydroxide-selective column, IonPac AS18, that was specifically designed for the determination of common inorganic anions, such as sulfate, was used. This column is suitable for well-characterized pharmaceutical formulations and provides a higher throughput analysis than the IonPac AS11-HC column. This analysis can be used to complement a separate quantitation of the aminoglycoside free base to determine the molar ratio. The free base can be separated on a CarboPac PA1 column combined with integrated pulsed amperometric detection [12]. The latter analysis in combination with the former results revealed that paromomycin sulfate is prepared in a 1:2 molar ratio. IC can also be used to determine trace concentrations of undesirable impurities in pharmaceutical preparations that can originate from the raw materials used in the manufacturing process, as shown in Figure 13.3 for Humatin®, which is a commercially available paromomycin sulfate drug product [11]. The column used for this

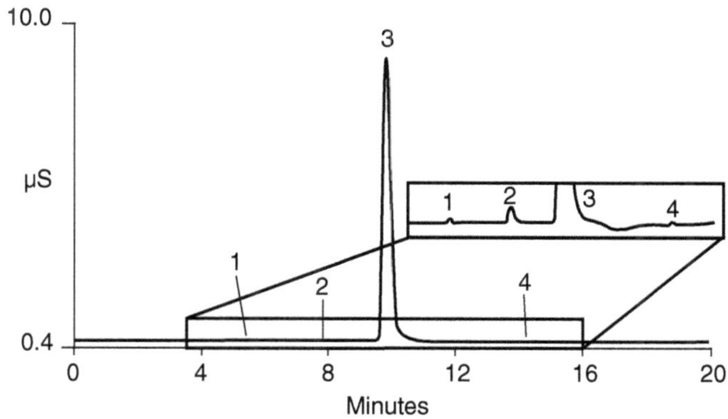

Figure 13.2. Separation of sulfate and anionic impurities in USP-grade paromomycin sulfate. Column: IonPac AS18 (2 × 250 mm); eluent: 22 mM KOH 0–7 min, then 22–40 mM 7–8 min, then 40 mM 8–20 min; eluent source: eluent generator cartridge; flow rate: 0.25 mL/min; column temperature: 30°C; detection: suppressed conductivity; injection volume: 5 μL; analyte concentrations: (1) 0.016% chloride, (2) carbonate, (3) 23.6% sulfate, (4) 0.040% phosphate.

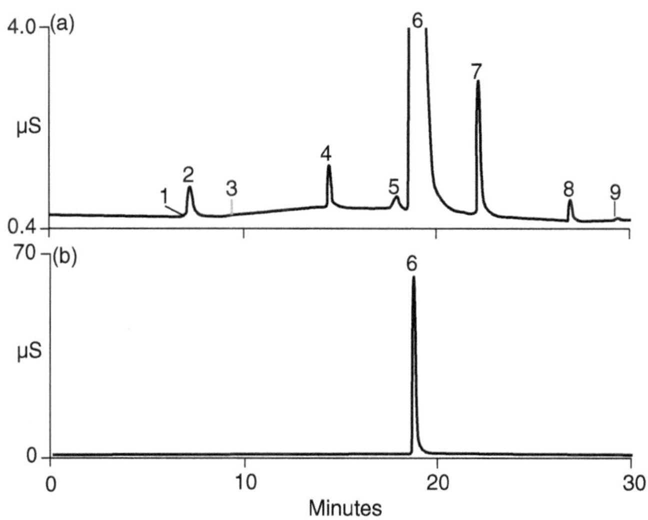

Figure 13.3. Separation of sulfate counterion and anionic impurities in Humatin®. Column: IonPac AS11-HC (2 × 250 mm); eluent: 1 mM KOH from 0–5 min, then 1–5 mM 5–9 min, 5–38 mM 9–20 min, 38–60 mM 20–25 min, 60 mM 25–30 min; eluent source: eluent generator cartridge; flow rate: 0.38 mL/min; column temperature: 30°C; detection: suppressed conductivity; injection volume: 5 μL; sample: (a) 2.5 mg/mL Humatin, (b) 0.25 mg/mL Humatin; analytes: (1) unknown, (2) 0.080% acetate, (3) unknown, (4) 0.025% chloride, (5) carbonate, (6) 24.7% sulfate, (7) 0.23% phosphate, (8) 0.0035% pyrophosphate, (9) unknown.

analysis, IonPac AS11-HC, is a high-capacity, hydroxide-selective anion-exchange column that provides high resolution separations of inorganic anions, low molecular weight organic anions, and polyvalent anionic species. Therefore, this column is best suited for the analysis of pharmaceutical samples that are not well-characterized. As a result, longer analysis times are typical for most separations compared to the example shown in Figure 13.2.

Another example is shown in Figure 13.4. It shows the separation of the chloride and bromide counter-ions of pseudoephedrine and dextromethorphan, respectively, in a multisymptom cold and flu formulation. These anions were separated using a carbonate/bicarbonate eluent in less than 10 min and provided accuracy within 10% of the label value [13]. This analysis represents a fairly simple sample matrix for determining common inorganic anions where a carbonate-based eluent was a suitable choice. Although this method offers the benefit of shorter run times, it does not allow the separation of polyvalent anions, such as citrate. To separate citrate in the sample matrix a hydroxide eluent is recommended.

Counter-ion assays also have an important role in investigating drug counterfeiting and in product tampering cases (see chapter 12 of this book). IC can be used as one of several analytical tools to determine the source and cause of the contaminants. Common household items contain corrosive mineral acids and bases that could be used in product tampering. In one example, liquid cold medication was investigated for tampering with NaOH, which caused severe burns to the lips, mouth, and esophageal tracts [14]. The initial analysis revealed a sample pH of 14 and sodium content of 10.9%. The active ingredient of the cold medicine was dextromethorphan HBr. The bromide counter-ion could be easily determined by anion-exchange chromatography. In this investigation, the counter-ion was a good indication of sample dilution as other ingredients can degrade under alkaline conditions. Comparison of the bromide response in the control and suspect sample indicated very little sample

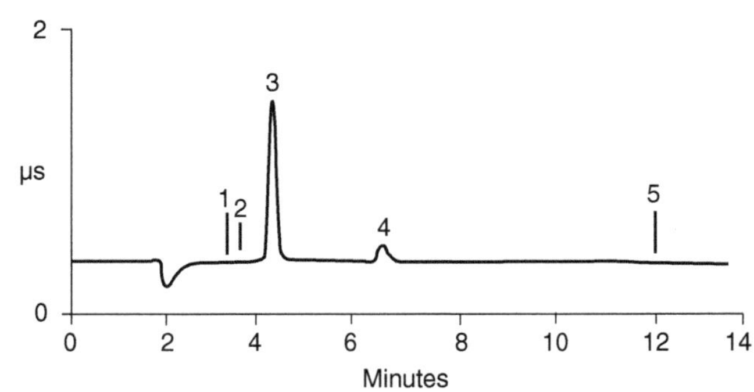

Figure 13.4. Separation of chloride and bromide counterions in a multisymptom cold and flu medication. Column: IonPac AS14 (4 × 250 mm); eluent: 3.5 mM sodium carbonate/0.8 mM sodium bicarbonate; flow rate: 1.2 mL/min; detection: suppressed conductivity; injection volume: 10 μL; sample preparation: 1:100 dilution with deionized water; analytes: (1) unknown, (2) unknown, (3) 3 ppm chloride, (4) 1 ppm bromide, (5) 0.08 ppm sulfate.

dilution and it was speculated that solid NaOH was added to the cold medicine. IC is also an important analytical technique in the investigations of drug authenticity. In one example the drug's active ingredient was substituted for a less expensive active ingredient [15]. The suspect drug was then repackaged as the more expensive drug for increased profit margin. However, the legitimate drug contained an equimolar ratio of phosphate and the active ingredient, whereas the suspect drug was found to contain low phosphate concentration and equimolar concentrations of chloride and the active ingredient. Thus, IC was a valuable tool in this investigation to identify that the suspect drug counter-ion concentration did not match the counter-ion concentration in the legitimate drug product.

Another IC application in the pharmaceutical industry is the determination of trifluoroacetic acid (TFA), which is commonly used when manufacturing peptides (Figure 13.5) [16–18]. TFA is also a common eluent additive in reversed-phased separations of peptides and proteins. The IC application is to assay for TFA to monitor its removal at different stages of purification. Figure 13.6 compares the separation of TFA in a crude synthetic peptide sample to the same sample after purification with gel permeation chromatography (GPC). As these chromatograms illustrate, GPC was ineffective in removing TFA from the peptide sample [19].

Some pharmaceutical compounds are less water soluble and therefore may be particularly challenging to analyze as the compound can potentially precipitate within the chromatography system and therefore cause excess backpressure and column contamination. This challenge can be overcome by adding sufficient amount of solvent to the eluent or precipitating the pharmaceutical then filtering the resulting solution prior to analysis. Another and perhaps simpler approach to resolve this issue is to preconcentrate the sample dissolved in 100% organic solvent on an anion-exchange

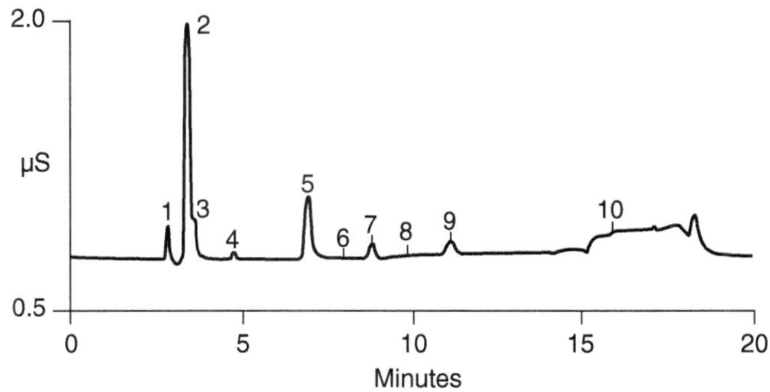

Figure 13.5. Determination of trace concentrations of trifluoroacetate (TFA) in a commercial peptide preparation. Column: IonPac AS18 (4 × 250 mm); eluent: 22 mM KOH 0–6 min, then 28 mM KOH 6–12 min, 50 mM KOH 12–15 min, 22 mM KOH 15–20 min; eluent source: eluent generator cartridge; flow rate: 1 mL/min; column temperature: 30°C; detection: suppressed conductivity; injection volume: 50 μL; analytes: (1) system peak, (2) acetate, (3) formate, (4) 15 ng/mL chloride, (5) carbonate, (6) unknown, (7) 140 ng/mL sulfate, (8) 14 ng/mL nitrate, (9) 100 ng/mL TFA, (10) 25 ng/mL phosphate.

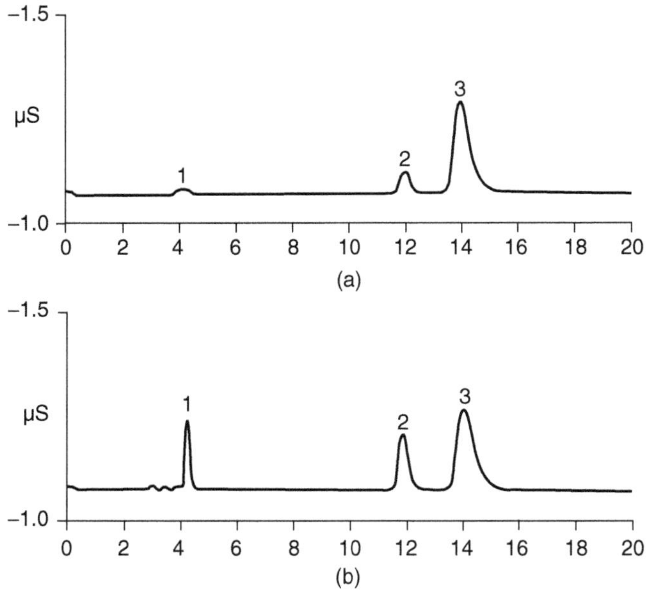

Figure 13.6. Separation of TFA in (a) crude in-process peptide, and (b) GPC purified in-process peptide. Column: IonPac AS14 (4 × 250 mm); eluent: 3.5 mM sodium carbonate/0.8 mM sodium bicarbonate; flow rate: 1.2 mL/min; detection: suppressed conductivity; injection volume: 10 μL; peptide concentration: 40 μg/mL; analyte concentration, (A): (1) 0.02% chloride, (2) 1.0% sulfate, (3) 19.6% TFA; (B): (1) 0.83% chloride, (2) 2.6% sulfate, (3) 16.7% TFA.

concentrator column and then eliminate the matrix with deionized water prior to analysis. This procedure was used successfully to determine inorganic anionic impurities in a proprietary pharmaceutical compound, which has poor water solubility (Figure 13.7) [20]. The IonPac AS15 is a high-capacity, hydroxide-selective column specifically developed for the rapid and efficient separation of trace inorganic anions in matrices of varying ionic concentration. Although the total ionic concentration of this sample was relatively low, the fluoride concentration was significantly higher than other close eluting anionic impurities and therefore this example demonstrates the suitability of the AS15 column to resolve anions at disparate concentration ratios. Fluoride is present in some raw materials used during the preparation of pharmaceuticals. Calcium salts are known to be highly contaminated with fluoride due to their manufacturing process. Phosphate is also expected to be present because phosphate buffers are used during the preparation of the final formulation. The sensitivity of IC is needed to determine trace concentrations of such ionic impurities. This is illustrated in the above example, which is found to have LODs in the sub-μg/L to μg/L range. The low detection limit is an advantage of selecting a hydroxide eluent over a carbonate eluent, as previously discussed.

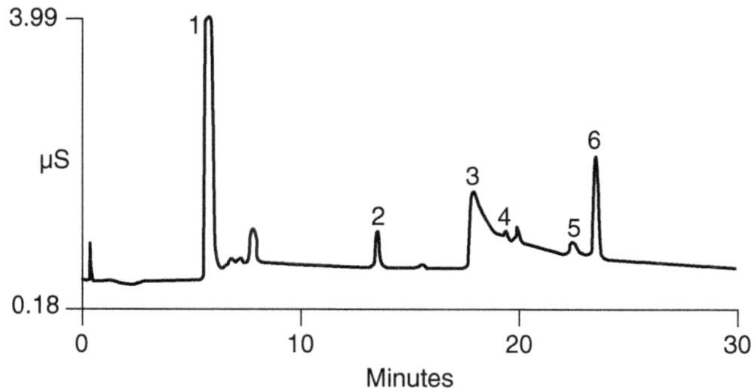

Figure 13.7. Separation of anionic impurities of a proprietary pharmaceutical compound, which has poor solubility in water. Column: IonPac AS15 (2 × 250 mm); eluent: 10 mM KOH 0–8 min, then 10–40 mM KOH 8–14 min, 40–60 mM KOH 14–20 min, 60 mM KOH 20–30 min; eluent source: eluent generator cartridge; flow rate: 0.40 mL/min; column temperature: 30°C; detection: suppressed conductivity; injection volume: 100 µL; matrix elimination volume: 1 mL (deionized water); sample: 0.30 mg/mL pharmaceutical compound in 100% methanol; analytes: (1) 975 µg/L fluoride, (2) 33.3 µg/L chloride, (3) carbonate, (4) 6.5 µg/L sulfate, (5) 33.8 µg/L nitrate, (6) 339 µg/L phosphate.

13.4 FUTURE DIRECTIONS

A few monolithic IC columns have been developed recently. These columns offer the benefit of faster run times [21]. The porous monolithic stationary phases enable the use of higher flow rates to significantly reduce the analysis time. These columns can potentially prove beneficial in the determination of active ingredient counter-ions, which allow researchers to rapidly obtain results and therefore improve productivity. Another important area is in the development of capillary IC systems. These systems offer the benefit of a significant reduction in sample size, which is particularly important for costly pharmaceutical compounds, and also for the reduction of waste generation. Further research in these areas is needed to determine their feasibility for routine counter-ion analysis in the pharmaceutical industry.

13.5 CONCLUSION

IC with suppressed conductivity detection is a well-established technique for the determination of a wide range of anions and cations. Therefore, IC has become an invaluable analytical tool in the pharmaceutical industry for the determination of counter-ions in drug products. Determination of the counter-ion concentration is critical for several reasons, such as establishing production of the correct salt form, determining the stoichiometric ratio, determining mass balance, and for determining completeness of salt formation. IC has also proven to be a valuable technique in drug authenticity investigations based on the presence or absence of an expected drug to counter-ion ratio. The introduction of Reagent-Free IC systems with improved hydroxide-selective columns has further enhanced the screening of pharmaceutical counter-ions and unintended ionic impurities, by allowing gradient elution with accurately produced pure hydroxide eluents to determine a wide range of inorganic and low molecular weight organic anions in a single analysis. Future developments in IC, such as monolithic columns and capillary systems, should provide additional benefits for counter-ion analyses.

13.6 ACKNOWLEDGEMENTS

We would like to gratefully acknowledge Edward Kaiser and Valoran Hanko for their contributions to portions of the work presented here.

REFERENCES

1. Haynes DA, Jones W, Motherwell WDS. Occurrence of pharmaceutically acceptable anions and cations in the Cambridge structural database. J. Pharmaceut. Sci. 2005;94:2111–2120.
2. Kumar L, Amin A, Bansal AK. Salt selection in drug development. Pharmaceutical Technology 2008;32(3):128–142.
3. De Borba BM, Later DW, Rohrer JS, Pohl CA, Jackson PE. Determination of common inorganic anions in environmental waters using a hydroxide-selective column. LCGC 2004;22:138–148.
4. Weiss J. Handbook of ion chromatography. Volume 1, Experimental parameters affecting retention when applying direct conductivity detection. New York: VCH; 2004.
5. Dionex Corporation Product Manual, IonPac AG11-HC and AS11-HC (2008) Document No. 031333–08.
6. Liu Y, Avdalovic N, Pohl C, Matt R, Dhillon H, Kiser R. An on-line, high-purity acid and base eluent generation system for ion chromatography. Am. Lab. 1998;November:48C–54C.
7. Murray HW, Berman JD, Davies CR, Saravia NG. Advances in leishmaniasis. The Lancet 2005;366:1561–1577.
8. Flanigan TP, Ramratnam B, Graeber C, Hellinger J, Smith D, Wheeler D, Hawley P, Heath-Chiozzi M, Ward DJ, Brummitt C, Turner J. Prospective trial of paromomycin for cryptosporidiosis in AIDS. Am. J. Med. 1996;100:370–372.
9. Fahey, T. Cryptosporidiosis. Primary Care Update for OB/GYNS 2003;10:75–80.
10. The European Agency for the Evaluation of Medicinal Products: Veterinary Medicines and Information Technology Unit. Committee for Veterinary Medicinal Products: Paromomycin Summary Report. London, UK; 2000 January. Report nr EMEA/MRL/718/99-FINAL. Available from: EMEA, London, UK.
11. Dionex Corporation Application Note 190, Determination of sulfate counterion and anionic impurities in aminoglycoside drug substances by ion chromatography with suppressed conductivity detection.

12. Dionex Corporation Application Note 186, Analysis of paromomycin by HPAE-IPAD.
13. Dionex Corporation Application Note 116, Quantification of anions in pharmaceuticals.
14. Kaine LA, Jackson DS, Heitkemper DT. Ion chromatography analysis of corrosive materials in product tampering cases. J. Forensic Sci. 1999;44:413–416.
15. Kaine LA, Heitkemper DT, Jackson DT, Wolnik KA. Use of ion chromatography for the verification of drug authenticity. J. Chromatogr. A 1994;671:303–308.
16. Dionex Corporation Application Note 160, Determination of residual trifluoroacetate in protein purification buffers and peptide preparations by Reagent-Free™ Ion chromatography.
17. Kabakoff B, Blank A, Heinsohn H. Final Program. International Ion Chromatography Symposium; 1997 Sept 14-17; Santa Clara, CA. p 165.
18. Weiss J. Handbook of ion chromatography. Volume 2, Ion chromatography in the pharmaceutical industry. New York:VCH;2004.
19. Dionex Corporation Application Note 115, Determination of trifluoroacetic acid (TFA) in peptides.
20. Dionex Corporation Application Note 220, Determination of inorganic anion impurities in a water-insoluble pharmaceutical by ion chromatography with suppressed conductivity detection.
21. Pohl C. Recent developments in ion-exchange columns for inorganic ions and low molecular weight ionizable molecules. Supplement to LCGC 2010:24–31.

14

SAMPLE PREPARATION TECHNIQUES IN ION CHROMATOGRAPHY FOR PHARMACEUTICAL APPLICATIONS

Jay Gandhi

Metrohm USA, Inc., Riverview, FL

14.1 INTRODUCTION: SAMPLE PREPARATION METHODS

There are numerous sample preparation methods for the wide range of problems to be solved. The apparatus and the time required for sample preparation vary significantly. The working steps range from relatively simple measures such as sample dilution, filtration, or pH adjustment, to complicated and sometimes even multi-step procedures such as analyte preconcentration and matrix elimination. Table 14.1 summarizes the most important sample preparation methods for ion chromatography (IC) and other chromatographic techniques. The sample preparation techniques are arranged according to their use for liquid, solid, and gaseous samples.

The choice of the sample preparation method depends on the one hand on the physical state and the composition of the sample. On the other hand it depends to a considerable extent on the analyte ions to be determined. The criteria that are ultimately important, however, are the (possible) choice of the separation principle and separation conditions for the IC analysis and, for many users, the availability of particular instrument configurations (e.g., chemical conductivity suppression, gradient elution, type of detector).

Applications of Ion Chromatography for Pharmaceutical and Biological Products, First Edition.
Edited by Lokesh Bhattacharyya and Jeffrey S. Rohrer.

TABLE 14.1. Sample Preparation Techniques for IC.

A-Liquid Samples	B-Solid Samples	C-Gaseous Samples
Filtration	Drying	Absorption in liquids
Dilution	Homogenization	Adsorption on solid-phases
pH adjustment	Dissolution	Membrane sampling
Standard addition	Extraction/leaching	Chemical conversion
Derivatization	Digestion	
Liquid-liquid extraction	Ashing	
Solid-phase extraction	Combustion	
Distillation		
Microdiffusion		
Membrane separation (dialysis/ultrafiltration)		

Sample pretreatment for the determination of anions frequently requires different measures from those used for the determination of cations. It is also clearly more difficult to carry out sample preparation when a large number of ions are to be determined simultaneously than when only a few or even a single ion is of interest. The target concentrations of the ions to be determined may, under certain circumstances, be critical for the choice of a particular sample preparation technique. The properties of the separating column used with respect to its resolution, capacity, pH stability, and compatibility with organic solvents as modifiers in the eluent can also be decisive for selection of the type of sample pretreatment.

Apart from the unwanted increase in analysis time and costs associated with sample preparation, as the number of steps that need to be carried out increases, the risk of analytical errors also increases. Possible errors due to contamination, loss of analyte or conversion of the analyte may falsify the results of qualitative and quantitative analyses. This must be accounted for when choosing the preparation steps and, if possible, avoided by taking suitable measures.

Sample preparation costs can increase as the complexity of the sample composition increases. Measured against the large number of IC applications there are only a few types of sample that can be directly analyzed without sample preparation. If samples are already present in aqueous form then sample preparation may have various targets.

Adjustment to a particular pH value is only necessary on rare occasions as many eluents will determine the pH of the injected sample, preventing interference with the separation.

For liquid samples, the sample preparation can be achieved using techniques such as gravity filtration, filtration through a sorbent under either gravity or pressure or vacuum (to speed filtration), or filtration under centrifugal force (centrifuge). However, sample preparation of aqueous samples can also be integrated as an automated sample preparation step into IC instrumentation. Figure 14.1 demonstrates general steps that can be easily automated along with IC instrumentation.

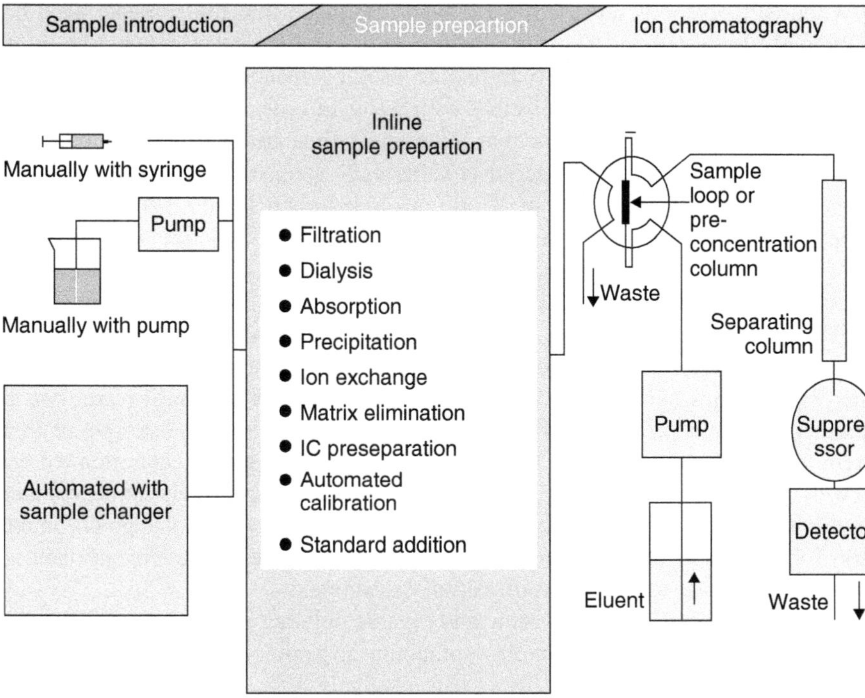

Figure 14.1. Integration of general sample preparation techniques with IC.

The basics of sample preparation and the possible applications of important separation methods are summarized and described in the following sections.

14.2 SOLID-PHASE EXTRACTION

For the isolation and preconcentration of analyte ions as well as the separation of interfering matrix components, various versions of SPE are particularly suitable. This normally involves passing the sample solution through a small column (SPE cartridge) that is filled with a sorbent material suitable for the separation, which either retains the analyte ions or the interfering matrix components. In the case of analyte preconcentration, the analyte ions are eluted from the SPE cartridge in a subsequent step and the eluate is then analyzed by IC. If matrix components are retained on the solid phase then the determination is carried out on the solution leaving the SPE cartridge.

Many SPE cartridge manufacturers offer ready-to-use cartridges filled with sorbent material as well as devices designed to pass the sample solution through the sorbent bed either manually, semi-automatically, or automatically. A practical and efficient way of executing automated sample preparation by SPE is with a column-switching technique. In this case, matrix elimination occurs in a flow system functioning in the low-pressure range that is connected in-line with the IC instrument.

When SPE methods are used, care should be taken that the sorbents used have a sufficiently high degree of purity as there is a serious contamination risk, particularly in trace determinations. Hydrophobic packing materials require conditioning with an organic solvent that is miscible with water in order to ensure good wettability of the sorbent. Finally, ensure that the contact time between the components to be separated and the sorbent material is sufficiently long. Regardless of the type of sorbent used and the practical execution, an optimization of the SPE that accounts for the sample's matrix composition is absolutely essential. This includes determination of reproducibility and recovery by spiking experiments as well as checking for contamination and analyte loss [1].

SPE can be used for matrix modification and the separation of interfering substances. The neutralization of acidic and alkaline samples can be carried out with ion-exchange resins in basic or acidic form, respectively. Ion-exchange resins can also be used for the selective removal of particular interfering ions. An example of this is the removal of chloride or sulfate from samples containing high concentrations of these ions with cation-exchangers in the Ag^+- and Ba^{2+}-forms, respectively, or the exchange of multivalent cations (which could cause interference in anion-exchange chromatography) for alkali metal ions. Interfering concentrations of carbonate/bicarbonate can be eliminated by the use of H^+-form cation exchangers.

Non-specific sorption on inorganic and organic polymer phases has achieved great importance in the analysis of samples containing organic substances. Depending on their polarity and molecular size these can adversely affect the IC analysis in various ways and, in the worst case, cause irreversible damage to the stationary phase. Hydrophobic packing materials such as polystyrene-divinylbenzene, polymethacrylate resins, or chemically modified silica gels, which are used in HPLC as reversed phases, are particularly suitable for the separation of hydrocarbons, surfactants, aromatic compounds and long-chain organic (fatty) acids from the ions of interest. Polyvinylpyrrolidone has been shown to be suitable for use as the sorbent phase for the separation of humic materials, tannins, lignins, as well as organic dyes, phenolic materials, aldehydes, and aromatic acids. Other, less commonly used sorbents are carbon materials (charcoal, graphitic carbon), inorganic oxides and silicates (Al_2O_3, Mg silicate, zeolites) as well as a wide variety of other organic polymers that have proved to be advantageous for specific separation problems.

Weak Cation exchanger SPE can also be used for the preconcentration of heavy metals as complexed compounds and their subsequent determination by cation-exchange chromatography. Alternatively, chelate-forming ion-exchangers can be used for this purpose. These are extraction reagents covalently bound to organic polymer carriers that selectively bind individual metal ions or groups of metal ions. In a second step the metal cations are eluted with a small volume of mineral acid and can then be directly injected into the IC instrument.

14.3 TRACE PRECONCENTRATION FOR IC ANALYSIS

Trace analysis problems occur in many scientific disciplines and the demand for correspondingly sensitive methods—including ion analysis—has led to various attempts to improve IC detection performance. With the currently available instrument configurations in IC equipped with conductivity detection, numerous ions can be routinely determined without any great difficulty even in concentration ranges below 1 mg/L. An increase in the injection volume can increase the sensitivity, but only for samples with a very low ionic strength, unless the analyst chooses a higher capacity column.

By preconcentration of the ions to be determined in a prechromatographic step (Figure 14.2), it is possible to achieve considerable improvements in the sensitivity and often, at the same time, better matrix elimination. This is of particular importance for the preconcentration of trace components in a difficult matrix. Whereas in individual cases as selective a preconcentration as possible of individual ions in the presence of very high concentrations of other ions may be asked for (e.g., iodide in seawater). In numerous other cases a multi-component preconcentration is required. This applies, for example, to the quality control of ultrapure water used in the semiconductor and pharmaceutical industries or to the monitoring of the boiler water in power plants.

In principle, the trace preconcentration methods are the same as those mentioned before in connection with sample preparation. The use of SPE with appropriate sorbents is very versatile and is by far the most frequently used method for this purpose. The selectivity can be influenced by the type of sorbent so that particular ions (alone

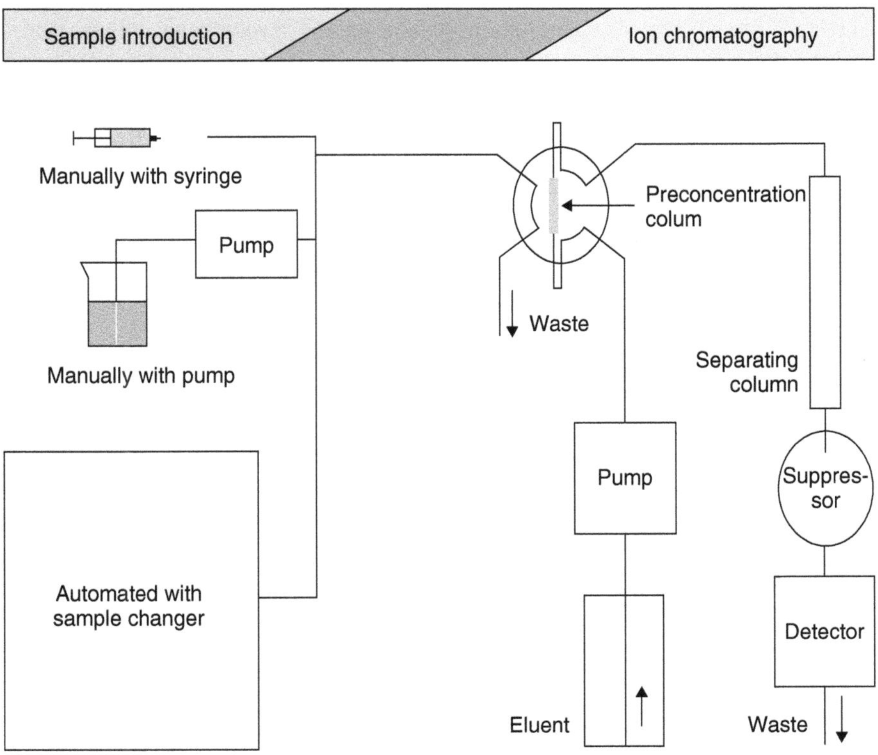

Figure 14.2. Sample preparation with preconcentration technique.

or together with others) can be preconcentrated and separated from the other matrix components. In principle, very high preconcentration factors can be achieved by loading the sorbent phase with relatively large volumes of the sample and eluting with a very small volume of the extraction solvent. However, there is also the risk that, even if the capacity of the sorbent is adequate with respect to the matrix, not all the ions to be determined will be retained quantitatively. For example, numerous anions in pure water can be preconcentrated simultaneously and without many problems on anion-exchange columns and then determined by anion IC, whereas samples with a higher total ionic strength function as an eluent during preconcentration themselves. This means that only ions with a high affinity to the anion exchanger (e.g., sulfate) will be retained completely. A further general problem results from the fact that a sufficiently strong elution agent must be used for eluting the preconcentrated ions; this again represents a new matrix that must be compatible with the IC analysis.

From the perspective of the apparatus required, trace preconcentration can be carried out in a conventional off-line procedure with ready-to-use cartridge systems. Although this off-line preparation can be easily automated, the numerous possible sample contacts with the various vessels and the atmosphere significantly increase the risk of contamination. Integration of a preconcentration step in the analytical procedure of an IC determination (inline preconcentration) is very easy to realize for certain problems by including a small preconcentration column in the injection loop of the IC system. The sample solution is then passed through the column and, on switching the valve; the pre-concentrated components elute and are separated on the column. A limitation is that the sorbent material must be stable at high pressures and the interaction between the analyte and the preconcentration column must not be too strong as the chromatographic eluent must be able to elute the analyte again. As an alternative to the position in the injection valve, the pre-concentration column can also be included in a secondary low-pressure flow system and the eluate transferred to the injection loop by directly coupling it to the IC instrument.

14.4 TRACE PRECONCENTRATION WITH MATRIX REMOVAL FOR IC ANALYSIS

Matrix elimination is the simplest form of inline sample preparation. In this case only the sample loop in the sample valve needs to be exchanged for a preconcentration column. A defined amount of the sample is then pumped through the preconcentration column in accordance with the flow chart shown in Figure 14.2. Only the analyte ions are retained on the column; the matrix is washed away. This method is particularly suitable whenever traces of analyte ions need to be determined as, in this case, matrix elimination and preconcentration are combined. By using the sample loop in Valve 1 a small amount of the sample can be dosed accurately and, in addition, the matrix fraction remaining in the dead volume of the preconcentration column can be rinsed out with water. Typical examples of applications are the determinations of anions or cations in 30% hydrogen peroxide (Figure 14.3) or in organic solvents.

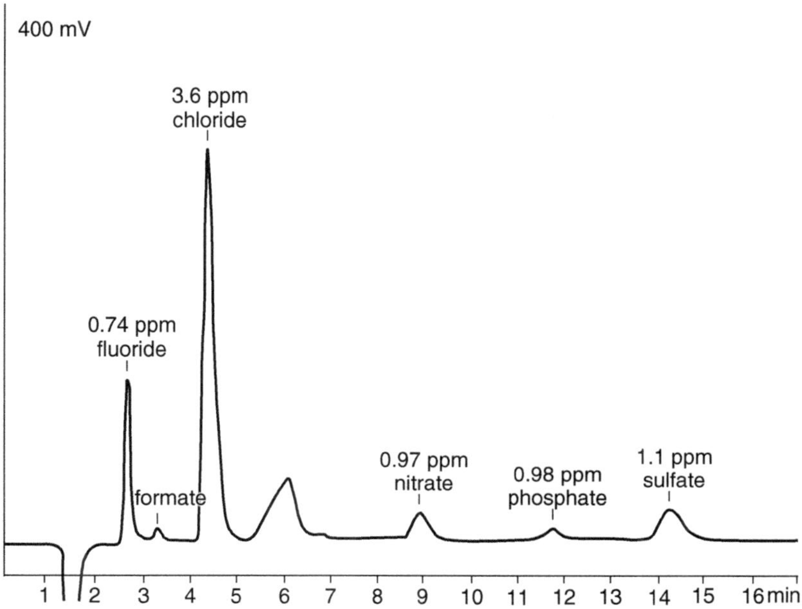

Figure 14.3. Anion analysis of a hydrogen peroxide sample (0.1 mL injected); fluoride 0.03 mg/L, chloride 0.32 mg/L, nitrate 0.23 mg/L and sulfate 2.8 mg/L. Column: ASUPP5-150, elution: isocratic with 3.2 mM $Na_2CO_3 + 1.0$ mM $NaHCO_3$ at 0.7 mL/min, sequential suppressed conductivity detection.

Hydrogen peroxide solutions of up to 50% can be analyzed directly with this technique. This method has been used in the semiconductor industry for a long time, even though interferences occurred—with the separation columns then available—in the sulfate determination (di- and tricarboxylic acids as well as an oxidation product of EDTA); this is not the case when high-performance columns such as the Metrosep A SUPP 5 are used, as can be seen from Figure 14.3 [2].

The determination of anions in organic solvents can also be carried out by the matrix elimination method. In this case the ionic contaminants can be determined in polar solvents (acetone, methanol, ethanol) as well as in very non-polar solvents (halogenated hydrocarbons such as dichloromethane), as only they are retained on the preconcentration column. Even aggressive solvents such as N,N-dimethylformamide can be analyzed. However, care must be taken that the volume in the preconcentration column is sufficiently large, as otherwise there is the risk that some of the ions may pass through the preconcentration column due to poor analyte/sorbent kinetics.

14.5 ULTRAFILTRATION

Classical filtration and microfiltration are not normally regarded as being membrane techniques but are directly linked to ultrafiltration with respect to their use in sample preparation. In all cases a pressure gradient affects the transfer of the solvent and its dissolved fraction through the filter. Particulate constituents are retained and, depending on the pore size of the filter or the ultrafiltration membrane, colloidal fractions and dissolved molecules with a high molecular weight are also retained. Ultrafiltration is, however, regarded as being a filtration technique in which the membranes have pore sizes that are very much smaller than those used in membrane filtration. The characteristic value for the filtration effect is also no longer the pore size, but is now the molecular separation limit. This refers to the molecular weight of a substance that can just (or just not quite) pass through the membrane. The smaller pore sizes also require much higher pressure gradients.

Considering the apparatus and time required, ultrafiltration is clearly more complicated and time-consuming than membrane filtration. Ultrafiltration also costs more. In certain cases, ultrafiltration is also carried out with a series of membranes in order to achieve a fractionation according to molecular weight. The high pressure that is required is obtained either by suction with a vacuum pump or by an ultracentrifuge. Special miniature devices are commercially available in which the membrane separates the sample vessel and the eluate collection vessel. These are placed in the centrifuge as a complete unit and make ultrafiltration much easier. It is also possible to carry out continuous ultrafiltration. In this case the sample solution is passed tangentially along the membrane (cross flow) and the eluate is removed continuously (Figure 14.4). As an

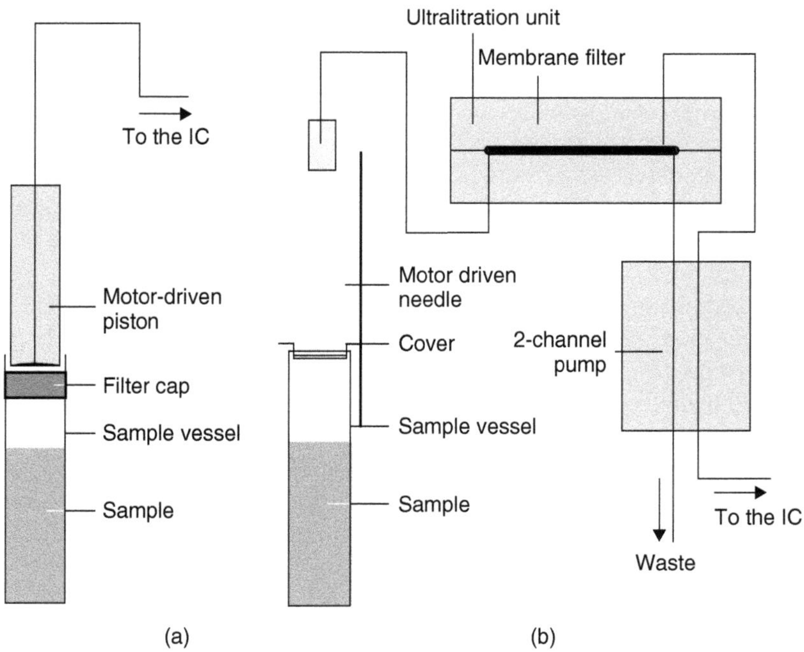

Figure 14.4. (a) Typical IC setup for single use disposable in-line filter. (b) Typical setup for multi-use filter membrane with cross-flow filtration.

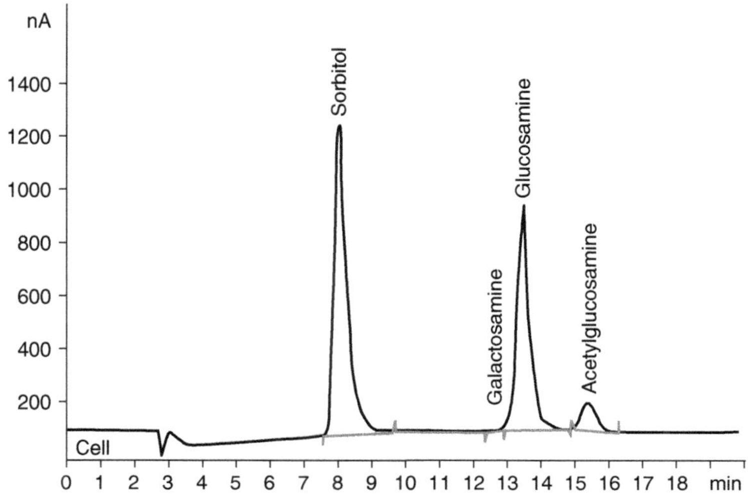

Result :	Sorbitol mg/L	Galactosamine mg/L	Glucosamine mg/L	Acetylglucosamine mg/L
	4.23	0.04	4.57	0.10

Figure 14.5. Example of pharmaceutical product analysis using in-line ultrafiltration. Electro Chemical Detector waveform used for detection: E1 = +0.05V, E2 = +0.75V, E3 = −0.15V, t1 = 0.40s, t2 = 0.20s, t3 = 0.40 s, sample time = 100ms, Detector Range = 0–5 μA full scale, 0% offset.

alternative, hollow fiber modules can also be used. These have a much larger exchange area, resulting in a higher permeate flow combined with a longer working life.

Within the context of IC analysis, ultrafiltration has, until now, mostly been used only for the determination of ions in samples containing large amounts of protein (blood, serum). However, in light of its extraordinarily good matrix elimination, ultrafiltration has great potential for other applications. In combination with HPLC, ultrafiltration has, for example, achieved great importance in monitoring fermentation processes. Figures 14.5 and 14.6 are typical examples of the use of ultrafiltration technique in pharmaceutical applications [3–5].

14.6 DIALYSIS

In principle the term "dialysis" is used for very different membrane separation techniques. However, the same basic procedure applies for all dialysis techniques: a membrane separates the sample solution (donor) from an acceptor solution that takes up the substances that are transported through the membrane. There is a great variety and variability in the geometric arrangements of the dialysis cells, in the dimensions of the liquid compartments, and in the operating procedures used in the various dialysis techniques. In addition to the usually very tedious batch techniques, continuously operating flow-through dialyzers are now used. Dialysis modules exist in a sandwich

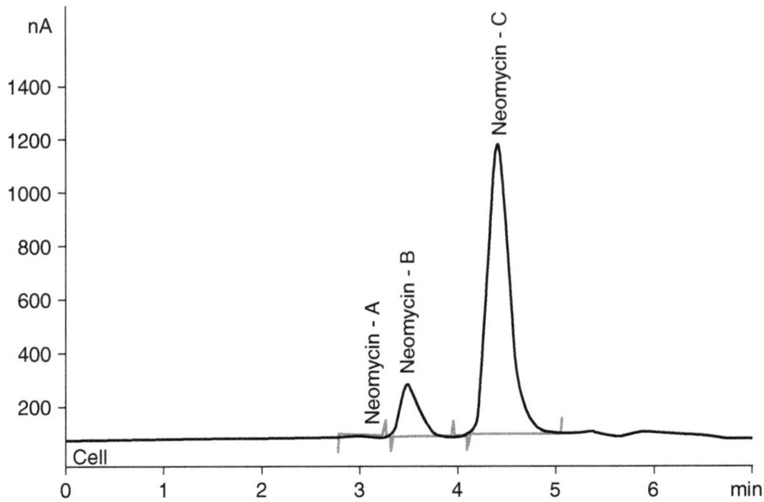

Result :	Neomycin - A wt %	Neomycin -B wt %	Neomycin - C wt %
	0.4	0.4	0.57

Figure 14.6. Example of neomycin powder analysis using in-line ultrafiltration. Electro Chemical Detector waveform used for detection: E1 = +0.05 V, E2 = +0.75 V, E3 = −0.15 V, t1 = 0.40 s, t2 = 0.20 s, t3 = 0.40 s, sample time = 100 ms, Detector Range = 0–5 μA full scale, 0% offset.

configuration for flat membranes and hollow fiber dialyzers. Miniaturized disposable dialyzers are also commercially available.

In passive dialysis ions or molecules are separated because of a concentration gradient across a neutral (nonreactive) membrane. The selectivity results from the pore size of the dialysis membrane used. This dialysis method is mainly used for the same purpose as ultrafiltration. However, its decisive advantage can be seen in the fact that it is far easier to carry out a continuous dialysis and it is therefore easier to automate. There is also no transfer of solvent through the dialysis membrane, which means that the membrane is not clogged even though it is subjected to samples with higher particle loading and/or high-molecular sample constituents.

In *Donnan* dialysis ion-exchange membranes are used; depending on the charge on the membrane, either anions or cations are transported. The driving force is the attempt to achieve an equal charge between the ions in the sample solution and the ions in the acceptor solution that diffuse in the opposite direction. There are two different Donnan dialysis applications used in combination with IC; these are matrix standardization and analyte preconcentration. Matrix standardization can be carried out by the selective addition or deliberate separation of ions with a particular charge. The selective separation of the ions by Donnan dialysis has only found a very limited application as difficulty exists in that the separated ions must be determined in the acceptor solution, which necessarily has a high ion concentration. The problems associated with this can be avoided to some extent by the use of a carbonate/hydrogen carbonate solution as the acceptor for separations by anion chromatography. Reports have also been made about the isolation and preconcentration of transition metal cations followed by their determination by cation-exchange chromatography [3,6,7].

By applying an electrical potential difference across an ion-exchange membrane, the transfer of ions can be realized without having to use a salt solution as the receiver flow. This is known as electrodialysis. Until recently, it has hardly been used as a sample preparation technique because of the more complicated apparatus required and the long dialysis times, but with newer technologies it can be used in IC as one of the possible means of conductivity suppression.

Dialysis is particularly well suited for inline sample preparation. Only the Metrohm inline version will be briefly treated here as dialysis is described in detail. The combination of dialysis cell, peristaltic pump, and an additional valve in an ion chromatograph allows the inline method to be carried out. The dialysis cells are made of polymethylmethacrylate or PEEK. The dialysis membrane has a standard diameter of 4.7 cm, so that all commercially available dialysis membranes can be used. The channels are arranged spirally. The sample flow is pumped in counter-flow to the acceptor solution.

The flow diagram for inline sample preparation by dialysis is shown in Figure 14.7. With the technique shown here a 100% dialysis rate is achieved as

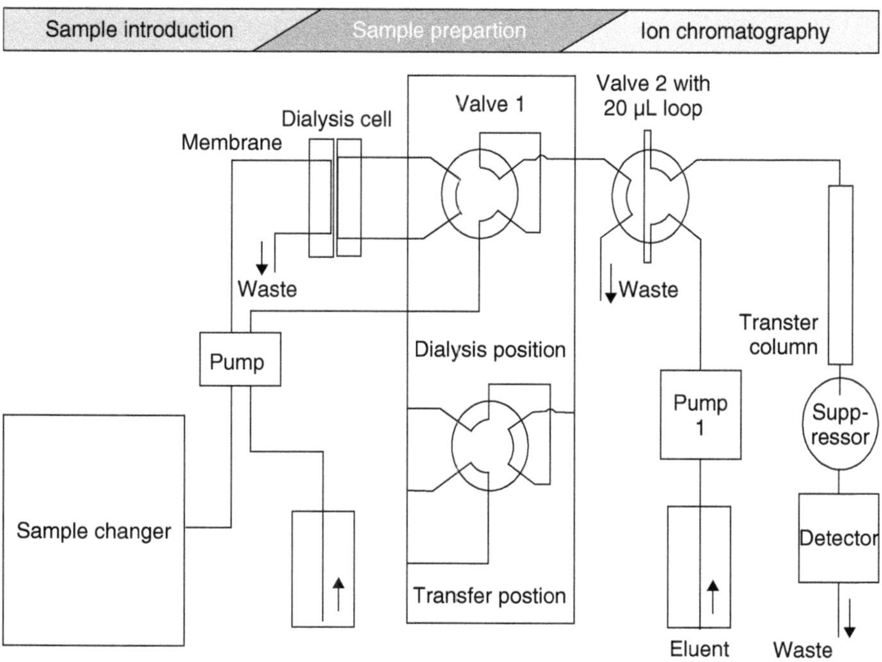

Figure 14.7. Typical schematic of dialysis sample preparation technique.

during dialysis (Figure 14.7, valve position: Dialysis) the acceptor circuit is closed, i.e. the acceptor solution is stationary on one side of the dialysis membrane while the sample solution flows past on the other side. This means that 100% of the analyte ions are always present on the sample solution side. The dialysis equilibrium is only achieved when the same concentrations are present on both sides of the membrane. Only then Valve 1 is switched to the transfer position (lower valve position in Figure 14.7) and the acceptor solution is transferred into the sample loop by Valve 2. Then, as in all other methods, the chromatogram is recorded.

14.7 APPLICATIONS EXAMPLES

Figure 14.8 shows an example of in-line dialysis application, in which phosphite and phosphate are quantitated in the presence of a large amount of the active pharmaceutical ingredient (API), pamidronic acid.

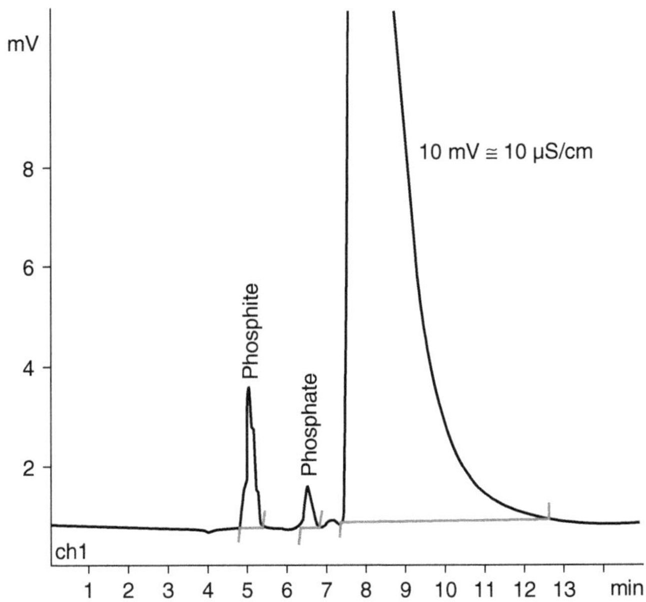

Result :	Phosphite wt %	Phosphate wt %
	0.029	0.775

Figure 14.8. Example of the Pamidronic Acid analysis using in-line dialysis (Column ASUPP3-4 mm ID x 250 mm length at 35°C).

14.8 SUMMARY

In this chapter, various sample preparation techniques are illustrated. With modern IC instrumentation and integrated sample preparation techniques, pharmaceutical products and active pharmaceutical ingredients (API) analysis can be made simple and robust. These techniques can also be integrated into clinical applications. As time progresses, demands for more precise, robust, and rugged sample preparation techniques will increase.

REFERENCES

1. Koulman A, Woffendin G, Narayana VK, Welchman H, Crone C, Volmer DA. High—resolution extracted ion chromatography, a new tool for metabolomics and lipidomics using a second-generation orbitrap mass spectrometer. Rapid Commun. Mass Spectrom. 2009;23:1411–1418.
2. Kerth J, Jensen D. Determination of trace anions in hydrogen peroxide. J. Chromatogr. 1995;706:191–198.
3. Steinbach A, Willie A. Ion chromatographic analysis of carbohydrates in essential and non-essential foodstuffs. Food Engineering & Ingredients. 2008;October: 33–36.
4. Metrohm Application Team. Azide in Irbesartan according to USP PF 30(1) Metrohm Application Note # AW IN6-0606-112005, Riverview FL: Metrohm USA.
5. Gryta M. Scaling diminution by heterogeneous crystallization in a filtration element integrated with membrane distillation module. Pol. J. Chem. Tech. 2009;11:60–65.
6. Monograph: Sample Preparation Techniques for Ion Chromatography. Riverview FL: Metrohm USA.
7. Monograph: Practical Ion Chromatography. Riverview FL: Metrohm USA.

15

VALIDATION OF ION CHROMATOGRAPHIC METHODS

Shreekant V. Karmarkar

Baxter Healthcare, Round Lake, IL

15.1 INTRODUCTION

Validation of a method demonstrates that the method is suitable for its intended purpose. Although the ICH guidance document, Q2(R1), entitled "Validation of analytical procedures: Text and methodology", is considered the main guidance document on method validations, it is essential to review other ICH guidance documents as well for several reasons (Table 15.1). For example, the guidance documents ICH Q1A(R2) and Q1B provide necessary information on the stability testing of drugs and the relevant factors that identify an analytical method as a stability-indicating method, including stress testing, storage conditions for drug substances and drug products during stability studies, and photostability testing. Likewise, ICH Q3A(R2) and Q3B(R2) provide information on the classification of impurities (process impurities vs. degradation products), and reporting, identification, and qualification thresholds of impurities as determined by the maximum daily dose. These sources of information provide guidance in designing validation studies of analytical methods and determining relevant validation characteristics to be addressed during validation studies. The reporting

Applications of Ion Chromatography for Pharmaceutical and Biological Products, First Edition.
Edited by Lokesh Bhattacharyya and Jeffrey S. Rohrer.

threshold has significance in targeting limit of quantitation (LOQ) and limit of detection (LOD) of an analytical method. For the US products, it is also useful to review guidance documents from USP and FDA [1–4]. For example, the CDER guidance document entitled "Guidelines for submitting samples and analytical data for methods validation" [3] includes useful information on characterization of a reference standard and also includes examples of common problems that can delay successful validation. The guidance documents from USP, FDA, and ICH are summarized in Table 15.1 including a short summary of information relevant to analytical validation in each of these documents.

This chapter will cover relevant information essential for method validation, evaluation of validation characteristics as dictated by the method type and the nature of the intended use, strategy of forced degradation studies, similarities and differences between IC and other types of HPLC methods as they relate to method validation, examples of IC method validation in the literature with additional discussion on method validation, and finally the method revalidations.

15.2 ESSENTIAL INFORMATION NEEDED FOR METHOD EVALUATION AND VALIDATION

Although this chapter is focused on the validation aspects of an IC method, it is worthwhile noting that the method validation is only the last step of a process that starts with initial feasibility study and goes through a formal evaluation step (pre-validation experiments). Once the intended purpose of the method is fully understood, as described below, validation-like experiments should be conducted during method evaluation.

For pharmaceutical applications, IC methods are used for the analysis of drug substance, drug product, and excipients, and also for other purposes such as analysis of process water and container leachables. In order to fully understand the intended purpose of the IC methods, especially the methods that are intended for analysis of drug products, the analyst should carefully study necessary information on a number of related areas as listed below:

a) Formulation components and pH range at release and through end of shelf life: this information is needed in preparing test solutions for analytical validations. Typically the test solutions are made to contain excipients at 110% of their nominal concentration to maximize any potential interference.
b) Maximum daily dose: to understand the reporting, identification, and qualification thresholds.
c) Target specifications: the knowledge of product specifications is extremely important (e.g., release and end of shelf life specifications) in setting up analytical expectations. For example, assay method accuracy and precision will need to be tighter for a product that requires assay limits of 95.0 ± 5.0% than for a product with assay limits of 90.0 ± 10.0%.

TABLE 15.1. Guidance Documents for Method Validations and Related Topics From USP, FDA, and ICH.

Source	Title and Reference	Pertinent Information on Analytical Validations
USP	<1225> Validation of compendial procedures [1]	The document provides information on description of data elements for method validation. Four categories of test procedures are described along with a list of data elements for each of the four categories.
USP-PF	<1086> Impurities in drug substances and drug products [2]	Major revision to an earlier version of the monograph has been proposed in the USP- PF 35(5) issue. Classification of impurities is provided as organic, inorganic, and residual solvents. The discussion on impurities in drug substance and drug products is very helpful for developing specific methods.
FDA	Guidelines for submitting samples and analytical data for methods validation—CDER guidance document, February 1987 [3].	Key items in this document include a) integrity of reference standard, b) expectations on method validation details in the report, and c) examples of common problems that can delay successful validation.
FDA	Reviewer guidance: validation of chromatographic methods—CDER guidance document, November 1994 [4]	The document describes validation characteristics and data to be included in the validation report. For example, it recommends recovery data, at least in triplicate, at each level (80, 100, and 120% of label claim) for method accuracy and it also states that the mean is an estimate of accuracy and the RSD is an estimate of sample analysis precision.
ICH	Q1A(R2): Stability testing of new drug substances and products, February 2003 [5]	Recommended stress conditions should be considered for evaluating method specificity. It is imperative to understand recommended storage conditions for the drug substance and drug product because the analytical methods are needed for analyzing samples exposed to these storage conditions.
	Q1B: Photostability testing of new drug substances and products, November 1996 [6]	The guidelines provide exposure conditions for light stressing of drug substance and drug product.

(*continued*)

TABLE 15.1. (*Continued*)

Source	Title and Reference	Pertinent Information on Analytical Validations
ICH	Q2(R1); Validation of analytical procedures: Text and methodology, November 2005 [7]	As the name indicates this is the main guidance document on method validation and it provides a) validation characteristics per the test category (similar to USP <1225>), b) revalidation, c) glossary of validation terms, and d) methodology.
	Q3A(R2): Impurities in new drug substances, October 2006 [8]	For effective validation, the analyst should read several subsections on a) classification of impurities (organic, inorganic and residual solvents), b) reporting of impurities based on maximum daily dose to rationalize the LOQ and LOD targets, c) number of decimal places (e.g., two decimal places when reporting below 1.0%), d) listing of impurities in the validated analytical procedures, and e) glossary.
	Q3B(R2): Impurities in new drug products, June 2006 [9]	Similar to Q3A(R2) document on drug substance, this guidance document is worth reading for discussions on a) process impurities vs. degradation products, b) reporting of impurities based on maximum daily dose to rationalize the LOQ and LOD targets, c) specificity samples by appropriate stress conditions, d) reporting of impurities, and e) glossary.
	Q5C: Quality of biotechnology products: Stability testing of biotechnological/biological products [10]	The document is similar to Q1A(R2), for guidance on storage conditions and testing of biotechnological and biological products.
	Q5E: Comparability of biotechnological/biological products subject to changes in their manufacturing process [11]	It is important to understand the proposed manufacturing process change when selecting the analytical methods. Also, it is necessary to evaluate appropriateness of intended use of existing methods for the process change being proposed.
	Q6A; Specifications: Test procedures and acceptance criteria for new drug substances and new drug products: Chemical substances [12]	It is essential to know the acceptance criteria for drug substance and drug product when setting performance targets for the methods being validated.

d) Container enclosure and target leachables: the information is needed to resolve any container leachables peaks from the target analyte peak in the IC method.
e) Product processing (aseptically filled vs. terminally sterilized product) and secondary packaging for the product (e.g., aluminum over pouch for injectable products): the information is necessary to determine force degradation strategy of the test solutions.
f) Sensitivity of drug molecule and excipient components to heat, light, pH, and oxidative conditions: this information is essential in force degrading the test solutions and in identifying potential analyte degradants.

A list of similar type of information can be put together prior to developing IC methods intended for analysis of other types of pharmaceutical materials. For example, when developing an IC method for analysis of water in container enclosure for the container leachable assessment, the relevant information that the analyst would seek includes (a) processing of water samples in the container, i.e. aseptically filled or terminally sterilized, (b) known extractables from the container system, and (c) storage conditions (time and temperature). Likewise, when developing an IC method for the analysis of a drug substance, it is essential to know the degradation pathways and, when possible, structures and standards of known degradants.

15.3 METHOD TYPES AND VALIDATION CHARACTERISTICS

Like all other pharmaceutical methods, IC methods could be used for various purposes, e.g., assay, identification of active drug substance, quantitation of impurities or absence of a specific impurity. Validation characteristics change depending on the method type as summarized in Table 15.2.

The IC methods intended for the analysis of drug substances and drug products are expected to be stability indicating, i.e. the method can accurately and precisely quantify the decrease of the active pharmaceutical ingredient content, alone or in the drug product, due to degradation. To develop such stability indicating methods, ideally the analyst would possess a drug substance or drug product sample at the end of shelf life to demonstrate specificity of methods intended for drug substance or drug product analysis through its shelf life (release and stability testing). Alternatively, again ideally, the analyst would know structures and have standards available for all impurities. Quite often, however, neither the end of the shelf life samples nor structures or standards for impurities are available when the method is developed during the development of a drug substance or drug product, as the analytical methods typically proceed on parallel paths to development of the manufacturing process. Therefore, it is necessary to create test samples that are representative of the drug substance or product at the end of shelf life or perhaps present the worst case scenario. Such samples are created using forced degradation, also called meaningful degradation, conditions. The FDA and ICH guidance documents provide very little information on how much stress is adequate in stress testing. Overstressing a molecule can lead to degradation profiles that are not representative of real storage conditions and perhaps no longer relevant to method development [13]. As Reynolds et al. [13] have summarized, it is the amount of stress that is more important and not necessarily the extent of degradation.

Karmarkar et al. [14] have described optimum forced degradation conditions for injectable products as summarized in Figures 15.1 and 15.2. The recommended conditions are expected to impart more stress than the product would be subjected to during manufacturing and storage, but not in such an excess that the samples are no longer relevant for the method development. Therefore, if the degradation is less than 10% even after stressing the solutions for injectable products, per the recommended forced degradation conditions (Figures 15.1 and 15.2), further stressing (e.g., heat stressing a test solution for a terminally sterilized product to 3X sterilization cycle if 2X sterilization cycle yields less than 10% degradation) could only make the solutions irrelevant for method development purposes.

15.4 SIMILARITIES AND DIFFERENCES IN IC AND OTHER TYPES OF HPLC METHODS

Reversed-phase HPLC methods are the workhorse of the pharmaceutical industry. Naturally there is an abundance of literature available on validation of various reversed-phase methods. Also, most likely the analysts would be more experienced in validating other types of HPLC methods than they would be validating IC methods. IC and other types of HPLC methods share many operating principles (Table 15.3), but there are also differences in the two techniques. This affects how the IC methods

TABLE 15.2. Recommended Validation Characteristics for the Various Types of Tests.

Type of analytical procedure Characteristics	Assay	Testing for Impurities		Identification
		Quantitative	Limit	
Accuracy	+	+	−	−
Precision				
Repeatability	+	+	−	−
Intermediate precision	+ (1)	+ (1)	−	−
Specificity (2)	+	+	+	+
Detection limit	−	− (3)	+	−
Quantitation limit	−	+	−	−
Linearity	+	+	−	−
Range	+	+	−	−

Notes:
−Signifies that this characteristic is not normally evaluated.
+Signifies that this characteristic is normally evaluated.
(1)In cases where reproducibility has been performed, intermediate precision is not performed.
(2)Lack of specificity of one analytical procedure could be compensated by other supporting analytical procedure(s).
(3)May be needed in some cases.

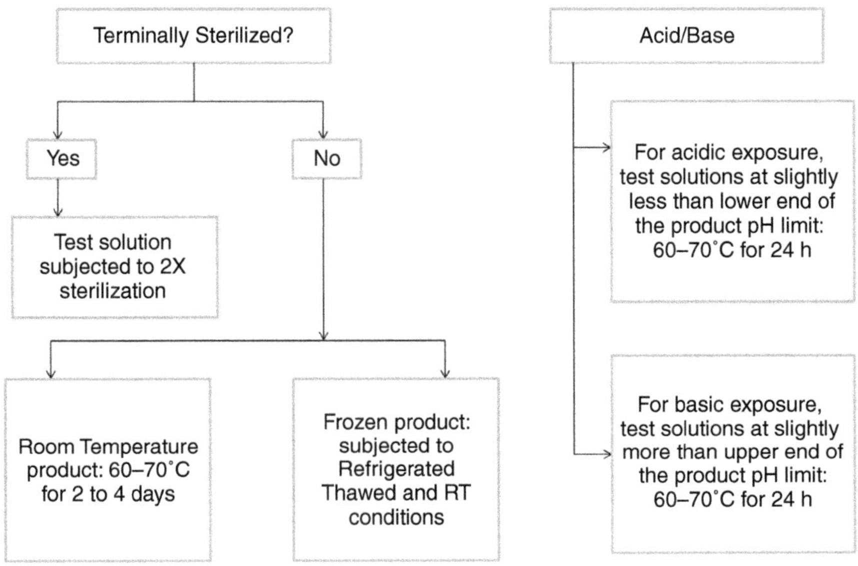

Figure 15.1. Recommended forced degradation scheme with respect to heat and acid/base for injectable drug products.

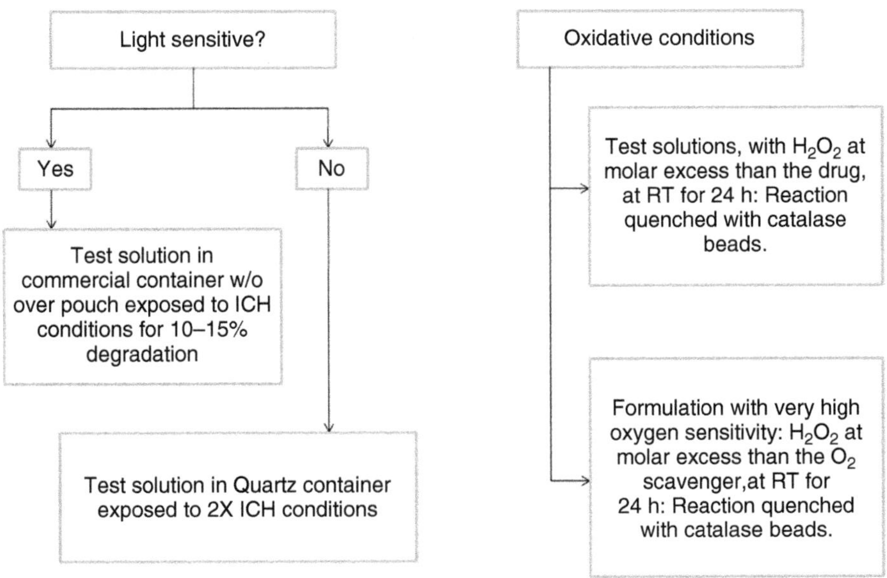

Figure 15.2. Recommended forced degradation scheme with respect to light and oxidative conditions for injectable drug products.

TABLE 15.3. Similarities and Differences Between IC and Other Types of HPLC as Applicable for Method Validations.

Item	IC and HPLC	Discussion Related to Method Validation
Separation modes	**IC:** Ion exchange, ion exclusion, and ion pair.	Selectivity differences among the three IC separation modes could be explored in assessing method specificity.
	HPLC: Reversed phase, normal phase, size exclusion, affinity, and chiral. The HPLC methods also sometimes use ion exchange and ion pair separations.	Ruggedness: Both techniques need to consider column to column and lot to lot variability.
Band broadening	Both techniques typically follow Knox equation.	Initial method development: The information on band broadening is useful in optimizing IC operating conditions.
Mobile phase (often called eluent in IC)	**IC:** Ionic mobile phases are commonly used. Solvents are sometimes added to enhance separation.	Robustness: Experiments to assess minor changes in mobile phase composition are needed for both IC and HPLC techniques.
	HPLC: Various solvents with or without buffers for pH adjustment	
Water quality	**IC:** Virtually all IC methods require very high purity water, i.e. conductivity transparent. Reagent quality is very crucial for trace level IC methods.	LOQ and LOD: The water quality needs must be met to achieve low level detection and quantitation of impurities in IC methods.
	HPLC: In virtually all cases, water quality in terms of UV transparency is important.	
Suppressor for enhancing S/N ratio	**IC:** Most commonly used for conductometric detection of anions and cations. Direct conductometric detection is suitable when detection at ppb levels is not necessary.	Robustness: Experiments are necessary to evaluate the effects of minor changes to suppressor settings on performance of suppressed IC methods.
	HPLC: None	

Detection	**IC:** Conductivity is the most commonly used detector. Also used are UV detection, post-column reaction followed by VIS detection, and pulsed amperometry. **HPLC:** UV, diode array, MS, RI, fluorescence, and post-column reaction followed by VIS detection.	Detector considerations for accuracy, precision, linearity, range, LOQ, LOD, and robustness experiments are applicable for both techniques. Specificity: • Suppressed conductivity, a bulk property detector, does not lend itself to peak purity detection like a UV detector in HPLC. • To assess specificity of IC methods, other approaches are worth considering, e.g., use of UV and conductivity detectors in series and three-D amperometric detectors.
Instrumental setup	**IC:** Resembles HPLC setup. Due to ionic mobile phases, system components are preferably made of inert materials. **HPLC:** Most of the system components are made of passivated SS 316, other metals, and plastic material	Robustness: Appropriate experiments on minor variations in instrumental parameters, e.g., flow rate and gradient composition are necessary with both techniques.
Computer based software	The computer based software is essentially same with both techniques.	No special considerations in terms of computer based software are needed to validate the IC methods.

can be validated. For instance, robustness with respect to mobile phase, instrumental parameters, and lot-to-lot column variability, and detector considerations for method accuracy, precision, and linearity are common to both techniques. The nature of IC technology, however, necessitates further considerations for a couple of parameters; a) reagent quality as it affects validation in terms of LOQ and LOD, and b) bulk property detection that affects validation in terms of method specificity.

Most IC detection methods require high quality water, i.e. low conductivity water (as opposed to UV-transparent water required for most HPLC methods) of 18.0 megaΩ-cm or better resistivity. It is impossible to attain trace level quantitation or detection without the high quality water in IC methods with suppressed conductivity or pulsed amperometric detection. It is highly recommended to follow instrument manufacturers' guidelines on minimizing ionic (e.g., sodium, chloride, or carbonate), biological (e.g., bacteria or mold), or organic (e.g., glycerol contamination from filters) contamination and obtaining high purity reagents [15,16]. System suitability tests, such as no peak above a certain threshold in the blank injections and a limit on background signal from pulsed amperometric or suppressed conductivity detection, are typically applied to ensure contamination-free operation prior to analyzing the samples.

The other major difference is in the detection techniques impacting method specificity assessment. Quite a large number of HPLC methods use UV detection for which method specificity in terms of spectral peak purity, in most cases, can be determined using photo-diode array detectors. It is, however, not possible to determine purity of peaks using IC methods with conductivity detection (commonly used detector in IC), because of the bulk property nature of the detector. In some cases, when the coeluting peak absorbs in the UV/VIS region, peak purity assessment could be augmented using an UV/VIS detector placed in series with a conductivity detector. Certain compounds containing oxidizable groups (e.g., amines and partially oxidized sulfur) can be oxidized at a given potential, and therefore IC methods with pulsed amperometric detection can impart specificity to a certain degree by adjusting applied voltage. Use of high performance anion exchange chromatography with three-dimensional detection has been reported in the literature for peak identification and estimation of percent purity of leucine [17].

15.5 EXAMPLES OF IC METHOD VALIDATION

Comprehensive validation information, specifically with respect to the common validation characteristics (e.g., accuracy, linearity, specificity, sensitivity, repeatability, intermediate precision, robustness and reproducibility), is summarized in Table 15.4 for the various pharmaceutical applications of IC. As shown in Table 15.4, IC methods have been validated for API (e.g., Alendronate in IV solutions and tablet formulations), excipients and inactive ingredients (e.g., acetic and lactic acids in LVP IV solutions), impurities and degradation products (e.g., monoethylsulfate, a degradant of Indinavir sulfate drug substance and galactosamine in total hexosamine for heparin), and process streams (e.g., inorganic cations in culture media and anions/cations in pharmacopeial grade water).

TABLE 15.4. Examples of Validation Data For Quantitative Ion Chromatography Methods.

	Performance Parameters						
Application	Accuracy	Repeatability/ Intermediate Precision	Linearity	Specificity	LOQ/LOD	Reproducibility	Ref.
			A. Active Ingredient Analysis				
Alendronate in IV solutions and tablet formulation by suppressed conductivity detection	Spiked drug into placebo, at 80, 90, 100, 110 and 120% of formulation level. Mean recovery, 100.2%. Demonstrated equivalence versus an HPLC/ fluorescence method.	Injection, $n = 10$ at 0.05 mg/mL, 1% RSD by peak height. Total method, $n = 10$ at 2.5 mg/mL 1.1% RSD or less.	Coefficients of determinations of 0.999 or greater for range of 40–160% of assay level. Non-zero intercepts observed.	Tested against formulation placebo and known thermal decomposition products.	Not evaluated.	Evaluated by performing testing on four analytical systems.	18
Biphosphonate drugs with indirect UV detection	Spiked drug into placebo, at 50, 75, 100, 125 and 150% of formulation level. Mean recovery for three drugs, 100±1% at sample level of 0.05 mg/mL.	Injection to injection, $n = 10$ at 0.05 mg/mL (25 μL injection) or 0.4 mg/mL (50 μL injection). %RSD less than 1%. Total method, 1–2% RSD for $n = 10$.	Coefficients of determinations of 0.999 or greater for range of 20–200% of formulation level. Areas better than heights, non-zero intercepts observed.	Tested against formulation placebo and known thermal decomposition products.	LOD of 0.001 mg/mL at S/N of 4 for all analytes	Evaluated via two analysts/two system testing, testing using five columns, and examining performance on one column after at least 500 injections.	19

(*continued*)

TABLE 15.4. (*Continued*)

	Performance Parameters						
Application	Accuracy	Repeatability/ Intermediate Precision	Linearity	Specificity	LOQ/LOD	Reproducibility	Ref.
Disodium clodronate in bulk materials pharmaceuticals by UV with post-column derivatization	Spiked drug into placebo capsules and tablets at 80, 100 and 120% of nominal formulation level (300–1000 mg). Mean recovery = 99.5–100.7%.	Six samples of bulk material and pharmaceuticals assayed. % RSD from 0.8% to 1.3% for 60–820 mg of drug.	50-175% of nominal analyte conc. (0.02–0.07 mg/mL), $r^2 = 0.9999$ ($n = 6$ at each level).	Tested against formulations degraded by acid, base, peroxide, heat, UV light) and against mixture of known impurities.	Not evaluated.	Stability of sample solutions = 60 hours at RT. Tested two columns with system suitability tests.	20
SPE removal of non-polar compounds, ion exclusion separation and UV detection of citrate and acetate in medical fluids	Recovery, compared with non-SPE treated, ranged from 99.9 to 100.3% for 80 to 120% concentration	% RSD ranging from 0.0 to 0.3%.	$r^2 \geq 0.999$, with % y-intercept ≤ 0.6	The SPE treatment did not introduce citrate or acetate.	Not evaluated.	Two analysts on separate systems. Pooled data, 99.9 to 100.2% recovery, and % RSD ≤ 0.3	21
Paromomycin using IC with separation on CarboPac PA1 column followed by PAD detection	Recovery of paromomycin at 0.5, 1.0, and 2.5 μM ranged from 96–106%, 98–107%, and 95–103%, respectively.	% RSD for intra-day analysis was <1% and for between-day analysis was <2%	Linear response in the range of 1.25 to 10 μM.	Paromomycin isomer peaks were well resolved from impurities and baseline dips.	Not evaluated.	Ruggedness demonstrated from no difference in analytical results with two different instruments, two different column lots and two different eluent cartridges.	38

Citric acid/citrate and phosphate in pharmaceutical dosage forms using IC with suppressed conductivity detection	Recovery of spikes in nine different dosage forms was within 95–105%.	The % RSD for intra-day and inter-day analyses was <1 and 2%, respectively.	Linear response in the range of 0.2–100 μg/mL for citrate and 0.2–60 μg/mL for phosphate	Matching retention times of citrate and phosphate in standards and samples and dilution parallelism (slope and dilution bias) were used to demonstrate specificity.	LOQ of 0.2 μg/mL and LOD of 0.06 μg/mL.	Method robustness demonstrated in terms of analyst, equipment, column lot and eluent preparation (electrolytically vs. manually prepared).	39
Citrate and phosphate in pharmaceutical solutions by IC with suppressed conductivity detection	Recovery of phosphate and citrate in the range of 100 ± 3%.	The % RSD for replicate analyses not more than 1.5%.	Instrument response is linear in the range of 10–30 μg/mL citrate and 6–18 μg/mL phosphate	Matrix blanks (with or without autoclaving) did not contain any peaks at the retention times of phosphate and citrate	Not determined	Contamination from CO_2 in the eluent was minimized by He sparging.	40
			B. Excipients and Inactive Ingredients				
Methane sulfonic acid in intermediates and drug substances by direct conductivity	Determined by mass balance calculations in batch release applications.	Injection precision, 1–2% RSD. Method precision, 1–3% RSD	$r^2 = 0.9997$ over the range of 0.1–5 mg/mL.	Extensive investigation of elution characteristics.	Not evaluated.	Not evaluated.	22

(continued)

TABLE 15.4. (*Continued*)

Application	Performance Parameters						Ref.
	Accuracy	Repeatability/ Intermediate Precision	Linearity	Specificity	LOQ/LOD	Reproducibility	
Alkyl sulfonic acids (for example, methansulfonic acid, MSA) by suppressed conductivity, total method performance (sample extraction)	Spiked samples at 6 levels between 80 and 120% of allowable limit, mean recovery $(n = 6) = 102.9\%$.	Injection precision $(n = 5)$ at the specification limit was 5% RSD.	$r^2 = 0.9999$ over application range.	Resolution demonstrated versus other sulfonic acids and chloride.	LOQ of 40 ppm (μg/g)in sample for MSA	Stock solutions of analytes stable for 1 week at RT. Analyzed 4 samples across 2 days, two columns and fresh mobile phase preparations. % RSD = 7.5%.	23
Acetic and lactic acids in LVP IV solutions by ion exclusion with direct conductivity	Autoclaved formulation blanks spiked with 80, 100 or 120% of product specification levels. Acetate recovery 99–101%, 1500–7900 mg/L. Lactate, 99–103%, 680–3800 mg/L.	For 5 preparations in 4 formulations, 0.35–1.2% RSD at levels from 840–6600 mg/L.	50 to 150% of sample dilution target with 5 standards, (150–450 mg/L for sodium acetate trihydrate, 50 to 150 mg/L for sodium lactate). $r^2 = 0.9997$, other data provided.	Autoclaved formulation blanks examined. Examined a test sample cocktail containing 13 dextrose impurities, related substances, "foreign sugars" and decomposition products.	LOQ = 9.9 mg/L for sodium acetate trihydrate and 3.6 mg/L for sodium lactate.	Performed accuracy assessment with two runs (different analysts and columns). No difference in performance noted. Also examined robustness and response stability.	24
Cyanamide as a synthesis residual via PAD	Bulk drug substance spiked with 3 to 25 ppm additional analyte, % recovery from 89–106%.	Ten injection of sample at 5.4 ppm, % RSD = 4.6%. Method precision tested with 8 preparations of one lot, % RSD = 6.4%.	Assessed over the range of 15–150 ng/mL (7 levels, duplicate injections per level), $r^2 = 0.9982$.	Absence of response noted in degraded bulk drug substance.	Minimum quantifiable limit = 3 ppm.	Sample solutions stable for 53 hours at RT.	25

Methanesulfonic acid as a synthesis impurity with suppressed conductivity	Batch of drug spiked with MSA levels of 0.1–2.0% (by weight), duplicate injections at 7 levels, % recovery 98–107%.	Six preparations of bulk drug, % RSD = 0.51%. Day to day reproducibility of 2.9–4.0% RSD.	Assessed over a range of 1–20 ppm, $r^2 > 0.999$ ($n = 10$)	Absence of response noted in formulation placebos.	LOD = 0.3 ppm.	Assessed as day to day reproducibility.	26
Oxalic acid, oxamic acid and oxamide as synthetic impurities by ion exclusion with UV detection	Examined by comparing calibration curves obtained in water matrix versus API. Less than 5% difference in slope, water versus API.	For six replicate preparations, % RSD = 9.3% at 2 ppm for oxalic acid; 4.1% at 1 ppm for oxamic acid and 3.4% at 0.6 ppm for oxamide.	$r^2 = 0.9999$ over range of 0.4–24 ppm for oxalic and oxalic acids; 0.1 to 6.2 ppm for oxamide. Six concentration, replicate injections.	Blank API matrix repetitively injected with no interfering peaks.	LOQ of 0.2 to 0.6 ppm.	Standards and samples stable for 24 hours at RT. Method tested in two different laboratories with new and aged columns.	27
Sulfate and sulfamate, decomposition products of topiramate, by suppressed conductivity	Examined by analyzing tablets spiked with the analytes over the range of 0.24–1.0 mol%. Mean recovery of 103%.	1.1% RSD (10 injections) for sulfamate and 1.5% RSD for sulfate at 0.5 mol%.	0.04 to 27 mol% for sulfamate, 0.1 to 30 mol% for sulfate, $r^2 > 0.999$ but systematic skew in calibration curve noted.	Formulation placebo (degraded and fresh) examined for absence of interfering peaks.	LOQ = 0.05 mol% for sulfamide, 0.1 mol% for sulfate.	Samples and standards stable for at least 6 days at RT. Performed robustness assessment.	28
Galactosamine in total hexosamine for Heparin	Evaluated from recovery of galactosamine spiked in heparin raw material in the range of 0.3 to 1.5%	% RSD values ranged from 0.1% to 1.5%	% Galactosamine linear in the range of 0.4% to 2.1% in heparin samples with correlation coefficient of 0.999	No peak in the diluent blank and in the unspiked solutions at the retention time of galactosamine. Formation of Galactosamine peak was observed in heparin samples spiked with OSCS and DS that were hydrolyzed.	LOQ = 0.29% LDO = 0.004%	The method is robust at column temperature of 30 ± 2°C, and eluent concentration of 14 ± 1 mM KOH. Hydrolysis step was robust at temperature of 95–110°C and hydrolysis time of 6 ± 1 hours.	a[1]

(*continued*)

TABLE 15.4. (*Continued*)

	Performance Parameters						
Application	Accuracy	Repeatability/ Intermediate Precision	Linearity	Specificity	LOQ/LOD	Reproducibility	Ref.
Monoethylsulfate in Indinavir sulfate drug substance	Monoethylsulfate spiked at three levels, mean recovery = 97.1%. Recovery was 89.2% using capillary electrophoresis method	Method precision: 4.5% RSD; intermediate precision: 4.9% RSD	Linearity studied in the range of 75 to 1200 ng/mL, R^2 value was 0.9997	Blank injections of indinavir sulfate and known impurities of indinavir sulfate showed no interference with monoethyl sulfate	LOD of 24 ng/mL and LOQ of 74 ng/mL	Solutions were stable for 22 hours	29
Human salivary anionic analysis for chloride, phosphate, nitrate, sulfate, thiocyanate, lactate, and acetate (ion exclusion method for lactate and acetate)	Recovery of anions in five-fold diluted saliva ranged from 86.2% for nitrate to 120.1% for chloride	Coefficient of variation ranged from 1.2% for thiocyanate to 9.7% for nitrate	Log-log plots of concentration and area counts were linear.	Ion exclusion method was used for lactate and acetate since they could not be separated on the anion exchange column. Recovery of spiked anions in diluted saliva matrix and analysis of water blank were performed to demonstrate specificity.	Not established	Necessary to filter saliva samples through 0.45 μm membrane filter and also ultra-filtration. Pre-rinsing of filter membrane was necessary to minimize any ionic contaminants leaching from the membrane.	30

N-methyl-pyrrolidine (NMP) in Cefepime using cation exchange IC	Recovery of spikes at three concentrations (0.26, 0.52, and 1.0%) ranged from 97.8 to 102.0%.	% RSD for Intra-day and between-day precision ranged from 1.3 to 1.5%	The method was linear in the range of 0.45 to 200 μg/mL NMP	NMP peak was well separated from Cefepime, arginine, and common cations in blank, or other sources	LOQ of 0.10 μg/mL and LOD of 0.03 μg/mL.	No significant increase in NMP concentration in Cefepime stored at -17°C for 96 h.	37
			D. Process Streams				
Inorganic cations in culture media by suppressed conductivity	Spike recoveries determined in chemically defined and complex media formulations.	Inter-day ($n = 6$) and intra-day ($n = 12$) precision assessed at concentrations ranging from 0.5 to 50 ppm. At > 1 ppm, inter-day %RSD <1%, intra-day <2%.	$r^2 > 0.9999$ using polynomial model, 0.5 to 25 ppm for Na and K, 1 to 50 ppm for ammonium, Ca, Mg.	Specificity considered from a theoretical perspective based on the separation and detection methods.	LOD 0.5 to 1 ppm.	Assessed via intra-day precision.	31
Hydroxylamine in waste streams by PAD	Mean recoveries of analyte spiked into waste water of 69.4% at 0.05 ppm and 93.3% at 0.5 ppm.	Triplicate injections at 0.05 ppm and 0.5 ppm had %RSD of 2.8% and 1.5%.	$r^2 = 0.999$ over range of 0.01 to 2.0 ppm.	Examined versus a mixture of n-methyl-hydroxyamine analogs at 1 ppm.	LOQ = 0.015 ppm.	Standards stable for 12 hours at RT.	32
Determination of phosphate in in-vitro samples containing lanthanum carbonate and covering GI tract pH	Mean accuracy of 88.8% to 103.3% in solution with pH ranging from 1.2 to 6.8	Inter-day precision (% RSD) ranged from 1.2% to 2.1%	The method is linear in the range of 2 to 200 ppm with R^2 value ≥ 0.9998).	Resolution of phosphate peak from chloride, borate, and carbonate demonstrated method selectivity for phosphate	Not determined	Phosphate solution stable for 90 days. Method was robust in terms of flow rate (± 10%), injector volume (± 50%), KOH Eluent concentration (± 1 mM), and column temperature (± 5°C).	33

(*continued*)

TABLE 15.4. (*Continued*)

	Performance Parameters						
Application	Accuracy	Repeatability/ Intermediate Precision	Linearity	Specificity	LOQ/LOD	Reproducibility	Ref.
Chloride, nitrate, sulfate in pharmacopoeial grades of water with suppressed conductivity	Recovery in solutions at 75%, 100% and 125% of nominal standard concentration. Mean accuracy = 105.1% for Cl; 104.5% for nitrate; 105.3% for sulfate.	Repeatability assessed by six injections on one day at the specification limit. %RSD ranged from 0.7 to 5.0%. Reproducibility assessed by duplicate injections on six different days by different analysts. % RSD ranged from 4.0% to 12.4%.	Five standards over range of 25%–150% of the pharmacopoeial limit (0.1 ppm Cl, 0.2 ppm nitrate, 1 ppm sulfate); $r^2 > 0.99$.	Tested against other common inorganic anions.	LOQ = 0.05 ppm, Cl; 0.004 ppm, nitrate; 0.04 ppm sulfate.	Standards and samples stable for > 7 days at RT. See method reproducibility.	34
Ammonium, magnesium, calcium in pharmacopoeial grades of water with suppressed conductivity	Recovery in solutions at 75%, 100% and 125% of nominal standard concentration. Mean accuracy = 97.9% for ammonium; 98.5% for magnesium; 98.4% for sulfate.	Method Repeatability assessed by six injections on one day at the specification limit. %RSD = 2.7% for ammonium, 1.7% for magnesium, 0.4% for calcium. Method reproducibility assessed by duplicate injections on six different days by different analysts. % RSD for ammonium = 5.4%, 1.7% for magnesium, 2.6% for calcium.	Duplicate injections for 5 standards over range of 25%–150% of the pharmacopoeial limit (0.2 ppm ammonium, 1.0 ppm magnesium, 2 ppm calcium); $r^2 > 0.99$.	Tested against other common inorganic cations.	LOQ = 0.02 ppm, ammonium; 0.25 ppm, magnesium; 0.35 ppm calcium.	Standards and samples stable for > 7 days at RT. See method reproducibility.	34

[1]Data from author's laboratory work and partially published in reference 35.

Experimental design for method specificity is one of the key elements of a validation protocol. Among the papers included in Table 15.4, selected papers on validation of methods intended for three different purposes, a) methods for assay of a drug molecule, b) methods on sample pre-treatment prior to IC determination, and c) methods on impurities, are further discussed below to illustrate experimental design for method specificity evaluation.

15.5.1 Assay of Disodium Clodronate in Drug Substance and Drug Product [20]

The IC method was intended as a stability indicating assay method for bulk material (drug substance), solid dosage samples (capsules and tablets), and injectable products. The anionic analyte was separated on an anion-exchange column and derivatized with an acidic iron (III) solution prior to UV detection at 300 nm. To evaluate method specificity, a 4 mg/mL solution was exposed to extreme conditions (acid, base, peroxide, heat, and UV light). Forced degradation with basic conditions (refluxing the sample in 0.1 M NaOH solution for 1 h yielded maximum degradation of disodium clodronate. The clodronate peak was well resolved from the degradation products tetraprotic acids (standards for these compounds were available); orthophosphate, phosphonoformate, hydroxymethylenebisphosphonate, and monochloromethylenebisphonate. The method was validated (see Table 15.4 for summary data) in terms of accuracy, precision, linearity, robustness and reproducibility (solution stability, column to column, and lot to lot variability).

15.5.2 Solid-phase Extraction for Sample Cleanup with an Existing Validated Ion Exclusion Chromatographic Method for Citrate and Acetate [21]

Late-eluting compounds, resulting from a process change, necessitated a runtime of 180 min with an existing validated method for citrate and acetate in medical fluid samples. An SPE clean up was developed to remove undesired late-eluting compounds to shorten the runtime to 35 min (Figure 15.3). Method specificity was demonstrated by absence of peaks in the injections of water and placebo blanks at the retention times of citrate and acetate. Quadruplicate injections of SPE treated samples, with a runtime of 35 min, were made, and it was confirmed that the SPE fully removed the late-eluting peaks as there was no indication of carried over peaks in these multiple injections. The method was validated in terms of accuracy for recovery versus untreated samples, precision, linearity, range, SPE cartridge ruggedness, stability of SPE treated samples, and robustness as summarized in Table 15.3.

15.5.3 Verification of USP Method for Galactosamine in Heparin [35]

An anion-exchange HPLC method was proposed in USP-PF, and subsequently adopted in USP-NF [36], to separate glucosamine and galactosamine after acid hydrolysis to

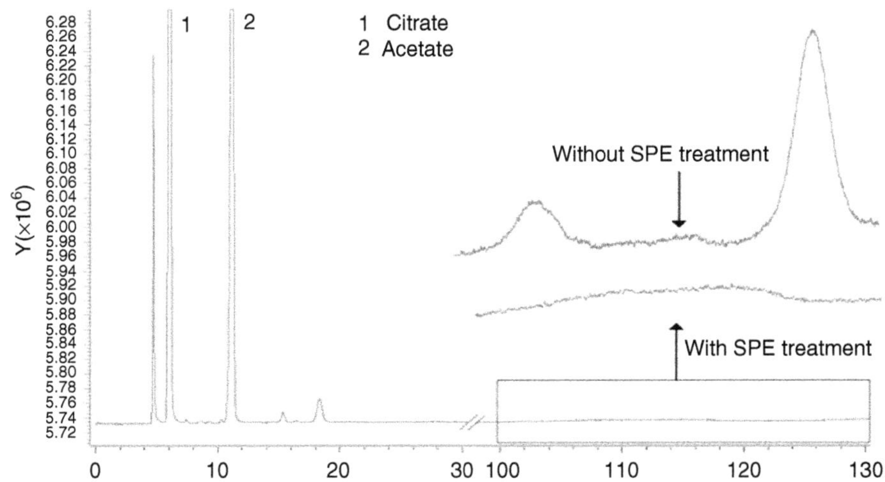

Figure 15.3. Chromatogram illustrating removal of late-eluting peaks in a medical fluid sample treated by SPE. An 8-mL sample was pushed through the SPE cartridge and discarded. Another 1 mL was then pushed through the cartridge and collected for the IEC analysis. Chromatographic conditions, Mobile phase; 50 mN H_2SO_4 0.8 mL/min, columns; 4.6 × 30 mm cation H+ guard cartridge and 7.8 × 300 mm HPX-87H analytical column, 20 μL injection, 60° C column temperature, and UV detection at 210 nm.

test for the limit of galactosamine in total hexosamine in heparin samples. Method specificity was demonstrated from lack of any peaks at the galactosamine retention time in an HCl blank, and presence of only glucosamine peak in the unspiked samples. Heparin sample spiked with over-sulfated chondroitin sulfate or dermatan sulfate and subjected to acid hydrolysis showed the presence of galactosamine formed due to hydrolysis of spiked impurities (Figure 15.4). The method was verified in terms of accuracy, precision, linearity, range, LOQ, LOD, robustness, and solution stability (see Table 15.4 for summary data).

15.6 REVALIDATION OF IC METHODS

The methods need to be maintained in a validated state as discussed in ICH Q2(R1) guidance document. Revalidation may be necessary in the following circumstances:

a) Changes in the synthesis of drug substance,
b) Changes in the composition of finished product, and
c) Changes in the analytical procedures.

The degree of revalidation requirements depends on the nature of the changes. Certain other changes may require revalidation, e.g., smaller LOQ because of tighter reporting threshold requirements.

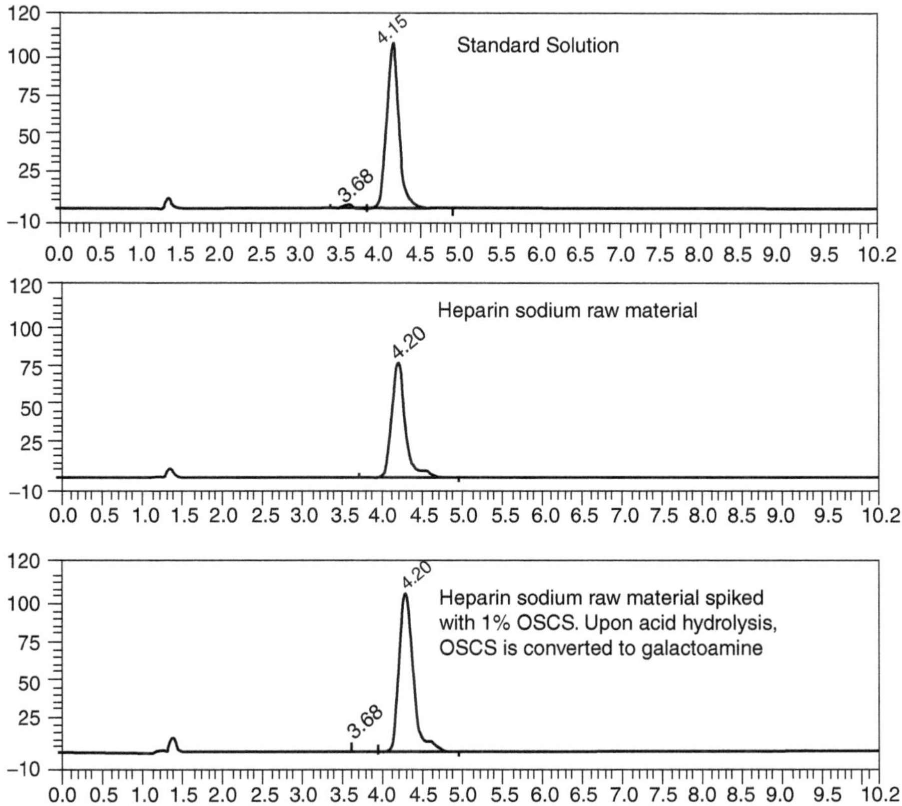

Figure 15.4. Chromatograms obtained for standard solution (top), unspiked heparin raw material (middle), and heparin raw material spiked with OSCS and acid hydrolyzed (bottom). Chromatographic conditions, Eluent A: water; Eluent B: 100 mM KOH (14%B from 0 to 10 min, 100%B at 10.1 min through 20 min, 14% B at 20.1 min through 35 min), 0.5 mL/min, columns; 3.0 × 30 mm Dionex BioLC AminoTrap column with Dionex CarboPac PA20, 3.0 × 150 mm analytical column, 10 μL injection, 30°C column temperature, and pulsed amperometric detection (+0.1 V at 0.00 s, integration from 0.20 to 0.40 s with +0.1 V, −2.0 V at 0.41 s, −2.0 V at 0.42 s, +0.6 V at 0.43 s, −0.1 V at 0.44 s, and −0.1 V at 0.50 s).

15.7 SUMMARY

The chapter has covered various aspects of IC method validation. Along with the ICH guidance document, Q2(R1) entitled "Validation of analytical procedures: Text and methodology", it is extremely useful to review other ICH, USP, and FDA documents as they provide necessary information related to method validation not found in the ICH Q2(R1) document. For IC methods that are intended for the analysis of drug substances and drug products, the forced degradation strategy should be carefully designed so that the forced degraded test solutions are representative of the drug substance or product at the end of shelf life or perhaps present worst case scenario. However, overstressing a molecule can lead to degradation profiles that are not representative of real storage conditions and perhaps not relevant to method development. Since IC and other forms of HPLC technologies share many operating principles, the validation approaches used with the former methods are also readily applicable to IC methods. Special considerations are, however, needed for reagent quality when IC methods are used for trace level analysis. Also, compared with HPLC methods with UV detection, peak purity determination is not possible in IC methods with conductivity detection. IC methods have been validated for analysis of API, excipients, degradants, and process waters.

REFERENCES

1. United States Pharmacopeial Convention. General Chapter: <1225>, Validation of Compendial Procedures, 2009. USP 32 - NF 27, Rockville, Maryland.
2. United States Pharmacopeial Convention. Pharmacopeial Forum, In-Process Revision to General Chapter: <1086>, Impurities in Official Articles. USP - PF 35(5), 2009, Rockville, Maryland.
3. FDA-CDER, Guidelines for submitting samples and analytical data for methods validation, February 1987.
4. FDA-CDER, Reviewer guidance: Validation of chromatographic methods—CDER guidance document, November 1994.
5. ICH Harmonised Tripartite Guideline, Q1A (R2), Stability testing of new drug substances and products, 2003.
6. ICH Harmonised Tripartite Guideline ICH Q1B, Stability testing: Photostability testing of new drug substances and drug products, 1996.
7. ICH Harmonised Tripartite Guideline ICH Q2(R1), Validation of analytical procedures: Text and methodology, 2005.
8. ICH Harmonised Tripartite Guideline ICH Q3A(R2), Impurities in new drug substances, 2006.
9. ICH Harmonised Tripartite Guideline ICH Q3B(R2), Impurities in new drug products, 2006.
10. ICH Harmonised Tripartite Guideline ICH Q5C, Quality of biotechnology products: Stability testing of biotechnological/biological products, 1995.
11. ICH Harmonised Tripartite Guideline ICH Q5E, Comparability of biotechnological/biological products subjected to changes in their manufacturing process, 2004.
12. ICH Harmonised Tripartite Guideline ICH Q6A, Specifications: Test procedures and acceptance criteria for new drug substances and new drug products: Chemical substances, 1999.
13. Reynolds D, Facchine K, Mullaney J, Alsante K, Hatajik T, Motto M. Available guidance and best practices for conducting forced degradation studies. Pharmaceutical Technology 2002; February: 48–56.
14. Karmarkar S, Miller R, Yang X, and Garber R. Design of Forced Degradation Experiments to Demonstrate Specificity of Stability Indicating Methods for Pharmaceutical Injection Products, Presented at the LC-GC's PharmSep, Symposium for Separation Scientists in Pharmaceuticals, Philadelphia, September 21-22, 2009.

15. Dionex Corporation. Technical Note 48: Determination of trace anions in high-purity water by high-volume direct injection with the EG40.
16. Dionex Corporation. Technical Note 71: Eluent preparation for high-performance-anion-exchange chromatography with pulsed amperometric detection.
17. Dionex Corporation. Technical Note 63: Peak identification and estimation of percent purity using HPAE with 3-D Amperometry.
18. Tsai EW, Ip DP, Brooks MA. Determination of alendronate in pharmaceutical dosage formulations by ion chromatography with conductivity detection. J. Chromatogr. 1992;596:217–224.
19. Tsai EW, Chamberlin SD, Forsyth RJ, Bell C, Ip DP, Brooks MA. Determination of biphosphonate drugs in pharmaceutical dosage formulations by ion chromatography with indirect UV detection. J. Pharm. Biomed. Anal. 1994;12:983–991.
20. Kosonen JP. Determination of disodium clodronate in bulk material and pharmaceuticals by ion chromatography with post-column derivatization. J. Pharm. Biomed. Anal. 1992;10:881–887.
21. Karmarkar S, Koberda M, Momani J, Kotecki D, Garber R. Validated ion exclusion chromatographic method for citrate and acetate in medical fluid. J. Chromatogr. A 2004;1039:147–153.
22. Lord BS, Stringham RW. Liquid chromatographic determination of organic acids used as pharmaceutical counterions. Anal. Chem. 1996;68:1067–1070.
23. Annable PL. Determination of alkyl sulfonic acids in pharmaceuticals by ion chromatography. J. Chromatogr. A. 1996;724:199–206.
24. Warnock M, Nair L, Sadain S, Jenke DR. Evaluation of an ion exclusion/direct conductivity method for quantitating acetic and lactic acids in pharmaceutical LVP base formulations. J. Liq. Chromatogr. & Rel. Technol. 2002;25:541–560.
25. Nair JB. Determination of trace levels of cyanamide in a novel potassium channel activator bulk drug by pulsed electrochemical detection. J. Chromatogr. A. 1994;671:367–374.
26. Jagota NK, Nair JB, Kurtulik PT. Ion chromatography of methansulfonic acid in pharmaceuticals. J. Pharm. Biomed. Anal. 1995;13:1291–1295
27. Yang L, Liu L, Olsen BA, Nussbaum MA. The determination of oxalic acid, oxamic acid and oxamide in a drug substance by ion exclusion chromatography. J. Biomed. Pharm. Anal. 2000;22:487–493.
28. Micheel AP, Ko CY, Guh HY. Ion chromatography method and validation for the determination of sulfate and sulfamate ions in topiramate drug substance and finished product. J. Chromatogr. A. 1998;709:166–172.
29. Prasanna JS, Sharma H K, Mukkanti K, Sivakumaran M, Pavan Kumar KSR, Jagadesh Kumar V. Validation of a sensitive ion chromatography method for determination of monoethylsulfate in Indinavir sulfate drug substance. J. Pharm. Biomed. Anal. 2009;50:1065–1069.
30. Chen Z, Darvell B, Leung V. Validation of ion chromatography for human salivary anionic analysis. Archives of Oral Biology, 2004;49:855–862.
31. Robinett RSR, George HA, Herber WK. Determination of inorganic cations in fermentation and cell culture media using cation-exchange liquid chromatography and conductivity detection. J. Chromatogr. A. 1995;718:319–327.

32. Fernando PN, Egwu IN, Hussain MS. Ion chromatographic determination of trace hydroxylamine in waste streams generated by a pharmaceutical reaction process. J. Chromatogr. A. 2002;956:261–270.
33. Samy R, Faustino P, Adams W, Yu L, Khan M, Yang Y. Development and validation of an ion chromatography method for the determination of phosphate-binding of lanthanum carbonate. J. Pharm. Biomedical Anal. 2010;51:1108–1112.
34. Waterworth JP. Validation of ion chromatographic methods for the trace analysis of ions in pharmacopeial grades of water. J. Chromatogr. A. 1997;770:99–104.
35. Moore E. Heparin impurities testing, 3rd workshop on the characterization of Heparin products,, Rockville, Maryland: US Pharmacopeial Convention; July 27–28, 2009.
36. USP-PF IRA, USP32-NF27. Heparin Sodium, 2009. Rockville, Maryland: US Pharmacopeial Convention; 2009.
37. Dionex Corporation. Application Note 199: Determination of N-methylpyrrolidine in Cefepime using a reagent-free ion chromatography system.
38. Dionex Corporation. Application Note 186: Analysis of paromomycin by HPAE-PAD.
39. DeBorba BM, Rohrer JS, Bhattacharyya L. Development and validation of an assay for citric acid/citrate and phosphate in pharmaceutical dosage forms using ion chromatography with suppressed conductivity detection. J. Pharmaceutical Biomedical Anal. 2004;36:517–524.
40. Jenke D, Sadain S, Nunez K, Byrne F. Performance characteristics of an ion chromatographic method for the quantitation of citrate and phosphate in pharmaceutical solutions. J. Chromatogr. Sci. 2007;45:50–56.

16

PHARMACOPEIAL APPLICATIONS OF ION CHROMATOGRAPHY

Ravi Ravichandran

Senior Scientist, US Pharmacopeial Convention, Rockville, MD

16.1 INTRODUCTION

This chapter provides an overview of the use of ion chromatography (IC) in the *US Pharmacopeia–National Formulary* (*USP–NF*), the *European Pharmacopoeia* (*EP*), the *British Pharmacopoeia* (*BP*), and the *Japanese Pharmacopoeia* (*JP*). *USP–NF* is a publication containing two official compendia: the *US Pharmacopeia* (*USP*) and the *National Formulary* (*NF*). The latter contains monographs for excipients. *BP* and *EP* contain similar information except that *BP* contains dosage form monographs and *EP* contains primarily drug substance monographs and only a few monographs for dosage forms (e.g., radiopharmaceuticals). Pharmacopeias are published periodically and are typically updated at least once during the period between two successive publications. For example, *USP–NF* is published annually and is updated twice a year via two *Supplements*. Publications that are official at the time of the preparation of this manuscript are *USP 32–NF 27* and its two *Supplements, EP 6th Edition* and its *Supplements,* and *JP 15*. Monographs in pharmacopeias are legally enforceable specifications (i.e., sets of tests, procedures, and acceptance criteria) for drug substances, dosage forms, and excipients, together with other requirements. *USP–NF* is the largest

Applications of Ion Chromatography for Pharmaceutical and Biological Products, First Edition.
Edited by Lokesh Bhattacharyya and Jeffrey S. Rohrer.

of the pharmacopeias in the world and contains more than 4000 monographs for small-molecule and biological drug substances and dosage forms, excipients (in *NF*), dietary supplements (*DS*), and radiopharmaceuticals. All pharmacopeias are published under stated authorities (e.g., *EP* is published by the European Union's European Directorate for the Quality of Medicines and Healthcare, EDQM). For *USP–NF*, that authority rests with the nongovernmental United States Pharmacopeial Convention (USP). Official text is prepared by USP's standards-setting body, the Council of Experts, and is published by USP's Board of Trustees.

This chapter is divided into three major sections: The first section deals with a description of ion-exchange columns that are referenced in pharmacopeias. The second section provides an overview of general chapters related to IC that are often referenced in official monographs. The last section deals with the actual applications of IC in official monographs. The chapter also includes two tables of specific monographs in which IC has been used in *USP–NF,* as well as in other pharmacopeias. This chapter is not an exhaustive review of IC applications in all pharmacopeias but is meant to provide an introduction to the readers. For further information, readers are directed to *USP* Chapter *Ion Chromatography* <1065>, updates in USP's journal of standards development and compendial revision, *Pharmacopeial Forum,* and standard sources [1].

16.2 COLUMNS IN PHARMACOPEIAS

IC is a specialized HPLC technique in which the separation is based on an ion-exchange mechanism. In addition, there are a few monographs in which the separation mode is ion exclusion, which also uses ion-exchange columns [2]. The ion-exchange packings used in monographs are referenced as reagents, and detailed descriptions of the packing materials are presented in the Reagents sections of the pharmacopeias [3–5]. In *USP–NF,* the column packings are described in the Chromatographic Columns subsection. *USP–NF* classifies packings with an L-number based on their column chemistry and characteristics. For example, *USP* text assigns separate classifications to columns based on type of supports such as silica (e.g., L9) or various polymers (e.g., L47). The columns are further classified by their separation modes, such as weak or strong ion exchangers. *USP 32–NF 27* includes a total of 70 different classes of liquid chromatographic supports with varying column chemistries, including silica-based and polymer-based supports [3]. Of these, 23 different supports can be used to carry out separations based on ion exchange and ion exclusion, and thus can be used in IC. *EP* has 11 types of liquid chromatographic supports dispersed throughout the *Reagents* section that can be used for IC [4], and JP has only five types of such supports [5]. Details about the specific column used in a given monograph are usually found in the knowledge database associated with that monograph.

16.3 GENERAL CHAPTERS IN PHARMACOPEIAS

Each pharmacopeia has specific sections termed *General Chapters*. General chapters provide information about, give definition to, or describe a particular subject or technique, and provide specific details regarding certain tests. Some chapters contain mandatory requirements applicable to many official articles, but others are considered informational in nature and are generally not intended to convey compendial requirements. Further discussion of the content of general chapters is beyond the scope of this chapter.

USP includes five general chapters that refer to ion-exchange separations in some form. General chapter *Assay for Citric Acid/Citrate and Phosphate* <345> is cited in more than 15 official *USP–NF* monographs [6]. Before 2005, citric acid and/or citrate content in several dosage forms were quantified using classical titrations. Dosage forms containing citric acid and/or citrate tend to be buffered and contain multiple excipients that require extensive sample preparation to isolate the citric acid or citrate. In 2005, USP proposed <345> as a new general chapter. The separation of citrate and phosphate is achieved using an L61 anion-exchange column. The mobile phase used in the procedure is 20 mM sodium or potassium hydroxide, and detection is accomplished using the suppressed conductivity mode. Subsequently monographs for all the dosage forms containing citrate and phosphate salts were revised, which led to an increase of IC applications in *USP–NF*.

General chapter *Ion Chromatography* <1065>, an informational chapter, provides an overview of IC, including information about stationary phases, mobile phases, and detectors used in IC [7]. General chapters *Biotechnology-dervived Articles* <1045>, *Biotechnology-derived Articles—Amino Acid Analysis* <1052>, *Biotechnology-derived Articles—Peptide Mapping* <1055>, and *Impurities in Official Articles* <1086> refer to the use of ion-exchange chromatography, although none of them make explicit reference to IC [8–11].

EP and *JP* include dedicated chapters for amino acid analysis [12,13]. In these chapters, the separation by ion exchange is followed by post-column derivatization reaction using either ninhydrin or *o*-phthalaldehyde (OPA) for spectroscopic detection. Ion-exchange separations in combination with post-column derivatization have been extensively used in all the three pharmacopeias for the compositional analysis of peptide drugs such as oxytocin. Although monographs that require amino acid analysis are found in the three pharmacopeias, they will not be discussed in this chapter.

16.4 IC APPLICATIONS IN PHARMACOPEIAL MONOGRAPHS

USP–NF contains several monographs that use ion-exchange separation for quantification. The majority of these monographs require the use of ion-exchange and ion-exclusion separation for assay of the drug substance or formulation components (e.g., amikacin), but others use ion-exchange chromatography for measuring trace levels of ionic impurities (e.g., limit of free sulfates in dental-type silica). Some monographs in *EP* use IC almost exclusively to measure levels of ionic impurities. There appear to be no monographs in *JP* that require the use of IC.

An instructive example of the definitional issues that arise is the case of the vecuronium bromide monograph [14]. Vecuronium bromide does not absorb appreciably even in the wavelength range of 200–210 nm. The assay procedure for vecuronium bromide uses typical reversed-phase separation with UV detection because of the high concentration of the analyte. Impurities are present in such low levels that UV detection can be difficult. The impurities of interest all carry a positive charge, which makes conductivity detection a suitable alternative. In the official monograph, the impurities of interest are separated using reversed-phase HPLC and are quantified using suppressed conductivity detection.

16.4.1 Conductivity and Amperometric and Refractive Index Detectors

IC applications can be classified into two major groups: The first group involves the separation of ions (cations or anions) using an ion-exchange or ion-exclusion mechanisms with an aqueous mobile phase and non-absorbance detection. The monographs that fall into this category are listed in Table 16.1. The table indicates the name of the monograph, the purpose of IC in the monograph, the type of column, detection, and the pharmacopeia in which it appears. The majority of applications use conductivity detection. The table also includes applications that use ionic separation mechanisms and detectors such as refractive index (RI) and pulsed amperometric detection (PAD). Although most of the separations use acidic or basic mobile phase with anions or cations to enable elution, a group of monographs uses water as the mobile phase. The columns used in such separations effect the separation by selective retention of the analytes by a combination of ion-exchange (or ion-exclusion), size-exclusion, and complexation mechanisms. Such a mode of separation has been termed Ion-moderated Partition chromatography [15]. Because the columns have been described as strong cation-exchange columns, they are included in this discussion. In almost all cases, RI is used as the detection mode. Monographs for organic acids, amino acids, carbohydrates, and sugar alcohols are examples of this category. Both *EP* and *USP* include monographs that fall into this classification.

A class of monographs uses PAD in conjunction with ion-exchange separation. These are antibiotics containing a carbohydrate functional group. Carbohydrates are very weak acids that can undergo partial deprotonation under strong alkaline conditions ($pH > 12$), resulting in anions that can be separated on an anion-exchange column. Carbohydrates can be electrochemically oxidized at the surface of a gold electrode at very alkaline conditions. Thus ion-exchange separation with PAD detection is a logical choice for compounds that contain a reducing functional group associated with a carbohydrate moiety. Although ELSD and MS detectors offer different types of selectivity, IC separations using these selective detectors have yet to be included in the pharmacopeias.

The information in Table 16.1 indicates that the majority of the applications use suppressed conductivity as the detection mode. Of the monographs listed only 10 or so use conductivity detection in a non-suppressed mode. Suppressed conductivity detection is well suited for the analysis of low-level ionic impurities. Thus one would expect IC with suppressed conductivity to be useful in detecting low levels of impurities. Etidronate disodium uses IC with suppressed conductivity to monitor phosphite and phosphate impurities at 0.1% (w/w). However, the Etidronate

TABLE 16.1. Monographs in *USP* and *EP* That Use IC with Nonspectroscopic Detection (Other Than Absorbance).[1]

Monograph	Type of Analysis	Column Type/Details	Detection Mode	Source
Ademetionine Disulfate Tosylate	Assay of sulfate	L46 IonPac AS14A	S-CD	*DS*
Amikacin	Assay of amikacin	L47 CarboPac MA1	PAD	*USP*
Amikacin Sulfate	Assay of amikacin	L47 CarboPac MA2	PAD	*USP*
Amikacin Sulfate for Injection	Assay of amikacin	L47 CarboPac MA3	PAD	*USP*
Anticoagulant Citrate Dextrose Solution	Assay of citrate	L61 IonPac AS11	S-CD	*USP*
Anticoagulant Citrate Phosphate Dextrose Adenine Solution	Assay of citrate and phosphate	L61 IonPac AS11	S-CD	*USP*
Anticoagulant Citrate Phosphate Dextrose Solution	Assay of citrate and phosphate	L61 IonPac AS11	S-CD	*USP*
Anticoagulant Sodium Citrate Solution	Assay of citrate	L61 IonPac AS11	S-CD	*USP*
Bethanechol Chloride	Assay of bethanechol chloride, related compounds	L55 IC-Pak C M/D	NS-CD	*USP 32*
Bethanechol Chloride Injection	Assay of bethanechol chloride and Limit of 2-hydroxypropyltrimethyl ammonium chloride	L53 IonPac CS 14	NS-CD	*USP*
Bethanechol Chloride Tablets	Assay of bethanechol chloride and dissolution	L55 IC-Pak C M/D	NS-CD	*USP*
Calcium Gluconate	Limit of oxalate	L12 HPIC Anion S4 Analysis	S-CD	*DS*
Calcium Gluconate for Injection	Limit of oxalate	Strong Anion-exchange Resin Bio-Rad AG1-X2	S-CD	*EP*
Cefepime Hydrochloride	Limit of *N*-methylpyrrolidine	L17 TSK IC SW cation	NS-CD	*EP*
Cefepime Hydrochloride Monohydrate	Limit of *N*-methylpyrrolidine	Strong Cation-exchange Resin R TSK IC SW Cation	NS-CD	*EP*

(*continued*)

TABLE 16.1. (*Continued*)

Monograph	Type of Analysis	Column Type/Details	Detection Mode	Source
Citric Acid, Magnesium Oxide, and Sodium Carbonate Irrigation	Assay of citrate	L61 IonPac AS11	S-CD	*USP*
Clodronate Disodium Tetrahydrate	Impurities–substituted phosphonates	Anion Exchange Resin R Dionex AS11-HC	S-CD	*EP*
Cranberry Liquid Preparation	Content of dextrose and fructose	L19		*DS*
Dental-type Silica	Limit of free sulfates	Anion-exchange Resin	S-CD	*BP*
Dextran 40 in Dextrose Solution	Assay of dextrose	L17 Aminex HPX 87H	RI	*NF*
Enoxaparin Sodium Injection	Free sulfate content	L61 IonPac AS11	S-CD	*USP*
Erythritol	Assay of erythritol	L17	RI	*NF*
Erythromycin Onintment	Assay of erythromycin	L50 OmniPac PAX 500	PAD	*USP*
Etidronate Disodium	Assay of etidronate	L23 IC Pak Anion	RI	*USP*
Etidronate Disodium	Limit of phopshite	L61 Ion Pac AS 11Anion	S-CD	*USP*
Etidronate Disodium Tablets	Assay of etidronate, dissolution	L23 IC Pak Anion	RI	*USP*
Fenolopam Mesylate	Limit of iodide	Ion-exchange Column	S-CD	*USP*
Ferumoxides Injection	Assay of citrate	L48 HPIC-IonPac AS5	S-CD	*USP*
Fludeoxyglucose F-18 Injection	Limit of 2-chloro-2-deoxy-*d*-glucose	L46 CarboPac PA1 (USP) CarboPac PA-100(EP)	PAD	*USP, EP*
High-fructose Corn Syrup	Assay of fructose	L19 Aminex HPX87	RI	*NF*
Inositol	Assay of inositol	L19 CARBOsep CHO	RI	*NF*
Irbesartan	Limit of azide	L31 IonPac AS10	S-CD	*USP*
Isomalt	Assay of isomalt and impurities	L19 Aminex HPX 87	RI	*NF*
Kanamycin Injection	Assay of kanamycin	L47 CarboPac MA6	PAD	*USP*
Kanamycin Sulfate	Assay of kanamycin	L47 CarboPac MA4	PAD	*USP*
Kanamycin Sulfate Capsules	Assay of kanamycin	L47 CarboPac MA5	PAD	*USP*
Lactitol	Assay of lactitol, impurities	L34 Aminex A7 Resin	RI	*NF*
Magnesium Carbonate and Ciric Acid for Oral Solution	Assay of anhydrous citric acid	L61 IonPac AS11	S-CD	*USP*

Magnesium Carbonate, Citric Acid, and Potassium Citrate for Oral Solution	Assay of anhydrous citric acid	L61 IonPac AS11	S-CD	*USP*
Magnesium Citrate for Oral Solution	Assay of citrate	L61 IonPac AS11	S-CD	*USP*
Magnesium Citrate Oral Solution	Assay of anhydrous citric acid	L61 IonPac AS11	S-CD	*USP*
Maltitol	Assay of maltitol	L34 Aminex Fast Carbohydrate Column	RI	*NF*
Maltitol Solution	Assay of maltitol	L34 Aminex Fast Carbohydrate Column	RI	*NF*
Maltose	Assay of maltose	L58 Aminex HPX 87N	RI	*NF*
Molsidomine	Limit of morpholine and methanesulfonic acid	Resin for Reversed-phase Ion Chromatography R IonPac CS14	S-CD	*EP*
Multiple Electrolytes Injection Type 2	Assay of citrate	L61 IonPac AS11	S-CD	*USP*
Nadroparin Calcium	Limit of free sulfates	Anion-separation Column	S-CD	*EP, BP*
Oil- and Water-soluble Vitamins with Minerals Oral Solution	Assay of fluoride	L17	NS-CD	*DS*
Oil- and Water-soluble Vitamins with Minerals Tablets	Assay for fluoride	L17	NS-CD	*DS*
Olsalazine Sodium	Limit of acetate and methanesulfonic acid	Resin for Reversed-phase Ion Chromatography R MPIC-NSI	S-CD	*EP, BP*
Oral Rehydration Salts	Assay for citrate	L61 IonPac AS11	S-CD	*USP*
Pamidronate Sodium for Injection	Assay of pamdironate	L23 Super-sep IC Anion	RI	*USP*
Pamidronate Sodium	Assay of pamidronate	L23 Super-sep IC Anion	RI	*USP*
Paricalcitol Injection	Content of propylene glycol and alcohol	L17 Aminex HPX 87H	RI	*USP*

(*continued*)

TABLE 16.1. (*Continued*)

Monograph	Type of Analysis	Column Type/Details	Detection Mode	Source
Potassium and Sodium Bicarbonates and Citric Acid Effervescent	Assay of anhydrous citric acid	L61 IonPac AS11	S-CD	*USP*
Potassium Citrate and Citric Acid Oral Solution	Assay of citrate	L61 IonPac AS11	S-CD	*USP*
Potassium Citrate Extended-release Tablets	Assay of citrate	L61 IonPac AS11	S-CD	*USP*
Potassium Perchlorate Capsules	Assay of perchlorate	L23 IC-Pak Anion HR	NS-CD	*USP*
Potassium Perchlorate	Assay of perchlorate	L23 IC-Pak Anion HR	NS-CD	*USP*
Rocuronium Bromide	Limit of chloride	Anion-exchange Resin R IonPac AS9-SC	S-CD	*EP, BP*
Sodium Fluoride Acidulated Phosphate Topical Solution	Assay of fluoride	L 46 Transgenomic AN1	S-CD	*USP*
Sorbitol	Assay of sorbitol	L34 Aminex Fast Carbohydrate Column	RI	*NF*
Sorbitol Sorbitan Solution	Assay of sorbitol	L34 Aminex Fast Carbohydrate Column	RI	*NF*
Streptomycin Injection	Assay of streptomycin	L46 CarboPac PA1	PAD	*USP*
Streptomycin for Injection	Assay of streptomycin	L46 CarboPac PA1	PAD	*USP*
Streptomycin Sulfate	Assay of streptomycin	L46 CarboPac PA1	PAD	*USP*
Tagatose	Assay of tagatose	L19 Aminex HPX 87	RI	*NF*
Topiramate	Limit of sulfate/sulfamate	L46 IonPac AS11	S-CD	*USP*
Topramate Tablets	Limit of sulfate/sulfamate	L17- PRP-X100	NS-CD	*PF*
Trehalose	Assay of trehalose	L58 Aminex HPX 87N	RI	*NF*
Tricytrates Oral Solution	Assay of citrate	L61 IonPac AS11	S-CD	*USP*

[1]Abbreviations: *DS* = dietary supplement monograph; *NF* = *National Formulary; PF* = *Pharmacopeial Forum;* S-CD = suppressed conductivity detection; NS-CD = non-suppressed conductivity detection; RI = refractive index detector.

Disodium and Etidronate Disodium Tablets monographs use RI detection for the assay of the main component. The rationale here is that because etidronate disodium is a major component, analysts do not need to use a suppressed conductivity detector. The Topiramate and the Topiramate Tablets monographs are other examples of IC using suppressed conductivity detection to measure low levels of sulfate and sulfamate ions. All IC applications in *EP* and *BP* use suppressed conductivity detection.

No monographs use IC for the analysis of small inorganic cations. Whenever the quantification of such cations is required, classical wet chemical procedures or atomic spectroscopic procedures (both absorption and emission procedures) appear to be the methods of choice for pharmacopeial applications. Lithium Carbonate is one such monograph [16]. This monograph uses a flame test for identification, flame photometry for limit of sodium (0.1%), a tedious visual color test for the limit of calcium (0.15%), and acid–base titration for carbonate content. All of these individual methods can be replaced by a single IC method.

16.4.2 Spectroscopic Detectors

Several drug substances contain an amine functional group (primary, secondary, or tertiary) that can be protonated using an acidic mobile phase that allows retention on an ion-exchange column. The hydrophobic character of the molecules often requires use of a large amount of organic modifier. Most of these molecules contain chromophores that absorb at 210–300 nm, and thus UV detection is prevalent. Table 16.2 contains a list of such monographs in *USP*. *EP* and *JP* do not appear to contain any monographs that use an ion-exchange mechanism in combination with spectroscopic detectors, with the exception of the amino acid analysis that was described earlier in this chapter.

16.5 CONCLUSION

IC is a useful technique that has not yet found wide acceptance in pharmacopeial applications. Still, during the past 5–10 years USP has seen a notable increase in monographs that use IC assays. This is not surprising because many drug substances are hydrophobic, chromophore-containing organic compounds that are best analyzed by reversed-phase separation with UV detection. Analytes requiring extensive sample treatment before analysis are exceptions to this observation, and reversed-phase separations in combination with suppressed conductivity detection have been used (e.g., the Vecuronium Bromide monograph discussed above). Because all pharmacopeias depend on industry to submit validated methods approved by regulatory agencies, development and implementation of additional useful IC applications rest largely with the pharmaceutical industry.

TABLE 16.2. *USP* Monographs That Use IC with Spectroscopic Detectors.

Monograph Name	Type of Analysis	Column Type/Details	Detection λ
Anticoagulant Citrate Phosphate Dextrose Adenine Solution	Assay of adeneine	L9 Partisil SCX	254 nm
Acetaminophen, Dextromethorphan Hydrobromide, Doxylamine Succinate, and Pseudoephedrine Hydrochloride Oral Solution	Assay of dextromethorphan hydrobromide, doxylamine, and pseudoephedrine	L9 Zorbax SCX	220 nm
Aprotenin	Limit of *N*-pyroglutamyl aprotetin	L52 TSK Gel SP-5 PW	210 nm
Butoconazole Nitrate Vaginal Cream	Assay of butoconazole	L9 Partisil SCX	225 nm
Calcitonin Salmon	Related petides and other impurities	L9	214 nm
Cinoxacin	Assay of cinoxacin	L12 Zipax SAX	254 nm
Ciprofloxacin Injection	Content of lactic acid	L17 Aminex HPX 87H	208 nm
Cisplatin	Limit of trichloramminoplatinate	L14	209 nm
Foscarnet Sodium	Limit of phosphite and phosphate	L23 IC Pak Anion	290 nm
Fumaric Acid	Limit of maleic acid	L17 Aminex HPX 87H	210 nm
Ganciclovir	Assay of ganciclovir and impurities	L9 Partisil SCX	254 nm
Garlic FluidExtract	Content of *S*-ally-*L*-cysteine	L17 TSK GEL AminePak	Fluorescence
Hydroxyzine Hydrochloride Tablets	Assay of hydroxyzine hydriochloride	L9 Partisil SCX	232 nm
Hydroxyzine Pamoate capsules	Assay of hydroxyzine pamoate, dissolution	L9 Partisil SCX	232 nm
Hydroxyzine Pamoate Oral Suspension	Assay of hydroxyzine pamoate	L9 Partisil SCX	232 nm

Isoetharine Mesylate	Assay of isoetharine mesylate	L9 Partisil SCX	254 nm
Malic Acid	Content of fumaric and maleic acid	L17 ORH 80L	210 nm
Meclizine Hyrochlroide Tablets	Dissolution	L9 Partisil SCX	230 nm
Metformin Hydrochloride	Impurities	L9 Partisil SCX	218 nm
Multiple Electrolytes Injection Type 1	Assay for acetate and gluconate	L17 Aminex HPX 87H	210 nm
Multiple Electrolytes Injection Type 2	Assay for acetate	L17 Aminex HPX 87H	210 nm
Oxymetazoline Hydrochloride	Assay of oxymetazoline	L9 Partisil SCX	280 nm
Oxymetazoline Hydrochloride Nasal Solution	Assay of oxymetazoline	L9 Partisil SCX	280 nm
Oxymetazoline Hydrochloride Opththalmic Solution	Assay of oxymetazoline	L9 Partisil SCX	280 nm
Oxymorphone Hydrochloride Suppositories	Assay of oxymorphone	L12 Zipax SAX	254 nm
Pseudoephedrine Hydrochloride, Carbinoxamine Maleate, and Dextromethorphan Hydrobromide Oral Solution	Assay	L9 Partisil SCX	225 nm 257 nm
Raclopride C11 Injection	Chemical purity	L9 Partisil SCX	254 nm
Ribavirin	Assay of ribavirin	L17-Aminex Fast Acid Analysis	207 nm
Ribavirin Inhalation Solution	Assay of ribavirin, impurities	L17-Aminex Fast Acid Analysis	207 nm
Risedronate Sodium	Assay of risedronate and impurities	L48 IonPac AS7	263 nm
Sodium Acetate C11 Injection	Chemical purity	L9	210 nm
Thiothixene	Limit of (*E*)-thiothixene	L9	254 nm

REFERENCES

1. Gjerde DT, Fritz JS, eds. Ion Chromatography, 2nd ed. Heidelberg (Germany): Alfred Huethig Verlag;1987.
2. Weiss, J, Jensen D. Modern stationary phases for ion chromatography. Anal. Bioanal. Chem. 2003;275:81–98.
3. USP 32–NF 27, Reagents. Rockville: United States Pharmacopeial Convention; 2009:877.
4. European Pharmacopoeia, Reagents 4.1.1. 7th ed. Strasbourg (France): Council of Europe; 2007:389.
5. Japanese Pharmacopoeia, 9.4.2 Solid Supports/Column Packings for Chromatography. 15th ed. Tokyo (Japan): Ministry of Health, Labour, and Welfare; 2006:259.
6. USP 32–NF 27, Assay for Citric Acid/Citrate and Phosphate <345>. Rockville: United States Pharmacopeial Convention; 2009:143.
7. USP 32–NF 27, Ion Chromatography <1065>. Rockville: United States Pharmacopeial Convention; 2009:514–516.
8. USP 32–NF 27, Biotechnology-derived Articles <1045>. Rockville: United States Pharmacopeial Convention; 2009:426–436.
9. USP 32–NF 27, Biotechnology-derived Articles—Amino Acid Analysis <1052>. Rockville: United States Pharmacopeial Convention; 2009:481–490.
10. USP 32–NF 27, Biotechnology-derived Articles—Peptide Mapping <1055>. Rockville: United States Pharmacopeial Convention; 2009:496–500.
11. USP 32–NF 27, Impurities in Official Articles <1086>. Rockville: United States Pharmacopeial Convention; 2009:546–549.
12. European Pharmacopoeia, Amino Acid Analysis 2.2.56. 7th ed. Strasbourg (France): Council of Europe; 2007:89–96.
13. Japanese Pharmacopoeia, Amino Acid Analysis. 15th ed. Tokyo, Japan: Ministry of Health, Labour, and Welfare; 2006:1655–1662.
14. USP 32–NF 27, Vecuronium Bromide. Rockville: United States Pharmacopeial Convention; 2009:3850.
15. Snyder LR, Kirkland JJ, Dolan JW, eds. Introduction to Modern Liquid Chromatography. 3rd ed. New York: Wiley; 2009:626–628.
16. USP 32–NF 27, Lithium Carbonate. Rockville: United States Pharmacopeial Convention; 2009:279.

PART III

BIOTECHNOLOGY APPLICATIONS

17

GLYCAN ANALYSIS BY HIGH PERFORMANCE ANION EXCHANGE CHROMATOGRAPHY WITH PULSED AMPEROMETRIC DETECTION

Lokesh Bhattacharyya

Division of Biological Standards and Quality Control, Office of Compliance and Biologics Quality, Center for Biologics Evaluation and Research, Food and Drug Administration, Rockville, MD

17.1 INTRODUCTION

The term glycan refers to mono-, oligo-, and polysaccharides. Many proteins and lipids contain glycan moieties covalently linked to them. Such proteins and lipids are called glycoproteins and glycolipids, respectively, and together, they are called glycoconjugates. The glycan residues of glycoproteins are linked to the polypeptide chains through the serine or threonine residues (*O*-linked) or asparagine residues (*N*-linked). The glycan moieties are integral parts of the protein structures and are associated with the functional properties of proteins. They play critical roles in numerous biological events, including protein folding, protein uptake and trafficking, inter-cellular interaction and communication, host-cell binding, immune response, blood-group antigenic determinants, cancer markers, and many others. The glycan moieties also play critical roles in protein stability and adverse antigenic reactions of proteins. Thus, there has been a strong interest in elucidating the composition, linkage, and three-dimensional structures of glycan moieties of glycoproteins with an objective to understand their

Applications of Ion Chromatography for Pharmaceutical and Biological Products, First Edition.
Edited by Lokesh Bhattacharyya and Jeffrey S. Rohrer.

molecular binding properties, mechanisms of action and the structure-function relationship.

The development of recombinant glycoproteins for human therapeutic use has led to an increasing need for understanding the therapeutic implications of the functional roles of their glycan moieties, and hence for methods to elucidate the glycan structures. However, until the late 1980s, there was no convenient, low cost, and robust method available that would provide the necessary structural information. Colorimetric methods have been used to determine total glycan content of glycoproteins. Combined with exo- and endoglycosidase digestions, colorimetric methods also provided information on specific monosaccharide content and linkage. Paper and high voltage electrophoresis and thin-layer chromatography were used to obtain data on monosaccharide composition as well as to purify oligosaccharides present in glycoproteins. Methods of chemical analysis, such as methylation analysis (Smith Degradation), have been used successfully to obtain linkage information of glycoside bonds in glycans.

A few available liquid chromatographic methods employed silica-based aminobonded or polymer-based metal-loaded cation exchange columns, with refractive index (RI) or low-wavelength ultraviolet (UV) detection. These methods required higher sample concentrations and were not suitable for the separation of oligosaccharides with closely related structures that arise due to microheterogeniety of glycosylation. In addition, RI and low-wavelength UV detection methods were sensitive to eluent and sample matrix components, permitted only limited use of gradients, and often required stringent sample clean-up prior to chromatography. More sensitive methods required pre-derivatization of carbohydrates, thus requiring multiple steps and the possible introduction of errors or uncertainties in the results. Conventional anion exchange chromatography has been used for the separation of sialylated, phosphorylated, or sulfated glycans. However, the technique was not suitable for the analysis of neutral carbohydrates. This is primarily because available column packings were not compatible with mobile phases that are sufficiently alkaline to ionize neutral sugars.

Until the late 1980s, the general methods that were available for monosaccharide composition analysis include gas-liquid chromatography of monosaccharide derivatives, such as alditol acetate and silyl derivatives of monosaccharides. Derivatization is necessary because the monosaccharides are nonvolatile. The alditol acetate method involved several steps after the hydrolysis to release the oligosaccharides from proteins and cleave the glycosidic bonds, including reduction with $NaBH_4$, filtration, concentration of the supernatant, repeated distillation with dry methanol to remove borate, and acetylation followed by extraction before the chromatography step. The procedure was laborious and resulted in significant loss of sample. Although silylation involved a single-step derivatization procedure after hydrolysis, the products were unstable due to sensitivity to moisture and, therefore, the results were often sub-optimal. Although informative, these methods were limited to monosaccharide composition analysis and showed poor accuracy. Another disadvantage was the complexity of the chromatograms due to the formation of anomeric derivatives. This made interpretation of the results more difficult. Derivatives of alditols are, however, free from this problem.

For the structural analysis of glycan moieties, the most meaningful method available was ^{1}H and ^{13}C NMR. Although powerful and informative, NMR had some draw-backs. (1) The high cost of purchasing and maintaining the equipment, which was later addressed partially by opening centralized NMR facilities (such as, the Complex Carbohydrate Research Center in Georgia). (2) NMR required relatively large amounts of sample. (3) Difficulty in the interpretation of results—sufficient data was not available to comprehensively assign signals to appropriate residues and linkages. Thus, interpretation of NMR results required significant expertise and experience. As more data became available, databases were developed correlating signals and coupling data (such as, ^{1}H-^{1}H or ^{1}H-^{13}C cross correlation) to glycan structures. (4)The high degree of microheterogeniety of glycans rendered interpretation of data more difficult. No suitable method was available to purify different oligosaccharides of a glycoprotein, which have closely related structures.

The break-through came with the introduction in the late 1980s of the IC-based High pH Anion Exchange Chromatographic (HPAEC) methods with electrochemical detection (Pulsed Amperometry Detection, PAD) for the analysis of mono- and oligosaccharides derived from glycoproteins [1–5]. The method was known popularly as the Dionex method because Dionex, Inc. marketed the instrument required for the analysis, including the chromatographic system, columns and accessories, and the detection system. Later, Metrohm and Shodex also marketed IC systems for carbohydrate analysis. Now more commonly called High Performance Anion Exchange Chromatography (still HPAEC), or HPAE chromatography, the technique provides a general method for the analysis of mono- and oligosaccharide moieties of glycoproteins and polysaccharides, which is affordable, simple enough for an analyst with knowledge and experience in chromatography to operate, and robust enough to provide reproducible accuracy and precision. In addition, HPAEC methods may be validated according to current analytical requirements [6]. The introduction of HPAEC permitted wide-spread investigation into the structure and functional properties of glycans, both in academics and in the biotechnology industry, and contributed to the explosion of information on glycan structures and functions that we witnessed during the 1990s.

HPAEC-PAD is free from many of the problems associated with carbohydrate analysis techniques that employ indirect detection or require derivatization. The use of HPAEC-PAD eliminates: (a) requirement for large samples, (b) the need to select an appropriate tag from the large number of tags available to label carbohydrates for their detection, (c) derivatization steps required for labeling, (d) sample cleanup associated with tag usage, (e) recovery problems encountered when using tags, and (f) any need for re-*N*-acetylation of amino sugars after hydrolysis to release mono- and oligosaccharides from glycoproteins, followed by the removal of excess reagents required in re-*N*-acetylation. The sensitivities of HPAEC-PAD for underivatized carbohydrates are orders of magnitude greater than those obtained by other chromatographic methods. For example, LODs for monosaccharides are in the hundred femtomoles range [7]. Neutral and cationic components in the matrix elute in, or close to, the void volume of the column. Therefore, even if such species are oxidizable, and thus detected by PAD, they do not interfere with the analysis of carbohydrates of interest.

17.2 HIGH PERFORMANCE ANION EXCHANGE CHROMATOGRAPHY

HPAEC permits direct quantitation of underivatized carbohydrates at low picomole levels with minimal sample preparation and clean-up. HPAEC takes advantage of the weakly acidic nature of carbohydrates to give highly selective separations at high pH using an anion exchange stationary phase.

Table 17.1 shows the dissociation constants of some of the neutral monosaccharides that commonly occur in glycoproteins. Neutral under physiological conditions, the monosaccharides (and oligosaccharides) are ionized at high pH and can be separated as oxyanions by anion exchange chromatography on quaternary-ammonium-bonded pellicular resins stable to operating conditions of high pH [1]. As the epimerization of sugars is very rapid at high pH, individual anomeric forms of monosaccharides are no longer resolved. Thus, each monosaccharide yields a single peak. The elution order of monosaccharides correlates with their pKa values, and excellent separations have been obtained for neutral, amino-, and acidic monosaccharides derived from hydrolysis of many glycoproteins and glycolipids [1–5,7–10].

Classical silica-based columns cannot be used under these conditions because such stationary phases dissolve at high pH. Carbohydrates are separable on porous polystyrene divinylbenzene anion exchangers but tend to exhibit broad peaks. However, nonporous microbead pellicular anion exchange column packings utilize uniform, small (< 10 micron diameter), nonporous beads onto which much smaller functionalized latex beads (< 0.1 micron diameter) are agglomerated. Thus the carbohydrate analyte interacts with the functional groups at the surface of the latex microbeads, eliminating the need to get into and out of pores, resulting in rapid mass transfer, eliminating peak broadening, and providing high resolution of peaks. Microbead resins also exhibit high pH stability (pH 0–14), and excellent mechanical stability that permits back pressure tolerance.

Monosaccharides undergo a number of reactions at high pH that have the potential to interfere with HPAEC. However, in most cases these reactions are slow at room temperature and do not appear to occur to any detectable extent over the time course of the chromatography presumably because of the use of low concentrations of alkali (NaOH or KOH) and short analysis time. For example, when glucose is left in 150 mM sodium hydroxide for four days at room temperature, there is no evidence of the presence of any mannose or fructose in the solution [10]. Most oligosaccharides and glycopeptides also are stable under these conditions. For example, the *N*-linked

TABLE 17.1. Dissociation Constants of Some Monosaccharides.

Monosaccharide	pKa in Water at 25°C
Mannose	12.08
Xylose	12.15
Glucose	12.28
Galactose	12.39

glycopeptides from soybean lectin (Man9-Asn) and asialofetuin (mixture, complex type) showed no detectable change in NMR spectra after overnight incubation in 100 mM NaOH at room temperature [11].

17.2.1 Eluent

HPAEC requires the use of high-quality water, free of ions and biological contamination, and as little dissolved carbon dioxide as possible. Biological contamination is often the source of unexpected glucose and other peaks in the chromatogram after exposure to the alkaline eluent. Water used in HPAEC should be of high resistivity ($\geq$18 MΩ). Alkali solutions, which are eluents used in HPAEC, absorb carbon dioxide from the atmosphere and form carbonate in solution. Over time, the accumulation of carbonate can be significant, particularly if adequate precautions are not taken. Carbonate, being a divalent anion at pH $\geq$ 12, the pH typically used in HPAEC, binds strongly to the columns and interferes with carbohydrate binding, causing a decrease in column capacity and ultimately, selectivity, loss of resolution and efficiency, and poor day-to-day reproducibility. Thus, it is critical to degas water to be used to prepare eluent solutions either by sparging with an inert gas, such as helium, or under vacuum with sonication, to remove carbon dioxide as much as possible, before sodium hydroxide (or potassium hydroxide) is added to it to prepare the eluent solution. Commercially available sodium hydroxide pellets are covered with a thin layer of sodium carbonate due to exposure to atmospheric carbon dioxide and should not be used. A 50% (w/w) sodium (or potassium) hydroxide solution is much lower in carbonate content because most of the carbonate formed is "salted out" by the high concentration of sodium hydroxide and remains either as a layer at the top or is precipitated to the bottom of the container. The amount of carbonate in the eluent can be minimized by taking appropriate care in alquoting solutions from the middle of the container without stirring. Once eluents have been prepared, it is important to keep them blanketed under an inert gas at a positive pressure during chromatography. It is also important to use eluents that are not older than about a week.

Eluent generators have been introduced to substantially eliminate this problem. Eluent generators automatically prepare hydroxide eluents of precise concentrations, without significant levels of carbonate contamination, through the electrolysis of water. Thus, the only source of carbonate in the mobile phase can be the carbonate or carbon dioxide in water used to supply the eluent generator.

The trace of carbonate that may be present or formed in the eluent can be removed by using an anion trap column (ATC) installed between the eluent delivery system and the pump. An ATC also removes trace amounts of other anions present in the eluent. When an eluent generator is used, an ATC may be used before the eluent generator to remove trace amount of carbonate from the feed water.

Borate is a known contaminant in laboratory water supplies. Borate contamination of eluents can come from degraded deionized water systems requiring maintenance or due to leaching from glass eluent bottles. When present, borate forms anionic complexes with carbohydrates having vicinal *cis* hydroxyl groups, such as sugar alcohols (alditol) and mannose, resulting in, peak tailing. An online guard column (for example,

BorateTrap column, Dionex) that can be used to remove borate from eluent streams prior to the injector is particularly useful in monosaccharide analysis [12].

To eliminate carbonate, borate and other contaminating anions from water, the Continuously Regenerated Anion Trap Column (CR-ATC) has been introduced, which works particularly well when an eluent generator is used. A CR-ATC is installed after the eluent generator and eliminates the need to use an ATC and a borate-trap column, and unlike these guard columns, does not need periodic regeneration.

In order to separate monosaccharides that are charged at neutral pH (e.g., sialic acids) or large oligosaccharides (e.g., branched N-linked oligosaccharides from glycoproteins), a stronger eluent is necessary. This is accomplished by using sodium hydroxide/sodium acetate eluent. The quality of sodium acetate has been reported to be a source of problems [13].

17.3 PULSED AMPEROMETRIC DETECTION

Pulsed amperometric detection (also referred to as pulsed electrochemical detection, PED) is a sensitive electrochemical technique that detects only those compounds that are oxidizable at the employed detection voltage. It permits detection of carbohydrates with excellent signal-to-noise ratios down to approximately 10 picomoles without requiring derivatization. Carbohydrates are detected by measuring the electrical current generated by their oxidation at the surface of a gold electrode (hence, amperometry). The products of the oxidation reaction also poison the surface of the electrode, which needs to be cleaned between measurements. This is accomplished by raising the potential to a level sufficient to oxidize the gold surface, which causes desorption of the carbohydrate oxidation products. The electrode potential is then lowered to reduce the electrode surface back to gold. Such a series of potentials is called a waveform because positive and negative potentials are applied alternately at the electrode within short intervals. PAD thus employs a repeating sequence of potentials. The step from one potential to the next produces a current that is not part of the analyte oxidation current. So, current from carbohydrate oxidation is measured at the first potential but after a delay to allow the current to stabilize. The carbohydrate oxidation current is measured by integrating the analyte oxidation current. Current integrated over time is charge, so the detector response is measured in Coulomb (C).

However, recent advances in electrochemical detection technology permit the use of fast waveforms in conjunction with reductive desorption (rather than oxidative cleaning) to clean the surface of gold electrodes in pulsed amperometry [14]. The importance of using such fast waveforms is that the time during which gold oxide is formed is minimized [15]. It is during the conversion of gold oxide to gold that gold oxide gets slowly chipped off the gold electrode surface. Preventing the formation of gold oxide during electrode cleaning prevents the resulting loss of gold from the gold working electrode that occurs with oxidative cleaning. As recession of the working electrode surface slowly and progressively occurs with oxidative cleaning, the liquid channel that passes over the gold electrode thickens, slowly increasing in volume. The consequences of an increase in channel thickness is that there is a decrease in the velocity of fluid flow over the electrode surface resulting in a decrease in detector response for a given concentration of analyte. By eliminating working electrode recession with a fast four-potential waveform, peak area reproducibility improves significantly [16]. This waveform enabled the design, production, and ultimately the use of disposable thin gold working electrodes (approximately 3000 angstroms thickness). A detailed description of the fabrication of thin disposable gold electrodes was reported [17]. Further details of detectors used in PAD are discussed in Chapter 3 of this book.

17.4 INSTRUMENTATION

The set-up of the HPLC system required for HPAEC-PAD is very similar to that used in a typical HPLC system but also involves significant differences. As with a typical HPLC system, the components include an autosampler, a high-pressure pump, an injection system with sample loop of suitable size (typically, 10–250 μL), one or more guard columns, as discussed above (also see later), an analytical column, a flow-through detector, and a processing system ranging from a data-processing integrator to a computerized system management unit, including software to run the system using pre-programmed method and schedule (sequence) files, perform data acquisition, and process data to produce the final results. A schematic diagram of the instrument set-up is shown in Figure 1.3 of Chapter 1 of this book and will not be repeated here. However, the set-up for HPAEC-PAD also has some very unique features. It requires an eluent management module to spurge an inert gas through water used to prepare the eluent to make it free of carbon dioxide as much as possible. Alternatively, it requires a system for vacuum degassing. Recent systems use an eluent generator that generates an eluent of accurate concentration of potassium hydroxide continuously and reproducibly through electrolysis of water to produce hydroxide ion that pairs with potassium ion migrating across a cation-exchange membrane from an aqueous solution of a potassium salt contained in a cartridge. The rate of formation of potassium hydroxide and its concentration are maintained by regulating current controlled by software.

As mentioned above, the detector consists of a flow-through system with gold electrode, which is coupled with Ag/AgCl reference electrode for the measurement of current.

17.5 MONOSACCHARIDE COMPOSITION ANALYSIS

Monosaccharide composition analysis is a valuable tool in the characterization of glycoproteins and polysaccharides. Glycosylation of proteins can vary under subtle conditions, including variations in fermentation and purification conditions of recombinant glycoproteins. Such variations can drastically affect the conformations and efficacies of therapeutic proteins [*cf*. 18]. Thus, monosaccharide composition analysis provides a rapid method for detecting even minor variations in glycosylation and may be used for process monitoring and process validation, particularly where variation in glycosylation is suspected or established, as was determined for monoclonal antibodies. For glycoproteins and polysaccharides, including polysaccharide and polysaccharide-protein conjugate vaccines, monosaccharide composition analysis often serves as part of an identity test of either individual active components or the final formulated product and is sometimes called carbohydrate fingerprinting analysis (*cf*. Chapter 22 of this book). Thus, monosaccharide composition analysis can provide valuable data supporting quality control of the manufacturing processes and final products, and sometimes constitutes part of lot-release testing.

17.5.1 Standards

Carbohydrates are hygroscopic in nature because water can form very strong hydrogen bonds with several hydroxyl groups present in carbohydrates. Thus, bound water can account for a significant percentage of their weight. Consequently, the monosaccharide standards must be stored over strong desiccating agents, such as phosphorus pentoxide, and monitored over time. When a stable vial weight is attained, the individual monosaccharides are weighed, dissolved in water, and can be stored frozen until needed. Mixtures of monosaccharide standards are available commercially. For example, a mixture of six monosaccharide standards that contain 100 ± 6 nmoles (label claim) each of fucose (Fuc), galactosamine (GalNH2), glucosamine (GlcNH2), galactose (Gal), glucose (Glc), and mannose (Man) is available from Dionex, Inc. These six monosaccharides, and sialic acids, are commonly found monosaccharide residues in mammalian glycoproteins.

17.5.2 Hydrolysis

In most instances, monosaccharide can be liberated from glycoproteins by hydrolyzing with 2–4 N trifluoroacetic acid (TFA) at 100°C for 4–6 hours in a 1.5-mL screw cap microfuge tubes. Hydrolysates are centrifuged under vacuum to dryness and the dried samples are reconstituted in an appropriate volume of water such that 10–25 μL injections contain up to several hundred picomoles of each component monosaccharide. For the purpose of analysis of amino sugars, the hydrolysis may be conducted with 4–6 N HCl at 100°C under vacuum. The latter hydrolysis condition is more effective for glycans that are high in amino sugar content. However, neutral sugars are destroyed under such hydrolysis conditions.

17.5.3 Column

Several columns are commercially available for monosaccharide composition analysis. Two types of columns that are widely used are available from Dionex, Inc.—CarboPac PA and CarboPac MA series columns.

The HPAEC-PAD technique was introduced with the CarboPac PA 1 column, which was (and still is) particularly well-suited for the analysis of monosaccharides, including sialic acids. Subsequently, an improved version, the CarboPac PA 10, was introduced. The CarboPac PA 10 column showed similar selectivity as the CarboPac

PA 1 column but resulted in faster separation compared to latter. For example, the mixture of six monosaccharides mentioned above can be separated baseline resolved on a CarboPac PA 1 column in about 20 minutes when eluted isocratically with 15 mM NaOH at 1 mL/min flow rate, whereas the same mixture takes about 15 min to be separated on a CarboPac PA 10 column under similar elution conditions. The most recent introduction in this series is the CarboPac PA 20, which shows somewhat different selectivity. This column can produce baseline resolution of the same mixture of six monosaccharides in about 10 minutes when eluted isocratically with 10 mM NaOH at a flow rate of 0.5 mL/min (Figure 17.1).

Separations of monosaccharides (and neutral oligosaccharides) may be achieved also using a CarboPac MA 1 column by isocratic elution with 400–600 mM NaOH at a flow rate of 0.4 mL/min at ambient temperature (lower Panel of Figure 17.4). However, the separation takes a longer time, about 30 min. The CarboPac MA 1 column is suitable for the analysis of alditols (reduced sugars). The alditols are present in several polysaccharides, including polysaccharides used in vaccines (such as, Hemophilius Influenza b, meningococcal, and pneumococcal polysaccharides), as the integral part of their backbone structures and play critical roles in maintaining their conformations. In addition, for samples containing monosaccharides, disaccharides and alditols, such as in fermentation or culture media, and oligosaccharides, it may be advantageous to use the CarboPac MA 1 column, as it permits simultaneous separation of all analytes [10].

Similar columns are also available from other manufacturers, for example Hamilton RCX-30 and Metrosep Carb1 from Metrohm. The selection of column depends on several of factors, including adaptability to the HPLC system used, nature of the sample, user's need and, of course, the cost. Thus, users should make an appropriate selection of columns considering their need and other factors.

17.5.4 AminoTrap Guard Column

Glycoproteins with low percentages of glycosylation represent a challenge for monosacchride composition analysis because of matrix interference. For such glycoproteins, an on-line guard column that removes interfering matrix amino acids and peptides or moves them out of the elution window may be added between the injector and the separator column. An example of such a guard column is the AminoTrap guard column from Dionex. Figure 17.1 illustrates the improvement in the quality of chromatogram when an AminoTrap guard column is used. Inclusion of an AminoTrap guard column moves interfering matrix amino acids and peptides out of the elution window producing more stable baseline, and accurate and reproducible results.

17.5.5 Sialic Acid Assay

Sialic acids (neuraminic acids) occupy terminal positions on many mammalian glycan moieties. The amino group of neuraminic acid is linked to either an *N*-acetyl or an *N*-glycolyl group, resulting in *N*-acetylneuraminic acid (Neu5Ac)

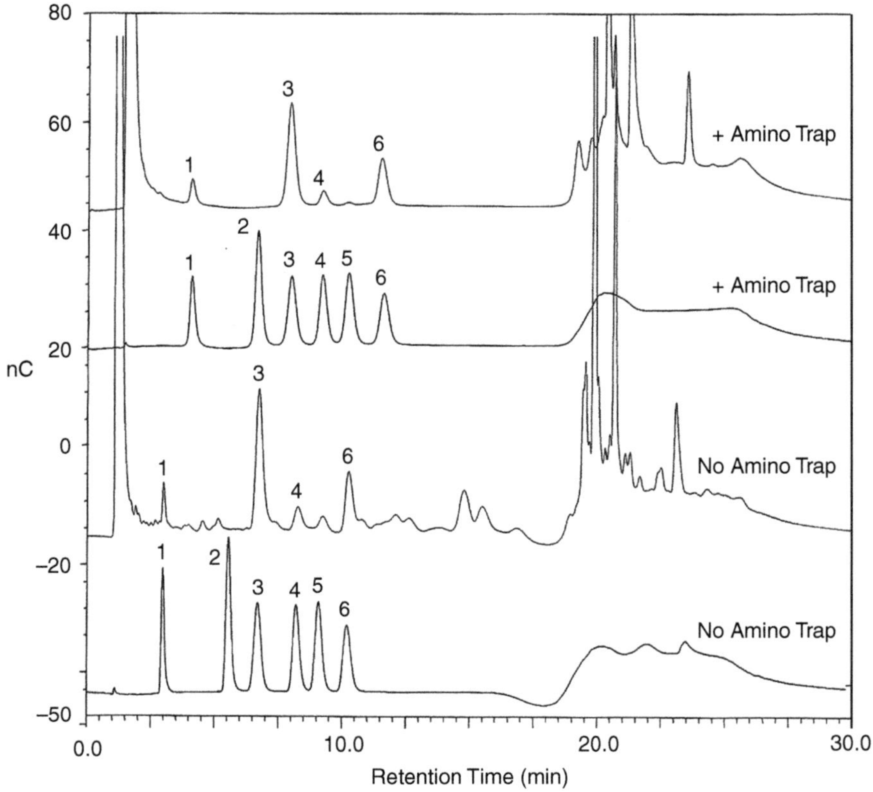

Figure 17.1. Monosaccharide analysis of a protein hydrolysate. Top two tracings are monosaccharide profiles on CarboPac PA 20 in the presence of a 3 × 30 mm AminoTrap cartridge (20 microgram hydrolysate injected), and a mixture of 6 monosaccharide standards (100 pmol), respectively. Column: 3 mm × 150 mm CarboPac PA 20; elution: 10 mM NaOH, isocratic; flow rate: 0.5 mL/min; Inj. Vol.: 10 μL; detection: PAD, 4-potential waveform [26], with a disposable gold electrode. Bottom two tracings are the same as above, respectively, except that the AminoTrap cartridge was not included. Note the cleaner baseline and absence of contaminant peaks in the top profile.

or *N*-glycolylneuraminic acid (Neu5Gc) residue, respectively. The elution of sialic acids from the CarboPac PA columns requires stronger eluents than those used for neutral sugars. This is usually accomplished by the addition of sodium acetate to the sodium hydroxide eluent. Sodium acetate accelerates the elution of strongly bound species without compromising selectivity and interfering with pulsed amperometric detection. Isocratic elution with 150 to 300 mM sodium acetate containing 100 mM sodium hydroxide provides well-resolved peaks (*cf*. Figure 20.6 of this book). It is necessary to maintain a constant concentration of sodium hydroxide in the eluent to maintain baseline stability. The analysis of neutral monosaccharides and sialic acids can be accomplished in a single run by eluting neutral monosaccharides with 10–20 mM sodium hydroxide and then changing the eluent to that required for the elution of sialic acids. However, the column needs to be re-equilibrated to the initial elution condition after every run.

17.6 OLIGOSACCHARIDE ANALYSIS

HPAEC-PAD has been widely used to characterize oligosaccharides. The method allows rapid, sensitive, and direct quantitation of underivatized samples. HPAEC-PAD not only separates oligosaccharides according to charge, but resolves oligosaccharides with the same charge according to size, composition, and linkage of monosaccharide units. Oligosaccharides can be identified through co-elution with standards or through retention time comparison and characterized further by more sophisticated techniques, such as NMR or on-line mass spectrometry. In the pharmaceutical and biotechnology industries, oligosaccharide analyses by HPAEC-PAD have been used: (1) during product development to identify oligosaccharides present in glycoproteins, isolate them, and elucidate their structures; (2) to monitor consistency of glycosylation patterns and the structures of oligosaccharides during process validation and subsequent production (process monitoring) of glycoproteins; (3) to monitor consistency of glycosylation and identify changes that may have resulted from alterations in cell culture conditions or during subsequent manufacturing process, as part of comparability studies; and (4) to identify and monitor changes in glycosylation that occur as a result of expression in different cell lines.

Figure 17.2 shows that bi, tri-, and tetra-sialylated branched fetuin oligosaccharides were separated not only by their sialic acid contents, but also by the

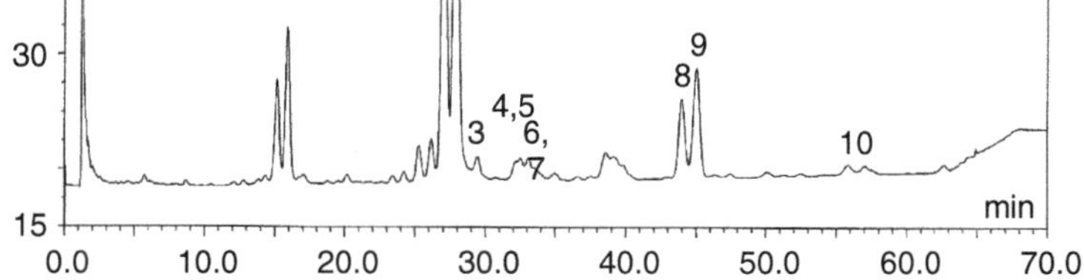

Figure 17.2. Separation of sialylated fetuin oligosaccharide alditols. Column: CarboPac PA 100 Analytical Column (4 × 250 mm) with guard column; elution: 50–200 mM sodium acetate gradient in 100 mM NaOH, over 0–60 min at 1.0 mL/min.; Inj. Vol.: 10 μL; detection: PAD, 4-potential waveform, with a disposable gold electrode.

isomeric α(2,3) and α(2,6) linkages of sialic acid residues. Oligosaccharides with more α(2,6)-linked sialic acids (versus α(2,3)-linked sialic acids) were eluted increasingly later and the ones with only α(2,3)-linked sialic acids were eluted last. Separation was also achieved for the isomeric neutral oligosaccharides. Oligosaccharides containing a terminal Galβ(1,3)GlcNAc sequence were more retained than those with a Galβ(1,4)GlcNAc unit (GlcNAc: *N*-acetylglucosamine). Empirical relationships between oligosaccharide structure and chromatographic retention for HPAEC-PAD have been described [19,20].

The typical run conditions for sialylated *N*-linked oligosaccharides involve elution under ambient condition from a CarboPac PA 100 column using 50–250 mM sodium acetate in 100 mM sodium hydroxide gradient. Such a gradient efficiently separated sialylated fetuin oligosaccharide alditols. As mentioned before, it is important to keep the sodium hydroxide concentration constant to maintain baseline stability. The development of two optimized gradients, one for the separation of sialylated *N*-glycans and a second for the separation of asialo-*N*-glycans, has been reported [21]. However, the users should evaluate and optimize the elution gradients depending on their need for separation.

For the separation of neutral *N*-linked oligosaccharides, better resolution was observed in the separation of a set of 12 neutral *N*-linked oligosaccharides by increasing the NaOH concentration from 100 mM to 250 mM. A 20 minute gradient from 0 to 80 mm sodium acetate in the presence of 250 mM NaOH gives a good separation of *N*-linked neutral oligosaccharides (Figure 17.3). The comparison of peak positions of oligosaccharides 1 and 2, 3 and 4 as well as 7 and 8 shows that fucosylation reduces oligosaccharide retention on a CarboPac PA 100 column. The separation of oligosaccharides 2, 5, and 12 shows that the larger a high mannose oligosaccharide, the later it elutes on the same column.

The examples discussed above used the CarboPac PA 100 column. However, more recently, it was shown that a lower hydroxide eluent concentration can improve the resolution of neutral *N*-linked oligosaccharides from monoclonal antibodies using CarboPac PA 200 column [22]. The applications are discussed in more detail in this book in Chapter 20. In addition to increased resolution, the CarboPac PA 200 column reduces eluent consumption and waste generation (optimum flow rate is 0.5 mL/min. compared to 1.0 mL/min. for the CarboPac PA 100), and requires less acetate to elute a given oligosaccharide compared to the CarboPac PA 100, allowing faster separations and making subsequent acetate removal easier. In certain instances, CarboPac PA 100 and CarboPac PA-200 showed different separation profiles of oligosaccharides [23].

The choice of hydroxide concentration for neutral oligosaccharide separations depends on the structures of oligosaccharides to be separated. The neutral oligosaccharides shown in Figure 17.3 are diverse both in size and composition. On the other hand, the neutral oligosaccharides of monoclonal antibodies presented in Chapter 20 of this book are closely related in size and structure. However, many glycoproteins contain neutral oligosaccharides that vary widely in size, monosaccharide composition and structure. Under such circumstances, a higher hydroxide concentration provides a better separation of oligosaccharides.

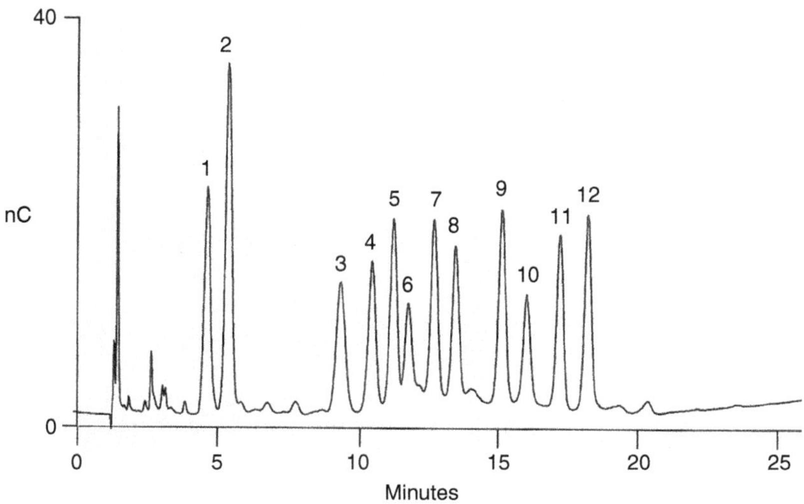

Figure 17.3. Separation of neutral oligosaccharides. Column: CarboPac PA 100 Analytical Column (4 × 250 mm) with guard column; elution: 0–4 min isocratic with 250 mM NaOH; gradient; 0–80 mM sodium acetate in 250 mM NaOH, from 4.1–24 min at 1.0 mL/min.; Inj. Vol.: 10 μL; detection: PAD, 4-potential waveform, with a disposable gold electrode. Peaks are as follows: 1. FucMan3GlcNAc2; 2. Man3GlcNAc2; 3. asialo-agalacto biantennary oligosaccharide with core Fuc; 4. asialo-agalacto biantennary oligosaccharide; 5. Man5GlcNAc2; 6. asialo-agalacto triantennary oligosaccharide; 7. asialo biantennary oligosaccharide with core Fuc; 8. asialo biantennary oligosaccharide; 9. asialo triantennary oligosaccharide; 10. bisected hybrid biantennary oligosaccharide; 11. asialo tetraantennary oligosaccharide; 12. Man9GlcNAc2.

In 100 mM sodium hydroxide, a condition frequently used in HPAEC-PAD for elution, there is small but measurable epimerization of terminal GlcNAc to *N*-acetylmannosamine. Thus, some highly retained oligosaccharides show peak complexity due to the epimerization of GlcNAc and GalNAc (*N*-acetylgalactosamine) at the reducing terminus, making the interpretation of results difficult. Thus, reducing oligosaccharides to the corresponding alditols prior to HPAEC-PAD is recommended.

17.7 GLYCAN MAPPING

Exoglycosidases are enzymes that hydrolyze non-reducing terminal monosaccharides of glycans and exhibit residue, anomeric, and linkage specificity. Thus, digestion with exoglycosidases has been employed to assess monosaccharide sequence and structure in several glycoproteins [*cf*. 24]. In this approach oligosaccharides are labeled at their reducing ends and the reactions are monitored by following labeled oligosaccharides. The digestion shortens the oligosaccharide chain length from the non-reducing end. Oligosaccharides, thus, retain the label. The release of monosaccharide residue(s) is concluded by the loss of molecular weight of the oligosaccharide, as determined by the

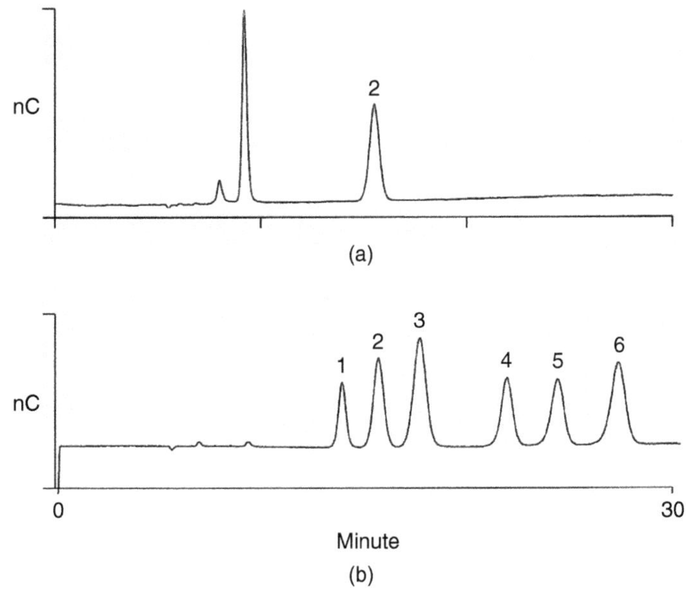

Figure 17.4. A glycoprotein impregnated on a PVDF membrane, is incubated with 0.1 U of Jack bean β-N-acetylhexosaminidase in 25 mM sodium citrate-phosphate buffer, pH 5.0 for 20 h at 37°C. 10 μL of the digest was injected directly to the column (a). Column: CarboPac MA 1 column (4 × 250 mm) column with guard column; elution: 480 mM, isocratic; flow rate: 0.4 mL/min; detection: PAD, 4-potential waveform, with a disposable gold electrode. (b) Chromatogram of a monosaccharide mixture under the same chromatographic conditions as above. Peaks: 1. Fuc; 2. GlcNAc; 3. GalNAc; 4. Man; 5. Glc; and 6. Gal.

peak shift in size-exclusion chromatography. The cleaved monosaccharide and its linkage are identified from the known specificity of the exoglycosidase used in the reaction. It is not possible to identify the other digestion products, the released monosaccharides, because they do not carry labels. Thus, the identification and hence the conclusion about the structure has been indirect. It is not possible to rule out the possibility that the release of oligosaccharide is due to a contaminating exoglycosidase activity. This could (and did) result in incorrect elucidation of the oligosaccharide structures. Direct identification and quantitative measurement of released monosaccharide(s) is critical to permit correct elucidation of sequence and linkage of monosaccharide residues of aglycan.

Application of HPAEC-PAD served this purpose by direct identification and quantitation of the exoglycosidase digestion product [25]. Protein bands separated by SDS-PAGE are electrotransferred onto a PVDF membrane. The membranes are then cut with each glycoprotein band on a piece of membrane and treated separately with exoglycosidases under appropriate pH, buffer and incubation conditions required for each exoglycosidase. At the end of each digestion, the membrane is washed with water and subjected to the digestion with another exoglycosidase. The solutions containing digestion products are dried, reconstituted with water and analyzed on a suitable CarboPac PA or CarboPac MA column for the identification and quantitation of the released monosaccharides. Where pure glycoproteins or glycopeptides are available, the PVDF membranes can be impregnated with them using dot-blot technique and analyzed as described above to obtain oligosaccharide linkage and the carbohydrate sequence of oligosaccharide chains (Figure 17.4). With appropriately purified glycopeptides such analysis could lead to the structure of different glycans at each glycosylation site of a glycoprotein.

REFERENCES

1. Hardy MR, Townsend RR, Lee YC. Monosaccharide analysis of glycoconjugates by anion-exchange chromatography with pulsed amperometric detection. Anal. Biochem. 1988;170:54–62.
2. Townsend RR, Hardy MR, Hindsgaul O, Lee YC. High-performance anion-exchange chromatography of oligosaccharides using pellicular resins and pulsed amperometric detection. Anal. Biochem. 1988;174:459–470.
3. Townsend RR, Hardy MR, Cumming DA, Carver JP, Bendiak B. Separation of branched sialylated oligosaccharides using high-pH anion-exchange chromatography with pulsed amperometric detection. Anal. Biochem. 1989;182:1–8.
4. Lee YC. High performance anion-exchange chromatography for carbohydrate analysis. Anal. Biochem. 1990;189:151–162.
5. Hardy MR, Townsend RR. Analysis of glycoprotein oligosaccharides using high-pH anion exchange chromatography. Glycobiology 1991;1:139–147.
6. ICH Harmonised Tripartite Guideline, Validation of analytical procedures: Text and methodology, Q2(R1), International Conference On Harmonisation of Technical Requirements For Registration Of Pharmaceuticals For Human Use, 2005.
7. Weitzhandler M, Barreto V, Pohl C, Jandik P, Cheng J, Avdalovic N. CarboPac™ PA20: a new monosaccharide separator column with electrochemical detection using disposable gold electrodes. J. Biochem. Biophys. Methods 2004;60:309–317.
8. Karlsson NG, Hansson GC. Analysis of monosaccharide composition of mucin oligosaccharide-alditols by high performance anion-exchange chromatography. Anal. Biochem. 1995;224:538–541.
9. Jahnel JB, Ilieva P, Frimmel FH. HPAE-PAD—a sensitive method for the determination of Carbohydrates. Fresenius J. Anal. Chem. 1998;360:827–829.
10. Dionex Corporation. Technical Note 20: Analysis of carbohydrates by anion exchange chromatography with pulsed amperometric detection.
11. Bhattacharyya L. Unpublished observation.
12. Weitzhandler M, Pohl C, Rohrer JS, Narayanan L, Slingsby R, Avdalovic N. Eliminating monosaccharide peak tailing in high pH anion-exchange chromatography with pulsed amperometric detection. Anal. Biochem. 1996;241:135–136.
13. Dionex Corporation, Technical Note 71: Eluent Preparation for High-Performance Anion-Exchange Chromatography with Pulsed Amperometric Detection.

14. Jensen MB, Johnson DC. Fast wave forms for pulsed electrochemical detection of glucose by incorporation of reductive desorption of oxidation products. Anal. Chem. 1997;69:1776–1781.
15. Clarke AP, Jandik P, Rocklin RD, Liu Y, Avdalovic N. An integrated amperometry waveform for the direct, sensitive detection of amino acids and amino sugars following anion-exchange chromatography. Anal Chem. 1999;71:2774–2781.
16. Rocklin RD, Clarke AP, Weitzhandler M. Improved long-term reproducibility for pulsed amperometric detection of carbohydrates via a new quadruple-potential waveform. Anal. Chem. 1998;70:1496–1501.
17. Cheng J, Jandik P, Avdalovic N. Development and characterization of microfabricated disposable gold working electrodes for high-performance ion chromatography and integrated pulsed amperometric detection. Anal. Chem. 2003;75:572–579.
18. Varki A. Biological roles of oligosaccharides: all the theories are correct. Glycobiology 1993;3:97–130.
19. Rohrer, JS. Separation of asparagine-linked oligosaccharides by high-pH anion-exchange chromatography with pulsed amperometric detection: empirical relationships between oligosaccharide structure and chromatographic retention. Glycobiology 1995;5:359–363 (1995).
20. Strang, A-M. Additional empirical relationships between oligosaccharide structure and high pH anion exchange chromatography. Glycobiology 1998;8: iii–v.
21. Hermentin P, Witzel R, Vliegenthart JFG, Kamerling JP, Nimtz M, Conradt HS. A strategy for the mapping of N-glycans by high-pH anion-exchange chromatography with pulsed amperometric detection. Anal. Biochem. 1992;203:281–289.
22. Grey C, Edebrink P, Krook M, Jacobsson SP. Development of a high performance anion exchange chromatography analysis for mapping of oligosaccharides. J. Chromatogr. B-Analytical Technologies in the Biomedical and Life Sciences 2009;877:1827–1832.
23. Dionex Corporation, Application Note 215: Separation of Asparagine-Linked (N-Linked) Oligosaccharides from Human Polyclonal IgG Using the CarboPac PA200 Column.
24. Kobata A. The carbohydrates of glycoproteins. In: Ginsburg V, Robbins PW, editors. The Biology of Carbohydrates, New York: John Wiley and Sons, Inc. 1984, p. 87–161.
25. Weitzhandler M, Kadlecek D, Avdalovic N, Forte G, Chow D, Townsend, RR. Monosaccharide and oligosaccharide analysis of proteins transferred to polyvinylidene fluoride membranes after sodium dodecyl sulfate-polyacrylamide gel electrophoresis. J. Biol. Chem. 1993;268:5121–5130.
26. Dionex Corporation, Technical Note 21: Optimal Settings for Pulsed Amperometric Detection of Carbohydrates Using the Dionex ED40 Electrochemical Detector.

18

MONOSACCHARIDE ANALYSIS OF GLYCOPROTEINS BY HIGH-PERFORMANCE ANION-EXCHANGE CHROMATOGRAPHY WITH PULSED AMPEROMETRIC DETECTION

Jeffrey S. Rohrer

Thermo Fisher Scientific, Sunnyvale, CA

18.1 INTRODUCTION

Although high-performance anion-exchange chromatography with pulsed amperometric detection (HPAEC-PAD) using a gold working electrode to determine carbohydrates was first described in 1983 [1], it was another five years before Hardy, Townsend, and Lee showed that HPAEC-PAD could be used to determine the monosaccharide composition of a glycoprotein [2]. This was an important step forward in glycoprotein characterization and HPAEC-PAD was rapidly adopted by university and industry laboratories for monosaccharide compositional analysis of a variety of glycoconjugates, including glycoproteins, glycolipids, and oligosaccharides. In fact, the application of HPAEC-PAD quickly expanded beyond monosaccharides to a wide range of carbohydrates. In this book there are chapters that show the application of HPAEC-PAD to oligosaccharide analysis of monoclonal antibodies [3,4], monosaccharide analysis of vaccines [5,6], sugar phosphate analysis of vaccines [6,7], oligosaccharide analysis of vaccines [6], aminoglycoside assay and purity analysis [8], sucrose analysis [9], hexosamine assay of heparin [10], and a chapter that reviews the principles and historical development of glycan analysis

Applications of Ion Chromatography for Pharmaceutical and Biological Products, First Edition.
Edited by Lokesh Bhattacharyya and Jeffrey S. Rohrer.

by HPAEC-PAD [11]. This chapter reviews the basic principles of HPAEC-PAD monosaccharide analysis, discusses how glycoprotein samples are prepared and analyzed by HPAEC-PAD, discusses why this technique is widely used, and briefly discusses how HPAEC-PAD glycoprotein monosaccharide analysis has been and continues to be used in the biotechnology and pharmaceutical industries.

18.2 HPAEC OF MONOSACCHARIDES

One usually does not consider monosaccharides such as glucose, mannose, and glucosamine as anions, but in solutions of pH 12 and higher these monosaccharides are oxyanions and can potentially be separated by anion-exchange chromatography. This is the result of monosaccharides having pK_as around 12. For example, the pK_a of glucose is 12.28. More monosaccharide pK_as, all greater than 12, can be found in Table 1 in the chapter of this book authored by Bruggink [4]. This table reveals an important point about monosaccharide acidity, that the hydroxyl group linked to the number one position of the pyranose ring, the anomeric hydroxyl, has the biggest impact on the monosaccharide's acidity. For example, when that hydroxyl group of glucose is reduced to form sorbitol, the pK_a of glucose increases from 12.28 to 13.60. If that same hydroxyl group is methylated to form α-methyl glucoside the pK_a increases from 12.28 to 13.71.

Figure 18.1 shows the separation of six monosaccharides commonly found in mammalian glycoproteins. The 10 mM sodium hydroxide eluent is sufficiently alkaline to ionize the monosaccharides and allow their separation on the HPAEC column,

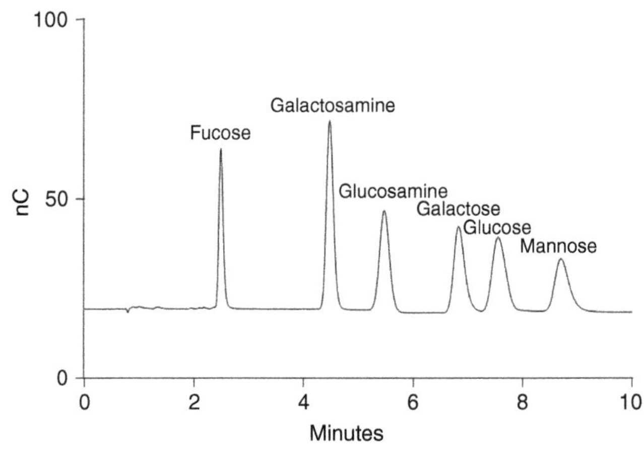

Figure 18.1. Chromatography of six monosaccharides commonly found in mammalian glycoproteins after acid hydrolysis. Monosaccharides were separated on a CarboPac PA20 column (3 × 150 mm) using 10 mM sodium hydroxide flowing at 0.5 mL/min. The column temperature was 30°C. There were 100 pmol of each monosaccharide in a 10 μL injection. Monosaccharides were detected by PAD with the four-potential waveform optimized for carbohydrate detection [30].

the CarboPac PA20. This separation can be rationalized based on the monosaccharide's acidity and the number of ionizable hydroxyl groups. The most acidic monosaccharide will bind to the column the tightest and therefore elute last. With a pK_a of 12.08, mannose is the most acidic of the monosaccharides in Figure 18.1 and it elutes last. Galactose and glucose elute 4th and 5th which represents least to most acidic with pK_as of 12.39 and 12.28, respectively. Galactosamine and glucosamine are simply galactose and glucose with the hydroxyl at the number 2-position of the hemiacetal ring replaced by an amine group. The loss of the hydroxyl that is second in acidity to the anomeric hydroxyl (number 1 position) results in these two monosaccharides eluting earlier than galactose and glucose, but in the same order (i.e. galactosamine followed by glucosamine). The first monosaccharide eluting in Figure 18.1 is fucose, which is a deoxy sugar and therefore has one less hydroxyl to ionize compared to glucose, galactose, and mannose. This order of elution is the same observed on earlier generation CarboPac columns used for the separation of monosaccharides, the CarboPac PA1 and PA10 columns.

Research groups from Germany and Iowa did a systematic study of small carbohydrate binding to the CarboPac PA1 column [12]. Each group determined the retention times of a collection of 93 mono-, di-, tri-, and tetrasaccharides, and small sugar alcohols and small glycosides using 100 mM NaOH flowing at 1.0 mL/min. After plotting the pK_a versus the capacity factor, they concluded that the retention of these small carbohydrates was governed primarily by their acidity.

To test the hypothesis that retention time is related to monosaccharide acidity the current author and Joseph Olechno attempted a HPAEC separation of deuterated glucoses [13]. Specifically we attempted to separate glucose, glucose-1-d (deuterium at the carbon to which the anomeric hydroxyl group is attached), and glucose-2-d (deuterium at the number 2 carbon on the hemiacetal ring). Figure 18.2 shows the

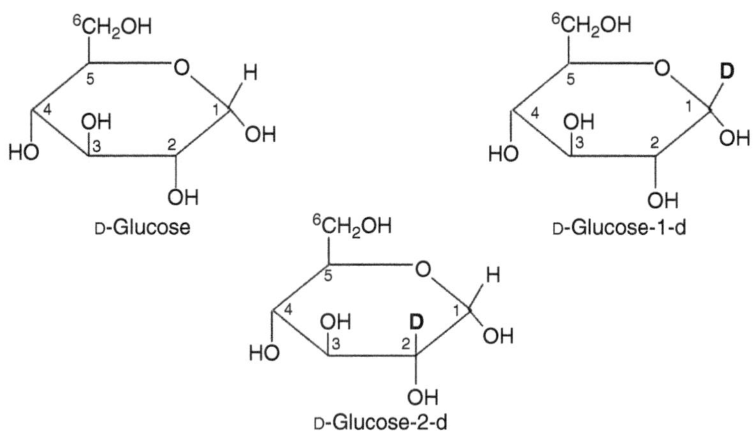

Figure 18.2. Structures of D-glucose and two deuterium-containing D-glucoses, D-glucose-1-d and D-glucose-2-d.

structures of these three monosaccharides. The secondary isotope effect suggests these compounds should have different acidities. We were able to separate glucose from glucose-1-d and glucose-1-d from glucose-2-d. However, we were unable to resolve glucose and glucose-2-d. If the pK_a is impacted by deuterium substitution the effect should be greatest with glucose-1-d, and the effect should be to make the glucose-1-d a weaker acid. Glucose-1-d does elute earlier than glucose, consistent with glucose-1-d being a weaker acid. We also plotted retention times of monosaccharides with known pK_as versus their pK_a and used a linear least squares fit to calculate a line. This line was used to estimate pK_as for the three compounds. This analysis yielded a pK_a for glucose of 12.28, which is the published value. The pK_as for glucose-1-d and glucose-2-d were 12.31 and 12.29 respectively. These data support the hypothesis that that HPAEC monosaccharide retention time is related to monosaccharide acidity.

There are other publications that have explored the impact of monosaccharide structure on CarboPac PA1 column retention. Publications that looked at the impact of loss of hydroxyl groups of glucose [14] and mannose [15] showed that as hydroxyl groups were lost, retention time decreased and that the loss of the anomeric hydroxyl (1-position) caused the greatest loss of retention time. This is consistent with the fact that the anomeric hydroxyl has the greatest impact on a monosaccharide's acidity. While monosaccharide acidity is perhaps the most important factor in monosaccharide retention on the CarboPac PA1 column there are other structural features that lead to less acidic monosaccharides eluting later than more acidic. Table 1 in Y. C. Lee's early review of HPAEC-PAD that shows the retention of 19 monosaccharides clearly reveals this, as the pentoses, xylose ($pK_a = 12.29$) and ribose ($pK_a = 12.21$), elute later than the hexose mannose ($pK_a = 12.08$) [16]. The remainder of the 19 monosaccharides that have known pK_as elute according to their acidity.

These studies also show that HPAEC with a CarboPac column is capable of resolving a wide range of monosaccharides. For example, one publication showed the separation of fucosamine, rhamnosamine, quinovosamine, and glucosamine from a capsular polysaccharide from a strain of *Vibrio vulnificus* [17]. There were other early examples of HPAEC-PAD used for monosaccharide compositional analysis of bacterial carbohydrates [18–20].

18.3 PAD OF MONOSACCHARIDES

PAD is discussed in detail in the chapter by LaCourse [21] and discussed briefly in other chapters so it will not be discussed in detail here. Instead the important points will be reviewed in this chapter as they relate to monosaccharide analysis. PAD detects monosaccharides and other carbohydrates by measuring the current due to oxidation of the carbohydrate in highly alkaline solution at a gold working electrode to which a potential (voltage) is applied. Only compounds oxidized at a gold electrode with the applied potential will be detected. Therefore PAD is a direct selective detection technique. This is important because nearly all monosaccharides lack a good chromophore and were often derivatized with a chromophore or fluorophore in order to be detected with sufficient sensitivity. PAD requires none of the labor and has none of the possible errors of sample derivatization. While monosaccharides can be detected without derivatization with either refractive index (RI) detection or by absorbance at a low UV wavelength (e.g., 200 nm) neither has the sensitivity or selectivity (RI is bulk property detector) to be applied to the determination of the monosaccharide content of a glycoprotein. PAD is a sensitive detection technique that detects pmol quantities of monosaccharides. Figure 18.1 shows that 100 pmol of each of the six monosaccharides are detected with sufficient sensitivity. A comparison of HPAEC-PAD to HPLC-RI for the determination of amino sugars concluded that HPAEC-PAD was nearly two times more sensitive, not subject to matrix interferences, highly precise and selective, and required less sample preparation [22]. Electrochemical responses for monosaccharides are nearly equivalent, though to quantify monosaccharides, typically calibration is performed for each monosaccharide using different concentrations of a mixed standard. In summary, the direct, selective, and sensitive detection of monosaccharides by PAD combined with the high-resolution separation of HPAEC with a highly alkaline mobile phase made PAD a popular detection technique for monosaccharides in a variety of samples including acid hydrolysates of glycoproteins.

18.4 PREPARATION OF GLYCOPROTEIN SAMPLES FOR HPAEC-PAD MONOSACCHARIDE ANALYSIS

To determine the monosaccharide composition of a glycoprotein, the monosaccharides must be first released from the glycoprotein. In other words, the chemical bonds between the polypeptide backbone and the attached oligosaccharide and between the monosaccharides in the oligosaccharides must be broken. This is typically accomplished by acid hydrolysis at elevated temperatures. While there are numerous published acid hydrolysis conditions, and ideally hydrolysis conditions should be optimized for the individual glycoprotein [23], the following conditions have proven to be good general conditions. To determine the neutral sugar content of a glycoprotein (i.e. mannose, fucose, galactose, etc.) low microgram quantities of the glycoprotein are hydrolyzed in 2M trifluoroacetic acid (TFA) for 4 h at 100°C. These conditions will also determine amino sugars (i.e. galactosamine, glucosamine, etc.), but their yield will only be 90–95%. For better recovery of amino sugars 6N HCl is used for 4 h at 100°C. These conditions destroy most of the neutral sugars. Figure 18.3 shows chromatography of human IgG (2 μg in each injection) that was hydrolyzed using both conditions. Note the small quantities of galactose and mannose in the HCl hydrolysate compared to the TFA hydrolysate. It is important that pure acids be used, and that the tubes in which the glycoprotein is hydrolyzed, yield no peaks that interfere with monosaccharide quantification [24]. The quality of the acid and the tubes can be determined by performing a blank hydrolysis and analyzing the sample in the same manner as the glycoprotein sample. It is difficult to avoid having a glucose contamination in the final sample. The step by step procedure for acid hydrolysis of a glycoprotein for monosaccharide analysis is described below.

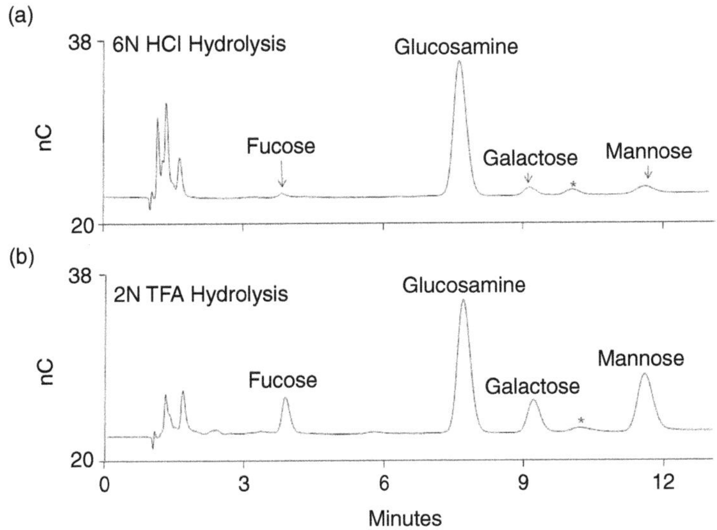

Figure 18.3. HPAEC-PAD analysis of human IgG monosaccharides. (a) 2 μg of a 6N HCl hydrolysis and (b) 2 μg of a 4N TFA hydrolysis. The monosaccharides were separated on a CarboPac PA20 column (3 × 150 mm) preceded by an AminoTrap guard column (3 × 30 mm) using 10 mM potassium hydroxide flowing at 0.5 mL/min. After 12 min at 10 mM potassium hydroxide the concentration was increased to 100 mM for 3 min, and then the concentration was reduced back to 10 mM prior to the next injection. The column temperature was 30°C and the injection volume was 10 μL for each sample. Monosaccharides were detected by PAD as in Figure 18.1. The potassium hydroxide was produced by an eluent generator. The asterisk shows the peak for glucose which is a ubiquitous contaminant of monosaccharide analysis.

1. Pipette 20 μL of each 1 mg/mL glycoprotein solution to a 1.5 mL polypropylene tube with a screw cap, add 320 μL deionized water, then add 60 μL concentrated (13 M) TFA, mix, and place in a heating block set at 100°C for 4 h. Note: hydrolyzing 20 μg of protein will yield enough materials for at least 10 injections from a lightly glycosylated (<3% by weight carbohydrate) protein for HPAEC-PAD monosaccharide analysis. Lower quantities of more glycosylated proteins can be used and lower volumes of glycoproteins prepared at concentrations greater than 1 mg/mL can be used with a concomitant increase in the volume of deionized water.
2. Remove the tubes from heating block, place them in a microfuge and briefly spin to cool and place all the liquid to the bottom of the tube.
3. Remove the lids and place the tubes in a SpeedVac to remove the volatile acid under vacuum. The SpeedVac should have an acid trap installed to protect the vacuum pump and no heat should be applied during drying.
4. After drying, the samples are reconstituted in 200 μL deionized water. The sample is ready for HPAEC-PAD monosaccharide analysis with a typical injection volume of 10 μL.

Note: Samples can be put in a freezer for future analysis either before or after adding deionized water. If you wish to have an internal standard, the sample can be dissolved in a solution of the internal standard that will match that in the external standard. This can be used to ascertain autosampler performance. 2-Deoxyglucose is an appropriate internal standard for this purpose.

For the 6N HCl hydrolysis add 400 μL of 6N HCl to the 20 μL of sample. Do not add deionized water. This will yield a final concentration of slightly less than 6N but it has been found to have a negligible impact on the final result. If more than 20 μL of sample must be used, it can be dried prior to adding the acid.

This general approach can be used for other glycoconjugates, such as oligosaccharides, glycopeptides, glycolipids, etc. albeit with possible changes in acid, acid concentration, hydrolysis temperature, and hydrolysis time.

While commonly used for monosaccharide analysis, acid hydrolysis of glycoproteins is far from a perfect technique. Even when conditions are optimized, acid hydrolysis is always a balance between release of the monosaccharide and its subsequent destruction. The conditions described above are a good approximation of this balance for many glycoproteins. Trying to hydrolyze a monosaccharide standard to correct for destruction will yield high results as the standard will be destroyed over the 4 h while the monosaccharides to be released from the glycoprotein are only subject to destruction after release. In other words, the acid-hydrolyzed standard will overcorrect for monosaccharide destruction in the acid-hydrolyzed glycoprotein.

While acid hydrolysis is required for all methods of glycoprotein monosaccharide compositional analysis, other methods also require sample derivatization. Prior to labeling the monosaccharide for absorbance or fluorescence detection, the amino sugars must be re-*N*-acetylated. Acid hydrolysis causes loss of the acetyl group of *N*-acetylglucosamine and *N*-acetylgalactosamine. Hardy et al. recognized that re-*N*-acetylation was not required for HPAEC-PAD [2]. They confirmed that no *N*-acetylglucosamine or *N*-acetylgalactosamine remained in the sample to interfere with the quantification of the other monosaccharides.

18.5 HPAEC-PAD MONOSACCHARIDE ANALYSIS

Acid-hydrolyzed glycoprotein samples prepared as described above are transferred to autosampler vials, leaving insoluble residue, if any, at the bottom of the tube, and the samples are now ready for analysis. Released monosaccharides can be separated on the CarboPac PA1, PA10, or PA20 column. The conditions described below are for the CarboPac PA20 column and references will be provided for the PA1 and PA10 conditions. The CarboPac PA100 and PA200 columns can also separate the six monosaccharides commonly found in mammalian glycoproteins, but both separations cannot tolerate small changes in eluent strength, and therefore are not as rugged as the separations on the other three columns. The CarboPac MA1 column also separates the same six monosaccharides, but glucosamine and galactosamine are not fully recovered from the MA1 resin and therefore the MA1 is only recommended for neutral monosaccharides and acetylated amino sugars [25].

Monosaccharides are separated in 15 min on the CarboPac PA20 column (0.3× 15 cm) and AminoTrap guard column (0.3 × 3 cm) with a 10 mM NaOH eluent flowing at 0.5 mL/min. The column is then washed with 200 mM NaOH for 5 min. The wash step is necessary to remove strongly bound carbonate from the eluent and sample as well as strongly bound amino acids and peptides from the sample. Failure to perform the column wash will eventually result in a loss of column capacity which will reduce monosaccharide retention times leading to monosaccharide coelution. After the wash step, the column is re-equilibrated at the starting conditions for 10 min prior to the next injection. Injection to injection time can be reduced to 25 min by using an eluent generator to produce the hydroxide eluent [26]. The use of eluent generation for HPAEC-PAD is described in the chapter on aminoglycoside determinations [8]. Briefly, the analyst adds deionized water to the system and the eluent generator (EG) produces pure, essentially carbonate-free, hydroxide eluent with concentrations up to 100 mM. The lack of carbonate in the eluent from EG reduces the need for column washing. The column wash solution concentration is reduced from 200 to 100 mM hydroxide and the wash time is reduced from five to three minutes. The initial separation time is also reduced from 15 to 12 min. Some reduction in separation time from 15 min may also be possible when the eluent is prepared manually. Manual eluent preparation requires that high purity 18 mΩ-cm deionized water and 50% NaOH solution are used, and adequate precautions are taken to minimize the introduction of carbonate in the eluent. Proper eluent preparation for HPAEC-PAD is described in Dionex Technical Note 71 [27].

The AminoTrap column is used in place of the CarboPac PA20 guard column for monosaccharide analyses of glycoproteins. Early eluting amino acids and small peptides can interfere with the detection and quantification of monosaccharides [28]. Though this effect is minimal for highly glycosylated proteins, it becomes significant for lightly glycosylated proteins (e.g., a monoclonal antibody). The AminoTrap delays the elution of amino acids and small peptides, until after the monosaccharides have eluted, and thus allows accurate monosaccharide quantification for even lightly glycosylated proteins.

The CarboPac PA20 column is preferred over the other columns used for HPAEC-PAD monosaccharide analysis because it delivers faster separations, requires a lower flow rate (less eluent prep and less waste generated) and a lower hydroxide concentration (less consumption of an EG cartridge if using eluent generation), and it has a greater toleration of eluent preparation errors (i.e. the six monosaccharides are baseline resolved over a wider hydroxide concentration range). Glycoprotein monosaccharide analysis with the CarboPac PA1 column uses 16 mM NaOH flowing at 1.0 mL/min with 50 min required per injection [29]. The CarboPac PA10 column also requires 50 min per injection, but uses an 18 mM NaOH eluent [28].

After separation the monosaccharides are detected by PAD at a gold working electrode. For the detection of carbohydrates a four-potential waveform designed for sensitive carbohydrate detection and good long-term reproducibility is used [30]. This waveform can also be used with disposable gold working electrodes [31]. These electrodes deliver better electrode to electrode reproducibility and eliminate the need to polish the working electrode.

To quantify a glycoprotein's monosaccharides, analyze two micrograms (per injection) of acid-hydrolyzed glycoprotein by HPAEC-PAD. This amount of sample is appropriate for lightly glycosylated proteins. For more heavily glycosylated proteins two micrograms can be used, but it is also possible to inject much less. Ten microliters is a convenient injection volume, but other volumes can be used. Quantification of the individual monosaccharides uses a calibration curve created by injecting 10 μL each of different concentrations of a mixture of the six monosaccharides in Figure 18.1. The calibration should cover the range expected in the sample. If that is unknown, a 20–500 pmol range is a good starting range. The standard mixture can be prepared as described by Hardy and Townsend [32], or purchased from Dionex. Usually the moles of each monosaccharide found in the sample are divided by the moles of protein injected to report the number of moles of each monosaccharide per mole of protein.

18.6 APPLICATION OF HPAEC-PAD GLYCOPROTEIN MONOSACCHARIDE ANALYSIS

HPAEC-PAD is widely used for monosaccharide analysis of glycoproteins in the pharmaceutical industry, essentially in the same manner as it is described here, in research and in later stages of the product pipeline. As long ago as 1990, Michael Spellman noted the use of HPAEC-PAD for carbohydrate analyses, including monosaccharide analysis, of recombinant glycoproteins [33]. More recently, Elizabeth Higgins reported the continued use of HPAEC-PAD monosaccharide analysis for recombinant glycoproteins [34]. Monoclonal antibodies are usually glycosylated and a 1997 FDA publication "Points to Consider in the Manufacture and Testing of Monoclonal Antibody Products for Human Use" suggests that carbohydrate content and composition could be used for in vitro evidence of product comparability [35]. HPAEC-PAD is commonly used for a product comparability analysis (i.e. how similar the current production of recombinant glycoprotein drug is to earlier productions of the glycoprotein, especially that which was used in the clinical trials) with respect to the glycoprotein's monosaccharide composition. HPAEC-PAD monosaccharide compositional analysis was one of the techniques used to compare two recombinant α-galactosidase A products (different manufacturers) and reveal difference in glycosylation [36]. As noted earlier in this chapter, HPAEC-PAD monosaccharide analysis has expanded well beyond glycoproteins to include vaccines and other glycosylated molecules. The enduring popularity of HPAEC-PAD glycoprotein monosaccharide analysis is due to the fact that it meets the required sensitivity for the analysis without sample derivatization. Direct detection reduces the necessary sample preparation and eliminates errors associated with derivatization.

REFERENCES

1. Rocklin RD, Pohl CA. Determination of carbohydrates by anion exchange chromatography with pulsed amperometric detection. J. Liq. Chromatogr. 1983;6:1577–1590.

2. Hardy MR, Townsend RR, Lee, YC. Monosaccharide analysis of glycoconjugates by anion exchange chromatography with pulsed amperometric detection. Anal. Biochem. 1988;170:54–62.
3. Grey C. Monitoring protein N-glycosylation consistency using HPAEC-PAD. In: Bhattacharyya, L and Rohrer, JS, editors. Applications of ion chromatography for pharmaceutical and biological products. New Jersey: John Wiley & Sons; 2012.
4. Bruggink C. Oligosaccharide analysis by high-performance anion-exchange chromatography hyphenated to integrated pulsed amperometric detection and on-line ion-trap mass spectrometry. In: Bhattacharyya, L and Rohrer, JS, editors. Applications of ion chromatography for pharmaceutical and biological products. New Jersey: John Wiley & Sons; 2012.
5. Ricci S, Fabbri B, Giannini S, Bernardini S, Morelli M. High performance anion-exchange chromatography with pulsed amperometric detection (HPAEC-PAD) analysis of GBS glycoconjugate vaccines: from fermentation to final product. In: Bhattacharyya, L and Rohrer, JS, editors. Application of ions chromatography for pharmaceutical and biological products. New Jersey: John Wiley & Sons; 2012.
6. Carinci V, Bernardini A, Campa C, Magagnoli C, Beccai F, D'scenzi S. Stability studies and characterization of meningococcal A, C, Y, and W_{135} glycoconjugate vaccine. In: Bhattacharyya, L and Rohrer, JS, editors. Applications of ion chromatography for pharmaceutical and biological products. New Jersey: John Wiley & Sons; 2012.
7. Del Grosso AV, Duong B. Applications of ion chromatography in biological product analyses. In: Bhattacharyya, L and Rohrer, JS, editors. Applications of ion chromatography for pharmaceutical and biological products. New Jersey: John Wiley & Sons; 2012.
8. Hanko VP, Rohrer JS. Ion chromatography analysis of aminoglycoside antibiotics. In: Bhattacharyya, L and Rohrer, JS, editors. Applications of ion chromatography for pharmaceutical and biological products. New Jersey: John Wiley & Sons; 2012.
9. Jackson DS. Ion chromatographic analysis of pharmaceuticals for authenticity and adulteration. In: Bhattacharyya, L and Rohrer, JS, editors. Applications of ion chromatography for pharmaceutical and biological products. New Jersey: John Wiley & Sons; 2012.
10. Karmarkar SV. Validation of ion chromatographic methods. In: Bhattacharyya, L and Rohrer, JS, editors. Applications of ion chromatography for pharmaceutical and biological products. New Jersey: John Wiley & Sons; 2012.
11. Bhattacharyya L. Glycan analysis by high performance anion exchange chromatography with pulsed amperometric detection. In: Bhattacharyya, L and Rohrer, JS, editors. Applications of ion chromatography for pharmaceutical and biological products. New Jersey: John Wiley & Sons; 2012.
12. Paskach TJ, Lieker HP, Reilly PJ, Thielecke, K. High-performance anion-exchange chromatography of sugars and sugar alcohols on quaternary ammonium resins under alkaline conditions. Carbohydr. Res. 1991;215:1–14.
13. Rohrer JS, Olechno JD. Secondary isotope effect: The resolution of deuterated glucoses by anion-exchange chromatography. Anal. Chem. 1992;64:914–916.
14. Koizumi K, Kubota Y, Ozaki H, Shigenobu K, Fukuda M, Tanimoto T. Analyses of isomeric mono-O-methyl-D-glucoses, D-glucobioses and D-glucose monophosphates by high-performance anion-exchange chromatography with pulsed amperometric detection. J. Chromatogr. 1992;595:340–345.
15. Li S, Mou S, Liao W, Lu D. The study of the relationship between retention and structure on D-mannose and its derivatives with high-performance anion-exchange chromatography. Carbohydr. Res. 1996;295:229–234.

16. Lee YC. High-performance anion-exchange chromatography for carbohydrate analysis. Anal. Biochem. 1990;189:151–162.
17. Reddy GP, Hayat U, Bush CA, Morris JG Jr. Evidence for the presence of multiple capsules in Vibrio vulnificus Strain B062316. Anal. Biochem. 1993;214:106–115.
18. Clarke AJ, Sarabia V, Keenleyside W, MacLachlan PR, Whitfield C. The compositional analysis of bacterial extracellular polysaccharides by high-performance anion-exchange chromatography. Anal. Biochem. 1991;199:68–74.
19. Ip CCY, Manam V, Hepler R, Hennessey Jr JP. Carbohydrate Compositional Analysis of bacterial polysaccharides: Optimized acid hydrolysis conditions for HPAEC-PAD analysis. Anal. Biochem. 1992;201:343–349.
20. Clarke AJ. Compositional analysis of peptidoglycan by high-performance anion-exchange chromatography. Anal. Biochem. 1993;212:344–350.
21. LaCourse WR. Pulsed electrochemical detection in ion chromatography. In: Bhattacharyya, L and Rohrer, JS, editors. Applications of ion chromatography for pharmaceutical and biological products. New Jersey: John Wiley & Sons; 2012.
22. Martens DA, Frankenberger WT Jr. Determination of aminosaccharides by high-performance anion-exchange chromatography with pulsed amperometric detection. Talanta 1991;38:245–251.
23. Fan JQ, Namiki Y, Matsuoka K, Lee YC. Comparison of acid hydrolytic conditions for Asn-linked oligosaccharides. Anal. Biochem. 1994;219: 385-378.
24. Dionex Corporation, Technical Note 53: Determination of glycoprotein monosaccharide composition by HPAE-PAD using online electrolytically generated eluents.
25. Dionex Corporation, Product Manual for CarboPac MA1, PA1, P10, PA100, p. 9.
26. Dionex Corporation, Technical Note 40: Glycoprotein monosaccharide analysis using high-performance anion-exchange chromatography with pulsed amperometric detection (HPAE-PAD).
27. Dionex Corporation, Technical Note 71: Eluent preparation for high-performance anion-exchange chromatography with pulsed amperometric detection.
28. Weitzhandler M, Pohl C, Rohrer J, Narayanan L, Slingsby R, Narayanan L, Avdalovic N. Eliminating amino acid and peptide interference in high-performance anion-exchange chromatography with pulsed amperometric detection glycoprotein monosaccharide analysis. Anal. Biochem. 1996;241:135–136.
29. Rohrer JS, Cooper GA, Townsend RR. Identification, quantification, and characterization of glycopeptides in reversed-phase HPLC separations of glycoprotein proteolytic digests. Anal. Biochem. 1993;212:7–16.
30. Rocklin RD, Clarke AP, Weitzhandler M. Improved Long-term reproducibility for pulsed amperometric detection of carbohydrates via a new quadruple-potential waveform. Anal. Chem. 1998;70:1496–1501.
31. Cheng J, Jandik P, Avdalovic N. Development and characterization of microfabricated disposable gold working electrodes for high-performance ion-exchange chromatography and integrated pulsed amperometric detection. Anal. Chem. 2003;75:572–579.
32. Hardy MR, Townsend RR. High-pH anion-exchange chromatography of glycoprotein-derived carbohydrates. In: Lennarz, WJ, Hart, GW, editors. Methods of Enzymology Guide to Techniques in Glycobiology, Vol. 230. San Diego: Academic Press; 1994. p 208–225.
33. Spellman MW. Carbohydrate characterization of recombinant glycoproteins of pharmaceutical interest. Anal. Chem. 1990;62:1714–1722.

34. Higgins E. Carbohydrate analysis throughout the development of a protein therapeutic. Glycoconjugate. J. 2010;27:211–225.

35. Zoon K. Points to consider in the manufacture and testing of monoclonal antibody products for human use. FDA CBER document UCM153152.pdf. 1997. Available at, http://www.fda.gov/downloads/BiologicsBloodVaccines/GuidanceComplianceRegulatoryInformation/OtherRecommendationsforManufacturers/UCM153182.pdf (Accessed 10/27/11).

36. Lee K, Jin X, Zhang K, Copertino L, Andrews L, Baker-Malcom J, Geagan L, Qui H, Seiger K, Barngrover D, McPherson JM, Edmunds T. A biochemical and pharmacological comparison of enzyme replacement therapies for the glycolipid storage disorder Fabry disease. Glycobiology 2003;13:305–313.

19

TWO-DIMENSIONAL ION CHROMATOGRAPHY FOR SIMULTANEOUS DETERMINATION OF AMINO ACIDS AND CARBOHYDRATES

Petr Jandik and Jun Cheng

Thermo Fisher Scientific, Sunnyvale, CA

19.1 INTRODUCTION

The chromatographic derivatization-free determination of amino acids was introduced in 1999 [1]. The amino acids and carbohydrates are separated by anion exchange and detected by integrated amperometry on a gold working electrode [1,2]. In the most widely used format, the total run time for each amino acid chromatogram is 70–80 min. This version of the method is capable of separating not only amino acids but also many other electroactive anionic compounds on a single column. To illustrate this point, Figure 19.1 shows a simultaneous separation of amino acids, carbohydrates, aminosulfonic acid and water soluble vitamins. Optimization of similar complex separations is a subject of several recent reports. Specialized gradient conditions for up to 75 compounds encountered in cell cultures and in fermentation broth media were developed by Hanko and Rohrer [3]. Mou et al. [4,5] published a report on optimized gradient conditions for the separation of 22 carbohydrates and amino acids from green tea samples. An additional tool for increasing selectivity of separation of complex amino acid and carbohydrate mixtures was developed with the help of bi-modal integrated amperometric detection [6,7]. The bi-modal amperometric detection

Applications of Ion Chromatography for Pharmaceutical and Biological Products, First Edition.
Edited by Lokesh Bhattacharyya and Jeffrey S. Rohrer.

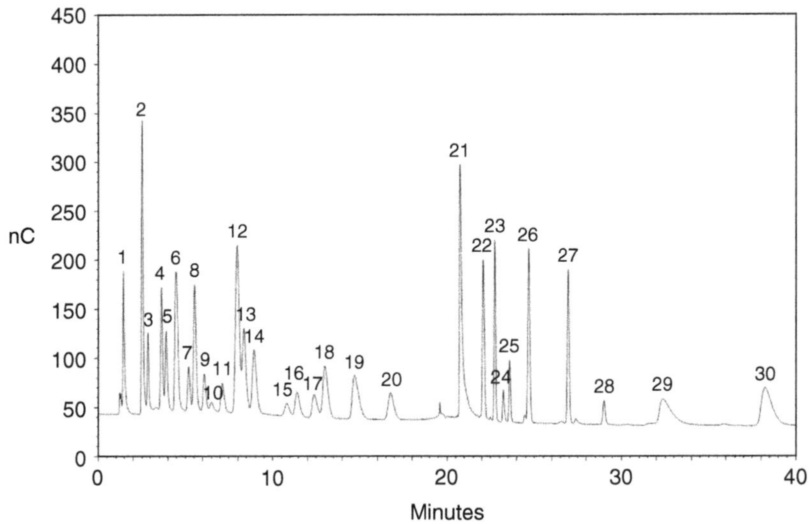

Figure 19.1. One-dimensional separation of amino acids, carbohydrates, and vitamins. Peaks: (1) arginine, (2) hydroxylysine, (3) lysine, (4) galactosamine, (5) glucosamine, (6) asparagine, (7) alanine, (8) threonine, (9) glycine, (10) fructose, (11) valine, (12) hydroxyproline, (13) serine, (14) proline, (15) sucrose, (16) isoleucine, (17) leucine, (18) methionine, (19) norleucine, (20) taurine, (21) histidine, (22) vitamin B6, (23) phenylalanine, (24) glutamate, (25) aspartate, (26) cystine, (27) tyrosine, (28) cysteic acid, (29) vitamin B2, (30) tryptophan; Separation column: AminoPac PA 10 (2 × 250 mm); Injection: 25 μL of 8 μM standard solution; see Table 19.1A for gradient elution conditions; column temperature 30°C; integrated amperometric detection on a gold electrode as described in reference 1.

gains selectivity by integrating the anodic current at two different potentials of the same amperometric waveform.

19.2 TWO-DIMENSIONAL SEPARATIONS

An important alternative to separations of amino acids in complex mixtures with other compounds by methods published in references [3–5] is two-dimensional chromatography, which is the subject of this chapter. In this approach, amino acid separations can be completed (including rinse and re-equilibration) with a reduced total run time of 25–40 min. Carbohydrates and other compounds are removed from the samples prior to amino acid analysis and separated in a second dimension on another column.

19.3 AMINO ACID SEPARATIONS WITH A REDUCED TOTAL RUN TIME

Examples of "regular" and shortened amino acid separations are included in Figure 19.2(a) and (b). Also shown in the same figure are the periods of time

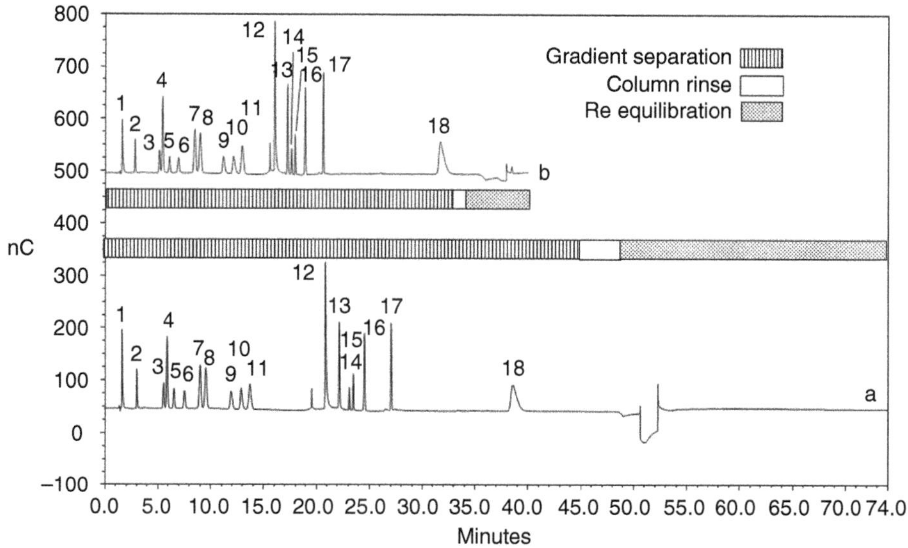

Figure 19.2. (a) Regular method and (b) short method for separations of amino acids. The AminoPac PA10 column was used for both chromatograms. Elution conditions are given in Tables 19.1A and 19.1B for the regular and short methods, respectively. Injection: 25 μL of 8 μM standard solution. Peaks: (1) arginine, (2) lysine, (3) alanine, (4) threonine, (5) glycine, (6) valine, (7) serine, (8) proline, (9) isoleucine, (10) leucine, (11) methionine, (12) histidine, (13) phenylalanine, (14) glutamate, (15) aspartate, (16) cystine, (17) tyrosine; column temperature 30°C; integrated amperometric detection [1].

required for the gradient separation, column cleanup with 200 mM sodium hydroxide, 0.10 M acetic acid (column rinse), and re-equilibration of the column to initial gradient conditions for both the regular and shortened separations. The corresponding gradient programs are compiled in Tables 19.1A and 19.2B. The reduction of the total run time for the chromatogram in Figure 19.2(b) is achieved mainly by reducing the resolution of methionine from histidine and by reducing the length of the re-equilibration time. The larger gap between methionine and histidine in Figure 19.2(a) is designed to accommodate nonproteinogenic amino acids such as norleucine, taurine, and others (see their separation in Figure 19.1) in case they are present in the sample. The longer re-equilibration time of Figure 19.2(a) allows for re-equilibration to lower initial hydroxide concentrations than those included in Table 19.1. The lowered initial hydroxide concentration is a principal tool for achieving simultaneous separations of a large number of carbohydrates along with amino acids. The detailed descriptions of the corresponding gradient programs can be found in references [3–5]. Without carbohydrates in the amino acid-containing sample, there is no need to lower the initial hydroxide concentration and a much shorter re-equilibration time of Figure 19.2B can be employed.

TABLE 19.1A. Typical Gradient Program for Anion-exchange Separation of Amino Acids.

Time (min)	%E1: Water	%E2: 0.25M NaOH	%E3: 1M Sodium Acetate	%E4: 0.1M Acetic Acid	Gradient Curve
0.0	76	24			
2.0	76	24			
8.0	64	36			8
11.0	64	36			
18.0	40	20	40		8
21.0	44	16	40		5
23.0	14	16	70		8
45.0	14	16	70		
45.1				100	5
47.1				100	
47.2	20	80			8
49.2	20	80			
49.3	76	24			5
74.0	76	24			

TABLE 19.1B. Gradient Program for Faster Anion-exchange Separation of Amino Acids.

Time (min)	%E1: Water	%E2: 0.25M NaOH	%E3: 1M Sodium Acetate	%E4: 0.1M Acetic Acid	Gradient Curve
0.0	76	24			
2.0	76	24			
6.0	64	36			8
8.0	64	36			
12.0	40	20	40		8
14.0	44	16	40		5
16.0	14	16	70		8
17.9	14	16	70		
18.0		16	84		8
32.0		16	84		
32.1				100	5
33.1				100	
33.2	20	80			8
34.2	20	80			
34.3	76	24			5
40.0	76	24			

19.4 PRE-SEPARATION OF CARBOHYDRATES FOLLOWED BY SIMULTANEOUS SEPARATIONS OF AMINO ACID AND CARBOHYDRATES ON TWO DIFFERENT COLUMNS

The pre-separation of carbohydrates from amino acids, the necessary first step in the two-dimensional chromatography of these two groups of compounds, has been

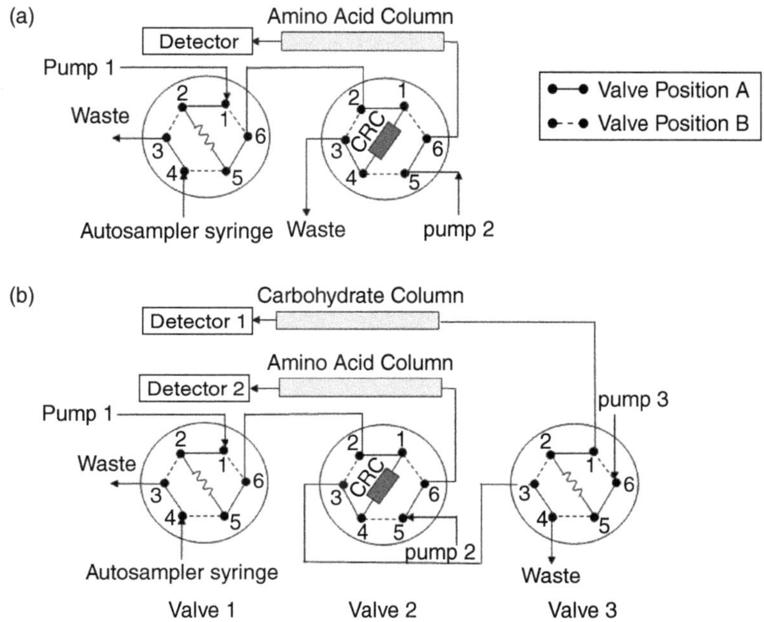

Figure 19.3. (a) The column switching scheme for one-dimensional separation of amino acids after carbohydrate removal. CRC: Carbohydrate Removal Column (2 × 5 mm column filled with a strong cation exchanger). Pump 1: Dionex AXP MS; 0.02% formic acid (transfer fluid) at 0.080 mL/min. Pump 2: Dionex DP pump; eluent and flow rate in Table 19.1A. Detector: Dionex ED in ICS3000 system. (b) The column switching scheme for simultaneous two-dimensional separation of amino acids and carbohydrates. Amino Acid Column: AminoPac PA10; One of the following columns can be used as a Carbohydrate Column: CarboPac PA20, CarboPac PA1, CarboPac MA1, CarboPac PA200; Pumps and eluents 1 and 2 as specified in (a). Pump 3: Dionex DP pump, eluent composition and flow rate as described in the following figures; Detectors 1 and 2: Dionex ED modules in the ICS3000 system. See the text for additional details.

developed as an extension of an on-line carbohydrate removal method [8]. With this technique, illustrated in Figure 19.3(a), the carbohydrates are removed by passing a sample through a cation-exchange cartridge in hydrogen form (Carbohydrate Removal Cartridge, CRC) with both valves of Figure 19.3(a) being in Position A (Figure 19.3). The amino acids (and also amino sugars, such as galactosamine or glucosamine) are cationic under acidic conditions and thus are retained on the CRC while carbohydrates and any other noncationic sample components are passing through the CRC to waste. After all carbohydrates are discarded to waste in this manner, Valve 2 is moved from Position A to Position B placing the CRC with retained amino acids into a stream of alkaline eluent from Pump 2. All amino acids are converted to anions, removed from CRC onto the anion-exchange column, and separated by the hydroxide/acetate gradient.

TABLE 19.1C. Gradient Program for Faster Anion-exchange Separation of Amino Acids without Determining Tryptophan.

Time (min)	%E1: water	%E2: 0.25M NaOH	%E3: 1M sodium acetate	%E4: 0.1M acetic acid	Gradient Curve
0.0	76	24			
2.0	76	24			
4.0	64	36			8
5.0	64	36			
9.0	40	20	40		8
12.0	44	16	40		5
13.0	14	16	70		8
17.0	14	16	70		
17.1				100	5
18.1				100	
18.2	20	80			8
19.2	20	80			
19.3	76	24			5
25.0	76	24			

The instrument configuration for the two-dimensional chromatography is illustrated in Figure 19.3(b). The first step of passing the carbohydrates through the CRC and retaining amino acids on it is identical with the carbohydrate removal depicted in Figure 19.3(a). However, instead of being discarded into the waste, carbohydrates are transferred to Valve 3 equipped with a sample loop. In the Position B of Valve 3 the carbohydrate fraction is entering the sample loop. At an appropriate time, Valve 3 rotates from Position B to Position A and the complete carbohydrate fraction is swept by the eluent from Pump 3 onto the carbohydrate column. All total chromatographic run times required for carbohydrate separations are either shorter or the same as those for amino acids separated by the shortened methods of Tables 19.1B or 19.1C. The simultaneous analysis of amino acid and carbohydrate content can thus be accomplished within the 25–40 min required for 1D separation of amino acids alone.

19.5 OPTIMIZATION OF A TWO-DIMENSIONAL SEPARATION SYSTEM FOR AMINO ACIDS AND CARBOHYDRATES

The key to optimizing the performance of the apparatus in Figure 19.3(b) is to find a proper timing for rotation of Valves 2 and 3 relative to the movement of Valve 1 marking the start of a chromatographic run. Note: the length of the 0.125 mm I.D tubing between port 3 of Valve 2 and port 3 of Valve 3 was 80 cm. Under the given conditions (transfer fluid flow rate of 80 μL/min, length of 0.125 mm I.D. tubing between port 6 of Valve 1 and port 2 of Valve 2: 50 cm, volume of 0.5 mm I.D. sample loop: 25 μL) and as determined in a previous report [8], an appropriate timing for the rotation of Valve 2 from A to B was 1.8 min after the start of the chromatographic run.

This delay time was found to fulfill both objectives of good recovery of amino acids retained on CRC and of a complete removal of carbohydrates prior to the injection of amino acids onto the amino acid column. The series of chromatograms in Figure 19.4 was obtained with 25 μL injections of 8 μM galactosamine and 8 μM glucose. Using the 1.8 min A to B switch of Valve 2, the rotation of Valve 3 was varied in 0.10 min increments between 0.7 and 1.8 min from the start of the chromatographic run. In doing so, a reproducible (peak area RSD = 0.7%) series of peaks of galactosamine was detected exiting the amino acid column (Figure 19.4a). This confirmed an acceptably reproducible transfer of amine compounds from CRC to the amino acid separation column. A good reproducibility of that transfer was also confirmed previously for 17 amino acids in the reference 8.

As illustrated in Figure 19.4(b), glucose peaks detected at the outlet of the carbohydrate column are reaching a maximum size when Valve 3 is rotated at 1.0 min after the start of the chromatographic run or 0.80 min before the rotation of Valve 2 from A to B. At times shorter and longer than 1.0 minute after the start, the peak sizes were decreasing gradually corresponding to the incomplete transfer of glucose. The rotation times of 1.8 and 1.0 min for Valves 2 and 3, respectively, were used in all subsequent experiments.

The optimization of the apparatus of Figure 19.3(b) was concluded by determination of optimum sample loop size for Valve 3. The loop sizes were varied between 25 and 40 μL and the loop sizes of 35 and 40 μL were found to produce the largest peaks of glucose (i.e. these volumes captured all the glucose passing through the CRC).

19.6 OPTIMIZED TWO-DIMENSIONAL SEPARATION OF A STANDARD MIXTURE CONTAINING AMINO ACIDS AND CARBOHYDRATES

The apparatus of Figure 19.3(b) can be used with a variety of columns available for carbohydrate separations. However, the amino acid separation is always performed with the AminoPac PA10 column. Two-dimensional separations of standard mixtures were successfully evaluated with four different carbohydrate column types (complete results are not shown). As an example, Figures 19.5 (a) and (b) show a two-dimensional separation of sugar alcohols, monosaccharides, and amino acids performed with the apparatus of Figure 19.3(b). The chromatogram of sugar alcohols and carbohydrates was generated with a CarboPac MA1 column connected to Valve 3. Figure 19.5(c) illustrates possible problems caused by excessive concentrations (50–100 μM) of sugar alcohols and monosaccharides in an one-dimensional amino acid separation. The peaks of arginine and lysine are obscured by unresolved peaks of sugar alcohols. The peaks of glucose and fructose are interfering with the peaks of alanine, threonine, and glycine, making their quantification either impossible or inaccurate. The comparison of chromatograms in Figures 19.5 (a) and (b) illustrate the completeness of removal of interfering peaks from the amino acid separation. Note: The amino acid retention times of Figure 19.5(a) are shifted by ca. 1.8 min in comparison with Figure 19.5(c) due to the delayed injection onto the amino acid column from Valve 2

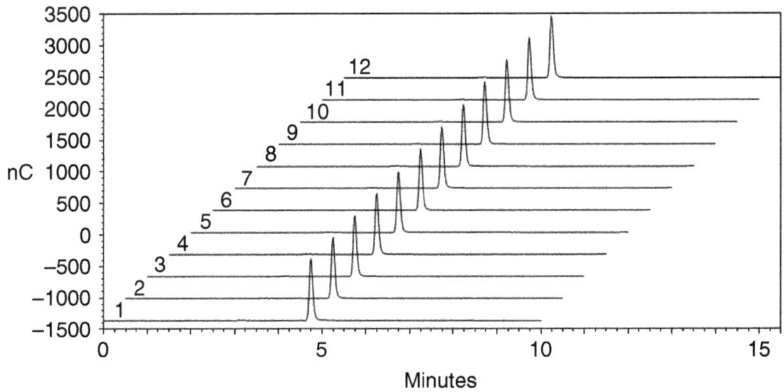

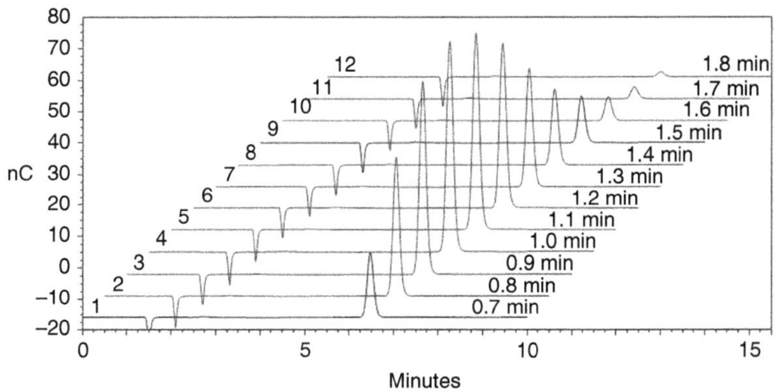

Figure 19.4. Last step in the preparation of apparatus of Figure 19.3(b) for two-dimensional separations, optimization of the timing of rotation of Valve 3 (see time intervals in (b)) by repeated injections of 25 μL of 8 μM solution of galactosamine and glucose. (a): Peaks of galactosamine on AminoPac PA10 column at 30 °C; 0.25 mL/min 50 mM NaOH from Pump 2; Pump 1: 0.02% formic acid at 0.080 mL/min; galactosamine peaks are detected by integrated amperometry [1]. (b): Peaks of glucose on CarboPac PA20 column at 30 °C; 0.5 mL/min 25 mM NaOH from Pump 3; Pump 1: 0.02% formic acid at 0.080 mL/min; glucose peaks are detected with quadruple pulse carbohydrate waveform [2]. See text and Figure 19.3(b) for additional details.

(at 1.8 min). The interfering compounds from the one–dimensional separation are well separated and can be reliably quantified using the chromatogram in Figure 19.5(b).

19.7 ANALYTICAL PERFORMANCE OF TWO-DIMENSIONAL SEPARATIONS

A good recovery of amino acids was previously reported for the apparatus of Figure 19.3(a) [8]. The recovery of amino acids was now re-evaluated for the

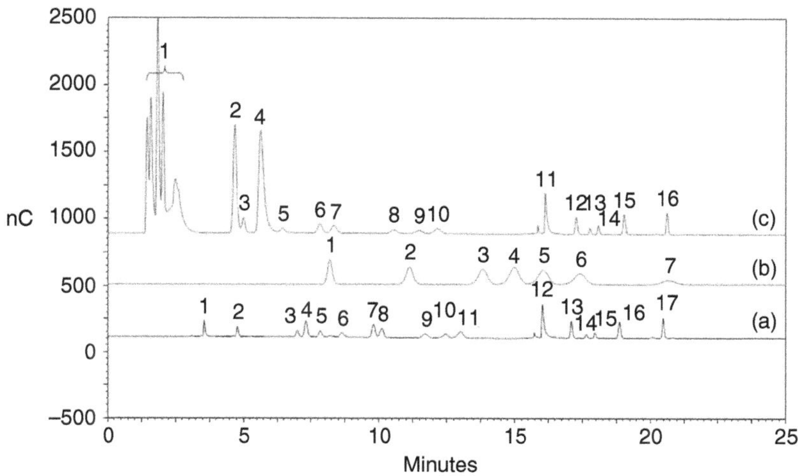

Figure 19.5. Comparing two-dimensional (a and b) and one-dimensional separations of a mixture containing amino acids, alditols, and monosaccharides. Injection: 25 μL of a standard mixture containing 8 μM concentrations of amino acids and 50—100 μM concentrations of carbohydrates. (a) Amino acids separated in the first dimension on an AminoPac PA10 column using eluent conditions of Table 19.1B. Peaks: (1) arginine, (2) lysine, (3) alanine, (4) threonine, (5) glycine, (6) valine, (7) serine, (8) proline, (9) isoleucine, (10) leucine, (11) methionine, (12) histidine, (13) phenylalanine, (14) glutamate, (15) aspartate, (16) cystine, (17) tyrosine; column temperature 30 °C; integrated amperometric detection [1]. (b) Carbohydrates separated in the second dimension on CarboPac MA1 column with 612 mM sodium hydroxide as an eluent at a flow rate of 0.4 mL/min. Peaks: (1) myo-inositol, (2) xylitol, (3) sorbitol, (4) dulcitol, (5) mannitol, (6) glucose, (7) fructose; column temperature 30° C; carbohydrate detection on a gold electrode with a four-potential waveform [2]. (c) Chromatogram of the same mixture of amino acids and carbohydrates on AminoPac PA10 without on-line removal of carbohydrates prior to the amino acid separation. Peaks: (1) alditols, (2) glucose, (3) threonine, (4) fructose, (5) valine, (6) serine, (7) proline, (8) isoleucine, (9) leucine, (10) methionine, (11) histidine, (12) phenylalanine, (13) glutamate, (14) aspartate, (15) cystine, (16) tyrosine; column temperature 30°C; integrated amperometric detection [1].

apparatus of Figure 19.3(b) where the average recovery of 17 proteinogenic amino acids was found to be 89%.

The following recoveries were determined for three of the carbohydrate columns:

CarboPac PA20: an average recovery of 92% for fucose, glucose, galactosamine, glucosamine, galactose, glucose, and mannose with 10 mM NaOH as an eluent pumped at 0.5 mL/min.

CarboPac PA1: an average recovery of 78% for glucose, fructose, lactose, and sucrose with 100 mM NaOH as an eluent pumped at 1 mL/min.

CarboPac MA1: an average recovery of 80% for myo-inositol, xylitol, sorbitol, dulcitol, mannitol, glucose, and fructose with 613 mM NaOH as an eluent pumped at 0.3 mL/min.

Good linearity of calibration in the first dimension was confirmed in one of the previous reports [8]. The linearity of calibration in the second dimension was confirmed by the authors in the range of 0.16 to 50 μM for glucose, fructose, lactose, and sucrose on the CarboPac PA20 column. The calibration in the second dimension was also tested using a CarboPac MA1 column with injections of seven standard solutions of glucose having concentrations from 0.5 to 50 μM. All calibration curves exhibited correlation coefficients greater than 0.999 in the evaluated concentration ranges.

19.8 EXAMPLES OF SAMPLE ANALYSIS

Combined chromatography after a sample pretreatment with an eluent containing sodium hydroxide and barium acetate for direct detection of proline and carbohydrates in wine samples was reported by Cataldi in 2003 [9]. The present technique allows generation of full amino acid profiles with simultaneous carbohydrate chromatography without a sample pretreatment. The wine samples can be diluted 100 to 500 times. The most prominent peak in the chromatogram of Figure 19.6(a) is the proline peak. The proline content in grapes increases with increased time on the vine. High levels of proline indicate high maturity of grapes and also a plant stress. Together with arginine, proline is one of the sources of nitrogen during fermentation. Unlike arginine, however, proline can not be used by the yeast without oxygenation of fermenting grapes. The determination of the arginine to proline ratio during the fermentation and in the final wine composition is thus of high interest to winemakers [10].

The CarboPac PA20 column selected as a carbohydrate column to generate the second dimension chromatogram presented in Figure 19.6(b) produced a peak of glycerin co-eluting with unknown weakly retained sample components. Four sugar alcohols are well separated from four monosaccharides in the chromatogram of Figure 19.6(b). An application of a two-dimensional separation with direct amperometric detection to cell culture media is shown in Figure 19.7. Most cell culture media contain relatively high concentrations of glucose which interfere with the alanine, threonine, and glycine peaks in one-dimensional amino acid separations (see also peaks 2 and 3 in Figure 19.5(c)). The only known method of avoiding such co-elution in one-dimensional chromatography consists of lowering the initial hydroxide concentration. As discussed above, lower initial concentrations of hydroxide make necessary longer re-equilibration times resulting in rather long total run times. The two-dimensional method eliminates glucose and other carbohydrates, if present, from the amino acid separation by transferring them to a carbohydrate column of choice.

The chromatogram of Figure 19.7(a) shows an amino acid composition of a cell culture media sample after the removal of glucose and other carbohydrates from the sample. Two of the frequently observed peaks in cell culture media, glutamine and asparagines, are not contained in this particular sample. If present, both compounds would elute well separated in the space between lysine and alanine. Another pair of frequently encountered cell culture media components is galactosamine and glucosamine. If present, both compounds would elute as a pair of well resolved peaks between lysine and glutamine. Hydroxyproline can be found in most cell culture media

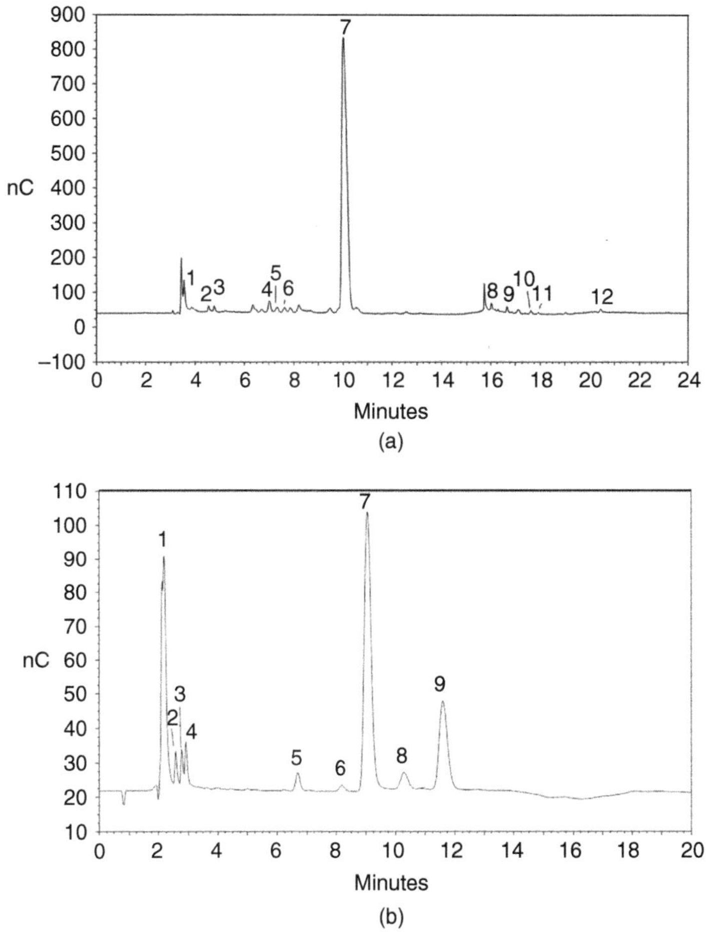

Figure 19.6. Amino acids and carbohydrates in red wine. Injection: 25 μL of a 500x diluted wine sample. (a) Amino acids separated in the first dimension on AminoPac PA10 column using eluent conditions of Table 19.1B. Peaks: (1) arginine, (2) unknown, (3) lysine, (4) alanine, (5) threonine, (6) glycine, (7) proline, (8) histidine, (9) phenylalanine, (10) glutamate, (11) aspartate, (12) tyrosine; column temperature 30°C; integrated amperometric detection [1]. (b) Carbohydrates separated in the second dimension on CarboPac PA20 column with 10 mM sodium hydroxide as an eluent at a flow rate of 0.5 mL/min. Peaks: (1) glycerol co-eluting with unknowns, (2) xylitol, (3) sorbitol, (4) mannitol, (5) unknown, (6) galactose, (7) glucose, (8) mannose; (9) fructose; column temperature 30 °C; carbohydrate detection on a gold electrode with a four-potential waveform [2].

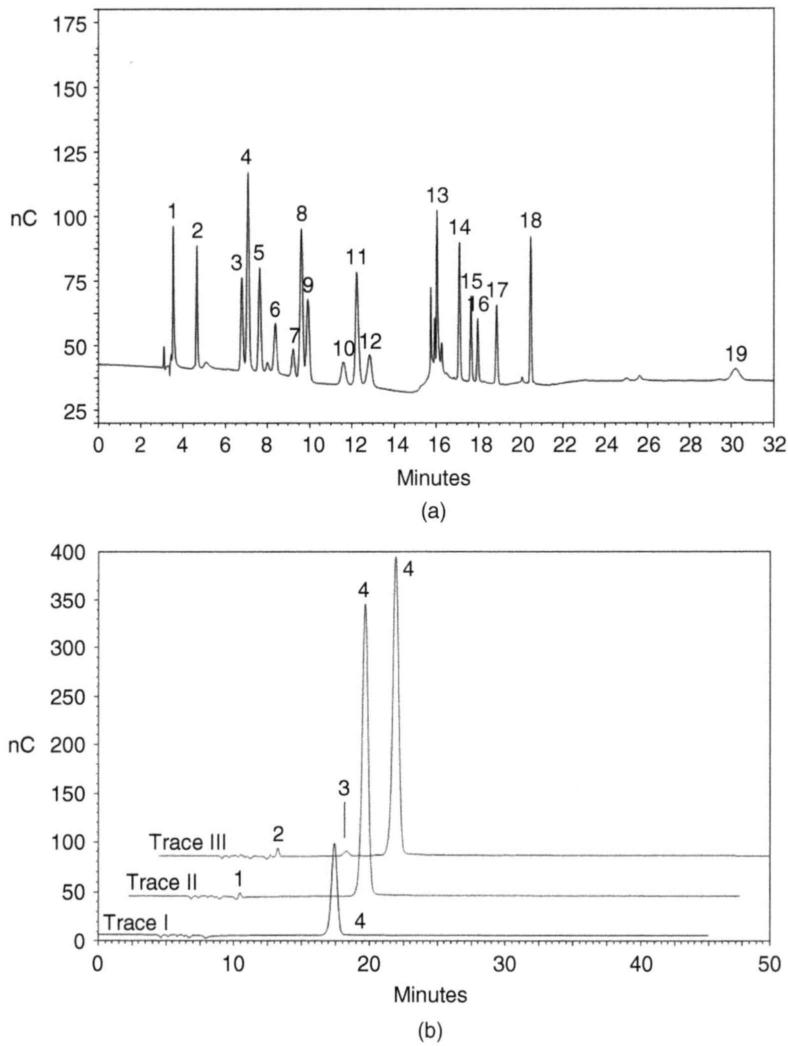

Figure 19.7. Two-dimensional separations of cell culture media samples. Injection: 25 μL of a 100x diluted cell culture media sample. (a) Amino acid profile obtained on AminoPac PA10 column using eluent conditions from Table 19.1B. Peaks: (1) arginine, (2) lysine, (3) alanine, (4) threonine, (5) glycine, (6) valine, (7) hydroxyproline, (8) serine, (9) proline, (10) isoleucine, (11) leucine, (12) methionine, (13) histidine, (14) phenylalanine, (15) glutamate, (16) aspartate, (17) cystine, (18) tyrosine; column temperature 30 °C; integrated amperometric detection [1]. (b) Three different cell culture media samples (traces I, II and III) separated in the second dimension on CarboPac MA1 column. Peaks: (1) glucose, (2) glycerol, (3) sorbitol, (4) glucose; carbohydrate detection on a gold electrode with a four-potential waveform [2].

and is shown in the chromatogram of Figure 19.7(a) as a well separated peak preceding serine. On the other hand, the presence of two unknown peaks eluting close to histidine in Figure 19.7(a) is somewhat atypical. Usually there are no unknown peaks in the vicinity of histidine in cell culture samples. The last peak of the separation eluting at ca. 30 min is that of tryptophan. With that peak the total run time of one analysis is 40 min (Table 19.1B). Without tryptophan, the total run time for obtaining the amino acid profile can be reduced to 25 min (Table 19.1C).

As shown in Figure 19.7(b), the CarboPac MA1 column, if selected as a carbohydrate column for the second dimension, provides interesting separations of not only glucose but also of analytes which would co elute with arginine in one-dimensional amino acid separations. While the sample separation in trace I of Figure 19.7(b) shows only glucose, the chromatograms of other two samples (traces II and III, Figure 19.7(b)) contain well-separated peaks of myo-inositol, glycerol, and sorbitol. Of the three, only myo-inositol was found in all cell culture samples we have analyzed so far.

19.9 CONCLUSION

A reduced total run time of 25–40 min that with the present method can include not only a full profile of amino acids, but also a simultaneous analysis of carbohydrates and related compounds, makes it easier to integrate the amino acid analysis into the monitoring of cell cultures [11,12]. An in-depth quantitative analysis of amino acids and carbohydrates, as it is offered by this two-dimensional technique, will provide more data which are required to gain an increased understanding of complex processes involved in the production of therapeutic proteins and of other important pharmaceutical compounds.

REFERENCES

1. Clarke AP, Jandik P, Rocklin RD, Liu Y, Avdalovic N. An integrated amperometry waveform for the direct, sensitive detection of amino acids and amino sugars following anion-exchange chromatography. Anal. Chem. 1999;71:2774–2781.
2. Rocklin RD, Clarke AP, Weitzhandler M. Long-term reproducibility for pulsed amperometric detection of carbohydrates via a new quadruple-potential waveform. Anal. Chem. 1998; 70:1496–1501.
3. Hanko VP, Rohrer JS. Determination of amino acids in cell culture and fermentation broth media using anion-exchange chromatography with integrated pulsed amperometric detection. Anal. Biochem. 2004; 324:29–38.
4. Yu H, Ding YS, Mou S, Jandik P, Cheng J. Simultaneous determination of amino acids and carbohydrates by anion-exchange chromatography with integrated pulsed amperometric detection J. Chromatogr. A 2002; 966:89–97.
5. Ding Y, Yu H, Mou S. Direct determination of free amino acids and sugars in green tea by anion-exchange chromatography with integrated pulsed amperometric detection. J. Chromatogr. A 2002; 982:237–244.

6. Jandik P, Clarke AP, Avdalovic N, Andersen DC, Cacia J. Analyzing mixtures of amino acids and carbohydrates using bi-modal integrated amperometric detection. J. Chromatogr. B 1999; 732:193–201.
7. Dionex Technical Note 63: Peak identification and estimation of peak purity using HPAE with 3D-amperometry (www.dionex.com).
8. Jandik P, Cheng J, Jensen D, Manz S, Avdalovic N. Simplified in-line sample preparation for amino acid analysis in carbohydrate containing samples. J. Chromatogr. B 2001; 758:189–186.
9. Cataldi TRI, Nardiello D. Determination of free proline and monosaccharides in wine samples by high-prefromance anion-exchange chromatography with pulsed amperometric detection (HPAEC-PAD). J. Agric. Food Chem. 2003; 51:3737–3742.
10. Enology Notes #104 at www.fst.vt.edu/extension/enology/EN/104.html
11. Larson T, Gawlitzek M, Evans H, Albers U, Cacia J. Chemometric evaluation of on-line high-pressure liquid chromatography in mammalian cell cultures: analysis of amino acids and glucose. Biotechnol. Bioeng. 2002; 77:553–563.
12. Genzel Y, Koenig S, Reichl U. Amino acid analysis in mammalian cell culture media containing serum and high glucose concentrations by anion exchange chromatography and integrated pulsed amperometric detection. Anal. Biochem. 2004; 335:119–125.

20

MONITORING PROTEIN *N*-GLYCOSYLATION CONSISTENCY USING HPAEC-PAD

Carl Grey

Department of Biotechnology, Lund University, Lund, Sweden

20.1 INTRODUCTION

Human proteins are processed after being expressed. Among these post translational modifications glycosylation is of particular interest, since it often affects both physicochemical properties such as solubility, folding and stability, as well as biological properties such as regulation of the activity, participation in cell-cell interactions, and determination of circulation half-life [1,2]. During the development and production of biopharmaceuticals, characterization and controlling of post translational modifications such as glycosylation are therefore of great importance. In recent years the amount of therapeutic proteins available on the market has grown steadily and there are many projects in development. The majority of these proteins are monoclonal antibodies (mAbs), which are glycoproteins. In general, the glycosylation should be similar to that of the corresponding human protein to ensure functionality, long circulation, and compatibility. Recombinant glycoproteins are therefore generally expressed in other mammalian cells, often Chinese hamster ovary (CHO) or mouse myeloma cell lines (NS0), whose glycosylation patterns resemble that of human proteins and where the differences have been shown not to have any

Applications of Ion Chromatography for Pharmaceutical and Biological Products, First Edition.
Edited by Lokesh Bhattacharyya and Jeffrey S. Rohrer.

effect on the activity in vivo [3]. However, an appropriate glycosylation is not the only consideration when producing biopharmaceuticals. The batch to batch consistency of the glycosylation is also important, since too large of a deviation may affect the protein function and other important properties. Even if the effect of a nonconsistent glycosylation is expected to be rather limited, regulatory authorities may still require that this is properly proven, which potentially could become quite expensive. Glycan heterogeneity is affected by several factors such as reactor and downstream processing conditions. For these reasons it might be advantageous to investigate the possibility of reducing potential sources of heterogeneity in the development of a new protein drug by e.g., removing glycosylation sites if the protein is heavily glycosylated and thereby making the production and purification less troublesome, i.e quality by design.

Clearly methods for characterizing and monitoring glycosylation are of great interest to the biopharmaceutical industry. Several analytical methods can be used for this purpose, in which mass spectrometry (MS) based methods are common for both quantitative analysis, using MALDI-TOF MS, and structural analysis using MS/MS based techniques. Other alternatives are the use of HPLC or capillary electrophoresis in combination with UV or fluorescence detection, in which most often chromophores are introduced to the sugars. Yet another attractive alternative is high performance anion exchange chromatography coupled with pulsed amperometric detection (HPAEC-PAD). Advantages with HPAEC-PAD are that it efficiently separates different oligosaccharides and the detection is possible without any derivatization. Structural isomers can be resolved and both acidic (containing sialic acid) and neutral oligosaccharides can be analyzed at the same time, which is not possible with MS. Further advantages with HPAEC-PAD are that it is better suited for quantitative analysis, has better precision, and is easier to automate. Although MALDI-TOF MS and HPAEC-PAD are based on completely different principles, the glycoprofile results have been shown to be similar [4]. A drawback with HPAEC-PAD is the need for analytical standards that are needed for oligosaccharide identification and which might be difficult to obtain. A combination of MS and HPAEC-PAD has been suggested, combining the strength of both approaches [5].

Protein glycans are divided into different classes, depending on the type of linkage between the protein and the sugar moiety. *N*- and *O*-glycosylation are the most common types, in which the carbohydrate is attached to an aspragine, or either a serine or a threonine amino acid residue respectively. C-glycosylation is yet another type, in which a single Man residue is linked by carbon-carbon bond to a tryptophan residue [6]. There are three types of *N*-linked (asparagine-linked) glycosylation; complex, high-mannose, and hybrid. They share a common pentasaccharide core consisting of two *N*-acetylglucosamine (GlcNAc) and three mannose (Man) residues, with one of the GlcNAc being linked to the asparagine (N) residue of the protein. The complex glycans contains more GlcNAc, galactose (Gal) and sialic acid residues in di- and sometimes higher antennary structures. The high-Man species contains additional mannose residues. Commonly occurring oligosaccharides of the complex and high-Man types are shown in Table 20.1. During the glycosylation process, which occurs in the endoplasmic reticulum and in the Golgi apparatus, a $Glc_3Man_9GlcNAc_2$ oligosaccharide (the core pentasaccharide + 3 Glc and 6 Man) is attached initially.

TABLE 20.1. Oligosaccharide Structures.

1. Man 3	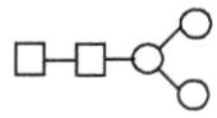
2. Man-5	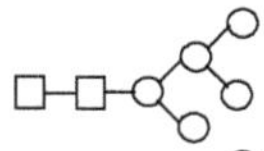
3. Man-6	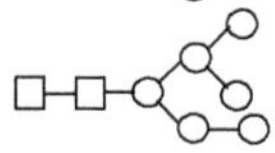
4. Man-9	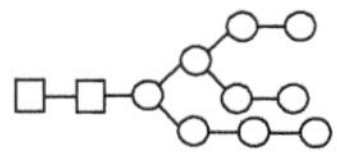
5. G0 (NGA2)	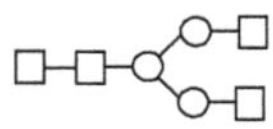
6. G2(NA2)	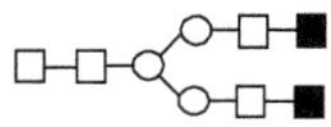
7. G0F (NGA2F)	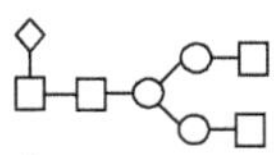
8. G1F (NA2G1F)	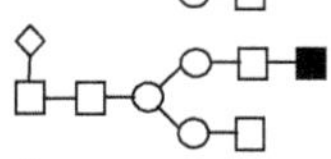
9. G2F (NA2F)	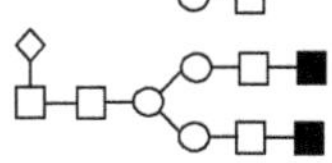
10. G2A1 (A1)	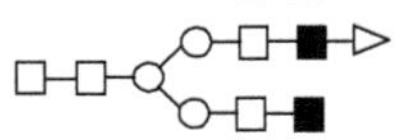
11. G2FA1 (A1F)	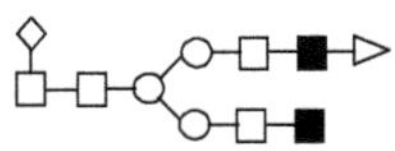
12. G2A2 (A2)	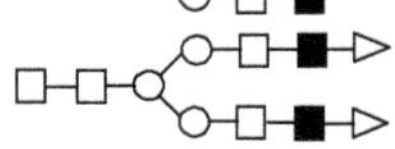
13. G2FA2 (A2F)	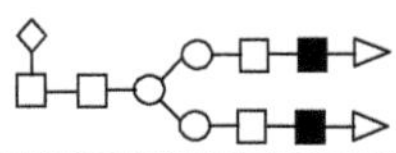

□ N-Acetylglucosamine (GlcNAc)
◇ Fucose (Fuc)
○ Mannose (Man)
■ Galactose (Gal)
◁ Sialic acid (Neu5Ac)

The different high-Man species are formed first by the removal of the three Glc, followed by trimming with α-mannosidases. If the glycan processing is stopped at this point high-Man type of glycosylation is observed which could affect both the biological properties and circulation half life of the protein in vivo [1]. The complex type is formed by sequential addition of GlcNAc, Gal, and sialic acid to the trimmed down core structure. The hybrid type of *N*-glycosylation is a mix of complex and high-Man type of glycosylation. Potential sites for *N*-glycosylation can be identified within the amino acid sequence of the protein, since a NXS/T motif is required for the initial glycosylation step, where X could be any amino acid except proline, and S and T are serine and threonine, respectively. However, glycosylation must be verified experimentally in each case since some of the potential sites are not glycosylated due to steric hindrances [7]. *O*-linked glycosylation (unlike *N*-glycosylation) lacks both a common core structure and a defined amino acid motif for the attachment. The methodology described in this chapter is suitable for analyzing *N*-glycosylation. If other glycosylations are to be analyzed other approaches must be used [8,9].

When beginning to work with a new protein it is important to perform an initial characterization using several methods. To reveal what kind of glycosylation is present in the protein and to confirm the number of glycosylation sites, protease cleavage followed by peptide identification with MS is often used. The HPAEC-PAD protocol described in this chapter (and MALDI-TOF MS) measures oligosaccharides released from all the glycosylation sites and will therefore only gives the average oligosaccharide composition of the protein. If there is heterogeneity in the glycosylation that is dependent on the glycosylation site, this will not be revealed. Characterization using several orthogonal methods, such as MS-peptide mapping, MS analysis of whole protein, HPAEC-PAD, and IEF (isoelectric focusing), is advantageous in the early stage of the characterization since each methods has different advantages as well as disadvantages and will provide overlapping but slightly different information. When the initial, and rather time consuming, characterization is completed, more simple analysis can follow using fewer methods, suitable for studying batch to batch variety or for evaluating changes in the production or in the downstream process.

20.2 DEGLYCOSYLATION

Removal of *N*-glycosylation can in most cases be achieved enzymatically using peptide-*N*-glycosidase F (PNGase F). The enzyme cleaves the linkage between the core GlcNAc and asparagine, and releases the intact oligosaccharide. PNGase F is quite nonspecific regarding the oligosaccharide composition and structure, and can therefore be used to release almost all N-linked glycans. It is important that the deglycosylation is complete to make sure that the released glycans are representative of the protein glycosylation and that no discrimination regarding site specific glycosylation occurs. The deglycosylation procedure should be evaluated individually for each new protein. Some glycoproteins are easily deglycosylated, for example, most mAbs are readily deglycosylated in their native form by PNGase F, whereas others are considerably tougher and require the use of denaturating conditions, reduction and alkylation of disulfide bonds, or the use of detergents that are compatible with PNGase F, e.g., RapiGest SF, an acid-cleavable anionic detergent [10,11]. Other factors that can be optimized are the length of the reaction time and amount of PNGase F used. A protocol suitable for mAb deglycosylation is shown in Figure 20.1 [4]. If a complete deglycosylation is not achieved using this protocol, it could be worthwhile to modify it by adding 100 μl of 1% (w/v) RapiGest SF solution and incubate for 30 min at

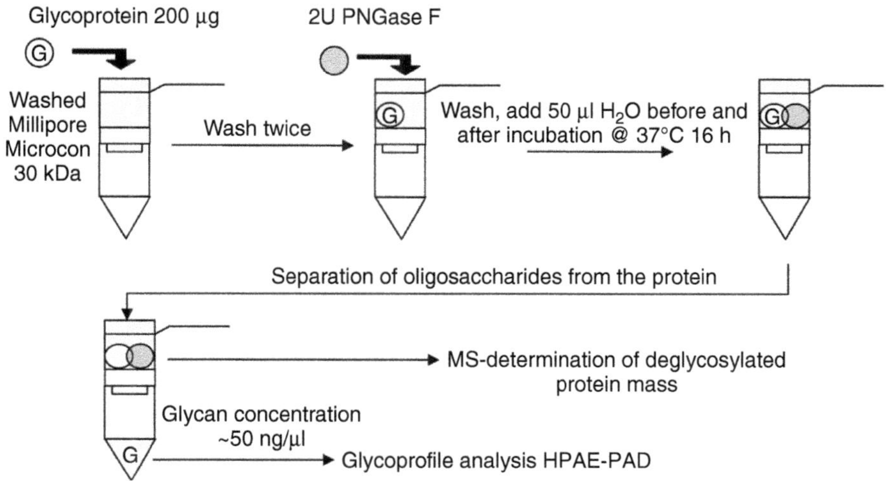

Figure 20.1. Deglycosylation protocol for mAbs, each wash is performed with 200 μl water followed by centrifugation 14,000 rpm for 15 min (or until the the filter is almost dry).

60°C and remove the solution by centrifugation, before the addition of the PNGase F. In Figure 20.2, the deglycosylation of a recombinant protein (no mAb), containing three glycosylation sites, using the protocol shown in Figure 20.1 with, and without the addition of RapiGest SF, is shown. Two out of the three sites were easily deglycosylated, but the third site clearly required the presence of the surfactant to be efficiently removed.

20.3 HPAEC-PAD

All results were obtained using a Dionex ICS3000 HPLC system, including SP pump, ED electrochemical detector and AS autosampler. The detector settings are shown in Table 20.2.

Columns: CarboPac PA200 column (3 × 250 mm).
Guard column (3 × 50 mm).
Reference electrode: Ag/AgCl
Detection mode: Integrated Amperometry
Data collection rate: 2 Hz
Column temperature: 30°C
Flow Rate: 0.5 mL/min.
Injection volume: 5–9 μl
Eluent A: 250 mM Sodium Acetate (NaAc)
Eluent B: 191 mM Sodium Hydroxide (NaOH) and 10 mM NaAc
Eluent C: MilliQ water, 18.2 MΩcm

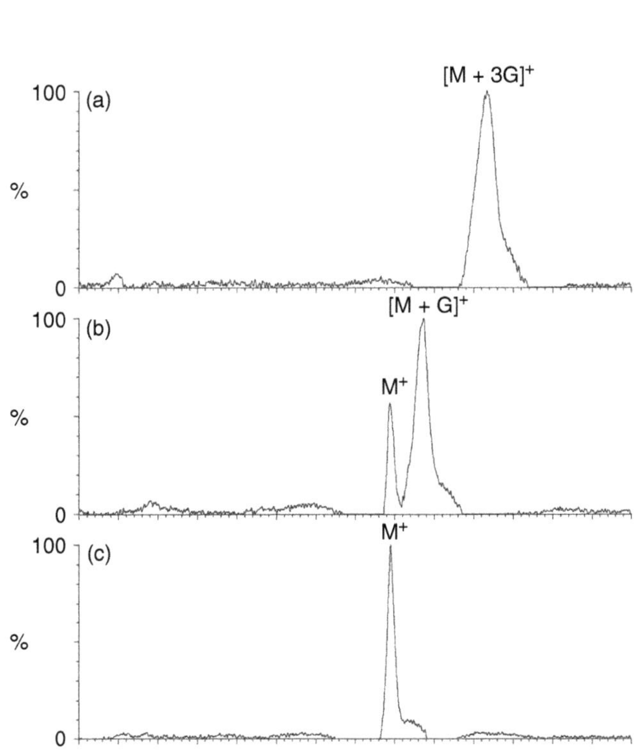

Figure 20.2. Whole protein analysis using MALDI-TOF MS to investigate the extent of deglycosylation. M is the molecular mass of the amino acid backbone and G is the average glycan mass. (a) Before deglycosylation treatment, (b) deglycosylation according to protocol in Figure 20.1, (c) deglycosylation as in (b) with the addition of RapiGest SF.

TABLE 20.2. Pulse Setting for ED-detector (Carbohydrates, Standard Quad).

Time/s	Potential/V	Integration
0.00	0.1	
0.20	0.1	Begin
0.40	0.1	End
0.41	−2.0	
0.42	−2.0	
0.43	0.6	
0.44	−0.1	
0.50	−0.1	

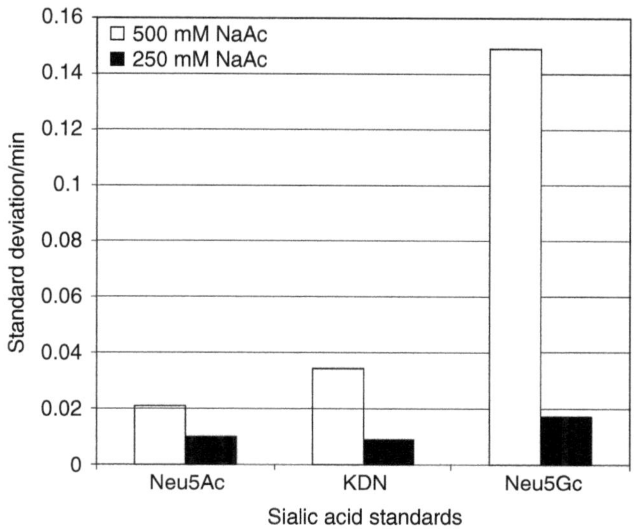

Figure 20.3. Effect of the stock concentration of the NaAc eluent on the standard deviation (17 replicates at 500 mM and 28 replicates at 250 mM) of the retention time obtained for three sialic acid standards (see Figure 20.6b).

The HPAEC-PAD analysis described was developed for the separation of oligosaccharides present in mAbs and other recombinantly expressed proteins in CHO and NS0 cells [4]. When preparing eluents it is important to use NaOH, NaAc, and water of high purity and to degas all eluents properly to avoid clogging of the column, fouling of the electrode, and to obtain a good signal to noise ratio. The eluents should be used within a week to ensure sufficient reproducibility and quality of the analysis. The eluent concentrations were chosen to give a good flexibility during method development. Other combinations giving the same gradient are of course possible, e.g., using only two eluent bottles. However, it could be worth noting that the reproducibility may be affected by such changes. It was found during the development of the method that the standard deviation for the retention times were improved quite drastically by lowering the concentration of the NaAc stock solution from the originally used 500 mM to 250 mM (Figure 20.3).

Three elution programs were created as shown in Table 20.3. Early on it was found that by lowering the concentration of NaOH to 55 mM, from the rather standardly used 100 mM, it was possible to better separate complex and high-Man oligosaccharides [4]. To compensate for the long retention times at this NaOH concentration, NaAc was added. The neutral oligosaccharides were eluted using isocratic conditions, whereas the larger neutral and the acidic glycans required considerably more NaAc and were therefore eluted using a NaAc gradient. Program 1 is the most extensive of the three and can handle both neutral and acidic oligosaccharides, as shown in Figure 20.4a. Program 2 also separates both types of glycans but the gradient is more aggressive, resulting in a faster elution of the oligosaccharides. But compared to Program 1 the

TABLE 20.3. HPAEC Programs.

Program	Time (min)	Eluent A (%)	Eluent B (%)	Eluent C (%)
1-Neutral and acidic	0.0	0	29	71
	12.0	0	29	71
	24.0	10	29	61
	42.0	48	29	23
	42.1	0	29	71
	50.0	0	29	71
2-Fast	0.0	2	29	69
	2.0	2	29	69
	12.0	48	29	23
	12.1	69	29	0
	15.0	69	29	0
	15.1	2	29	69
	23	2	29	69
3-Neutral	0.0	0	29	71
	12.0	0	29	71
	24.0	10	29	61
	24.1	48	29	23
	28	48	29	23
	28.1	0	29	71
	32.0	0	29	71

oligosaccharides are not as well resolved (especially the neutral glycans). Program 2 can be used when the identity of the individual glycans are either not a priority or difficult to confirm, which is often a problem for complex samples. Instead the glycans can be quantified in groups depending on the number of sialic acid residues present, as shown in Figure 20.4b. Program 3, shown in Figure 20.4c, is identical to the initial part of Program 1 but the NaAc gradient is stopped much earlier, making it shorter. This program can be used if only the neutral oligosaccharides are of interest or present in the sample. Many mAbs expressed in CHO often contain a very limited amount of acidic glycans and the time saved using this program could be considerable if the number of samples is large.

20.4 GLYCAN COMPOSITION AND METHOD VALIDATION

To obtain glycoprofiles with good precision the chromatograms have to be integrated consistently. Depending on the sample complexity, and if all major glycans can be properly identified, the composition could be either calculated for individual glycans or in groups depending on the number of sialic acid residues present. In the former case

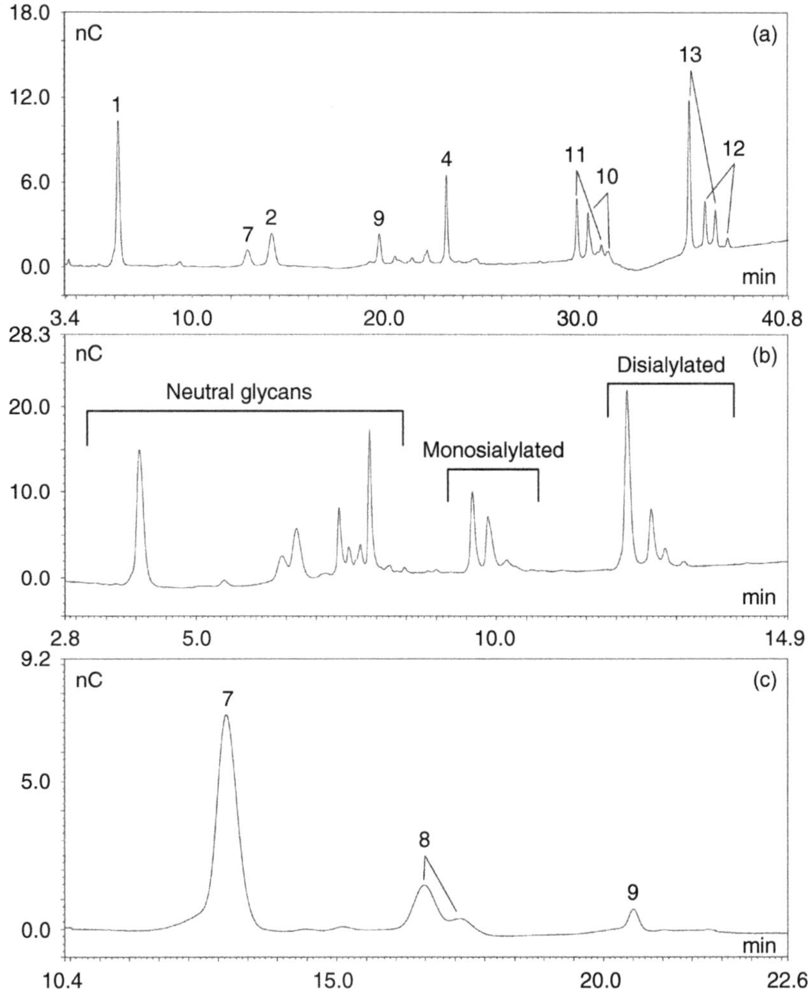

Figure 20.4. Separation of oligosaccharides standards (see Table 1) with HPAEC-PAD using (a) Program 1, (b) Program 2, and (c) and Program 3.

it would be possible to quantify the glycans in absolute terms, if quantitative standards are used. However, in most cases a relative quantification assuming an equal response factor for all the glycans will provide sufficient information. An example of monitoring batch variation of the glycosylation of protein recombinantly expressed in CHO cells using Program 1 is shown in Figure 20.5. The glycosylation was in this case rather complicated, including many different glycoforms with 0–3 sialic acid residues and a grouped type of integration was therefore chosen. In this example there are quite large differences in the glycosylation pattern between the batches, which would not be ideal if a constant glycosylation is required.

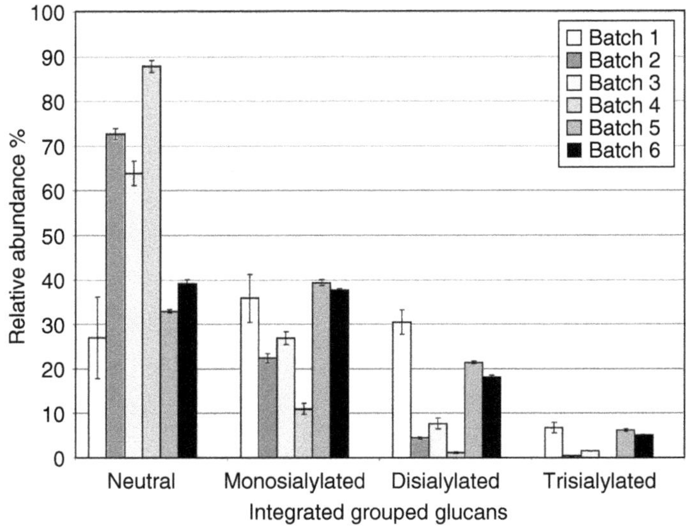

Figure 20.5. Batch to batch monitoring of glycosylation consistency.

A method validation of Program 1 showed that the analysis is highly reproducible and quite robust [4]. The response was found to be linear (0–120 pmol) and the limit of quantification (nine times the noise) was about 0.9 pmol for the glycan with the lowest response. Small variation in mobile phase composition affected the observed retention times but not the relative abundance of the glycans. The same results were obtained when new and one week old mobile phase were compared. Thus, if the identity of the individual glycans is important, oligosaccharide standards must be included in the analysis on a daily basis.

20.5 ANALYSIS OF SIALIC ACID

Sialic acid determination is a complement to the glycosylation profile analysis. The result does not give as detailed information as the glycosylation profile but on the other hand it is quantitative in absolute terms, giving the average amount of sialic acid per protein molecule. In addition, the analysis is quite fast and easy to perform. Sialic acid consists of a large family of *N*- and *O*-subsituted neuraminic acid, in which *N*-acetylneuraminic acid (Neu5Ac) and *N*-glycolylneuraminic acid (Neu5Gc) are among the most commonly occurring species [12]. A third type of sialic acid 3-Deoxy-D-glycero-D-galacto-2-nonulosonic acid (KDN) was used as an internal standard to eliminate variation in injection volumes and detector response. The release of sialic acid can be achieved by either acid hydrolysis or using enzymes. Both methods give approximately the same results, but the acid method is more robust, while the enzymatic method is done under milder conditions and therefore potentially more accurate. The separation of the three sialic acids was achieved in about 15 min using isocratic conditions. It was developed using the same eluents as in the glycoprofile analysis, in order to allow both assays to be run without changing eluents. Also, the method was developed using both the CarboPac PA200 column, used in the glycoprofile analysis, and the CarboPac PA20 column which is more optimal for monosaccharide analysis. The isocratic conditions were slightly modified for the two columns to achieve comparable retention times. The following conditions were used for the PA200 column: A: 20%, B: 29% and C: 51% (55 mM NaOH and 53 mM NaAc). For the PA 20 column: A: 30%, B: 29% and C: 41% (55 mM NaOH and 78 mM NaAc). Separation of sialic acid standards using the PA200 and the PA20 are shown in Figure 20.6. Both the resolution and the peak shape were somewhat better using the PA20 column, but the PA200 column can still be used successfully, allowing the sequential analysis of sialic acid and glycoprofile without the need to change the column.

The acid hydrolysis method can be performed in the following way: add about 50 μg of glycoprotein to 150 μl 0.1 M HCl. Add 50 μl 190 μM KDN (internal standard) and adjust the volume to 300 μl with water if necessary. Incubate the sample for 1 h at 80°C. Remove all solvent afterwards using a Speed Vac. Dissolve the released sialic acid in 200 μl of water and analyse with HPAEC-PAD. Prepare standard solutions, e.g., six levels, containing suitable concentrations, such as 15, 12, 9, 6, 3 and 0 μM of Neu5Ac (and/or Neu5Gc) instead of glycoprotein and incubate and treat the standards in the same manner as described above.

The enzymatic release of sialic acid can be done as follows: add about 50 μg glycoprotein to 100 μl 100 mM acetate buffer pH 5.5. Add 50 μl 190 μM KDN,

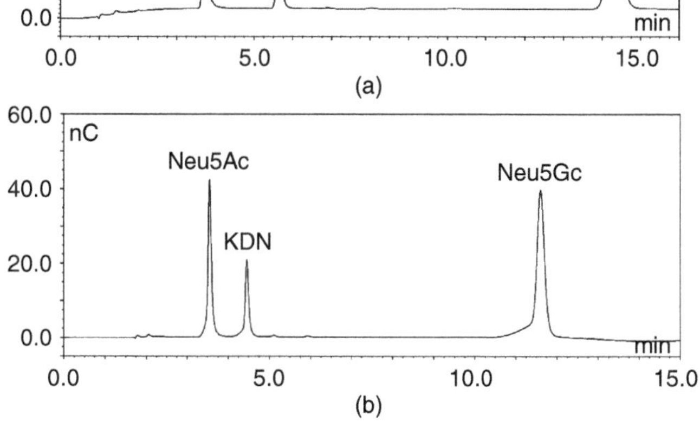

Figure 20.6. Separation of sialic acid standards using (a) the PA20 column and (b) the PA200 column.

0.015 U sialidase and adjust the volume to 300 μl with water if necessary. Incubate the sample 18 h at 37°C and analyze with HPAEC-PAD. Prepare the standard solutions, by treating them in the same manner as the unknown samples.

20.6 CONCLUDING REMARKS

The methods described above have been found to be well suited for both initial characterization and monitoring batch to batch variation of recombinantely expressed glycoproteins. Depending on the type of glycoprotein, one of the three elution programs can be chosen to optimize for either data quality or short analysis time. If the programs suggested here are not able to separate the glycans sufficiently, it is quite easy to modify the programs since three separate eluent bottles are used. For instance, it is possible to increase the gradient to elute more acidic species, such as tetrasialylated glycans, or to change the NaOH concentration to improve the selectivity of co-eluting glycans. In addition, the flexible set up allows different glycoproteins, using different elution programs, to be analyzed within the same sequence as well as performing additional analysis, e.g., the described quantitative sialic acid method, without interruption.

REFERENCES

1. Wright A, Morrison SL. Effect of glycosylation on antibody function: Implications for genetic engineering. Trends in Biotechnology 1997;15(1):26–32.
2. Lis H, Sharon N. Protein Glycosylation - Structural and Functional Aspects. European Journal of Biochemistry 1993;218(1):1–27.
3. Chadd HE, Chamow SM. Therapeutic antibody expression technology. Current Opinion in Biotechnology 2001;12(2):188–194.
4. Grey C, Edebrink P, Krook M, Jacobsson SP. Development of a high performance anion exchange chromatography analysis for mapping of oligosaccharides. Journal of Chromatography B 2009;877(20–21):1827–1832.
5. Bruggink C, Maurer R, Herrmann H, Cavalli S, Hoefler F. Analysis of carbohydrates by anion exchange chromatography and mass spectrometry. Journal of Chromatography A 2005;1085(1):104–109.
6. Hofsteenge J, Muller DR, Debeer T, Loffler A, Richter WJ, Vliegenthart JFG. New-Type of Linkage between a Carbohydrate and a Protein - C-Glycosylation of a Specific Tryptophan Residue in Human Rnase U-S. Biochemistry 1994;33(46):13524–13530.
7. Charlwood J, Bryant D, Skehel JM, Camilleri P. Analysis of N-linked oligosaccharides: progress towards the characterisation of glycoprotein-linked carbohydrates. Biomolecular Engineering 2001;18(5):229–240.
8. Rohrer JS, Miller HI. Detecting O-linked oligosaccharides on glycoproteins. Analytical Biochemistry 2003;316(1):131–134.
9. Medzihradszky KF, GilleceCastro BL, Townsend RR, Burlingame AL, Hardy MR. Structural elucidation of O-linked glycopeptides by high energy collision-induced dissociation. Journal of the American Society for Mass Spectrometry 1996;7(4):319–328.

10. Yu YQ, Gilar M, Kaska J, Gebler JC. Deglycosylation and sample clean-up method for mass spectrometry analysis of *N*-linked glycans. Lc Gc Europe 2005; 34–36.
11. Papac DI, Briggs JB, Chin ET, Jones AJS. A high-throughput microscale method to release, N-linked oligosaccharides from glycoproteins for matrix-assisted laser desorption/ionization time-of-flight mass spectrometric analysis. Glycobiology 1998;8(5):445–454.
12. Schauer R. Sialic acids: fascinating sugars in higher animals and man. Zoology 2004;107(1):49–64.

21

OLIGOSACCHARIDE ANALYSIS BY HIGH-PERFORMANCE ANION-EXCHANGE CHROMATOGRAPHY HYPHENATED TO INTEGRATED PULSED AMPEROMETRIC DETECTION AND ON-LINE ION-TRAP MASS SPECTROMETRY

Cees Bruggink

Thermo Fisher Scientific, Breda and Biomolecular Mass Spectrometry Unit, Department of Parasitology, Leiden University, Medical Center, RC Leiden, The Netherlands

21.1 INTRODUCTION

High performance anion-exchange chromatography (HPAEC) with integrated pulsed amperometric detection (IPAD) is a well accepted liquid chromatographic technique for characterization of oligosaccharides free in solution or after being released from glycoproteins or glycolipids [1,2]. The selectivity of anion-exchange chromatography for separating carbohydrates at a high pH is unique and its ability to separate isomeric carbohydrates is unequaled, and IPAD provides sensitive detection of underivitized glycans with a gold working electrode [3–5]. However, the characteristics, such as size and composition, of an unknown eluting carbohydrate are not easily derived from its retention time. This lack of information can be rectified by hyphenation with mass spectrometry (MS). Moreover, with tandem MS structural information can be obtained by fragmentation. Cleavage of glycosidic linkages provides information on sugar composition and sequence, while cross-ring fragments yield data on linkage

Applications of Ion Chromatography for Pharmaceutical and Biological Products, First Edition.
Edited by Lokesh Bhattacharyya and Jeffrey S. Rohrer.

positions between different sugar monomers [6–8]. In this chapter on-line hyphenation of a mass spectrometer (MS) to HPAEC-IPAD is described. Van der Hoeven et al. reported a detection limit of 17 pmol for maltoheptaose obtained with a 2 mm I.D. column and HPAEC-IPAD-MS [9]. While this is a sensitive technique, gaining additional sensitivity is important so that (i) only small amounts of precious samples need to be analyzed and (ii) low abundant carbohydrates can be detected for impurity analysis or for the study of new anabolic or catabolic pathways of glycans [10]. To obtain progress in biotechnical and pharmaceutical research better sensitivity, which can be obtained by further miniaturization of columns, is a requisite. This chapter reports results with a prototype capillary HPAEC-IPAD-MS system that delivers the desired higher sensitivity.

21.2 MECHANISM OF SEPARATION

Anion-exchange chromatography can separate anionic carbohydrates that contain a sialic acid, a carboxylate, a sulfate, or a phosphate group [1,2]. Neutral carbohydrates are very weak acids and they will dissociate in an eluent at a sufficiently elevated pH and interact with anion-exchange resins (Table 21.1). From selectivity research with small anions it is known that analyte charge is the dominant retention factor, followed by ionic size [11]. The property of analyte charge is nicely illustrated in Figure 21.1 that shows the separation of phosphorylated sugars. A monophosphorylated glucose (peak 4) elutes faster from the anion-exchanger than a diphosphorylated glucose (peak 13). The property of ionic size is illustrated in Figure 21.2 that shows the separation of fructo-oligosaccharides. The higher the degree of polymerization (DP), the longer the retention time. The separation of dihexose structural isomers in Figure 21.3 is a result of something more than charge and size. This chromatogram shows separations of different compositional isomers (i.e. peak numbers 2, 3, and 9) and linkage type and position type within the same compositional isomers (i.e. peak numbers 1, 4, 6, 7, and 9) of several dihexoses. An example of an isomeric *N*-linked oligosaccharide separation is shown in Figure 21.4, and an example of a baseline separation of two asialo-triantennaryglycopeptides only differing in the linkage of a galactose (either β 1—3 or β 1—4) was published by Hardy et al. [12]. For a more in depth study of HPAEC oligosaccharide separations the following review articles are recommended [1,12,13].

21.3 DETECTION

21.3.1 Integrated Pulsed Amperometric Detection (IPAD)

Underivatized carbohydrates can be sensitively detected with a gold working electrode, because aldehyde, ketone, and alcohol groups are oxidizable at 0.10 V when combined with a Ag/AgCl reference electrode [5,14]. Oxidation at this low potential is possible because gold serves as a catalyst in the oxidation reaction of the aforementioned

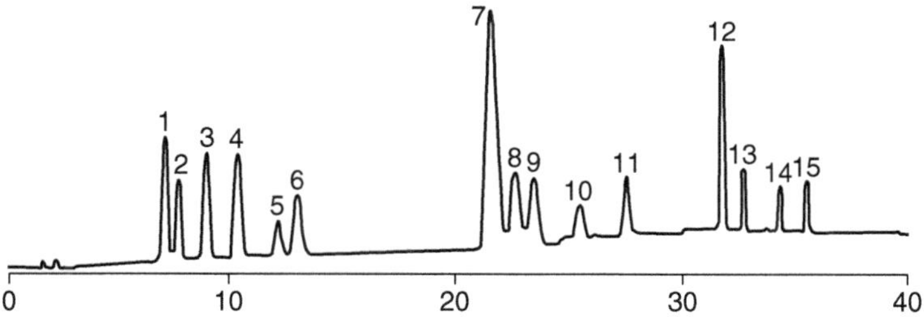

Figure 21.1. Separation of mono- and diphosphorylated monosaccharides with a CarboPac PA1 column using a sodium acetate gradient in 100 mM sodium hydroxide and IPAD detection at a gold electrode. Peaks and amount (μg) (1) α-D-Galactosamine-1-phosphate, 1.13; (2) α-D-Glucosamine-1-phosphate, 0.45; (3) α-D-Galactose-1-phosphate, 1.75; (4) α-D-Glucose-1-phosphate, 1.75; (5) α-D-Ribose-1-phosphate, 1.46; (6) β-D-Glucose-1-phosphate, 1.75; (7) D-Glucosamine-6-phosphate, 3.75; (8) D-Galactose-6-phosphate, 2.04; (9) D-Glucose-6-phosphate, 1.25; (10) D-Fructose-1-phosphate, 0.96; (11) D-Fructose-6-phosphate, 0.42; (12) α-D-Glucuronic acid-1-phosphate, 3.08; (13) α-D-Glucose-1,6-diphosphate, 1.06; (14) β-D-Fructose-2,6-diphosphate, 0.92; (15) D-Fructose-1,6-diphosphate, 0.92.

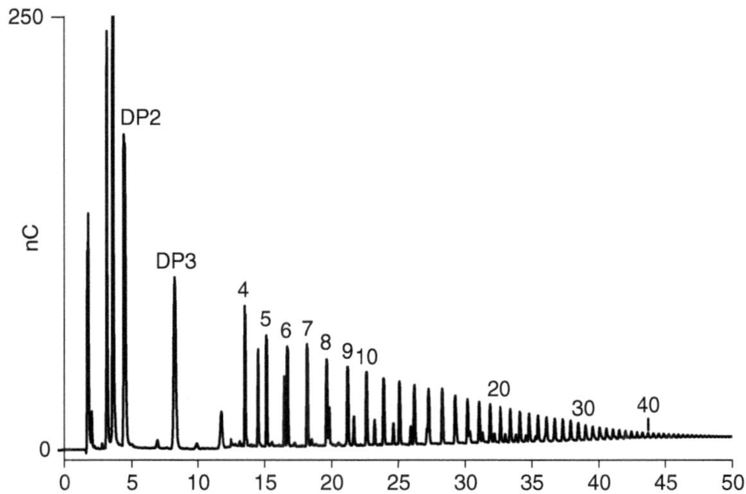

Figure 21.2. Separation of inulin with a CarboPac PA200 column using a sodium acetate gradient in 60 mM NaOH and IPAD detection at a gold electrode.

groups [3]. To keep the gold electrode active as a catalyst, cleaning and activation potentials are applied after the detection potential [5]. All of these potentials are only applied for less than a second with the sequence of potentials and their time periods referred to as a waveform. A modern waveform provides two data points per second and is stable over a long period of time [5]. Moreover, HPAEC-IPAD allows

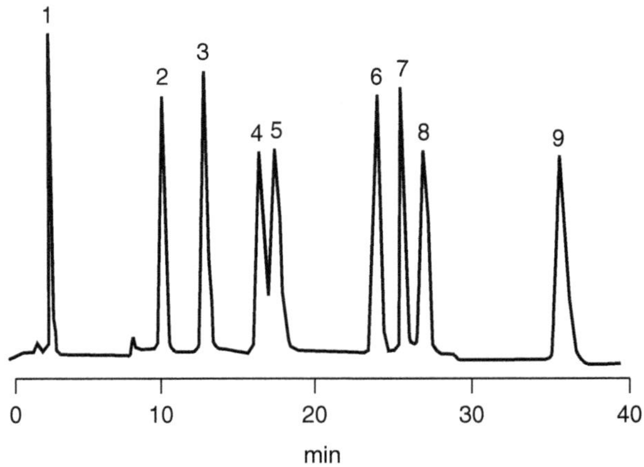

Figure 21.3. Separation of disaccharides with HPAEC-IPAD. Peaks (1) Glc(α1, α1)Glc (trehalose); (2) Fru(β2, α1)Glc (sucrose); (3) Gal(β1, 4)Glc (lactose); (4) Glc(α1, 6)Glc (isomaltose); (5) Glc(α1, 6)Gal (melibiose); (6) Glc(β1, 6)Glc (gentiobiose); (7) Glc(β1, 4)Glc (cellobiose); (8) Glc(α1, α3)Fru (turanose); (9) Glc(α1, 4)Glc (maltose).

TABLE 21.1. Dissociation Constants of Common Carbohydrates in Water at 25°C[1].

Carbohydrate	pK_a
Fructose	12.03
Mannose	12.08
Xylose	12.15
Glucose	12.28
Galactose	12.39
Dulcitol	13.43
Sorbitol	13.60
α-Methyl glucoside	13.71

[1] Reference: Lange's handbook of Chemistry 13th ed.

the analysis of glycans with reducing end, reduced terminus, and C_1-oxidation [10], which makes it more broadly applicable than methods that depend on reducing ends for a reductive amination reaction [15].

21.3.2 Mass Spectrometry

To obtain more information from complex samples, especially where standards are not available, mass spectrometric detection (MS) is indispensable. The ionization efficiency of carbohydrates with electrospray ionization is low resulting in relatively low signal intensity compared with peptides and proteins. Anionic carbohydrates can

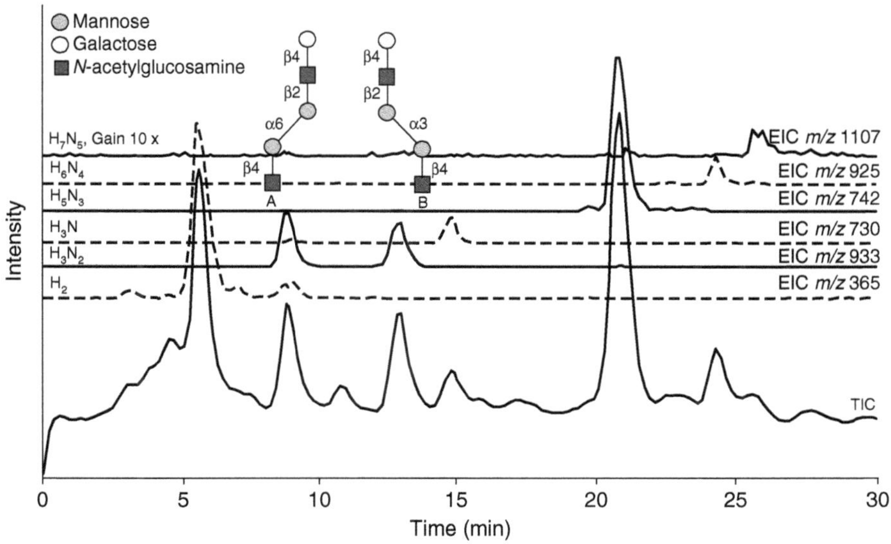

Figure 21.4. Separation of oligosaccharides with on-line HPAEC-MS in a urine sample of a G_{M1}-gangliosidosis patient [27] with a capillary CarboPac PA200, showing a separation of linkage isomers of asialo-monoantennary glycans (A and B). H = hexose, N = *N*-acetylhexosamine.

selectively be detected with electrospray ionization MS in the negative mode by loss of a proton. Neutral carbohydrates easily form adducts with a proton or a metal ion so that they can be detected in the positive mode [16].

21.4 SETUP OF HPAEC-IPAD-MS

Separation of carbohydrates with HPAEC is usually performed with sodium hydroxide in the eluent. To elute oligosaccharides from the anion-exchange column a sodium acetate gradient is executed, while typically keeping the hydroxide concentration constant. When combining IPAD and MS detections, a flow splitter after the analytical column is recommended (Figure 21.5). The detection volume in the amperometric cell is low enough but the volume of the cavity containing the reference electrode is high and causes extensive peak broadening. A nice feature of this setup is that the split ratio of the flow splitter can be chosen to compensate for the sensitivity difference between the IPAD and MS detections.

A mobile phase containing sodium hydroxide and sodium acetate is incompatible with the electrospray ionization (ESI) interface of the mass spectrometer. To convert the eluent into a compatible fluid for on-line MS, a membrane based desalter is used. There are several types mentioned in the literature such as the carbohydrate membrane desalter (CMD) [17] and the anion self-regenerating suppressor (ASRS) [18]. The latter is available in two formats to be used with columns having a 2 or 4 mm I.D. Such a membrane desalter can be considered as a cation-exchanger in the

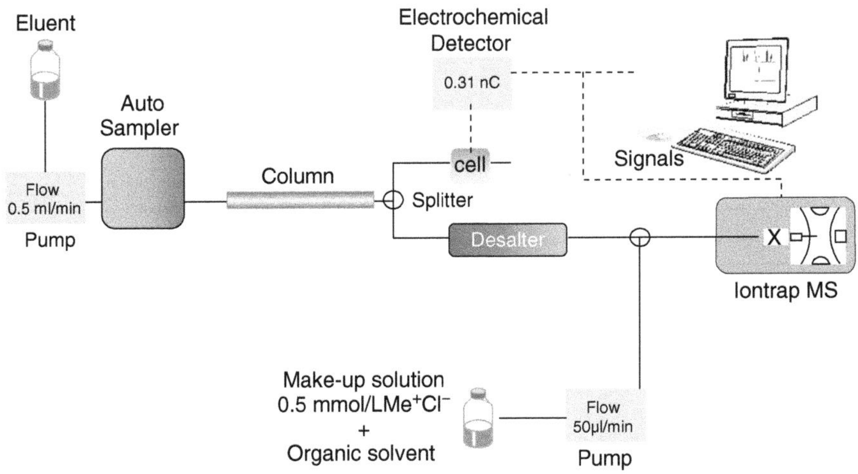

Figure 21.5. A schematic of instrumental setup for on-line HPAEC-IPAD-MS.

acidic form and exchanges sodium ions for hydronium ions. The sodium hydroxide is converted into water while sodium acetate is converted into volatile acetic acid. The CMD and ASRS are continuously regenerated by electrolysis of water, which is sometimes assisted with trifluoroacetic acid [17] to obtain more desalting capacity. After desalting the eluent, a make-up solution is pumped in via a tee to facilitate carbohydrate adduct and spray formation.

21.5 ADDUCT FORMATION

For the positive ion mode in MS detection adduct formation is accomplished by adding an acid or a metal salt to the make-up solution, see Table 21.2 for commonly used adducts. Mohr et al. [16] have studied the relative affinities of different alkali metal ions for carbohydrates, and have found that the affinity order is $Cs^+ > K^+ > Na^+ > Li^+ > H^+$. Although most efficient at producing ions, cesium is unable to ionize small carbohydrates [19]. Di- and trivalent metals are also capable of ionizing carbohydrates, but only singly charged quasimolecular ions are formed [20]. If the glycan contains fucose as a constituent and proton adducts are formed, rearrangement can occur and its position in the glycan changes [21,22], leading to structural misinterpretation. Lithium adducts are the most sensitive for relatively small sugars. Carbohydrate lithium adducts exhibit easy in-source fragmentation; though this is often viewed as a disadvantage because of signal intensity loss. It can be turned into an advantage, when using a single quadrupole MS detector, as glycosidic cleavages can be induced, leading to partial elucidation of a carbohydrate structure [18]. Sodium adducts are more stable to in-source fragmentation and are commonly used in MS detection. The stability of sodium adducts is such that with mild energy levels fragmentation can be induced with tandem MS, leading to structure elucidation.

TABLE 21.2. Adduct ions and their use in the ESI (+) mode.

Adduct Ion	Monoisotopic Mass	Remarks
Hydrogen	1.01	Fucose rearrangement possible
Lithium	7.02	In source fragmentation easy to induce
Sodium	22.99	Commonly used
Potassium	38.96	Stable adduct
Cesium	132.91	Only for larger oligosaccharides

21.6 MS DATA INTERPRETATION

Tandem MS can partly elucidate carbohydrate structures. The first piece of information is the mass of the carbohydrate. From this mass and knowledge of the masses of contributing monosaccharides (e.g., hexoses have a mass increment of 162, *N*-acetylhexosamines of 203 and others are listed in Table 21.3) possible structures can be proposed, though the anomeric configuration (i.e. whether a linkage is α or β) cannot be revealed from MS spectra.

Broadly accepted nomenclature for fragmentation of carbohydrates was proposed by Domon and Costello [23] and is depicted in Figure 21.6. At the left side of the figure is the non-reducing end of the oligosaccharide, while at the right side is the reducing end. Glycosidic cleavages containing the non-reducing end are called B and C fragments and Y and Z fragments contain the reducing end. A-ions are cross-ring fragments containing the non reducing part of the glycan, while X-ions are cross-ring fragments including the reducing part. These cross-ring fragment ions are accompanied by a subscript indicating the position of cleavage relative to the termini and a superscript in front indicating the cleavage position in the ring, starting with bond zero between the ring oxygen and the anomeric carbon with further counting in the

TABLE 21.3. Monoisotopic Mass Addition of Some Carbohydrates and water.

Residue	Mass Contribution
Water	18.01
Deoxyhexose	146.06
Hexose	162.05
Hexosamine	161.07
Uronic acid	176.03
N-acetylhexosamine	203.08
N-acetylneuraminic acid	291.10
N-glycolylneuraminic acid	307.09

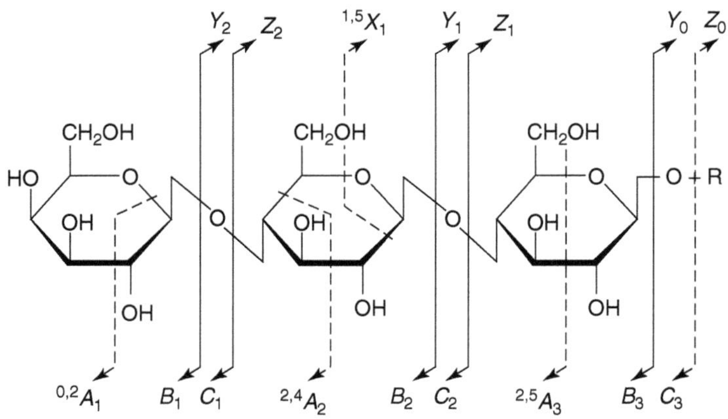

Figure 21.6. Glycan fragmentation types and their nomenclature as proposed by Domon and Costello [23].

TABLE 21.4. Mass Loss of Hexose (Hex) and *N*-acetylhexosamine (HexNAc) Caused by Cross-ring Fragmentation and Different Linkage Positions Between Carbohydrate Units in the Positive Ion Mode [6–8].

		Linkage			
Hex	HexNAc	1–6	1–4	1–3	1–2
−60	−101	^{0,2}A	^{0,2}A		
−90	−131	^{0,3}A			
−120	−161	^{0,4}A	^{2,4}A		^{1,3}A
−106	−147	^{3,5}A	^{3,5}A		
−78		^{0,4}A-H_2O			^{0,4}A-H_2O

clockwise direction (Figure 21.6). Linkage position information between the different sugar constituents is derived from cross-ring fragments. Table 21.4 lists mass losses observed depending on cross-ring fragmentation when tandem MS is used in the positive mode.

21.7 APPLICATION EXAMPLES

21.7.1 Released *N*-glycans With An Analytical Bore Anion-Exchange Column

N-glycans were released from a therapeutic glycoprotein and characterized by HPAEC-IPAD-MS. The system was a Dionex ICS-3000 ion chromatograph consisting of a low pressure gradient pump, an isocratic pump, a chromatography detector module equipped with an electrochemical detector and cell outfitted with a gold working electrode and Ag/AgCl as reference electrode, an injection valve, a dual zone temperature

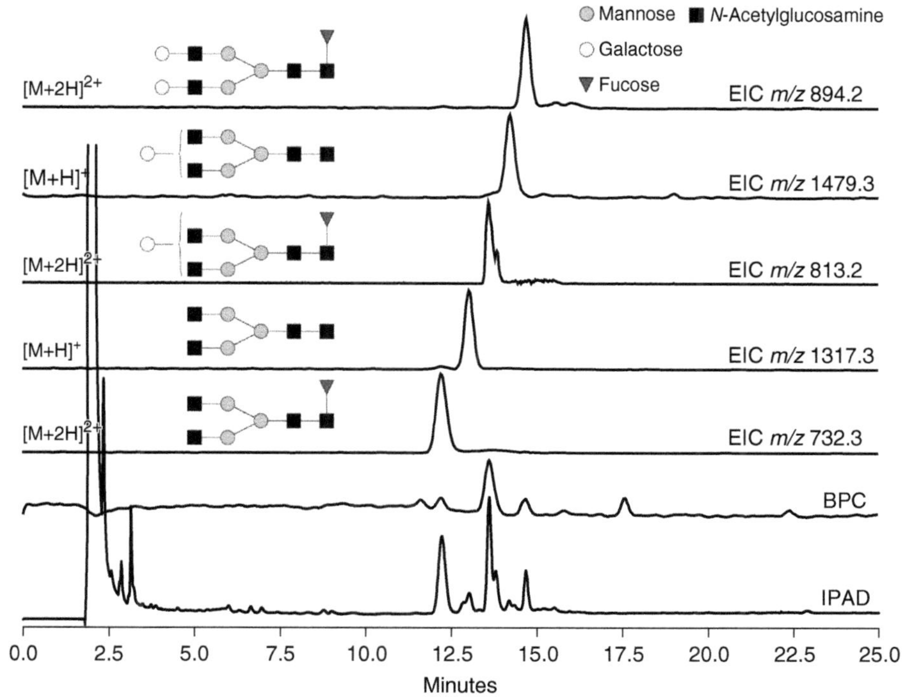

Figure 21.7. Separation of PNGase F-released *N*-linked glycans from a therapeutic glycoprotein with on-line HPAEC-IPAD-MS using a CarboPac PA200 column.

control compartment, and a cooled autosampler. The MS was a Bruker HCT Ultra ion-trap and an ESI interface. The separation was carried out on a CarboPac PA200 (3 × 250 mm) column from Dionex Corp. at 30°C and a flow rate of 360 μL/min. After the column a homemade PEEK splitter followed, which split 90 μL/min to the electrochemical cell and 270 μL/min to the desalter (ASRS-300 2 mm Dionex in the external water mode). The eluent leaving the desalter was combined with 90 μL/min 50% acetonitrile for MS detection (Figure 21.5). Eluent A was ultra pure water, B 200 mM NaOH, and C 400 mM NaOAc. The gradient was 0–5 min 50% A, 50% B, 0% C and from 5–65 min 0% A, 50% B, and 50% C. The IPAD waveform applied to the gold working electrode was 0.00–0.20 s 0.10 V, 0.20–0.40 s 0.10 V (data collection), 0.41–0.42 s–2.00 V, 0.43 s 0.60 V, 0.44–0.50 s–0.10 V versus an Ag/AgCl reference electrode [5]. The resulting separation is shown in Figure 21.7. The identification of different peaks are obtained from the MS and tandem MS. Where MS can only partly elucidate structure, structures were confirmed by running standard solutions of the different glycans. The chromatogram shows that the fucosylated glycans are eluting earlier from the column than their respective afucosylated glycan, which is consistent with the literature [24–26].

21.7.2 Glycans in a Urine Sample of a G_{M1}-gangliosidosis Patient With Capillary HPAEC-IPAD-MS

In clinical, pharmaceutical, and bio research, sensitivity is of paramount importance. Therefore use of an analytical column of narrower diameter is necessary. Research with a prototype capillary HPAEC-IPAD-MS in a clinical setting has been reported [27]. The column was a prototype CarboPac PA200 column (0.4 × 250 mm) and a prototype desalter regenerated with dilute sulfuric acid was used prior to the MS. To ensure that the system was functioning as expected, the gradient performance of this system was tested with an inulin solution and the expected sensitivity gain was tested with an asialo biantennary *N*-glycan. A sub-picomolar detection limit was achieved for both IPAD, 50 fmol, and MS detection, 110 fmol. With this capillary HPAEC-PAD-MS system, a 400 nL urine sample of a G_{M1}-gangliosidosis patient was characterized after first desalting with graphitized carbon SPE. The two isomeric glycans in the extracted ion chromatogram EIC m/z 933 are sodium adducts of monoantennary endo-β-*N*-acetylglucosamidase cleaved products of a complex type *N*-glycan structure (Figure 21.4). This separation is a representative example of the separation power of HPAEC chromatography for isomeric oligosaccharides. The carbohydrate constituents are the same and in the same order for both isomers. The difference lies in the linkage between the two mannoses being a 1–6 linkage for the first eluting peak and a 1–3 linkage for the second eluting peak. One may expect similar mass spectra and tandem mass spectra from both glycans. The expected exception is that the glycan with the 1–3 linkage between the two mannoses will not exhibit cross-ring fragment ions, while the one with the 1–6 link will show characteristic cross-ring fragment ions, as shown in Figure 21.8, $^{0,2}A_4$, $^{0,3}A_4$, and $^{0,4}A_4$ (see Table 21.4 for fragment designations). The detailed structures for both glycans are based on tandem MS linkage-specific fragmentation data in combination with the chromatographic data and what is known from the literature on *N*-glycan structures.

21.8 CONCLUSION

The two application examples in this chapter illustrate successful hyphenation of on-line MS with HPAEC-IPAD by adding a desalter in front of the ESI-MS. The unique separation power for isomeric carbohydrates is specifically shown by the detailed interpretation of the tandem MS data of the two baseline resolved isomeric glycan structures found in the urine sample of a G_{M1}-gangliosidosis patient. In the case of a chromatographic coelution of these glycans or direct infusion into the MS, the glycan having the 1–6 link would be reported, due to the detected cross-ring fragments, while the glycan with the 1–3 link will be missed, because cross-ring fragments will be absent (Table 21.4). This result shows the necessity of a high resolution separation prior to MS for characterization of a glycan sample. The same application example shows that using a prototype capillary HPAEC-PAD-MS system results in a sensitivity gain that is needed for bio research. Both application examples show that extending HPAEC-IPAD, a well accepted separation technique for profiling and characterizing oligosaccharides, with on-line tandem MS provides valuable structural information.

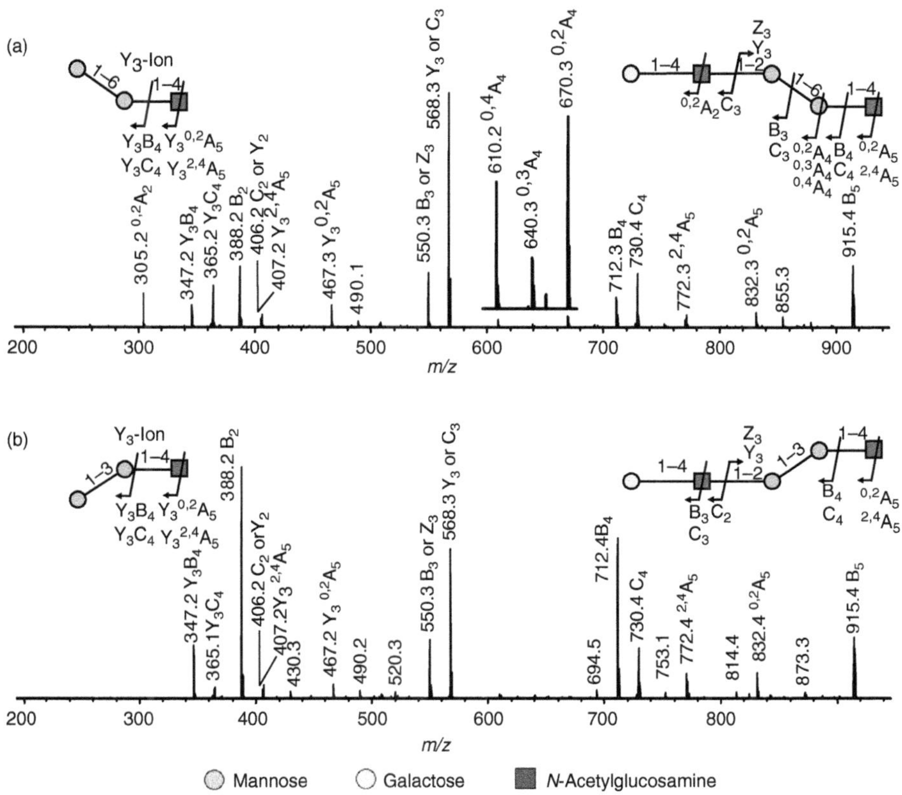

Figure 21.8. Fragment ion analysis of the H_3N_2 glycans as indicated in Figure 21.4 were subjected to MS^2 analysis in their sodiated form (m/z 933.5). The observed fragment ions are schematically interpreted according to the nomenclature of Domon and Costello [23].

REFERENCES

1. Lee, YC. High-performance anion-exchange chromatography for carbohydrate analysis. Anal. Biochem. 1990;189(2):151–162.
2. Townsend RR, Hardy MR, Hindsgaul O, Lee YC. High-performance anion-exchange chromatography of oligosaccharides using pellicular resins and pulsed amperometric detection. Anal. Biochem. 1988;174(2):459–470.
3. LaCourse WR, Johnson DC. Optimization of waveforms for pulsed amperometric detection of carbohydrates based on pulsed voltammetry. Anal.Chem. 1993;65(1):50–55.
4. Rocklin RD, Tullsen TR, Marucco MG. Maximizing signal-to-noise ratio in direct current and pulsed amperometric detection. J. Chromatogr. A. 1994;671(1-):109–114.
5. Rocklin RD, Clarke AP, Weitzhandler M. Improved long-term reproducibility for pulsed amperometric detection of carbohydrates via a new quadruple-potential waveform. Anal. Chem. 1998;70:1496–1501.

6. Harvey DJ. Electrospray mass spectrometry and fragmentation of N-linked carbohydrates derivatized at the reducing terminus. J. Am. Soc. Mass Spectrom. 2000;11(10):900–915.
7. Spina E, Sturiale L, Romeo D, Impallomeni G, Garozzo D, Waidelich D, Glueckmann M. New fragmentation mechanisms in matrix-assisted laser desorption/ionization time-of-flight/time-of-flight tandem mass spectrometry of carbohydrates. Mass Spectrom. 2004;18(4):392–398.
8. Weiskopf AS, Vouros P, Harvey DJ. Electrospray ionization-ion trap mass spectrometry for structural analysis of complex N-linked glycoprotein oligosaccharides. Anal. Chem. 1998;70(20):4441–4447.
9. van der Hoeven RAM, Tjaden UR, van der Greef J, van Casteren WHM, Schols HA, Voragen AGJ, Bruggink C. Recent progress in high-performance anion-exchange chromatography/ionspray mass spectrometry for molecular mass determination and characterization of carbohydrates using static and scanning array detection. J.Mass Spectrom. 1998;33(4):377–386.
10. Bruggink C, Poorthuis BJHM, Piraud M, Froissart R, Deelder AM, Wuhrer M. Glycan profiling of urine, amniotic fluid and ascitic fluid from galactosialidosis patients reveals novel oligosaccharides with reducing end hexose and aldohexonic acid residues. FEBS J. 2010;277:2970–2986.
11. Weiss J. Ion Chromatography, 3rd ed. Weinheim (Germany): VCH Verlag; 2004.
12. Hardy MR, Townsend RR. Separation of positional isomers of oligosaccharides and glycopeptides by high-performance anion-exchange chromatography with pulsed amperometric detection. Proc. Natl. Acad. Sci. U.S.A. 1988;85(10):3289–3293.
13. Cataldi TR, Campa C, De Benedetto GE. Carbohydrate analysis by high-performance anion-exchange chromatography with pulsed amperometric detection: The potential is still growing. Fresenius. J. Anal. Chem. 2000;368(8):739–758.
14. LaCourse WR. Pulsed electrochemical detection in high-performance liquid chromatography. New York: Wiley-Interscience; 1997.
15. Anumula KR. Advances in fluorescence derivatization methods for high-performance liquid chromatographic analysis of glycoprotein carbohydrates. Anal. Biochem. 2006;350(1):1–23.
16. Mohr MD, Bornsen KO, Widmer HM. Matrix-assisted laser desorption/ionization mass spectrometry: Improved matrix for oligosaccharides. Mass Spectrom. 1995;9(9):809–814.
17. Thayer JR, Rohrer JS, Avdalovic N, Gearing RP. Improvements to in-line desalting of oligosaccharides separated by high-pH anion exchange chromatography with pulsed amperometric detection. Anal. Biochem. 1998;256(2):207–216.
18. Bruggink C, Maurer R, Herrmann H, Cavalli S, Hoefler F. Analysis of carbohydrates by anion exchange chromatography and mass spectrometry. J. Chromatogr. A. 2005;1085(1):104–109.
19. Penn SG, Cancilla MT, Lebrilla CB. Collision-induced dissociation of branched oligosaccharide ions with analysis and calculation of relative dissociation thresholds. Anal. Chem. 1996;68(14):2331–2339.
20. Wong CKL, Chan T-W. Cationization processes in matrix-assisted laser desorption/ionization mass spectrometry: Attachment of divalent and trivalent metal ions. Mass Spectrom. 1997;11:513–519.
21. Franz AH, Lebrilla CB. Evidence for long-range glycosyl transfer reactions in the gas phase. J. Am. Soc. Mass Spectrom. 2002;13(4):325–337.

22. Harvey DJ, Mattu TS, Wormald MR, Royle L, Dwek RA, Rudd PM. "Internal residue loss": Rearrangements occurring during the fragmentation of carbohydrates derivatized at the reducing terminus. Anal. Chem. 2002;74(4):734–740.

23. Domon B, Costello CE. A systematic nomenclature for carbohydrate fragmentation in FAB-MS/MS spectra of glycoconjugates. J. Glycoconj. 1988;5:397–409.

24. Basa LJ, Spellman MW. Analysis of glycoprotein-derived oligosaccharides by high-pH anion-exchange chromatography. J. Chromatogr. 1990;499:205–220.

25. Wang WT, Zopf D. Liquid ion-exchange chromatography under pressure of milk oligosaccharides using a pulsed amperometric detector. Carbohydr. Res. 1989;189:1–11.

26. Hardy MR, Townsend RR. Separation of fucosylated oligosaccharides using high-pH anion-exchange chromatography with pulsed-amperometric detection. Carbohydr. Res. 1989;188:1–7.

27. Bruggink C, Wuhrer M, Koeleman CA, Barreto V, Liu Y, Pohl C, Ingendoh A, Hokke CH, Deelder AM. Oligosaccharide analysis by capillary-scale high-pH anion-exchange chromatography with on-line ion-trap mass spectrometry. J Chromatogr. B. 2005;829(1–2):136–143.

PART IV

VACCINE APPLICATIONS

22

APPLICATIONS OF ION CHROMATOGRAPHY IN BIOLOGICAL PRODUCT ANALYSES[1]

Alfred V. Del Grosso and Brandon Duong

Division of Biological Standards and Quality Control, Office of Compliance and Biologics Quality, Center for Biologics Evaluation and Research, Food and Drug Administration, Rockville, MD

22.1 INTRODUCTION

The term "Biological Product", as used in this chapter, refers specifically to products as defined in Section 351 (i) (1) of the Public Health Service (PHS) Act, ". . . a virus, therapeutic serum, toxin, antitoxin, vaccine, blood, blood component or derivative, allergenic product, protein (except any chemically synthesized polypeptide) or analogous product, or arsphenamine or derivative of arsphenamine (or any other trivalent organic arsenic compound), applicable to the prevention, treatment, or cure of a disease or condition of human beings". These products are regulated under the PHS Act under the regulatory jurisdiction of FDA's Center for Biologics Evaluation and Research (CBER) and include vaccines, allergenic extracts, antitoxins, human blood and blood derivatives, and tissue, cellular, and gene therapeutic products.

[1]The findings and conclusions in this chapter have not been formally disseminated by the Food and Drug Administration and should not be construed to represent any Agency determination or policy.

Applications of Ion Chromatography for Pharmaceutical and Biological Products, First Edition.
Edited by Lokesh Bhattacharyya and Jeffrey S. Rohrer.

For Biological Product analyses, important applications involving ion chromatography (IC) have been made in the areas of buffer ions, ionic impurities, monosaccharides, oligosaccharides, and small repeating saccharide units of polysaccharides. Products to which IC has been applied include vaccines, hematologic products, and allergenic extracts.

Topics in this chapter are limited to those products regulated by CBER. All applications described in this chapter are taken from published literature or are procedures developed by laboratories at the FDA's CBER. Inclusion of methods in this chapter does not imply endorsement by FDA-CBER.

22.2 VACCINES

High-performance anion-exchange chromatography with pulsed amperometric detection (HPAEC-PAD) has been widely adopted for the determination of saccharide content as a measure of product characterization and potency determination in bacterial capsular polysaccharide (PS) vaccines. Applications have been made to raw material polysaccharides, in-process materials, and bulk drug substances, as well as drug products in final formulations. Bacterial polysaccharide vaccines are routinely characterized and released largely on the basis of properties measured by physicochemical methods that include IC.

Vaccines based on bacterial capsular polysaccharides have been in use since the 1970s. Vaccines consisting of isolated and purified bacterial capsular polysaccharides from *S. Typhi, N. Meningitidis, S. Pneumoniae* and *Haemophilus influenzae* Type b were first developed and licensed. While these free PS vaccines effectively immunize adults and older children by means of a T-cell independent immune response, the response is short term. A particular shortcoming is that effective immunity is not developed in infants and children younger than two years of age as polysaccharides must be associated with a T-cell epitope for sufficient production of IgG antibodies. In order to produce safe and effective PS vaccines for infants and young children, polysaccharides derived from specific bacterial capsular serotypes have been synthetically conjugated with carrier proteins of diphtheria or tetanus toxoids, or other immunogenic carrier proteins. A PS conjugate vaccine was first licensed for *H. Influenzae* (Hib) in 1987, and then for Meningococcal serotypes in 1999, and for Pneumococcal serotypes in 2000 [1].

Steps involved in the manufacture of a PS conjugate, or glycoconjugate vaccine include activation of the purified PS serotype to create reactive groups followed by condensation with free amino groups of the protein, primarily epsilon amines of lysine, to produce the conjugate. PS activation is performed primarily using periodate oxidation to form reactive aldehydes or cyanylation using CDAP (1-cyano-4-dimethylaminopyridinium tetrafluoroborate) to form highly reactive cyanoesters [2].

Physicochemical characteristics that are required to be determined for manufacturing control and release of drug substance glycoconjugates or final formulated product include molecular weight dispersity, quantitation of unconjugated carrier protein, quantitation of individual serotype saccharides, determination of free saccharide, and determination of PS-protein ratio. Additionally, quantitation of free saccharide is an important determinant of product stability during the specified dating period for the product [2].

Because bacterial capsular polysaccharides are composed of polymers made of repeating units, the determination of specified sugars or other chemical moieties has been employed to determine concentration or quantity. Saccharide determinations of bulk PS serotypes and conjugates had been originally performed using colorimetric procedures for characteristic chemical functionalities; some manufacturers still use colorimetric methods for some or all steps of manufacturing and product release. In many cases, ion chromatographic procedures have been utilized due to required improvements regarding specificity or sensitivity, or for the purpose of improving analytical throughput.

22.2.1 Haemophilus Influenzae Type b (Hib)

The capsular PS of *Haemophilus influenzae* type b (Hib) is a polymer consisting of the repeating unit, → 5-D-ribitol-(1-1)-ß-D-ribose-3-phosphate. The PS is also termed polyribosylribitolphosphate or PRP. Determination of the PS content of Hib PS and the corresponding glycoconjugate vaccines was originally performed using the orcinol colorimetric assay to quantify the ribose moiety [3] or by the Chen colorimetric assay [4] to quantify on the basis of phosphorus content. Tsai et al. [5] first developed an HPAEC-PAD technique to quantitate PS in *Haemophilus influenzae* type b (Hib) PS and glycoconjugate vaccines using methodologies that had been previously reported for natural glycoconjugate proteins [6]. PS was depolymerized in 0.1M NaOH for 8 h at room temperature to yield ribitol-ribose-phosphate (RRP). Residual protein was removed from glycoconjugate vaccine preparations by passing the digest through a reversed-phase SPE cartridge.

Chromatography was performed using a Dionex CarboPac PA1 column with a mobile phase of 28 mM NaOH and 100 mM sodium acetate. Detection utilized a gold working electrode with a three-potential waveform:

$$E_1 = 0.05\text{V},\ t_1 = 360\text{ ms};\ E_2 = 0.8\text{ V},\ t_2 = 120\text{ ms};\ E_3 = -0.6\text{ V},\ t_3 = 420\text{ ms}.$$

Quantitation was against a purified Hib PS solution, standardized using the orcinol procedure [3], that was hydrolyzed in the same manner as the samples. Glucose-1-phosphate was used as an internal standard. The procedure was shown to measure PS content as low as 0.1 μg and was capable of quantifying Hib PS in the presence of high concentrations of lactose stabilizer in the formulated vaccine product (Figure 22.1).

Sturgess et al. [7] adapted Tsai's method to quantify total and unconjugated PRP in a vaccine product of Hib PS conjugated to an outer membrane protein complex (OMPC) from *N. meningitidis*. This procedure was applied to a final formulated product adsorbed onto aluminum hydroxide adjuvant. The aluminum hydroxide component was dissolved by incubating in 3% sodium citrate, 50 mM sodium phosphate, pH 7.2 for 3 min at 95°C. PRP hydrolysis was performed using 0.3 M NaOH at room temperature for 8 h; protein was subsequently removed using a 10 kDa MW polysulfone

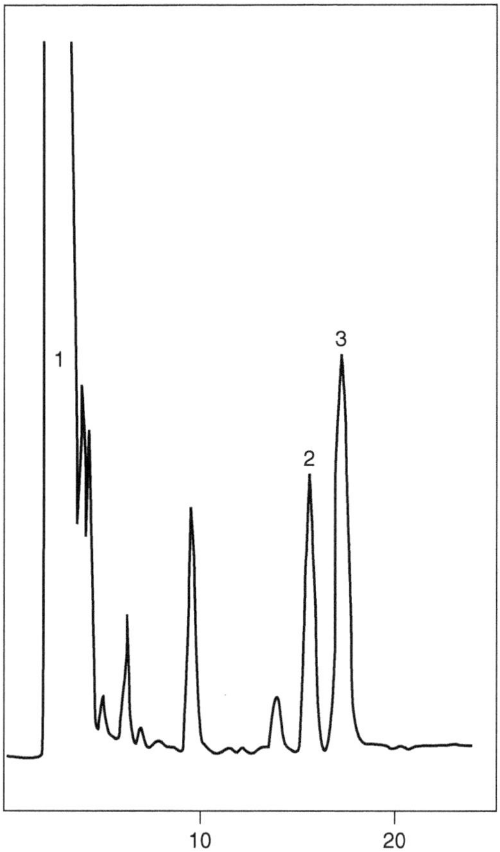

Figure 22.1. PS in Hib conjugate vaccine by HPAEC-PAD. Peak 1 is the breakthrough peak; includes lactose. Peaks 2 and 3 are glucose-1-phosphate (internal standard), and the PRP Hib PS subunit, respectively. Reprinted from Ref. 5 with permission from Elsevier.

cut-off membrane. Unconjugated PRP was determined following separation of the adjuvant by centrifugation at 625K × g and analysis of the supernatant for the free PS.

Sturgess's procedure used a Dionex CarboPac PA10 column with slightly modified mobile phase conditions of 25 mM NaOH and 100 mM sodium acetate with the addition of a "regeneration" step of 250 mM NaOH and 1.0 M sodium acetate followed by re-equilibration to the starting conditions. Glucosamine-1-phosphate was used as an internal standard. Otherwise, conditions were similar to those of Tsai et al. [5].

Stability studies of total and free PRP saccharide content in the formulated vaccine showed evidence of a catalytic reaction involving the phosphodiester bond of PRP and the aluminum hydroxide adjuvant that resulted in a modest but measurable decrease in conjugated PRP content from approximately 95 to 80%, which reached an asymptotic limit after about two years of normal storage. Clinical studies over this time period showed no significant impact on clinical effectiveness. The procedure was also applied to a combination vaccine containing a recombinant hepatitis B antigen along with Hib-conjugate.

Bardotti et al. applied HPAEC-PAD to the determination of PRP content in Hib glycoconjugate vaccine alone and also in a combination product that included vaccine components for Diphtheria, Pertussis, and Tetanus (DPT) [8]. Hydrolysis was carried out using trifluoroacetic acid (TFA) as hydrolysis of this combination product with NaOH resulted in interference from an unassigned peak that was unresolved from the RRP Hib-PS subunit. TFA hydrolysis resulted in complete depolymerization of PRP into ribitol and ribose. However, TFA hydrolysis of the DPT components resulted in a peak that interfered with ribose. Therefore ribitol was used for quantitation of PRP.

Sample preparation involved extraction of Hib conjugate from aluminum phosphate adjuvant with 10 mM phosphate followed by centrifugation, and then hydrolysis of the supernatant with 4 M TFA at 100°C for 2 h, followed by centrifugal evaporation and redissolution in water. Determination of unconjugated PRP PS was accomplished by ultrafiltration using a 30 kDa cut-off membrane; determination of free PS was quantifiable at levels as low as 2% of total saccharide.

Chromatography was on a Dionex CarboPac MA1 column (4 × 250 mm) [9] with MA1 guard column; mobile phase was isocratic 650 mM NaOH. PAD used a four-potential waveform [10]: $E_1 = 0.1$ V,$t_1 = 400$ ms;$E_2 = -2.0$ V,$t_2 = 20$ ms;$E_3 = -0.6$ V,$t_3 = 10$ ms,$E_4 = -0.1$ V,$t_4 = 60$ ms with integration from 200 to 400 ms at E_1. Figure 22.2 illustrates separations of total and free saccharide in the DPT-Hib combination vaccine along with acid hydrolysate of DPT alone showing the lack of interference.

Belfast described a multi-factor optimization in the development of a method for the determination of PRP from a Hib conjugate in a combination vaccine also containing Diphtheria, Tetanus, and Pertussis components [11]. Aluminum adjuvant was dissolved with 51 mM NaOH/63 mM Na citrate at room temperature for 11 min. Sample hydrolysis was with 0.3 M sodium hydroxide at room temperature followed by removal of protein with a 10 kMW cut-off filter. Unconjugated PRP was isolated by ultracentrifugation at 600K × g for 30 min.

Chromatography used Dionex CarboPac PA10 analytical and guard columns thermostated at 30°C and a mobile phase of 32 mM NaOH/120 mM Na acetate, with isocratic elution for 23 min at 1.2 mL/min followed by a regeneration step with 250 mM NaOH/1 M sodium acetate for 10 min. PAD conditions were not described in this publication. For the product under test, RRP was resolved from potential interferences. Limit of detection and quantitation was evaluated as 0.15 μgPRP/mL.

A 2007 publication [12] reported a collaborative study evaluating the saccharide content of a new WHO International Standard for *Haemophilus influenzae* b capsular PS. Results from colorimetric tests based on the determination of ribose or phosphorus were evaluated along with HPAEC-PAD assays that were performed by six different laboratories. Of the six laboratories using the HPAEC-PAD methods, five used alkaline hydrolysis for sample preparation with subsequent measurement of the ribitol-ribose-phosphate repeating unit, and one used the acid hydrolysis method as described by Bardotti and measurement of ribitol content for quantitation. HPAEC-PAD assays

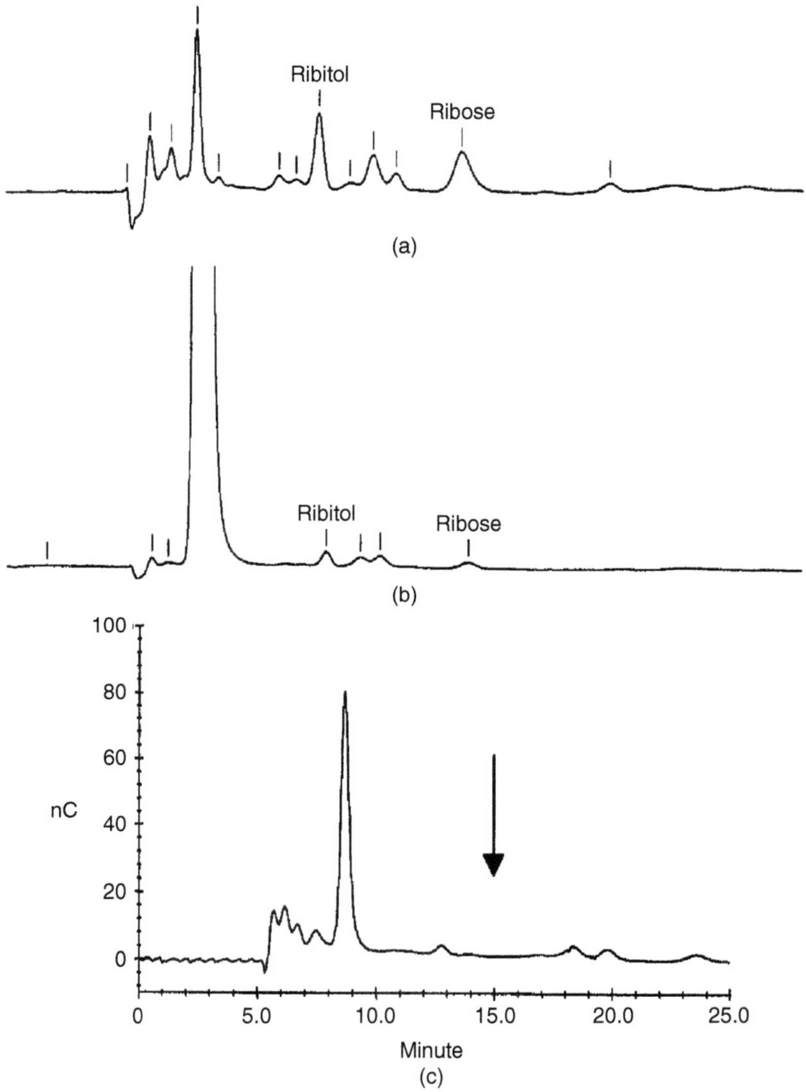

Figure 22.2. HPAEC-PAD of acid hydrolyzates of (a) total saccharide in combination vaccine DPT-Hib-CRM, (b) free saccharide from DPT-Hib-CRM combination vaccine, (c) DPT control showing absence of interfering peaks. Reprinted from Ref. 8 with permission from Elsevier.

used either PRP polysaccharide or PRP PS bulk conjugate as analytical standards. Participants assaying for phosphorus used sodium or potassium phosphate salts or ribose-5-phosphate as analytical standards while all participants that submitted results based on ribose assays used D-ribose as standard. Data analysis showed no significant difference in the values obtained by the different colorimetric and HPAEC-PAD assay

TABLE 22.1. PRP Saccharide Content Evaluation of WHO International Standard for *Haemophilus influenzae* Capsular Polysaccharide

PRP–Quantitation Procedure	Mean mg PRP/ampoule	95% Confidence Limits	No. of Labs.	% CV
Overall*	5.155	4.846 - 5.424	20	11.2
HPAEC-PAD	5.458	4.626–6.290	7	16.5
Phosphorus	5.077	4.847–5.307	5	3.6
Ribose	4.933	4.743–5.123	7	4.2

*Overall mean includes one result obtained by quantitative NMR–4.986 mg PRP/ampoule.

procedures, suggesting the suitability of the reference for use with the different procedures for quantitation of PRP. Mean results for each quantitation procedure along with limits at 95% confidence and precision as % CV are shown in Table 22.1.

Inter-laboratory variability was observed to be higher for the HPAEC-PAD assays (CV 16.5%) than the colorimetric ribose (CV 4.2%) and phosphorus assays (CV 3.6%). This was attributed to the higher variability inherent in the digestive procedures used with the PRP polysaccharides or conjugates in these procedures. Ultimately, PRP content of the standard was established using the result obtained by the ribose determinations; the ribose assay being chosen on the basis of traceability and specificity issues.

22.2.2 Meningococcal Polysaccharide Vaccines

Vaccines based on both free saccharides and glycoconjugates have been licensed for *N. Meningitidis* serotypes A, C, W135, and Y (MenA, MenC, MenW135 and MenY, respectively). Serotype A is a homopolymer of α-(1 $\rightarrow$ 6)-linked N–acetylmannoseamine phosphate in which 70% of the 3-hydroxyl groups are O-acetylated [13]. Quantitation of MenA PS has been performed using the Chen colorimetric method for phosphorus [4], but in vaccine samples phosphate from buffers and media may present a positive interference.

Ricci described an HPAEC-PAD method for MenA PS based on quantitation of the mannoseamine-6-phosphate subunit [14]. Samples were prepared by the hydrolysis of the polysaccharide or conjugate with 2 M TFA at 80°C for 3 h followed by lyophilization and redissolution.

Chromatographic procedures using both CarboPac PA1 and IonPac AS11 columns were investigated and compared. Using the PA1 column, optimum resolution was obtained with initial eluent conditions of 100 mM NaOH (6 min), followed by a 4 min gradient of 0 to 200 mM sodium acetate in 100 mM NaOH, followed by a 15 min gradient to 400 mM sodium acetate in 100 mM NaOH.

Due to the long time for analysis and re-equilibration a procedure was implemented using an AS11 column (4 × 250 mm) with AS11 guard, operated at room temperature with isocratic mobile phase of 11 mM NaOH. Pulsed amperometric detection used a standard four-potential waveform as previously described [10]. Results

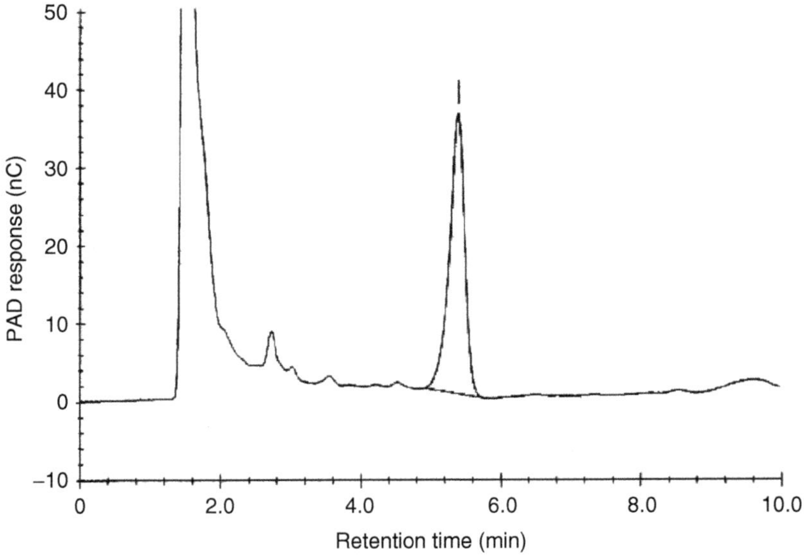

Figure 22.3. HPAEC-PAD of acid hydrolyzed Meningococcal A PS using an IonPac AS11 column with 11 mM NaOH eluent. Mannosamine-6-phosphate peak is indicated. Reprinted from Ref. 14 with permission from Elsevier.

obtained using the HPAEC-PAD procedure with mannoseamine-6-phosphate as an analytical standard correlated well with the standard colorimetric assay. Figure 22.3 illustrates the separation of a prepared MenA PS using the AS11 column and described conditions.

MenW135 is a heteropolymer of sialic acid alternating with galactose and MenY, a similar heteropolymer of sialic acid and glucose alternating units. Colorimetric assay procedures, including that of resorcinol [15] as a sialic acid determining reagent, have traditionally been used for quantitation. Bardotti et al. [16] described the quantitation of MenW135 and MenY by hydrolysis with 2 M TFA for 2 h at 100°C followed by centrifugal evaporation and redissolution in water. Chromatography was performed on a CarboPac PA1 column (4 mm × 250 mm) with PA1 guard column, mobile phase was 15 mM NaOH, isocratic for 15 min, followed by a 10 min regeneration step with 200 mM NaOH/50 mM sodium acetate. PAD used a gold working electrode with a three-potential waveform: $E_1 = 0.05\text{V}, t_1 = 400$ ms;$E_2 = 0.75\text{V}, t_2 = 200$ ms;$E_3 = -0.15\text{V}, t_3 = 400$ ms, with integration at 200–400 ms during $E_{1.}$. Galactose and glucose were used as analytical standards. MenW135 and MenY polysaccharides were quantitated on the basis of their galactose and glucose contents, respectively.

Meningococcal serotype C polysaccharide is composed of homopolymers of N-acetylneuraminic acid (sialic acid) linked (α 2 → 9) with O-acetylation at O-7 or O-8. MenC PS in glycoconjugate vaccines, as described by Ho et al. [17], was determined by HPAEC-PAD after hydrolysis by 100 mM HCl at 80°C for 2 h followed by 30 K MW cut-off filtration using Microcon-30 filters (Amicon). In the case of glycoconjugates adsorbed onto aluminum phosphate, the adjuvant was dissolved in 100 mM NaOH. For the determination of free MenC PS (unconjugated) 30K ultrafiltation was performed prior to hydrolysis.

Chromatography utilized an AminoTrap guard and a CarboPac PA1 analytical column; mobile phase was isocratic 100 mM NaOH/150 mM sodium acetate or 100 mM NaOH/80 mM sodium acetate. MenC PS of standardized sialic acid content was used as an analytical standard.

Lei et al. [18] described a procedure for determination of free PS in Meningococcal conjugate vaccines in which unconjugated PS was recovered after precipitation of the protein-PS conjugate with deoxycholate (DOC). One hundred microliters of 1% DOC was added to 1 mL sample, and then held at 0°C for 30 min. Fifty microliters of 1 M HCl was added followed by centrifugation at 6000 × g for 15 min. Supernatant was hydrolyzed with 1 M TFA for 1 h at 100°C followed by centrifugal evaporation and redissolution in 1.5 N NaOH containing 12.5 ppm glucosamine-1-phosphate internal standard. Protein was removed using a 10 K MW cut-off filter (Amicon).

Chromatography was performed using a CarboPac PA10 analytical column with guard column. The mobile phase was 9 mM NaOH for 10 min, followed by a 20 min linear gradient to 200 mM NaOH/1 M sodium acetate, and then 10 min regeneration with 200 mM NaOH. Detection was by pulsed amperometry using a gold working electrode; specifics of electrode potential were not described. The method was evaluated as showing good recovery and could quantitate free saccharide levels to 2%.

Kao and Tsai [19] developed procedures that utilized HPAEC-PAD and HPAEC with conductivity detection (HPAEC-CD) for the determination of monosaccharide units along with *O*-acetyl, *N*-acetyl and phosphate functional groups in Meningococcal and Typhoid Vi polysaccharides. *O*-acetyl moieties were hydrolyzed to free acetate with 10–20 mM NaOH at 37°C followed by removal of the polysaccharide using a 10K MW cut-off filter. Acetate was separated on an IonPac AS11-HC (4 × 250 mm) column with 2 mM NaOH mobile phase at 1 mL/min. Detection was by suppressed conductivity with an ASRS suppressor in self-regenerating configuration and 50 mA current. Sodium acetate was used as an analytical standard.

Phosphate was hydrolyzed from MenA polysaccharide using 38% hydrofluoric acid at 65°C for 3 h followed by vacuum drying using NaOH to trap HF. Samples were redissolved in water, redried, reconstituted, and 10 K filtered. Phosphate was determined using the same AS11-HC column as acetate, but with a mobile phase of 28 mM NaOH. Molar ratio of *O*-acetyl to phosphate in MenA PS was determined as 0.96.

Total acetyl content (*O*-acetyl plus *N*-acetyl) was determined in order to quantitate the *N*-acetyl content after subtraction of the *O*-acetyl content. This procedure was applied to polysaccharides from all four Meningococcal serotypes. Calcium salt was removed using a sulfonic acid functional SPE cartridge in order to avoid precipitation. The PS was then hydrolyzed with 2 N NaOH at 110°C for 4 to 6 h. After dilution and filtration, samples were separated on an IonPac AS15 column using a mobile phase of 3 mM NaOH.

Monosaccharides from MenY and W135 polysaccharides were hydrolyzed using 2 N TFA at 98°C for 1 h. Hydrolysates were lyophilized, reconstituted in water, and filtered using a 10 K cut-off filter, then glucose (Y) or galactose (W135) quantified using a CarboPac PA1 column (4 × 250 mm) using a mobile phase of 14 mM NaOH for 25 min followed by a regeneration step with 200 mM NaOH. Detection was by standard three-potential waveform described above with a gold working electrode. Molar ratios of *O*-acetyl to monosaccharides were found to be 0.83 (*O*Ac to Gal) for MenW-135 and 1.15 (*O*Ac to Glc) for MenY.

The described IC procedures for *O*-acetyl, *N*-acetyl, phosphate, and monosaccharides enabled the quantitation of *O*-acetyl content and its molar ratio to the other components. *O*-acetyl groups are an important factor in the immunogenicity of some capsular polysaccharides. Chemically, *O*-acetyl is a labile group and is hydrolyzed easily. Therefore, the extent of *O*-acetylation may be variable, even within the same manufacturing process. In addition, *O*-acetyl groups may dissociate while in refrigerated storage. For these reasons, *O*-acetyl quantitation may be an important measure of PS quality and stability.

22.2.3 Pneumococcal Polysaccharide Vaccines

More than ninety serotypes of *S. Pneumoniae* have been identified, each having different capsular PS structures and immunogenicity. Currently licensed vaccines contain up to 23 different PS serotypes in free PS vaccines and 13 different serotypes in the case of glycoconjugate vaccines. Specific serotypes are included in human vaccines largely on the basis of their frequency of infections in target populations.

Traditionally, characterization and quantitation of polysaccharides or bulk drug substance glycoconjugates have been based on colorimetric assays for characteristic saccharides or functional groups including hexoses [20], uronic acids [21], phosphate [22], sialic acids [15], and *O*-acetyl [22]. Quantitation of polysaccharides by physicochemical methods must be done at the bulk drug substance stage before formulation of the final combined vaccine drug product, due to the lack of any specific chemical identifiers unique to the individual serotypes. Repeating unit structures of 11 serotypes as described by Talaga [23] are shown in Figure 22.4.

The compositional differences of the various Pneumococcal serotypes require that hydrolysis conditions be optimized for the various repeating unit structures and glycosidic linkages. Talaga et al. [23] investigated three different hydrolysis procedures for analysis of Pneumococcal polysaccharides by HPAEC-PAD. These were:

1. *TFA Hydrolysis:* 2 N TFA 2 h at 110°C, followed by evaporation under nitrogen then reconstitution in water.
2. *Methanolysis and TFA Hydrolysis:* 1.1 N anhydrous methanol-HCl, 24 h at 80°C, followed by evaporation and TFA Hydrolysis per procedure 1).
3. *HF Hydrolysis and TFA Hydrolysis:* 16% v/v HF 2 h at 65°C, evaporation under nitrogen at 40°C, reconstitution in water followed by TFA hydrolysis per procedure 1.

Chromatography was performed on a CarboPac PA10 (4 mm × 250 mm) analytical column with PA10 guard. Elution gradients of NaOH and sodium acetate (NaAc) were

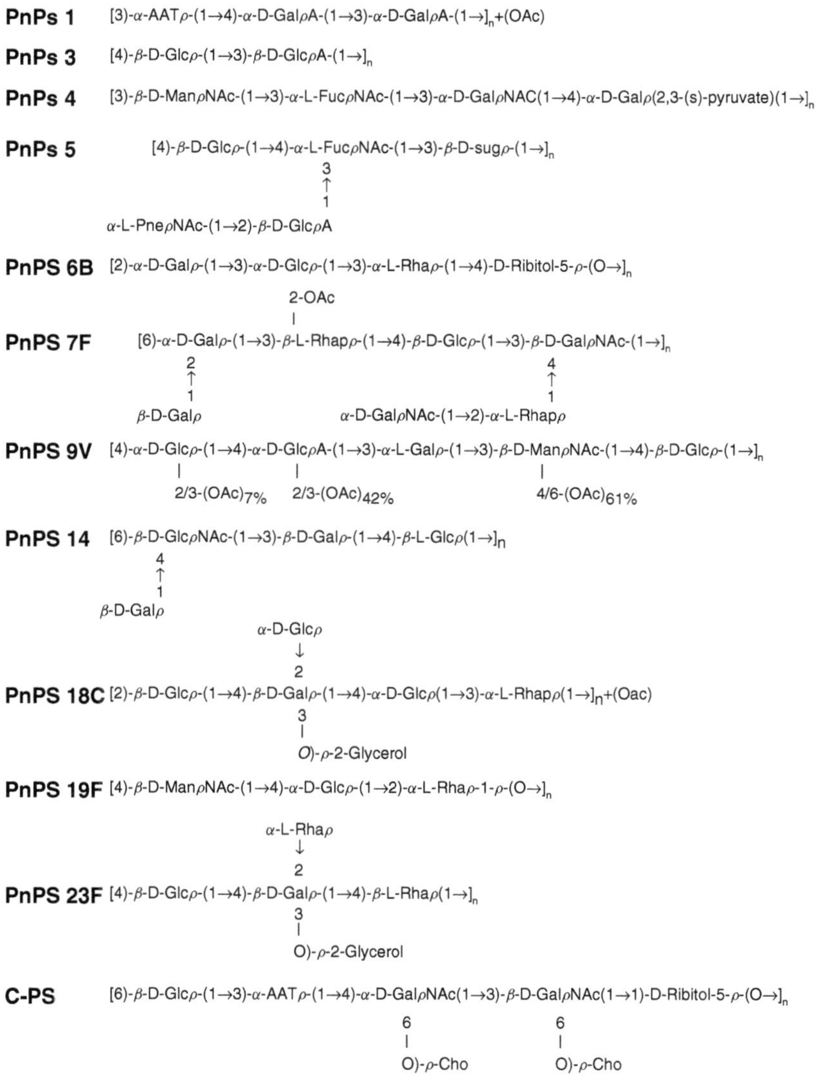

Figure 22.4. Structures of the repeating units of 11 pneumococcal PS used in the manufacturing of an 11 valent glycoconjugate vaccine product. Reprinted from Ref. 23 with permission from Elsevier.

applied at a flow rate of 1.0 mL/min. These were: 15 min isocratic 18 mM NaOH; 15.1–18 min 18 mM–100 mM NaOH; 18.1–35 min 100 mM NaOH/0–300 mM NaAc. Pulsed amperometric detection used a standard four-potential waveform [10] applied to a gold working electrode.

TFA hydrolysis was found to be satisfactory for serotypes 4, 6B, 7F, 14, 18C, 19F, and 23F based on identification of quantitative recovery of at least one monosaccharide to enable quantitation. Methanolysis followed by TFA hydrolysis was required for serotypes 1, 3, 5, and 9 V that contained uronic acids. HF followed by TFA hydrolysis was required for serotypes 6B, 18C, and 23F, which contain the sugar alcohols ribitol and glycerol. It was stated that determination of sugar alcohols would require the use of a Dionex CarboPac MA1 column, but operating conditions were not described. Figure 22.5 shows the chromatograms of four uronic acid-containing serotypes hydrolyzed by the two-step methanolysis—TFA hydrolysis procedure. Based on quantitative recovery after hydrolysis by TFA, the assay was performed using rhamnose as the standard for pneumococcal polysaccharides (PnPs) 6B, 18C, 19F, and 23F and galactose for PnPS 4, 7F, and 14. Similarly, glucuronic Acid (GlcA) was used as the standard for quantitation of PnPS 3, 5 and 9 V after methanolysis followed by TFA hydrolysis. In the case of PnPs 1, prepared by methanolysis followed by TFA hydrolysis, due to non-quantitative recovery of

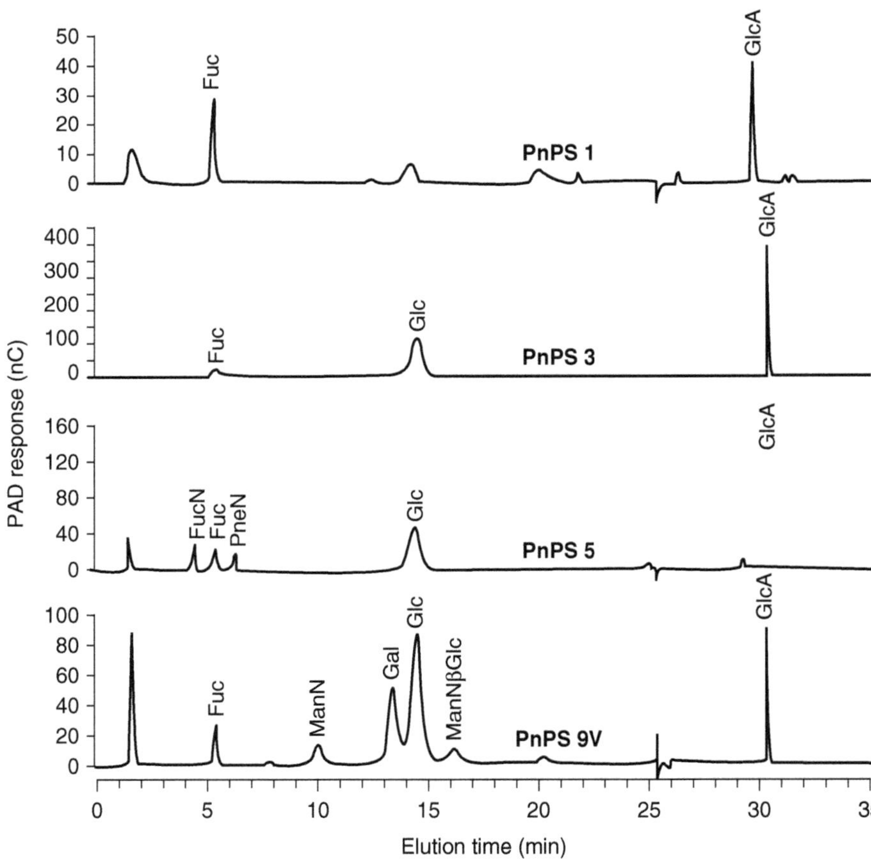

Figure 22.5. HPAEC-PAD of uronic acid containing polysaccharides after methanolysis followed by hydrolysis with TFA. Reprinted from Ref. 23 with permission from Elsevier.

galacturonic Acid (GalA), quantitation using a calibration curve with standardized PnPS 1 was performed.

Ip et al. compared TFA hydrolysis with a sequential HF and TFA hydrolysis for pneumococcal serotypes PnPS 6B, 14, 18C, and 23F [24]. Saccharide recovery from serotypes 6B, 18C, and 23F was significantly improved using the HF (65°C, 1 h) and TFA (98°C, 16 h) hydrolysis procedure. This was ascribed as resulting from efficient hydrolysis of the phosphodiester bond in the repeating units of these serotypes. Chromatography in this study used a CarboPac PA1 (4 × 250 mm) column with PA1 guard column, mobile phase for analysis was 1.0 mL/min 13 mM NaOH for 30 min followed by cleaning and regeneration NaOH/NaAc step gradients. Pulsed amperometry used a gold working electrode and the three-potential waveform.

Hepler and Ip [25] quantitated *O*-acetyl in *S. Pneumoniae* types 9 V and 18 C using IC and suppressed conductivity detection following alkaline hydrolysis with 2 mM NaOH at room temperature for 16 h. Chromatography utilized the Omnipac-PAX 500 guard and analytical columns (4 × 250 mm) along with an anion trap column (ATC-1) placed before the injection valve. Elution was isocratic with 0.98 mM sodium hydroxide in 2% methanol for 15 min at 1 mL/min, followed by 10 min step gradient of 50 mM NaOH, 2% methanol and 12.5 minute re-equilibration with the isocratic mobile phase. Eluent suppression for conductivity detection used a Dionex Anion Micro Membrane Suppressor (AMMS-II) with 50 mM sulfuric acid regenerant delivered at 10 mL/min.

As part of the study, the IC method was compared to a capillary ion electrophoretic (CIE) procedure with indirect photometric detection. The CIE method involved a 72cm × 75 μm fused silica capillary and a running buffer of 5 mM potassium hydrogenphthalate, 0.5 mM tetradecyltrimethylammonium bromide and 2 mM sodium tetraborate, pH 5.8. Samples were prepared in the same manner for the IC procedure. It was concluded that the CIE procedure offered advantages in terms of linear range (acetate concentrations up to 5 mM vs. 250 μM for IC) and time (21 min per run for CIE vs. 37.5 for IC). Additionally CIE was evaluated as capable of quantitating background levels of free residual acetate in unhydrolyzed samples while positive interference resulting from on-column hydrolysis was encountered with the alkaline mobile phase in the IC procedure. Determination of residual acetate may be important to avoid overestimation of *O*-acetyl content in PS samples that may contain free acetate.

The previously described HPAEC procedures for monosaccharide, *O*-acetyl, and phosphate in meningococcal polysaccharides by Kao and Tsai [19] were also applied to pneumococcal serotypes PnPS 9 V and 18C. For the determination of glucuronic acid content in Pneumo 9 V, hydrolysis with 2 N TFA at 98°C was performed for 8 h, otherwise sample preparation and chromatographic conditions were identical to those applied to the meningococcal polysaccharides. An *O*-acetyl to Glc ratio of 0.92 was determined for PnPS 9 V and an *O*-acetyl to phosphate ratio of 1.1 for PnPS 18C.

In addition, Kao and Tsai [19] used HPAEC to determine the pyruvic ketal component of PnPS 4 as pyruvate after acid hydrolysis. PS was hydrolyzed in 10 mM HCl with 1 mM propionate as an internal standard at 100°C for 20 min, then filtered through a 10K molecular weight cutoff cartridge following 10x dilution with water.

Chromatography utilized the same AS11-HC column with 2 mM NaOH eluent that was used for the determination of acetate in the same study. PnPS 4 was found to contain 0.82 nmol pyruvate per microgram PS or 9%w/w.

C-Polysaccharide (C-Ps) is common to all types of pneumococci. C-Ps consists of tetrasaccharides linked through ribitol phosphate diesters with phosophocholine (PCho) substituents (Figure 22.4, C-Ps). C-Ps is highly immunogenic, but anti-C-Ps antibodies are not protective [26]. As C-Ps is a common contaminant of capsular polysaccharides used in pneumococcal vaccines, levels of this impurity must be determined as a measure of serotype quality.

Talaga and Moreau determined C-Ps by HPAEC-PAD of ribitol obtained by a two-step hydrolysis with HF and TFA [27,28]. Hydrolysis steps and conditions were as previously described [23]. Chromatography utilized a CarboPac Dionex MA1 analytical column with 480 mM NaOH at 0.4 mL/min. Detection used a gold working electrode with a standard quadruple potential waveform [10]. C-Ps was quantitated in polysaccharides from Pneumococci types 6B, 7F, and 23F.

Vialle et al. [29] used HPAEC-PAD, along with MALDI-TOF mass spectrometry and NMR to identify a specific structure for the C-Ps constituent of *S. Pneumoniae* serotype 5. C-Ps from pneumococcal Type 5 was shown to contain galactose in place of glucose in its repeat unit. In these studies, Type 5 C-PS was digested with 48% HF. Analytical chromatography was performed using a CarboPac PA1 column (4 × 250 mm) with isocratic 75 mM NaOH eluent at 1 mL/min. Detection was by PAD on a gold working electrode with a quadruple potential waveform. Purification of oligosaccharides for MALDI-MS and NMR analyses used the same stationary phase in a semi preparative scale (9 × 250 mm) at 2 mL/min.

22.2.4 Oligosaccharide Profiling

The molecular size distribution of bacterial capsular oligosaccharides and polysaccharides is a critical parameter in the manufacturing of both PS and glycoconjugate vaccines. Average molecular size and dispersion have traditionally been performed by size-exclusion chromatography (SEC) using refractive index detection with a secondary calibration or by SEC-MALLS (multiple angle laser light scattering) which can yield absolute determination of molecular weight [30].

HPAEC-PAD has been utilized to profile bacterial polysaccharides by charge and size for the purpose of monitoring manufacturing process consistency. Ravenscroft et al. [31] described the use of the Dionex IonPac AS11 column with NaOH gradients for chromatographic profiling of polyphosphate functional Hib saccharides, MenA PS and oligosaccharides, and *O*-acetylated MenC oligosaccharides. Bardotti et al. [32] described profiling of Men Y and W135 oligosaccharides using IonPac AS11 and CarboPac PA100 columns (both 4 × 250 mm).

Elution from the AS11 was with a flow rate of 1.0 mL/min and a linear gradient of 5–100 mM over 50 min. Chromatography on the CarboPac PA100 used 10 mM NaOH as the mobile phase with isocratic elution at 1.0 mL/min for 15 min followed by a linear gradient of 0 to 500 mM sodium acetate in 100 mM NaOH for 50 min. Detection used a gold working electrode with the standard three-potential waveform. Figure 22.6

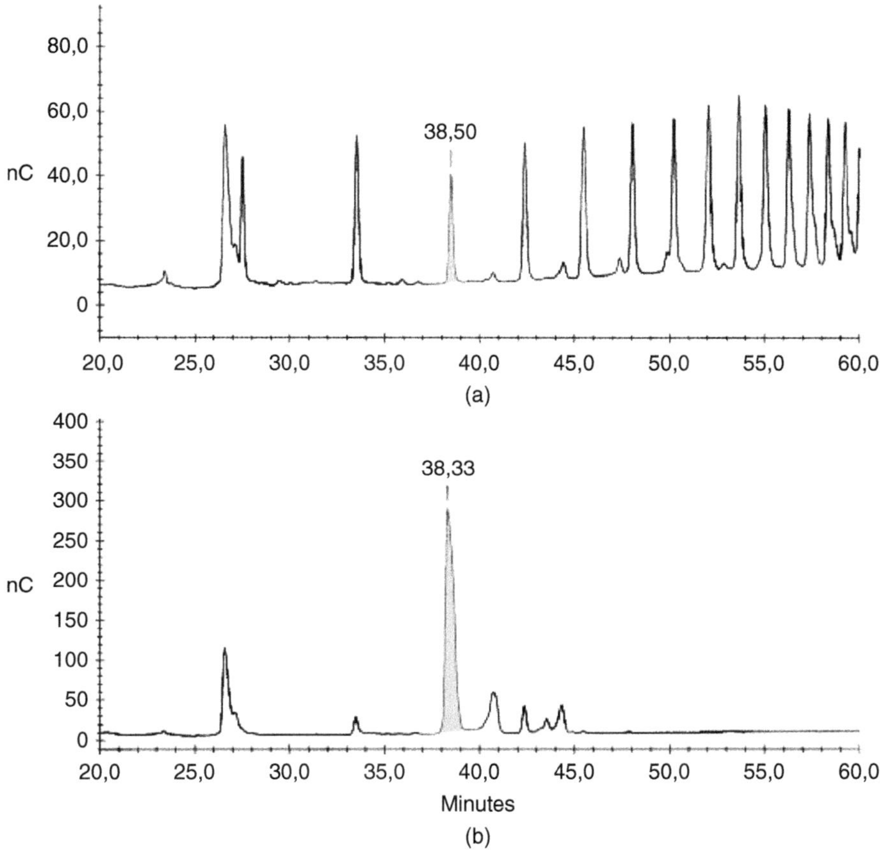

Figure 22.6. HPAEC-PAD profiling using a CarboPac PA100 column of: (a) oligosaccharides after acid hydrolysis of Men Y capsular PS (b) purified oligosaccharide of degree of polymerization (DP) = 4, Rt = 38 minutes. The numbers in the chromatograms indicate the retention times. DP4 peaks are highlighted. Reprinted from Ref. 32 with permission from Elsevier.

shows the profile of MenY oligosaccharides obtained by hydrolysis of capsular PS and that of a purified oligosaccharide fraction with degree of polymerization (DP) 4. Calibration using purified oligosaccharide fraction(s) characterized by electrospray ionization mass spectrometry (ESI-MS) and NMR spectroscopy allowed a correlation between the chromatographic peak numbers and the DP of the eluted oligosaccharides.

22.3 EXCIPIENTS IN BIOLOGICAL PRODUCTS

Analyses of ionic excipients in Biological Products such as anions and cations of common buffers and salts by IC with suppressed conductivity detection have typically utilized well-established procedures. Frequently such standard procedures are recommended by the IC column manufacturers. Applications involving excipients of Biological Products have seldom been the subject of product specific publications but several general publications, and column and instrument manufacturer's application notes, are relevant [33–35].

Saccharides including lactose, sucrose, and sorbitol are common excipients in lyophilized Biological Products, including vaccines and blood products. These stabilizers and bulking agents are frequently determined using standard IC techniques [36].

CBER has utilized IC analyses of vaccine excipients as evidence to support criminal investigations related to fraudulent dilutions of commercial vaccines. As an example, in a 2003 case involving suspected dilution of two viral vaccine products (here termed "vaccine 1" and "vaccine 2"); a laboratory investigation comparing levels of sodium, potassium, chloride, and phosphate as well as sucrose and sorbital from samples seized as evidence with normal samples was used to establish that the suspect samples had been diluted.

In these studies chloride and phosphate were separated on a Dionex IonPac AS4A-SC column with 1.8 sodium carbonate/1.7 mM sodium bicarbonate eluent. Sodium and potassium were separated on a Dionex CS12A column with an eluent of 20 mN methanesulfonic acid. Eluate conductivity in both cases was suppressed using SRS-Ultra suppressors (Dionex) in recycle mode. Table 22.2 shows results of this study. Lots are normal except those designated "Diluted", which are suspect samples.

Sucrose in "vaccine 1" and sorbitol in "vaccine 2" were determined using a Dionex CarboPac PA1 column with 0.2 N sodium hydroxide eluent and PAD using the standard triple waveform. Table 22.3 shows results for sucrose ("vaccine 1") and sorbitol ("vaccine 2").

Test results comparing potassium, phosphate, sucrose, and sorbitol indicated that the "vaccine 1" and "vaccine 2" samples under test for suspected dilution contained approximately 10% of the levels of these excipients that were found in the "normal" control samples. However, concentrations of sodium and chloride were similar or greater in the "Diluted" samples, indicating that these excipients may have been present

TABLE 22.2. Concentrations of Sodium, Potassium, Chloride and Phosphate Determined by IC in Normal and Diluted Samples of Two Viral Vaccines "Vaccine 1" and "Vaccine 2".

	Milligrams per Milliliter			
Sample	Sodium	Potassium	Chloride	Phosphate
"Vaccine 1" Lot 1	2.35	0.11	2.69	0.39
"Vaccine 1" Lot 2	2.20	0.10	2.62	0.40
"Vaccine 1" Lot 3	2.32	0.11	2.76	0.39
"Vaccine 1" Diluted	3.25	<0.02*	4.52	<0.15*
"Vaccine 2" Lot 1	4.13	0.18	2.91	6.14
"Vaccine 2" Lot 2	3.74	0.16	2.51	5.13
"Vaccine 2" Lot 3	3.81	0.16	2.65	5.45
"Vaccine 2" Diluted	3.35	<0.02*	4.11	0.59

*Observed, but less than limit of quantitation.

TABLE 22.3. Concentrations of Sucrose and Sorbitol Determined by HPAEC-PAD in Normal and Diluted Samples of Two Viral Vaccines "Vaccine 1" and "Vaccine 2".

	Milligrams per Milliliter		
Sample	Sucrose	Sample	Sorbitol
"Vaccine 1" Lot 1	34.7	"Vaccine 2" Lot 1	21.3
"Vaccine 1" Lot 2	36.4	"Vaccine 2" Lot 2	22.1
"Vaccine 1" Lot 3	37.3	"Vaccine 2" Lot 3	21.5
"Vaccine 1" Diluted	3.9	"Vaccine 2" Diluted	2.2

in an adulterant solution. These results were presented as evidence in Federal Court in Providence, Rhode Island in 2003 and resulted in a conviction.

Anticoagulant Sodium Citrate Solution along with Anticoagulant Citrate Dextrose Solution, Anticoagulant Citrate Phosphate Dextrose Solution, and Anticoagulant Citrate Phosphate Dextrose Adenine Solution are intended for use in the collection of blood or plasma for human infusion. These are regulated by FDA—CBER as blood related products. The current USP monographs for these products specify an IC method [38,39] for the determination of citrate and phosphate. In the case of Anticoagulant Sodium Citrate, the IC procedure replaced as of April 2009 a method based on potentiometric titration with 0.1 N perchloric acid for the assay of citrate. The IC procedure uses a USP L61 packing (Dionex IonPac AS11 250 × 4 mm with AG11 guard 50 × 4 mm). Mobile phase is 20 mM sodium or potassium hydroxide isocratic at 2.0 mL/min, detection is by suppressed conductivity and the column is maintained at 30°C [40].

Glycerol is frequently used in 50% v/v concentration in allergenic extracts and vaccine diluents, including smallpox vaccine, as a stabilizer and bacteriosatic/fungiostatic agent. Allergenic Products containing 50% or more (v/v) glycerin are exempt from the CFR requirement that products in multiple dose containers shall contain a (anti-microbial) preservative [41]. Glycerol has been determined in these products using an ion-exclusion chromatography (IEC) procedure using a 300 × 7.8 mm Aminex HPX-87H column (sulfonic acid functional polystyrene-divinyl benzene, hydrogen form) with 30 × 4.6 mm Micro-Guard Cation H guard column (Biorad Laboratories, Hercules, CA). Mobile phase was 0.01 N sulfuric acid and detection was by differential refractometry [42]. The column was maintained at 65°C in order to achieve optimal peak shape and chromatographic efficiency. Figure 22.7 illustrates a chromatogram of a Tuberculin Purified Protein Derivative diagnostic test antigen with ca. 50% v/v glycerin using this procedure. The IEC method showed equivalent or better precision when compared to potentiometric titration and gas chromatographic procedures. Slightly lower recoveries were obtained by IEC relative to the other procedures. The IEC procedure has also been successfully implemented using a 220 × 4.6 mm Brownlee Polypore H column with 30 mm guard (Perkin Elmer Inc., Waltham, MA) that enabled operation near ambient temperature (35°C) [43].

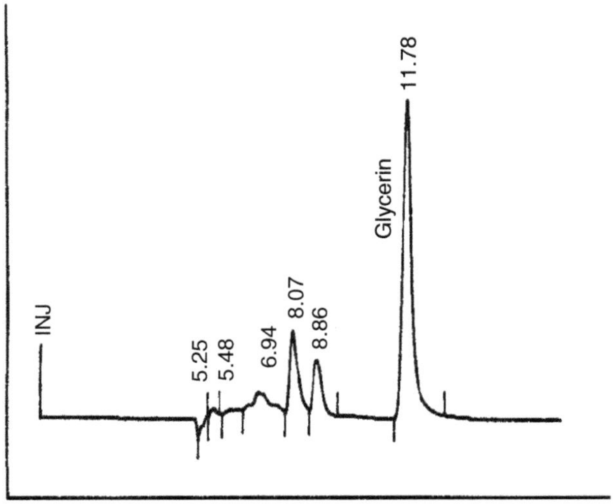

Figure 22.7. Ion exclusion chromatography of glycerin in a tuberculin purified protein derivative (PPD) diagnostic test antigen. Aminex HPX-87H (7.6 × 300 mm) column with guard, 0.6 mL/min 0.01 N H_2SO_4 mobile phase, refractive index detection. Reprinted from Ref. 42 with permission from A.O.A.C International.

22.4 QUANTIFICATION OF PROTEIN NITROGEN IN VACCINE REFERENCE ANTIGEN PREPARATIONS

Determination of total protein content is an important analytical parameter in the characterization or quantitation of a wide variety of biological samples ranging from foods to biotechnological products including vaccines and biopharmaceuticals. Several different analytical procedures have involved colorimetric and titrimetric as well as modern instrumental methodologies that are based on responses associated with protein peptide linkages or other components of amino acids [44]. Among these are copper binding assays, the most basic form of which is the Biuret procedure where analytical response is measured by the absorbance at 540 nm resulting from the complexation of Cu^{2+} with two peptide bonds. Popular sensitivity enhancements of the basic copper binding method include the Lowry method in which sensitivity is enhanced through the use of phosphomolybdotungstic acid chromogenic reagent [45] and the BCA or bininchonic acid procedure. Other methods include the Bradford dye binding assay and the o-phthalaldehyde fluorometric procedure [46]. Colorimetric and fluorometric methods offer advantages of sensitivity, low sample mass or volume requirements, speed and ease of use, but may be subject to interferences from detergents, solubilizing agents and various buffers that are frequently present in biological product samples.

The Kjeldahl procedure for nitrogen determination, originally introduced in 1883, has been one of the common procedures used to determine total protein content of biological products and reference antigens through the measurement of nitrogen content [47,48]. In its classical form the Kjeldahl method involves digesting a sample with sulfuric acid in the presence of potassium and copper sulfate, resulting in the liberation of nitrogen as ammonium sulfate. Alternately mercury salts have been used as a catalyst. Sodium hydroxide is added and the resulting ammonia distilled into a solution of boric acid. The distillate is then titrated with hydrochloric acid to a colorimetric endpoint visualized using methyl red-bromo-cresol green indicator. Chemical reactions involved in each process may be described by:

$$\text{Digestion}: \text{Protein} + H_2SO_4 \rightarrow (NH_4)_2SO_4 + CO_2 + SO_2 + H_2O$$

$$\text{Distillation}: (NH_4)_2SO_4(aq) + 2NaOH \rightarrow 2NH_3 \uparrow + Na_2SO_4 + 2H_2O$$

$$NH_3 \uparrow + H_3BO_3 \rightarrow NH_4{}^+ : H_2BO_3{}^- + H_3BO_3(\text{excess})$$

$$\text{Titration}: NH_4{}^+ : H_2BO_3{}^- + HCl \rightarrow NH_4Cl + H_3BO_4$$

Automatic analyzers for routine analysis using the Kjeldahl method are available from a number of manufacturers. Automated functions include sampling, digestion, distillation, and titration using either colorimetric or potentiometric endpoint detection. Other variations of the Kjeldahl method have involved digestion using microwave radiation [49] or the use of alternate digestion mediums including hydrogen peroxide in an acetic acid–sulfuric acid mixture [50].

Expression of the determined nitrogen content as protein is obtained by applying a conversion factor. As a common convention, a 16% contribution of nitrogen to the molar mass weight of a typical protein is assumed and measured nitrogen content is multiplied by 6.25 for the calculation of protein content. If the actual nitrogen content of a specific purified protein intended for quantitation is known, the exact molar ratio of protein/nitrogen may be applied directly.

In some cases precipitating agents may be used to isolate protein from potential interferants such as azide, Tris, or other nitrogen containing buffers. Precipitation with phosphotungstic acid (PTA) has been used as a standard method for the quantitation of protein in allergenic extracts [51,52]. In this procedure protein is expressed as protein nitrogen units (PNUs) in which one PNU/ml is equivalent to 1×10^{-5} mg of nitrogen in the material precipitated from 1 mL of allergenic extract.

Single-radial immunodiffusion (SRID) is based on the diffusion of antigen of interest into agarose gel containing specific antibodies, producing a zone of precipitation [53–55]. Potency of a test sample may be determined by comparison of zone diameter against those obtained from a calibrated reference of standardized antigen content.

The calibration of reference standards for immunoassay procedures typically relies on measurement of protein content to which immunological measurements may be correlated. In the standardization of influenza antigen reference standard materials, the measured total protein content of partially purified antigen preparation is adjusted on the basis of the relative density of the antigen-specific electrophoretic band obtained by SDS-PAGE relative to the total density of the observed electrophoretic protein bands. Determination of total protein has been performed variously using Kjeldahl, Lowry and other colorimetric methods and the OPA fluorometric method.

A significant limitation of the traditional Kjeldahl method is its sample size requirement relative to the Lowry and other colorimetric procedures. The USP general chapter < 461 > Nitrogen Determination [56] specifies a sample size equivalent to 2–3 mg nitrogen, equivalent to a minimum of ca. 12 mg protein, while the micro-Kjeldahl procedure described in AOAC 960.52 [48] specifies a sample requiring 3–10 mL 0.01 N HCl, equivalent to a minimum of 0.42 mg nitrogen or ca. 2.6 mg protein. Since the total protein content of influenza antigen solutions for use as immunoassay references is typically in the approximate range of 1–2 mg/mL, a minimum of a few mL is typically required for each determination. With replicate determinations, a total of several mL of sample would be required for accurate reporting of test results.

As a well established means of quantifying monovalent cations, including ammonium [57–59], IC may be utilized as a means of the final determination of nitrogen as ammonium in a variation of the Kjeldahl procedure. Jackson et al. [60] used polystyrene-divinylbenzene cation exchange columns with dilute nitric acid eluents and indirect (non-suppressed) conductivity detection to determine ammonium after Kjeldahl digestion of animal feed grain samples with acetic acid-sulfuric acid mixture in the presence of hydrogen peroxide catalyst. Chromatography involved an IC-Pak Cation (50 × 4.6 mm) (Waters Associates, Milford, MA) with a mobile phase of 2.0 mM nitric acid at 1.2 mL/min (Figure 22.8).

The same authors used a potassium sulfate-sulfuric acid digestion mixture with a mercury(II) catalyst for the processing of organic compounds and sludges for Kjeldahl analysis by IC. Chromatography of samples processed by this digestion method used a

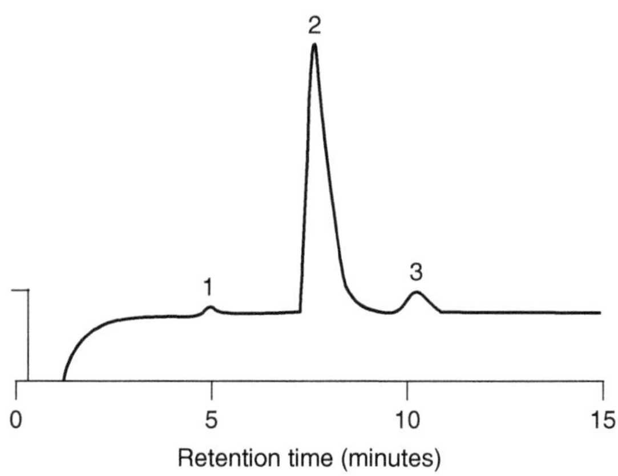

Figure 22.8. Chromatogram of an acetic acid-sulfuric acid-digested grain sample using the IC-Pak Cation column. Eluent, 2 mM nitric acid; flow-rate; 1.2 mL/min; injection, 100 μL. Sample preparation: 125-fold dilution of acetic acid-sulfuric acid-digested grain. Peak IDs: 1 = sodium; 2 = ammonium (3.55 ppm); 3 = potassium. Reprinted from Ref. 60 with permission from Elsevier.

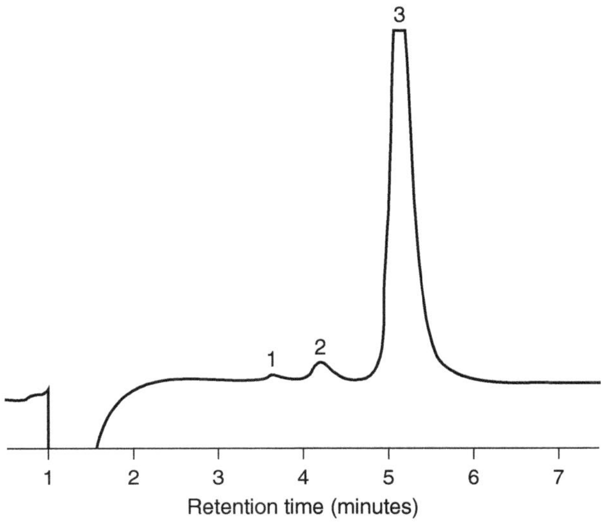

Figure 22.9. Chromatogram of a potassium sulfate-sulfuric acid-digested sample of caprolactam using the Protein-Pak SP-5PW column. Eluent, 25 mM nitric acid-5% acetone; flow-rate, 2.4 mL/min; injection, 100 μL. Sample preparation: 1000-fold dilution of potassium sulfate-sulfuric acid-digested caprolactam. Peak IDs: 1—sodium, 2—ammonium (0.45 ppm); 3—potassium. Reprinted from Ref. 60 with permission from Elsevier.

higher capacity Protein-Pak SP-5PW (75 × 4.6 mm) (Waters Associates, Milford, MA) with a mobile phase of 25 mM nitric acid-5% acetone at 2.4 mL/min (Figure 22.9). A higher capacity column was required to ensure the resolution of ammonium from the much larger quantity of potassium (ca. 20 mg/mL) in the final digest. While the amounts of original samples taken for digestion were not stated, solutions obtained from both digestion procedures typically contained 100–500 ppm ammonium and were diluted 100–1000-fold prior to the injection of 100 μL without further preparation.

Most current applications for analytical IC with conductivity detection incorporate devices for online suppression of eluent conductivity in order to minimize background conductivity (noise). In cation-exchange chromatography, stripping columns or, more currently, suppressors serve to convert the analyte's anion to hydroxide. Group I and II cations form strong bases, e.g., sodium hydroxide, and exhibit high conductivity responses and good linearity over a wide range. However ammonium and other weak bases are partially dissociated and show lower conductivity and a non-linear response with increasing analyte concentration [61]. Calibration of ammonium and other weakly dissociated amine bases resulting from the use of an eluent suppressor may be achieved by limiting the calibration range to a region of low concentration where adequate linearity may be obtained or by applying a polynomial line fit, usually quadratic across a wider range.

In order to improve detector response and linearity for ammonium, Dionex introduced a "Salt Converter-Cation Self-Regenerating Suppressor" Model SC-CSRS 300 [62]. Intended for use with methanesulfonic acid eluent (MSA), this device

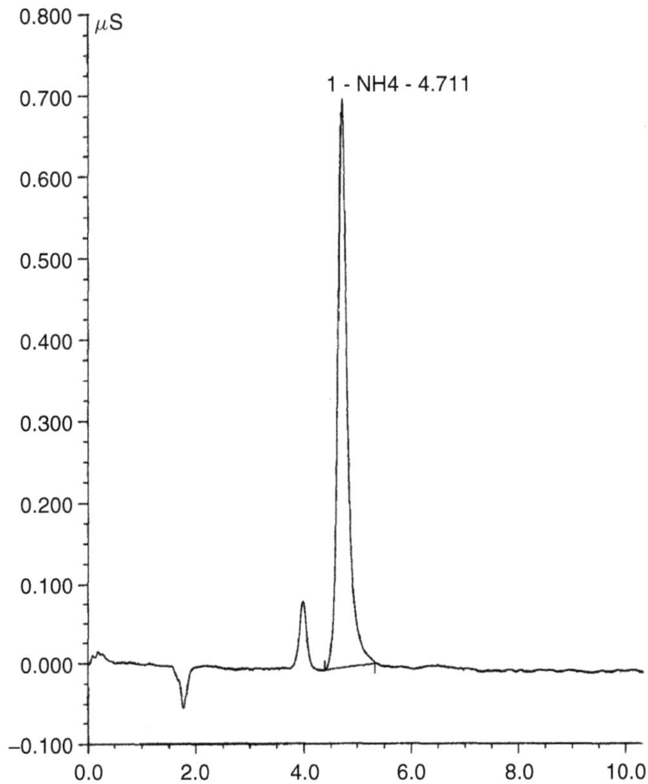

Figure 22.10. Chromatogram of 0.17 ppm ammonium chloride standard solution using Dionex IonPac CS12 4 × 250mm and CG12 4 × 50mm columns. Eluent 20 mM MSA, flow rate 1.0 mL/min; Injection, 50 μL. Eluent suppressor: SC CSRS-300 4 mm; conductivity detection.

incorporates a membrane type eluent suppressor device in which the MSA is largely replaced by hydroxide. This is followed by a second device in which the ammonium analyte is exchanged for the hydronium ion, resulting in the delivery of protonated MSA to the conductivity detector. Using this device with an IonPac CS12A 4 mm column (Dionex), 20 mM MSA eluent at 1.0 mL/min and 25 μL injection volumes, the manufacturer demonstrated excellent linearity in the range of 0.1–10 mg/L as $r^2 = 0.9999$.

With the purpose of reducing the sample size requirements for the Kjeldahl procedure as applied to the standardization of Influenza antigen reference standards, we investigated an approach that involved sample preparation, including distillation, following the classical Kjeldahl procedure with quantitation of ammonium in the distillate by IC. Initial development utilized an IonPac CS12 analytical column (4 × 250 mm) with CG12 guard (4 × 50 mm) and a mobile phase of 20 mM MSA at 1.0 mL/min used with a SC CSRS-300 suppressor system as previously described and a conductivity detector. Ammonium showed a retention time of 4.7 min and was adequately

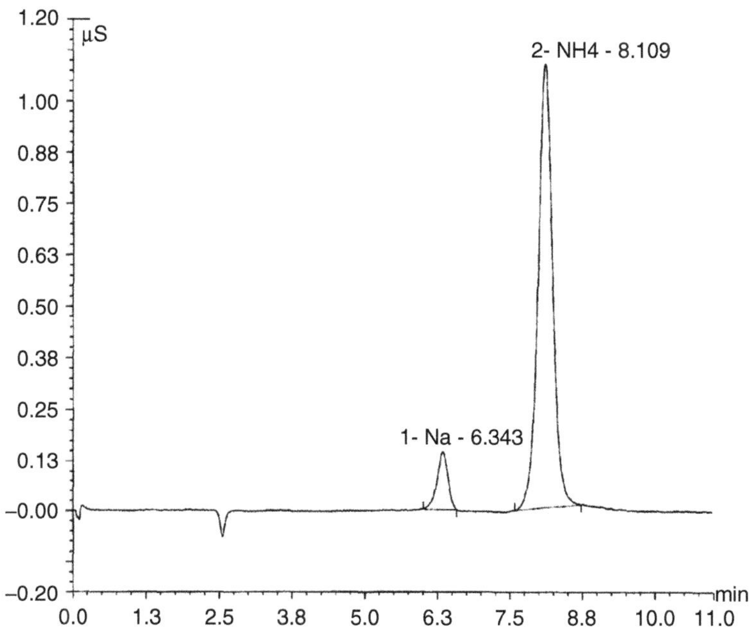

Figure 22.11. Chromatogram of 5.12 ppm ammonium chloride standard solution using Dionex IonPac CS16 5 × 250mm and CG16 5 × 50mm columns. Eluent 30 mM MSA at 1.25 mL/min; Injection, 10 μL. Eluent suppressor: CSRS 300 4 mm; conductivity detection.

resolved from sodium, the only other peak detected (Figure 22.10). Standards showed good linearity ($r^2 \geq 0.999$) in the evaluated range of 0.1–100 ppm.

As a second approach, the use of a CSRS-300 suppressor without the salt converter device was investigated. In this case an IonPac CS16 analytical column (5 × 250 mm) with CG16 guard (5 × 50 mm) was used with 30 mM MSA eluent at 1.25 mL/min [61]. Ammonium was eluted at 8.1 min and the resolution with sodium was superior to that obtained with the CS12 column system (Figure 22.11). Ten microliter injections of ammonium standards ranging from 0.5 to 10 ppm yielded a linear calibration with $r^2 = 0.9998$. Deviations from linearity were apparent with standard injections greater than 10 ppm (100 ng ammonium on column).

Kjeldahl determinations of a Bovine Serum Albumin NIST protein standard (BSA-927D), an in-house BSA Protein Nitrogen Reference material (Reference for Nitrogen Assay Lot 4) and representative lots of influenza reference antigen preparations of different viral strains have been performed using both the classical and ion chromatographic methods for ammonium quantitation. Kjeldahl digestions were performed using 0.3 mL or less of sample for products determined by IC while 3.0 mL of each sample was used with samples processed for use with the classical Kjeldahl titrimetric determinations.

IC quantitation of ammonium for the determination of total protein as nitrogen allows for approximately a 10-fold decrease in sample size requirements while providing results that are equivalent to the classical Kjeldahl method. Further evaluation and validation of the method are in progress.

REFERENCES

1. Finn A. Bacterial polysaccharide-protein conjugate vaccines. Br. Biomed. Bull. 2004;70: 1–14.
2. Frasch CE. Preparation of bacterial polysaccharide-protein conjugates: Analytical and Manufacturing Challenges. Vaccine 2009;27:6468–6470.
3. Brown A. Determination of pentose in the presence of large quantities of glucose. Arch. Biochem. 1946;11; 269–278.
4. Chen D, Toribara T, Warner H. Microdetection of Phosphorus. Anal. Chem. 1956;28:1756–1758.
5. Tsai C, Gu X, Byrd A. Quantification of polysaccharide in Haemophilus influenzae type b conjugate and polysaccharide vaccines by high-performance anion-exchange chromatography with pulsed amperometric detection. Vaccine 1994;12:700–706.
6. Hardy M, Townsend R, Lee YC. Monosaccharide Analysis of Glycoconjugates by Anion Exchange Chromatography with Pulsed Amperometric Detection. Anal. Biochem. 1988;170:54–62.
7. Sturgess A, Rush K, Charbonneau R, Lee J, West, D, Sitrin R, Hennessey, J. Haemophilus influezae type b conjugate vaccine stability: catalytic depolymerization of PRP in the presence of aluminum hydroxide. Vaccine 1999;17:1169–1178.
8. Bardotti A, Ravenscroft N, Ricci S, D'Ascenzi S, Guarnieri V, Averani G, Constantino P. Quantitative determination of saccharide in Haemophilus influenzae type b glyconconjugate vaccines, alone and in combination with DPT, by use of high-performance anion-exchange chromatography with pulsed amperometric detection. Vaccine 2000;18:1982–1993.
9. Dionex Corporation, Technical Note 20: Analysis of Carbohydrates by High Performance Anion Exchange Chromatography with Pulsed Amperometric Detection (HPAE-PAD).
10. Rocklin R, Clarke A, Weitzhandler M. Improved long-term reproducibility for pulse amperometric detection of carbohydrates via a new quadruple-potential waveform, Anal. Chem. 1998;70:1496–1501.
11. Belfast M, Lu R, Capen R, Jinglin Z, Nguyen M, Gimenez J, Sirin R, Mancinelli R. A practical approach to optimization and validation of a HPLC assay for analysis of polyribosylribitol phosphate in complex combination vaccines. J. Chromatogr. B 2006;832:208–215.
12. Mawas F, Bolgiano B, Rigsby P, Crane D, Belgrave D, Corbel M. Evaluation of the saccharide content and stability of the first WHO International Standard for Haemophilus influenzae b capsular polysaccharide. Biologicals 2007;35:235–245.
13. Bundle D, Smith I, Jennings H. Determination of the structure and conformation of bacterial polysaccharides by carbon 13 nuclear magnetic resonance. J. Biol. Chem. 1974;249:2275–2281.
14. Ricci S, Bardotti A, D'Ascenzi S, Ravenscroft N. Development of a new method for the quantitative analysis of the extracellular polysaccharide of Neisseria meningitidis serogroup A by use of high-performance anion-exchange chromatography with pulsed-amperometric detection. Vaccine 2000;19:1989–1997.

15. Svennerholm L. Quantitative estimation of sialic acids. II. A colorimetric resorcinol-hydrochloric acid method. Biochim Biophys Acta 1957;24:604–611.
16. Bardotti A, Averani G, Berti F, Berti S, Carinci V, D'Ascenzi, S, Fabbri B, Giannini, S, Giannozzi, A, Magagnoli, C, Proiei, D, Noelli, F, Rappuoli R, Ricci, S, Costantino, P. Physicochemical characterization of glyconconjugate vaccines for prevention of meningococcal diseases. Vaccine 2008;26:2284–2296.
17. Ho M, Lemercinier X, Bolgiano B, Corbel M. Monitoring Stability of Meningococcal Group C Conjugate Vaccines; Correlation of Physico-Chemical Methods and Immunogenicity Assays. In: Brown F, Corbel M, Griffiths E, editors. Developmental Biology, Volume 103: Physico-Chemical Procedures for the Characterization of Vaccines. Basel: Karger; 2000. p 139–150.
18. Lei Q, Shannon A, Heller R, Lamb D. Quantification of Free Polysaccharide in Meningococcal Polysaccharide-Diphtheria Toxoid Conjugate Vaccines. In: Brown F, Corbel M, Griffiths E, editors. Developmental Biology, Volume 103: Physico-Chemical Procedures for the Characterization of Vaccines. Basel: Karger; 2000. p 259–264.
19. Kao G, Tsai C. Quantification of O-acetyl, N-acetyl and phosphate groups and determination of the extent of O-acetylation in bacterial vaccine polysaccharides by high-performance anion-exchange chromatography with conductivity detection (HPAEC-CD). Vaccine 2004;22:335–344.
20. Morris D. Quantitative determination of carbohydrates with Dreywood's anthrone reagent. Science 1948;107:254–255.
21. Blumenkrantz N, Asboe-Hansen G. New method for quantitative determination of uronic acids, Anal. Biochem. 1973;54:484–489.
22. Hestrin S. The reaction of acetylcholine and other carboxylic acid derivatives with hydroxylamine, and its analytical application. J. Biol. Chem. 1949;180:249–261.
23. Talaga P, Vialle S, Moreau M. Development of a high-performance anion-exchange chromatography with pulsed-amperometric detection based quantification assay for pneumococcal polysaccharides and conjugates. Vaccine 2002;20:2474–2484.
24. Ip CY, Manam V, Hepler R, Hennessey J. Carbohydrate Composition Analysis of Bacterial Polysaccharides: Optimized Acid Hydrolysis Conditions for HPAEC-PAD Analysis. Anal. Biochem. 1992;201:343–349.
25. Hepler R, Ip C. Application of capillary ion electrophoresis and ion chromatography for the determination of O-acetate groups in bacterial polysaccharides. J. Chrom. A. 1994;680:201–208.
26. Nielsen S, Sorensen U, Henrichsen J. Antibodies against pneumococcal C-polysaccharide are not protective. Microb Pathog. 1993;14:299–305.
27. Talaga P, Moreau M. Quantification of C-polysaccharide in Streptococcus pneumoniae polysaccharides by high performance anion exchange chromatography with pulsed amperometric detection. In: Brown F, Corbel M, Griffiths E, editors. Developmental Biology, Volume 103: Physico-Chemical Procedures for the Characterization of Vaccines. Basel: Karger; 2000. p 27–34.
28. Talaga P, Bellamy L, Moreau M. Quantitative determination of C-polysaccharide in Streptococcus pneumoniae capsular polysaccharides by use of high-performance anion-exchange chromatography with pulsed amperometric detection. Vaccine 2001;19:2987–2994.

29. Vialle S, Sepulcri P, Dubayle J, Talaga P. The teichoic acid (C-polysaccharide) synthesized by Streptococcus pneumoniae serotype 5 has a specific structure. Carbohydr. Res. 2005;340:91–96.
30. D'Ambra A, Baugher, J, Concannon P, Pon R, Michon F. Direct and indirect methods for molar-mass analysis of fragments of the capsular polysaccharide of Haemophilus influenzae type b. Anal Biochem. 1997;250:228–236.
31. Ravenscroft N, D'Ascenzi S, Proietti D, Noelli F, Costantino P. Physicochemical Characterization of the Oligosaccharide Component of Vaccines In: Brown F, Corbel M, Griffiths E, editors. Developmental Biology, Volume 103: Physico-Chemical Procedures for the Characterization of Vaccines. Basel: Karger; 2000. p 35–47.
32. Bardotti A, Averani G, Berti F, Berti S, Galli C, Giannini, S, Fabri, B, Proietti D, Ravenscroft N, Ricci S. Size determination of bacterial capsular oligosaccharides used to prepare conjugate vaccines against Neisseia meningitidis groups Y and W 135. Vaccine 2005;23:1887–1899.
33. Dionex Corporation. Application Note 106: Ion Chromatography in the Pharmaceutical Industry.
34. Dionex Corporation. Application Note 116: Quantification of Anions in Pharmaceuticals.
35. Dionex Corporation. Application Note 164: Assay for Citrate and Phosphate in Pharmaceutical Formulations Using Ion Chromatography.
36. Dionex Corporation. Application Note 117: Application Quantification of Carbohydrates and Glycols in Pharmaceuticals.
37. Cassidy S, Demarest C, Wright, P, Zimmerman, J. Development and application of a universal method for quantitation of anionic constituents in active pharmaceutical ingredients during early development using suppressed conductivity ion chromatography. J. Pharm. Biomed. Anal. 2004;34:255–264.
38. USP32-NF27, Monograph for Anticoagulant Sodium Citrate Solution. Rockville: United States Pharmacopeial Convention; 2009: 1567.
39. USP32-NF27, General Chapter <345> Assay for Citric Acid/Citrate and Phosphate. Rockville: United States Pharmacopeial Convention; 2009: 143.
40. DeBorba B, Rohrer J, Bhattacharyya L. Development and validation of an assay for citric acid/citrate and phosphate in pharmaceutical dosage forms using ion chromatography with suppressed conductivity detection. J. Pharm. Biomed. Anal. 2004;36:517–524.
41. Title 21 Code of Federal Regulations Part 610.15a. General Biological Product Standards—Constituent Materials—Ingredients, preservatives, diluents, adjuvants.
42. Del Grosso A, May J. Gas Chromatographic, Liquid Chromatographic, and Titrimetric Procedures for Determination of Glycerin in Allergenic Extracts and Diagnostic Antigens: Comparative Study. J. Assoc. Off. Anal. Chem. 1987;70:825–828.
43. Del Grosso A. Unpublished results.
44. European Pharmacopeia 6th Edition 2010, Part 2.5.33, Total Protein.
45. Lowry O, Rosebrough N, Farr A, Randall R. Protein measurement with the Folin phenol reagent. J. Biol. Chem. 1951;193:265–275.
46. Benson, J, Hare, P. o-Phthalaldehyde: fluorogenic detection of primary amines in the picomole range. Comparison with fluorescamine and ninhydrin. Proc. Nat. Acad. Sci. 1975;72:619–662.
47. Kabat, E, Mayer, M. Experimental Immunochemistry, Charles C. Thomas, Springfield, IL 1961:476–479.

48. Method Number 960.52 Micro-chemical determination of Nitrogen, Official Methods of Analysis (OMA) 18th Edition, AOAC International, Gaithersburg, MD.
49. Korn M, Wagna P, Korn M., Sergio, L. Optimisation of focused-microwave assisted digestion procedure for Kjeldahl nitrogen determination in bean samples by factorial design and Doehlert design. Talanta 2005;65:710–715.
50. Cresceri L, Greenberg A, Trussell R. (Eds), Standard Methods for the Examination of Water and Wastewater, American Public Health Association, Washington, D.C., 17th ed. 1989:4–144.
51. May J, Sih J. Protein nitrogen unit precipitation procedure for Allergenic Extracts: Collaborative Study. J. Assoc. Off. Anal. Chem. 1981;64:1435–1438.
52. May J, Sih J, Miller J, Seligmann E. Optimization of parameters in protein nitrogen unit precipitation procedure for allergenic extracts. J. Allergy Clin. Immun. 1979;63:87–97.
53. Williams, M. Single-radial-immunodiffusion as an in vitro potency assay for human inactivated viral vaccines. Veterinary Microbiology 1993;37:253–262.
54. Wood J, Schild G, Newman R, Seagroatt V. An improved single-radial immunodiffusion technique for the assay of influenza haemagglutinin antigen: application for potency determinations of inactivated whole virus and subunit vaccines. J. Biol. Standardization 1977;5:237–247.
55. Harvey R, Wheeler J, Wallis, C, Robertson, J, Engelhardt, O. 2008 Quantitation of haemagglutinin in H5N1 influenza viruses reveals low haemagglutinin content of vaccine virus NIBRG-14(H5N1). Vaccine 2008;26:6550–6554.
56. USP 32–General Chapter <461> Nitrogen Determination Method II
57. Small H. 1989 Ion Chromatography; 1989 Plenum: New York, p175.
58. Haddad P, Jackson P. Ion Chromatography—Principles and Applications (J. Chrom. Library vol. 46) 1987 Elsevier, Amsterdam.
59. Gerde D, Fritz J. Ion Chromatography, 1987 Hüthig, Heidelberg, 2nd ed.
60. Jackson P, Krol J, Heckenberg A. Determination of total nitrogen in food, environmental and other samples by ion chromatography after Kjeldahl digestion. Journal of Chromatography 1991;546:405–410.
61. Bouyoucos A. Determination of ammonia and methylamines in aqueous solutions by ion chromatography. Anal. Chem. 1977;49:401–403.
62. Dionex Product Brochure: Salt-Converter-Cation Self-Regenerating Suppressor 300© 2008, Dionex Corporation

23

HIGH PERFORMANCE ANION-EXCHANGE CHROMATOGRAPHY WITH PULSED AMPEROMETRIC DETECTION (HPAEC-PAD) ANALYSIS OF GBS GLYCOCONJUGATE VACCINES: FROM FERMENTATION TO FINAL PRODUCT

Stefano Ricci, Barbara Fabbri, Sara Giannini, Silvia Bernardini, Marcello Morelli, and Sandro D'Ascenzi

Technology Development Department, Novartis Vaccines and Diagnostics S.r.L., Siena, Italy

23.1 INTRODUCTION

Group B streptococci (GBS) are the major cause of infections in neonates, pregnant women, and immunocompromised individuals. Immunity to GBS in neonates is associated with naturally acquired maternal antibodies to the type-specific capsular polysaccharides of these organisms. IgG class antibodies directed to these polysaccharides are passed transplacentally and protect the child from invasive GBS disease. The strains usually isolated from clinical cases belong to one of the major capsular types Ia, Ib, and III. In the early 1990s, reports of infection in newborn infants caused by GBS serotype V were published in the pediatric literature [1]. Moreover GBS type V has emerged as the most frequently identified serotype causing invasive GBS disease in nonpregnant adults [2]. A common feature of GBS type polysaccharides is that they are high-molecular-weight polymers with a repeating unit composed of glucose (Glcp), galactose (Galp), N-acetylglucosamine (GlcpNAc), and *N*-acetylneuraminic acid (NeupNAc, sialic acid).

Applications of Ion Chromatography for Pharmaceutical and Biological Products, First Edition.
Edited by Lokesh Bhattacharyya and Jeffrey S. Rohrer.

Capsular polysaccharides of serotypes Ia and V of GBS were purified at Novartis V&D Technology Development. Various analytical methods were developed in order to determine saccharide titer, residual protein content, impurities, molecular size and polydispersity [3]. HPAEC-PAD [4–6] methods were applied to determine the monosaccharide components of the GBS polysaccharides and also for rhamnose derived from group B carbohydrate, which is present as a process-related impurity [7]. The capsular polysaccharides of GBS are constituted by a repeating unit of five monomers: 2 units of β-D-Galp and 1 unit each of β-D-Glcp, β-D-GlcpNAc, α-D-NeupNAc linked differently depending on the serotype. In serotype V the composition of monosaccharide is the same as other serotypes, but with the difference that this serotype has a trisaccharide backbone with two side chains. In the development of the quantitative method for serotype Ia, Galp was identified as the best monosaccharide candidate for HPAEC-PAD analysis, while for serotype V, GlcpNAc was chosen because it is the only monosaccharide unaltered during the production process, apart from glucose, which was not considered as a good candidate because it is ubiquitous.

The next step in the production of GBS vaccines is the conjugation of the polysaccharide to a carrier protein. In general GBS purified polysaccharide, apart from serotype V, is linked to CRM197 carrier protein individually after sialic acid residue activation by an oxidation reaction. Galp is present in the backbone structure and in the lateral chain of the polysaccharide. The glycoconjugates produced are characterized by different analytical methods in order to evaluate saccharide titer, free (unbound) saccharide, dimensional analysis by size exclusion chromatography for Kd determination, where Kd = [(Te-T0)/(Tt-T0)] (Te = glycoconjugate retention time; T0 = elution time of high molecular weight fragment of λDNA HindIII; Tt = total time corresponding to sodium azide retention time), free sialic acid, and free (unconjugated) CRM197. Several methods were developed in the past on different glycoconjugates produced in our company [8,9]. In GBS glycoconjugates a solid phase extraction (SPE) method based on C4 cartridges was developed in order to measure the unbound polysaccharide. Galp residues, both in conjugated and unbound polysaccharides, are released by acid hydrolysis in TFA and determined by HPAEC-PAD. Figure 23.1 shows a schematic diagram of the steps involved in glycoconjugate manufacturing.

After concentrated bulk production, the glycoconjugates are formulated in a lyophilized presentation, which induces additional analytical challenges. The chromatographic method developed for bulk concentrate samples is not applicable to lyophilized product because of the presence of contaminant peaks derived from mannitol, the excipient used for lyophilization, and its impurities. Because of the huge amount of mannitol (and the small percentage of impurities) in comparison with the small amount of antigen/vaccine, the Galp peak detected by CarboPac PA1 chromatography was completely hidden by matrix peaks. As a consequence, another chromatography column was tested. The column identified was a CarboPac MA1, more selective for Galp in this sample type, enabling an optimal separation from matrix peaks. Generally, CarboPac PA series of columns are more suitable for the monosaccharide analyses of glycoconjugate vaccines because they need shorter

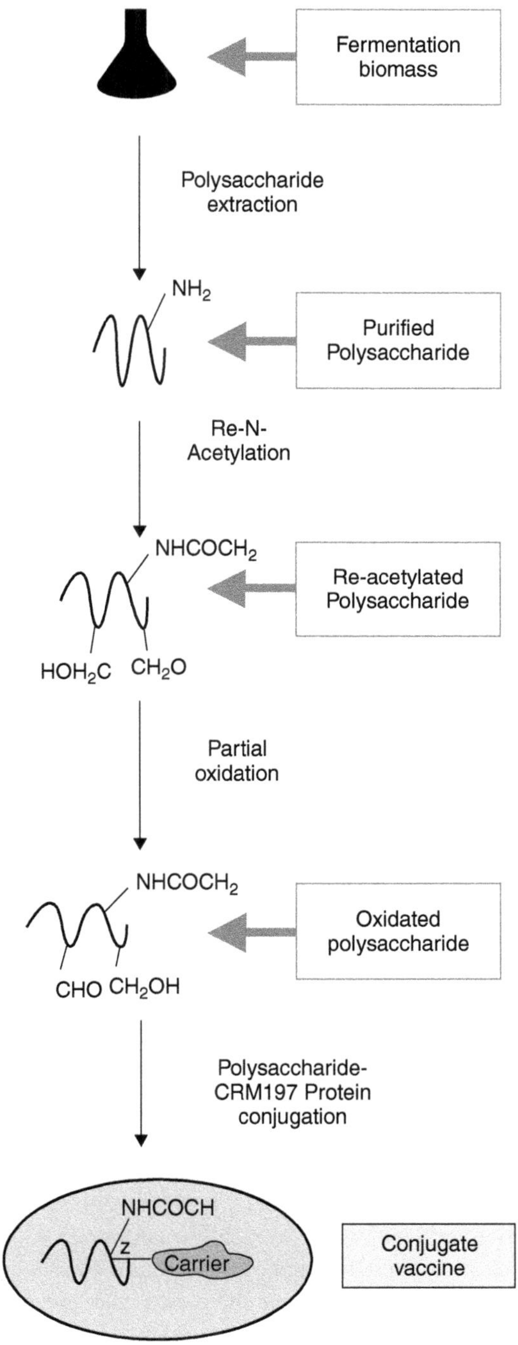

Figure 23.1. General scheme of glycoconjugate vaccine production.

TABLE 23.1. Chromatographic Elution Conditions for Sialic Acid Determination.

Time (min)	0.1 M NaOH	0.1M NaOH, 0.1 M Na Acetate	0.1M NaOH, 0.5 M3 $NaNO_3$
0.0	100	0	0
8.0	100	0	0
8.1	0	90	10
15.0	0	60	40
20.0	0	60	40
20.1	100	0	0
30.0	100	0	0

time and lower concentrations of NaOH. However, CarboPac PA1 did not have the selectivity for the current application. It was necessary to use CarboPac MA1 that uses more concentrated hydroxide eluent to achieve the required selectivity.

23.2 HPAEC-PAD ANALYSIS OF PURIFIED POLYSACCHARIDE

23.2.1 Saccharide Content by NeupNAc Quantification (Serotype Ia)

The saccharide content in purified polysaccharide samples is determined by HPAEC-PAD quantitative analysis of NeupNAc, because at this step the polysaccharide is still in the native form, with NeupNAc residues on the side chains. Samples are hydrolyzed in 100 mM HCl at 80°C for 90 min, neutralized with NaOH and injected into a Dionex system equipped with a CarboPac PA1 column and PA1 guard. The flow rate was 1.0 mL/min. All of HPAEC-PAD assays described in this work are based on a triple potential waveform for detection using the following settings: E1 = 0.05V, t1 = 400 ms; E2 = 0.75V, t2 = 200 ms; E3 = 0.15 V, t3 = 400 ms. Integration occurs from 200 to 400 ms during E1 application. Data are processed by Chromeleon® Software. The chromatographic conditions are reported in Table 23.1.

23.2.2 Free NeupNAc Determination (Serotype Ia)

NeupNAc is the residue involved in the conjugation chemistry and its presence in the structure is important for the immune response to the vaccine. Its importance in immunologic response and its labile chemical bond necessitated an analytical method for determination of free NeupNAc as a stability parameter. Chromatographic conditions are the same as applied for titer determination by NeupNAc HPAEC-PAD analysis. Polysaccharide samples are not pretreated and are loaded on column at ca.1 mg/mL. For glycoconjugate samples a pretreatment of ultrafiltration is applied in order to load only permeate material and avoid a protein overload of the column. Figure 23.2 shows a chromatogram of a free NeupNAc determination of a polysaccharide sample.

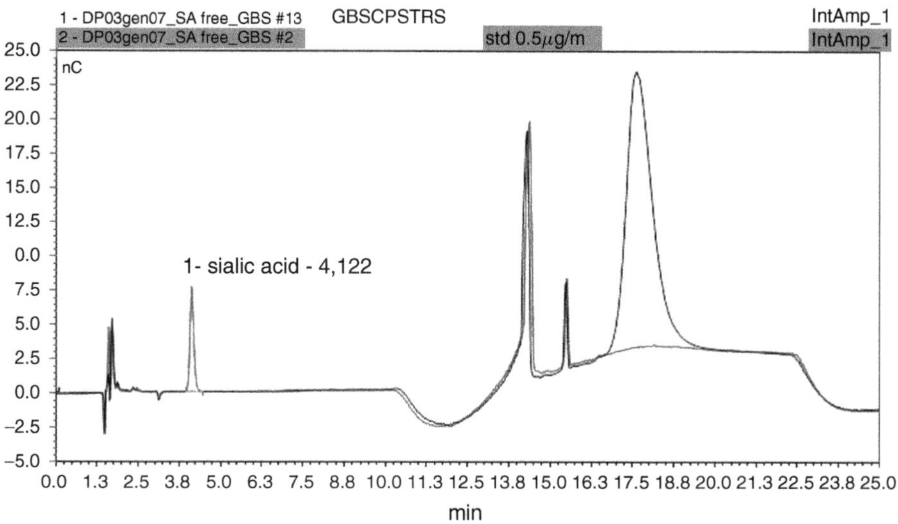

Figure 23.2. Overlay of standard sialic acid injection (blue chromatogram) and sample injection (black chromatogram). The sample does not contain any free sialic acid and the large peak in the regeneration step is the whole polysaccharide.

23.2.3 Saccharide Titer by GlcpNAc Quantification (Serotype V)

The saccharide titer for serotype V is determined by HPAEC-PAD quantitative analysis of GlcpNAc instead of Galp because in the conjugation step the sialic acid is completely removed and the conjugation is performed by activating the Galp residue. Samples are hydrolyzed in 4M TFA for 2 h at 100°C, evaporated in a SpeedVac, resuspended in H_2O, and injected into a Dionex system equipped with a CarboPac PA1 column and guard. Data are processed by Chromeleon® Software. The chromatographic conditions are reported in Table 23.2. During the optimization of the chromatographic method, the isocratic step was changed from 12 mM NaOH to 24 mM NaOH in order to improve the separation of the GlcpNAc peak from the nearby Galp peak. Figure 23.3 shows an example of a chromatogram obtained for GlcpNAc determination of GBS serotype V polysaccharide.

TABLE 23.2. Chromatographic Elution Conditions for Glucosamine Determination.

Time (min)	Deionized Water	0.5 M NaOH	0.1 M NaOH
0.0	76	0	24
20.0	76	0	24
20.1	0	100	0
25.0	0	100	0
25.1	76	0	24
40.0	76	0	24

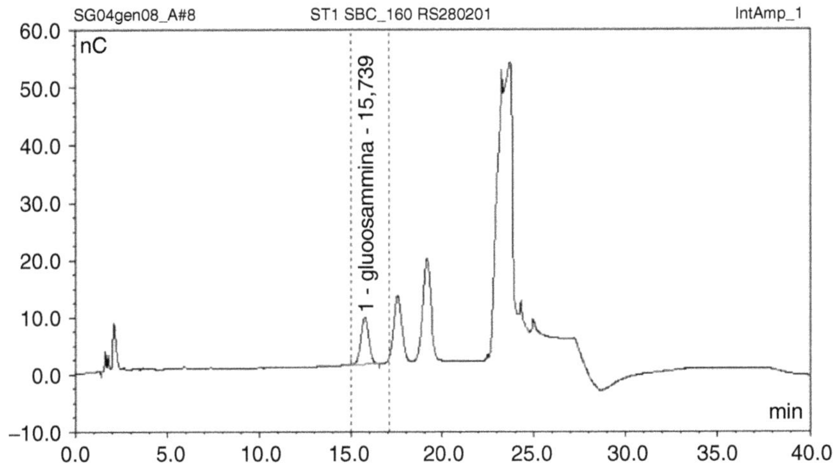

Figure 23.3. Sample of GBS polysaccharide serotype V hydrolyzed. The glucosamine peak is labeled.

TABLE 23.3. Chromatographic Elution Conditions for Rhamnose Determination.

Time (min)	Deionized Water	0.5 M NaOH	0.1 M NaOH
0.0	88	0	12
20.0	88	0	12
20.1	0	100	0
25.0	0	100	0
25.1	88	0	12
45.0	88	0	12

23.2.4 Rhamnose Determination

Rhamnose is a specific saccharide in the group B carbohydrate and it is determined as a possible contaminant residue after capsular polysaccharide purification. Samples and standards are hydrolyzed in TFA 2N for 3 h at 100°C, then evaporated in a SpeedVac and reconstituted with 450 μL of H_2O. The rhamnose standard curve range is 1.0–10.0 μg/mL. Chromatographic conditions: Column CarboPac PA1 with PA1 guard runs at a flow rate of 1.0 mL/min. The eluents settings are reported in Table 23.3. The elution of rhamnose is obtained by isocratic elution with 12 mM NaOH. The purified sample does not show a rhamnose peak or it is below the limit of detection. Due to the presence of other peaks close to the rhamnose peak, the assay includes a spike of standard added to the sample in order to avoid misinterpretation of the chromatogram and to be sure to identify the analyte peak.

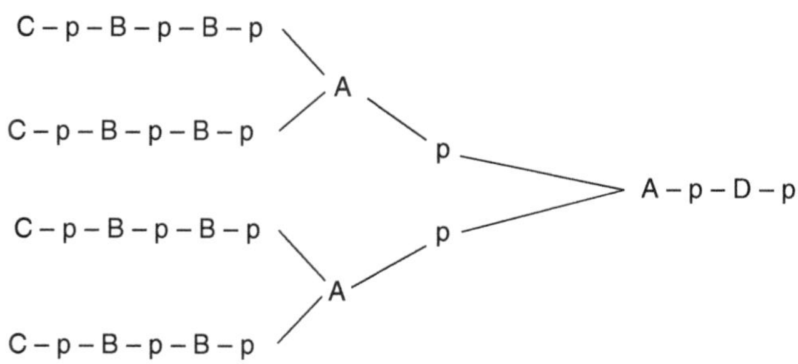

Figure 23.4. Tetra-antennary structure of group B carbohydrate (see the text for descriptions of A, B, C, and D).

Figure 23.4 shows the proposed tetra-antennary structure of group B carbohydrate [7]. A through D represent the four major component oligosaccharides whose structure is described below, and P represents phosphate. The sequences of the four oligosaccharides are as follows:

A: α-L-Rhap-(1→3)-α-D-Galp-(1→3)-β-D-GlcpNAc-(1→4)-α-L-Rhap-(1→2)-[α-L-Rhap-(1→3)-α-D-Galp-(1→3)-β-D-GlcpNAc-(1→4)-]-α-L-Rhap-(1→2)-α-L-Rhap-(1→1″)-D-glucitol(3″→1)-α-L-Rhap. **MW∼1790 Da**

B: α-L-Rhap-(1→2)-[α-L-Rhap-(1→3)-α-D-Galp-(1→3)-β-D-GlcpNAc-(1→4)-]-α-L-Rhap-(1→2)-α-L-Rhap-(1→1″)-D-glucitol(3″→1)-α-L-Rhap. **MW∼1278 Da**

C: α-L-Rhap-(1→2)-α-L-Rhap-(1→2)-α-L-Rhap-(1→1″)-D-glucitol(3″→1)-α-L-Rhap. **MW∼767 Da**

D: α-L-Rhap-(1→3)-α-D-Galp-(1→3)-β-D-GlcpNAc-(1→3)-α-L-Rhap-(1→3)-α-L-Rhap-(1→3)-α-L-Rhap-(1→3)-β-L-Rhap-(1→4)-D-GlcNAc. **MW∼1317 Da**

The group B carbohydrate also contains additional minor component oligosaccharides. Figure 23.5 displays chromatographic results of the rhamnose assay of the purified polysaccharide. Observe that there is no detectable rhamnose. To confirm that the assay can detect rhamnose (i.e. not column overload, etc.) the sample was spiked with rhamnose standard and Figure 23.5 shows that it is easily detected.

23.3 HPAEC-PAD ANALYSIS OF GLYCOCONJUGATES

23.3.1 Total and Free Saccharide Determination in Glycoconjugates

Separation of free/unbound polysaccharide is achieved using an SPE cartridge by loading 1.0 mL of the bulk concentrate sample, that was first diluted in 5 mM sodium

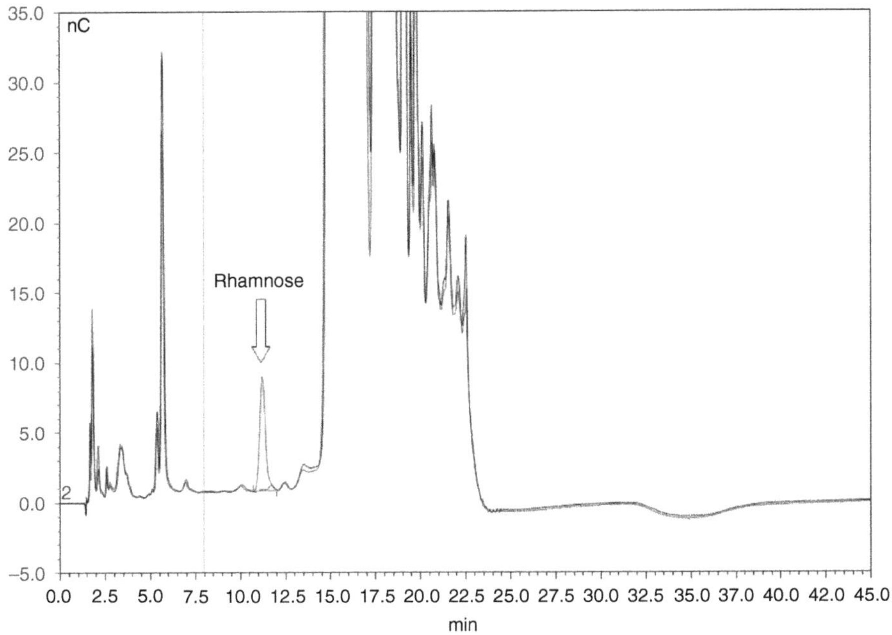

Figure 23.5. The sample is a purified GBS polysaccharide treated as described. The sample does not present a rhamnose peak. The blue chromatogram is obtained by adding rhamnose standard to the sample.

phosphate/NaCl 2.0% pH 7.2 to a final concentration of 20 μg/ml in protein, on a C4 cartridge (Isolute Vydac), and eluting the free/unbound polysaccharide with 1.0 mL of 5% acetonitrile/0.05% TFA. The sample for the determination of total saccharide titer is diluted in water. A Galp calibration standard curve is prepared ranging from 0.5 to 8.0 μg/mL. 450 μL of each sample and standard prepared as described above are combined with 450 μL 8M TFA and incubated in an oven at 100°C for 2 h. After hydrolysis the samples are evaporated in a SpeedVac, resuspended in 450 μL deionized water, filtered with a 0.45 μm filter, and injected into a Dionex ICS3000 system equipped with CarboPac PA1 column with guard. Chromatographic conditions are as follows: flow rate 1.0 mL/min; 20 min isocratic elution with 12 mM NaOH, 5 min washing with 500 mM NaOH, and 15 min column re-conditioning. The effluent is monitored using an electrochemical detector in the pulsed amperometric mode with a gold working electrode and an Ag/AgCl reference electrode. A triple-potential waveform was applied using the following settings: E1 = 0.05V, t1 = 400ms; E2 = 0.75V, t2 = 200ms; E3 = 0.15V, t3 = 400ms. Integration occurs from 200 to 400 ms during E1 application. The resulting chromatographic data are integrated and processed using Chromeleon software. Calibration curves, treated as samples, were set up with galactose in the range of 0.5–8.0 μg/mL. Figure 23.6 reports examples of chromatograms obtained for serotype Ia glycoconjugate.

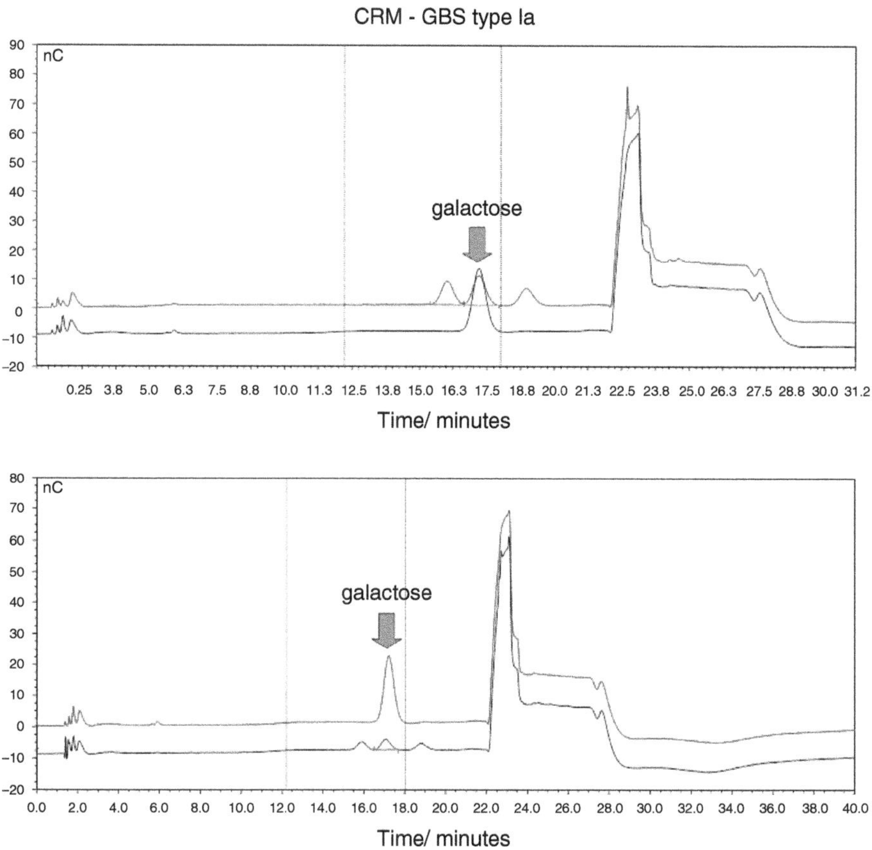

Figure 23.6. Galactose determination in concentrate bulk serotype Ia. Total and free saccharide chromatograms are reported. An overlay with galactose standard is reported in order to identify the galactose peak in the sample.

23.3.2 Total and Free Saccharide Determination in Lyophilyzed Glycoconjugates

The method developed for bulk concentrate samples is not applicable to lyophilized samples because of the presence of large amounts of the excipient (mannitol) and related interfering peaks. Another chromatographic method was developed using the CarboPac MA1, more specific for Galp, by delivering an optimal separation from matrix peaks. The chromatographic conditions are as follows: flow rate of 0.4 mL/min, eluent of 350 NaOH mM for 40 min. Detection is performed as described in the previous paragraph. Figure 23.7 shows examples of total and free saccharide determinations in lyophilized CRM-GBS serotype Ia. The total saccharide profile is shown in the upper panel of the figure, and the free saccharide chromatogram is in the lower panel. The chromatogram overlay with galactose standard highlights the huge amount of interfering substances present in the sample profiles in comparison with the standard

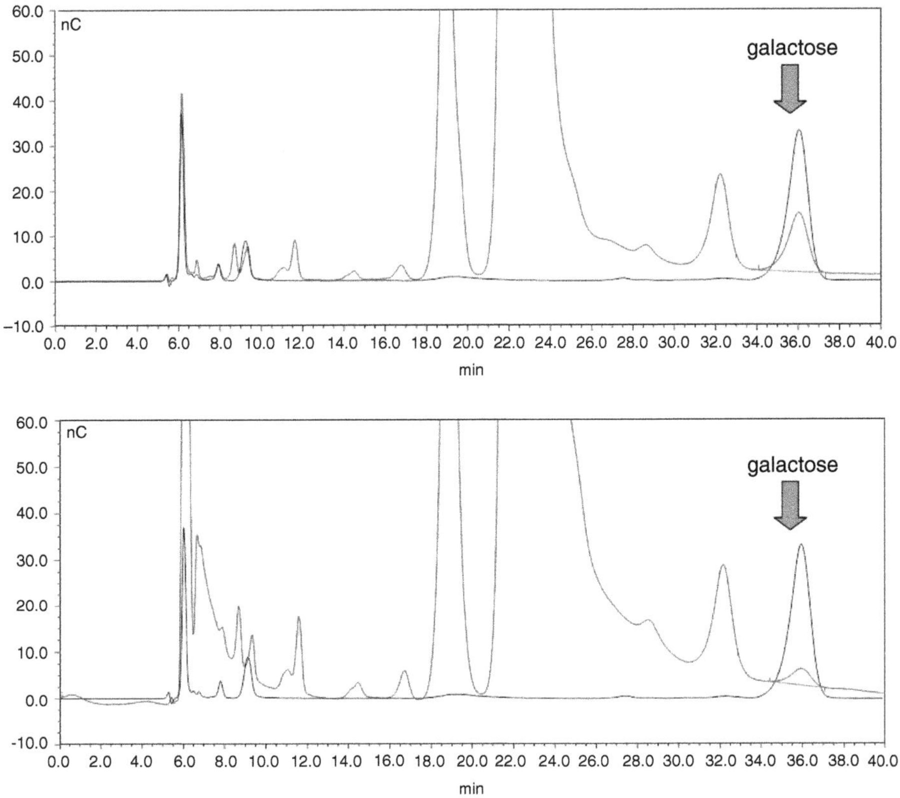

Figure 23.7. Determination of galactose in lyophilized CRM-GBS serotype Ia glycoconjugate by CarboPac MA1 chromatography. In each figure the black chromatogram is the galactose standard and the blue chromatogram is the sample. The upper panel is total saccharide determination and the lower panel is the free saccharide determination.

chromatogram. The interfering material originates from mannitol and its impurities. In terms of amount, the excipient is in the order of milligrams while the antigen is in the order of micrograms. During the development of the analytical method, attempts were made to remove mannitol by ultrafiltration, but sample recoveries were too low. For this reason, method optimization focused on chromatographic conditions.

23.4 CONCLUSIONS

The production of GBS-CRM glycoconjugates is a complex process starting from fermentation of the bacteria and passing through chemical treatment, purification steps, polysaccharide modification, and conjugation reaction. HPAEC-PAD is a suitable technology in order to follow the different steps of the process. In this work several applications of HPAEC-PAD are presented. The main goal of the technique is the titer determination of purified polysaccharide and final conjugates, bulk concentrate, and final formulated product. Depending on the serotype, a single monosaccharide is chosen for assay by HPAEC-PAD. For serotype Ia, NeupNAc is the analyte indicative of titer in purified polysaccharide, while for serotype V GlcpNAc is used, because of the removal of NeupNAc from the structure of the purified polysaccharide. Other applications of HPAEC-PAD are used for determination of oxidation percentage in serotypes Ia and V (not shown) and are based on the identification of specific molecules formed during the chemical steps preceding the conjugation reaction. After conjugation, titer determination is based on Galp for serotype Ia because of the modification applied on the NeupNAc in the conjugation step. For serotype V, titer determination is still based on the GlcpNAc molecule and the chromatographic conditions are suitable from purified polysaccharide to final lyophilized CRM-GBS product. In the final formulated product of serotype Ia Galp is still the analyte for titer determination, but because of the presence of a high concentration of mannitol, new chromatographic conditions had to be developed for this sample. Therefore, while the method applied on bulk concentrate is based on CarboPac PA1 column, the method for final formulated product is based on CarboPac MA1 column.

Other applications described in this chapter are for the analysis of rhamnose and free NeupNAc. Rhamnose content is determined in order to check the efficiency of purification of the polysaccharide. Free NeupNAc, determined in purified polysaccharide and also in bulk concentrate glycoconjugates, is checked as a stability parameter in serotype Ia conjugation because of its importance for the immunoresponse to the vaccine.

REFERENCES

1. Rench MA, Baker CJ. Neonatal sepsis caused by a new group B streptococcal serotype. Journal of Pediatrics 1993;122(4):638–640.
2. Farley MM. Group B streptococcal disease in nonpregnant adults. Clinical Infectious Diseases 2001;33:556–561.
3. Von Hunolstein C, D'Ascenzi S, Wagner B, Jelinkova J, Alfarone G, Recchia S, Wagner M, Orefici G. Immunochemistry of capsular type polysaccharide and virulence properties of type VI Streptococcus agalactiae (Group B Streptococci). Infection and Immunity 1993;61(4):1272–1280.
4. Ricci S, Bardotti A, D'Ascenzi S, Ravenscroft N. Development of a new method for the quantitative analysis of the extracellular polysaccharide of Neisseria meningitidis serogroup A by use of high-performance anion-exchange chromatography with pulsed-amperometric detection. Vaccine 2001;19:1989–1997.
5. Bardotti A, Averani G, Berti F, Berti S, Galli C, Giannini S, Fabbri B, Proietti D, Ravenscroft N, Ricci S. Size determination of bacterial capsular oligosaccharides used to prepare conjugate vaccines against Neisseria meningitidis groups Y and W135. Vaccine 2005;23:1887–1899.
6. D'Ascenzi S, von Hunolstein C, Dentini M, Alfarone G, Crescenzi V, Orefici G. Characterization of group B streptococcus type capsular polysaccharide. New perspective on

streptococci and streptococcal infections, XI Lancefield International Symposium, Siena, September 10–14, 1990, pp. 458–460.

7. Deng L, Kasper DL, Krick TP, Wessels MR. Characterization of the linkage between the type III capsular polysaccharide and the bacterial cell wall of group B Streptococcus. J. Biol. Chem. 2000;275:7497–7504.
8. Bardotti A, Ravenscroft N, Ricci S, D'Ascenzi S, Guarnieri V, Averani G, Costantino P. Quantitative determination of saccharide in Haemophilus influenzae type b glycoconjugate vaccines, alone and in combination with DPT, by use of high-performance anion-exchange chromatography with pulsed amperometric detection. Vaccine 2000;18(19):1982–1993.
9. Bardotti A, Averani G, Berti F, Berti S, Carinci V, D'Ascenzi S, Fabbri B, Giannini S, Giannozzi A, Magagnoli C, Proietti D, Norelli F, Rappuoli R, Ricci S, Costantino P. Physicochemical characterization of glycoconjugate vaccines for prevention of meningococcal diseases. Vaccine 2008;26(18):2284–2296.

24

STABILITY STUDIES AND CHARACTERIZATION OF MENINGOCOCCAL A, C, Y, AND W_{135} GLYCOCONJUGATE VACCINE

Valeria Carinci, Alberto Bernardini, Cristiana Campa, Claudia Magagnoli, Francesca Beccai, and Sandro D'Ascenzi

Technology Development Department, Novartis Vaccines and Diagnostics S.r.L., Siena, Italy

24.1 INTRODUCTION

Neisseria Meningitidis is a major cause of meningitis and sepsis in the world. The majority of meningococcal diseases are caused by the capsular groups A, B, C, Y, and W_{135}, although the precise epidemiological contribution varies from country to country. Prevention through vaccination is an important component for meningococcal disease containment.

Although polysaccharide meningococcal vaccines against groups A, C, Y, or W_{135} have been available since the 1960s, they are generally poorly immunogenic in children or require repeated doses and do not produce long-lasting immunity. This is because they are T cell-independent antigens not able to induce B cell memory. Chemical conjugation of purified poly- or oligosaccharides to a carrier protein results in a vaccine that induces a T cell response, which is immunogenic in young children and induces immunologic memory.

Applications of Ion Chromatography for Pharmaceutical and Biological Products, First Edition.
Edited by Lokesh Bhattacharyya and Jeffrey S. Rohrer.

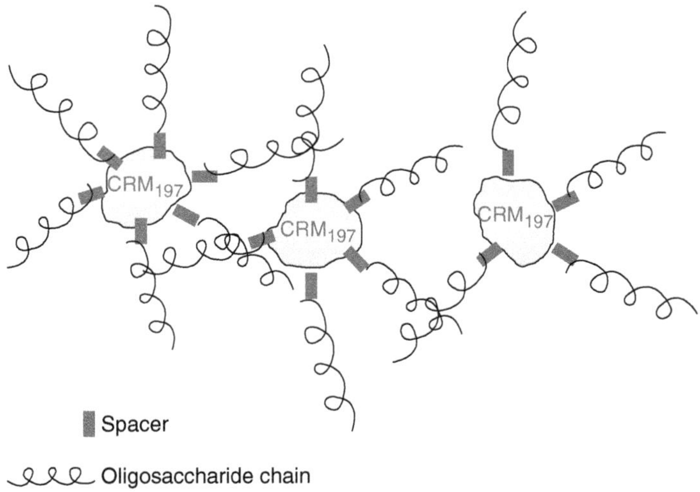

Figure 24.1. Schematic structure of glycoconjugate molecules: oligosaccharide chains covalently attached to a non-toxic mutant of the diphtheria toxoid CRM_{197}.

In our laboratories a tetravalent vaccine against serogroups A, C, Y, and W_{135} is produced from their respective capsular oligosaccharide covalently attached to a non-toxic mutant of the diphtheria toxoid CRM_{197} (Figure 24.1). The stability of the MenA (meningococcal A), C, Y, and W_{135} conjugates and the integrity of their saccharide and protein components are relevant factors for vaccine immunogenicity. Different analytical methods have been utilized to verify the quality of these vaccines to ensure their immunogenicity and compliance with manufacturing specifications [1–2]. One of these methods is an ion chromatography method, high performance anion-exchange chromatography coupled with pulsed amperometric detection (HPAEC-PAD) [3].

The molecular size and polidispersity of different lots of the bulk CRM_{197} conjugates of meningococcal A, C, Y, and W_{135} were evaluated by determining their average molar masses and distribution coefficients (K_D) using HPLC size exclusion chromatography (SEC) coupled with multi-angle laser light scattering (MALLS), refractive index (RI), or UV detection [4–6]. The structural stability and molecular integrity of the final tetravalent MenACYW vaccines were monitored by determining their total and free saccharide contents by HPAEC-PAD and a colorimetric method. The identification of the different meningococcal A, C, Y, and W_{135} serogroups in the final tetravalent formulation was obtained by capillary zone electrophoresis (CZE).

24.2 MATERIALS AND METHODS

24.2.1 Glycoconjugate Samples

In the characterization study, bulk concentrated lots of CRM_{197} conjugates of different meningococcal A, C, Y, and W_{135} polysaccharides and the final tetravalent formulation were used.

24.2.2 Chromatographic Condition for K_D Determination

SEC-HPLC analysis was performed using an Alliance Waters 2996 system coupled with a phododiode array detector (PDA). Empower Waters chromatography software was used to program runs and to perform data analysis. A TSK gel G4000SWXL TosoHaas column equipped with a TSKgel SWXL guard column was used at room temperature while the samples were maintained in the autosampler at 2–8°C.

24.2.3 Chromatographic Conditions for SEC/MALLS Analysis

The SEC/MALLS analysis was performed using an Alliance Waters 2996 system connected in series with a Dawn EOS multi-angle light scattering photometer (Wyatt Technology Corp.) to measure scattered light intensities and an Optilab DSP (Wyatt Technology Corp.) interferometric refractometer to measure concentrations. The refractive index (RI) signal is proportional to concentration while MALLS signal is proportional to concentration and molar mass. The *dn/dc* is the change in the refractive index (n) of a polymer solution relative to the change of the polymer concentration (c) and was experimentally determined using the same interferometric refractometer. Data were analyzed using ASTRA™ software (Wyatt Technology Corp.) [7–8].

Two high-performance columns with different size exclusion ranges were used: an Ultrahydrogel™ 1000 column (Waters) serially equipped with an Ultrahydrogel™ 250 and Ultrahydrogel™ guard column. The columns were run at room temperature while the samples were maintained in the autosampler at 2–8°C.

24.2.4 HPAEC-PAD Analysis of MenACYW$_{135}$ Vaccine

For the tetravalent vaccine the total and free saccharide resulting from MenA-CRM$_{197}$ component is determined by HPAEC-PAD. The method is based on trifluoroacetic acid (TFA) hydrolysis of the MenA-oligosaccharides and determination of the resulting mannosamine-6-phosphate (ManN6P).

24.2.5 Colorimetric Analysis of MenACYW$_{135}$ Vaccine

The cumulative total and free sialic acid resulting from meningococcal MenC, MenY, and MenW conjugates are determined by a colorimetric method. The assay is based on the reaction between sialic acid and resorcinol at 100°C in the presence of hydrochloric acid and copper (II) ions for 40 min; this leads to the formation of a blue-violet complex which exhibits a strong absorbance at 564 nm. The total sialic acid concentration is determined from a calibration curve obtained using a series of sialic acid standards.

24.2.6 Capillary Zone Electrophoresis (CZE)

The CZE analysis was performed using a ProteomeLab 800 CE System (Beckman Coulter) equipped with an UV detector. Data acquisition and analysis is accomplished by 32 Karat software (Beckman Coulter).

The following electrophoretic conditions were used:

1. Capillary: Uncoated fused silica, L = 80 cm, l = 72 cm, i.d. 50 μm (kept at 25°C).
2. Capillary pre-treatment – water (MilliQ) rinse (5 min at 50 psi) was followed by a 70/30 water/methanol rinse (10 min at 50 psi) then by a basic rinse (0.1M NaOH, 10 min at 50 psi). This was followed by a running buffer rinse (4 min at 50 psi). This procedure was repeated before each sample injection. The buffer solution was replaced at four runs interval.
3. Separation voltage and sample storage temperature: 15 kV (normal polarity) and 25°C.
4. Running buffer: Na_2HPO_4, $2H_2O$: 90 mM – H_3BO_3: 10 mM, pH 11.5.
5. Injection: Hydrodynamic, 0.5 psi for 30.
6. UV detection: 200 nm

24.3 RESULTS AND DISCUSSION

24.3.1 Dimensional Characterization of Meningococcal Glycoconjugate Using HPLC Size-Exclusion Chromatography (SEC/MALL)

Dimensional analyses of the glycoconjugates vaccine components are of extreme importance because they allow the analyst to monitor stability over time, monitor batch to batch consistency, and evaluate the molecular constitution in solution. The molecular sizes of different lots of meningococcal A-, C-, Y-, and W_{135}–CRM_{197} conjugates were evaluated by determining their average molar masses and distribution coefficients (K_D) using HPLC-SEC coupled with MALLS, RI, or UV detection. HPLC-SEC is a chromatographic technique for the separation of mixtures based on the molecular size of the components. K_D values were calculated according to the following formula:

$$K_D = (T_e - T_0)/(T_t - T_0)$$

where, T_0 is the time corresponding to the void volume of the column as determined by Blue Dextran 2000 and T_t is the time corresponding to the total permeation volume as determined by sodium azide. The elution time T_e of glycoconjugates vaccines was calculated from the peak of the absorbance curve measured at 214 nm. The K_D determination is a rapid and reproducible method for determining the molecular size distributions of conjugate vaccines but it is dependent on the elution behavior of these macromolecules on chromatography columns.

SEC with in-line MALLS and RI detector is a more complex technique that allows the analyst to obtain the absolute molecular mass of the conjugate according to light scattering equation:

$$I_{\text{scattered}} \propto Mc \left(\frac{dn}{dc}\right)^2$$

where the amount of light scattered is directly proportional to the product of the polymer molar mass M, concentration **c** and to the square of the specific refractive index increment *dn/dc* [9].

The *dn/dc* value has to be accurately measured by a differential refractometer. During *dn/dc* determination it is important to use the same solvent, wavelength, and temperature employed in chromatographic analysis to evaluate the molar mass by light scattering. Measured *dn/dc* values obtained for MenA, C, Y, and W_{135} – CRM_{197} conjugates are shown in Table 24.1.

The molecular weights (M_w) and polydispersity (M_w/M_n) values, and distribution coefficients (K_D) of different lots of bulk MenA -CRM_{197}, MenC -CRM_{197}, MenY -CRM_{197} and MenW -CRM_{197} conjugates were determined. The results are summarized in Tables 24.2–24.5. The data reported in the tables confirm the batch to batch consistency of the glycoconjugates.

The elution profiles from RI (smaller peak) and the 90° MALLS detectors (larger peak) of bulk MenA -CRM_{197}, MenC -CRM_{197}, MenY -CRM_{197} and MenW -CRM_{197} conjugates are shown in Figures 24.2–24.5, respectively. The vertical bars indicate the

TABLE 24.1. *dn/dc* Values.

MenA-CRM_{197}	MenC-CRM_{197}	MenY-CRM_{197}	MenW-CRM_{197}
0.163	0.190	0.201	0.159

TABLE 24.2. Dimensional Results of MenA - CRM_{197}.

MenA CRM_{197} (Lot #)	Mean Mw (g/mole × 10^5)	Polydispersity Mw/Mn	K_D
1A	1.10	1.1	0.40
2A	1.25	1.1	0.39
3A	1.30	1.1	0.38
4A	1.15	1.1	0.40
5A	1.20	1.1	0.37

TABLE 24.3. Dimensional Results of MenC - CRM_{197}.

MenA CRM_{197} (Lot #)	Mean Mw (g/mole × 10^5)	Polydispersity Mw/Mn	K_D
1C	8.48	1.0	0.38
2C	9.50	1.0	0.38
3C	9.23	1.0	0.38
4C	9.40	1.0	0.38

TABLE 24.4. Dimensional Results of MenY - CRM_{197}.

MenY CRM_{197} (Lot #)	Mean Mw (g/mole × 10^5)	Polydispersity Mw/Mn	K_D
1Y	1.03	1.1	0.36
2Y	1.00	1.1	0.37
3Y	1.14	1.2	0.35

TABLE 24.5. Dimensional Results of MenW - CRM_{197}.

MenW CRM_{197} (Lot #)	Mean Mw (g/mole × 10^5)	Polydispersity M_w/M_n	K_D
1 W	1..57	1.2	0.29
2 W	1.68	1.2	0.30
3 W	1.54	1.2	0.29

integration limits for the peaks and the trace with the larger peak is always from the MALLS detector. All MenA-CRM_{197} lots eluted as a single peak with a shoulder on the higher-molecular-mass side (Figure 24.2). This evidence, indicating the presence in the sample of a small amount of high-molecular-mass glycoprotein(s), was supported by polydispersity M_w/M_n values of around 1.1. The majority of MenC-CRM_{197} is

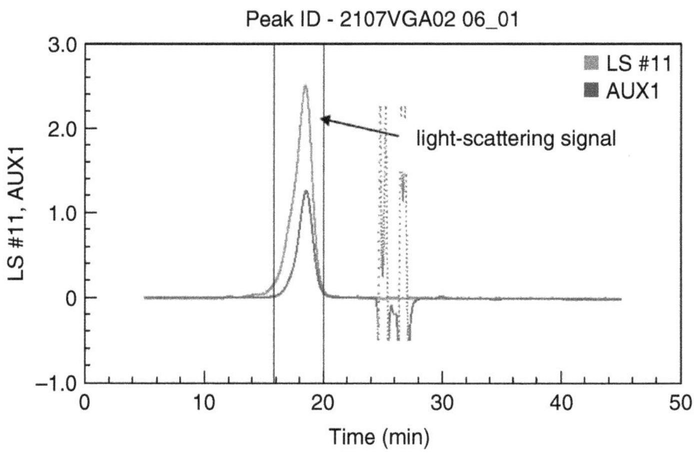

Figure 24.2. Typical SEC/MALLS elution profiles from light-scattering and RI detectors for MenA-CRM_{197}. The vertical bars indicate the integration limits for the peak. Samples were eluted at a flow rate of 0.8 mL/min in 0.2M sodium phosphate buffer pH 7.0 on tandem Ultrahydrogel™ 1000 and 250 columns.

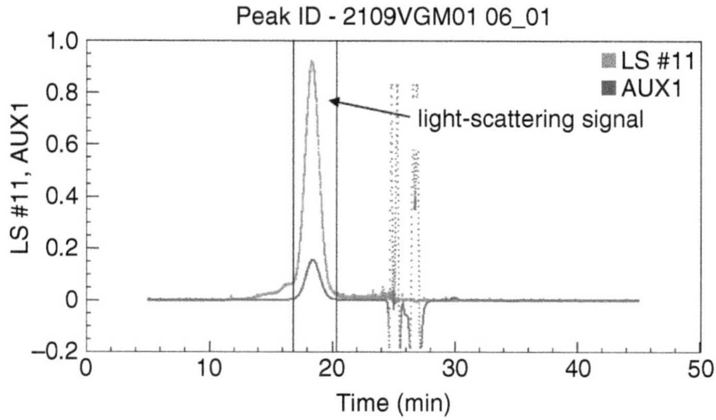

Figure 24.3. Typical SEC/MALLS elution profiles from light-scattering and RI detectors for MenC-CRM_{197}. The vertical bars indicate the integration limits for the peak. Conditions were the same as Figure 24.2.

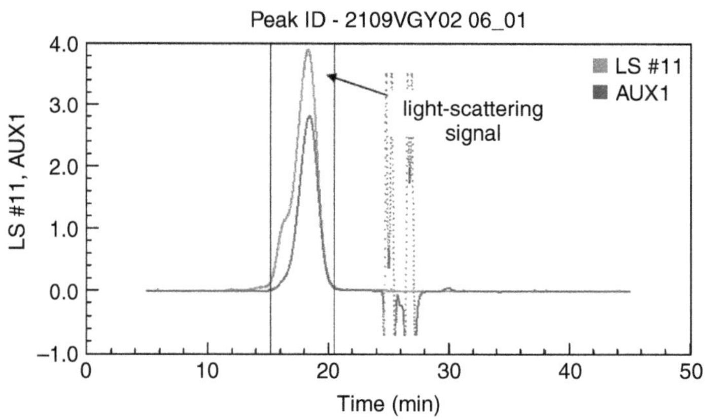

Figure 24.4. Typical SEC/MALLS elution profiles from light-scattering and RI detectors for MenY-CRM_{197}. The vertical bars indicate the integration limits for the peak. Conditions were the same as Figure 24.2.

eluted as a single symmetrical peak (Figure 24.3) corresponding with a monodisperse solution having an M_w/M_n value of 1.00. The chromatographic profiles of the MenY and MenW glycoconjugates show a single light-scattering peak with a shoulder on the higher-molecular-mass side which did not give a distinct RI signal, indicating the presence of a small amount of high molecular mass material.

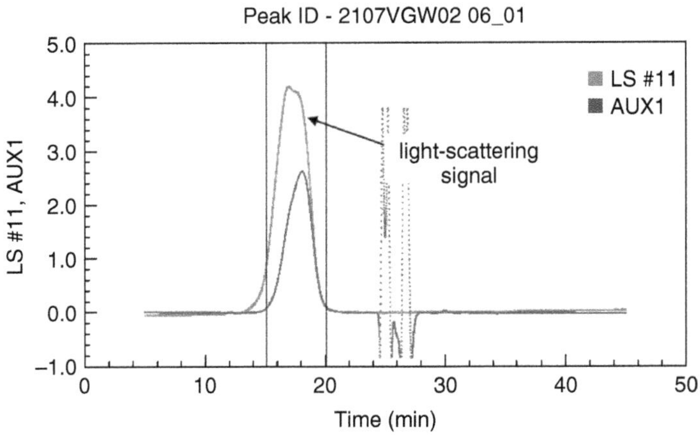

Figure 24.5. Typical SEC/MALLS elution profiles from light-scattering and RI detectors for MenW-CRM_{197}. The vertical bars indicate the integration limits for the peak. Conditions were the same as Figure 24.2.

24.3.2 Determination of Total and Free Saccharide Resulting from the MenA-CRM Component in the Tetravalent $MenACYW_{135}$ Vaccine

The capsular polysaccharide of serogroup A is a homopolymer of α-(1→6)-linked N-acetylmannosamine phosphate. We developed and validated a HPAEC-PAD analytical method for monitoring the concentration and stability of this conjugate in the final vaccine formulation with sucrose as stabilizer. The acidic hydrolysis of the MenA oligosaccharide is followed by chromatographic separation and quantification of the resulting ManN6P monomer from the area of the peak obtained using a CarboPac PA10 column coupled to a pulsed amperometric detector. Free saccharide was separated from meningococcal conjugates by solid phase extraction (SPE) using C4 cartridges. The concentration of free MenA saccharide (CRM unbound) indicates the stability of the product. The data obtained confirmed the stability of MenA component in tetravalent vaccine when stored at the recommended temperature for more than a year. The validation has shown that this method is highly specific for the quantification of the MenA in the tetravalent $MenACYW_{135}$ formulation. Figure 24.6 shows an overlay of the HPAEC-PAD chromatographic profiles of the blank, matrix, and the standard samples from the specificity test. No peaks were detected at the retention time of ManN6P in the chromatograms of blank and matrix samples. The standard curves were linear over the concentration range 1.0–8.0 μg/mL. In this range the correlation coefficient was greater than 0.99. Figure 24.7 shows an overlay of the HPAEC-PAD chromatograms of the different ManN6P standard points.

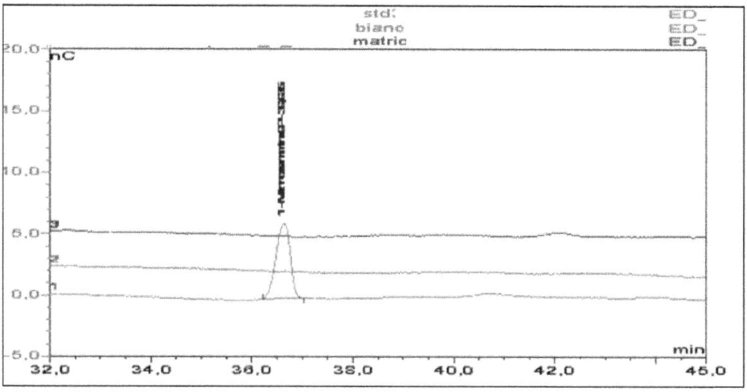

Figure 24.6. HPAEC-PAD chromatograms of blank, matrix, and standard on CarboPac PA10 column in the Specificity Test. The chromatographic analyses were performed with a flow rate of 1.0 mL/min and a run time of 55 min by a gradient mode using 100 mM NaOH and 1M sodium acetate/100 mM NaOH. The peak is the mannosamine-6-P standard (2.0 μg/mL).

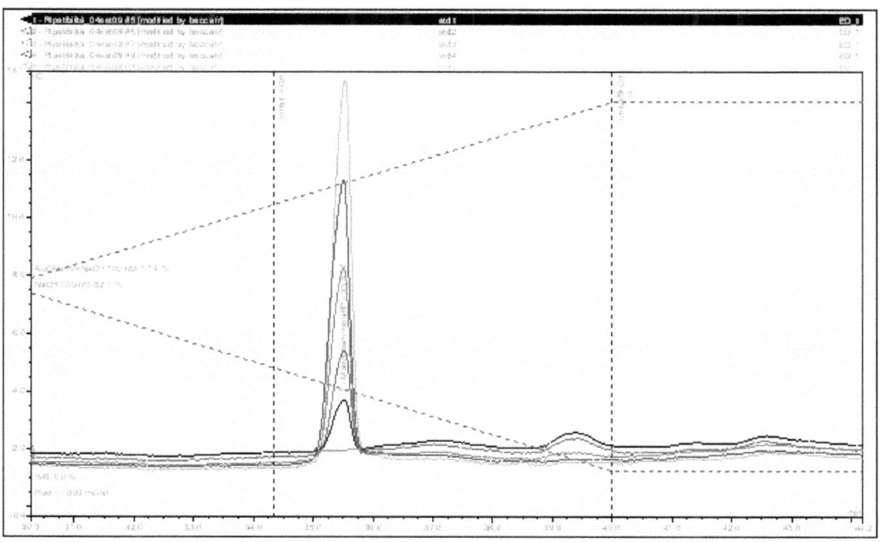

Figure 24.7. Overlay of HPAEC-PAD chromatograms of ManN6P standard points. The amount of total MenA saccharide is extrapolated using a calibration standard curve constituted of 5 Man-6-P standard points (1.0 μg/mL–2.0 μg/mL–4.0 μg/mL–6.0 μg/mL–8.0 μg/mL). The linearity is good over the range of concentrations tested with correlation coefficient greater than 0.99.

24.3.3 Determination of Total and Free Sialic Acid of MenC-CRM, $MenW_{135}$-CRM and MenY-CRM Components in Tetravalent $MenACYW_{135}$ Vaccine

The capsular polysaccharide of group C is a homopolymer of α-(2→9)-linked polysialic acid, the capsular polysaccharides of group Y consist of a repeating unit of →4-O-α-D-Glcp-(1→6)-β-D-NeuAc-2→, and the capsular polysaccharides of group W135 consist of a repeating unit of →4–O-α-D-Galp-(1→6)-β-D-NeuAc-2→. The capsular polysaccharides are hydrolyzed to oligosaccharide chains of defined length and conjugated with CRM_{197} protein. The quantification of sialylated single component by HPAEC-PAD is hindered because of the glucose formation from the sucrose added as a stabilizer. Therefore the determination of MenC, MenY, and $MenW_{135}$ total and free sialic acid is performed by the modified Svennerholm colorimetric method [10]. The method is based on the reaction between sialic acid and resorcinol at 100°C in the presence of hydrochloric acid and copper (II) ions for 40 min; this leads to the formation of a blue-violet complex which exhibits a strong absorbance at 564 nm. The removal of sucrose by ultrafiltration from reconstituted MenACYW samples is a necessary step because of its interference at 564 nm. Spectra of aqueous solutions of sucrose at different concentrations have shown the high absorption and therefore interference of sucrose at 564 nm (Figure 24.8). The separation of free sialic acid

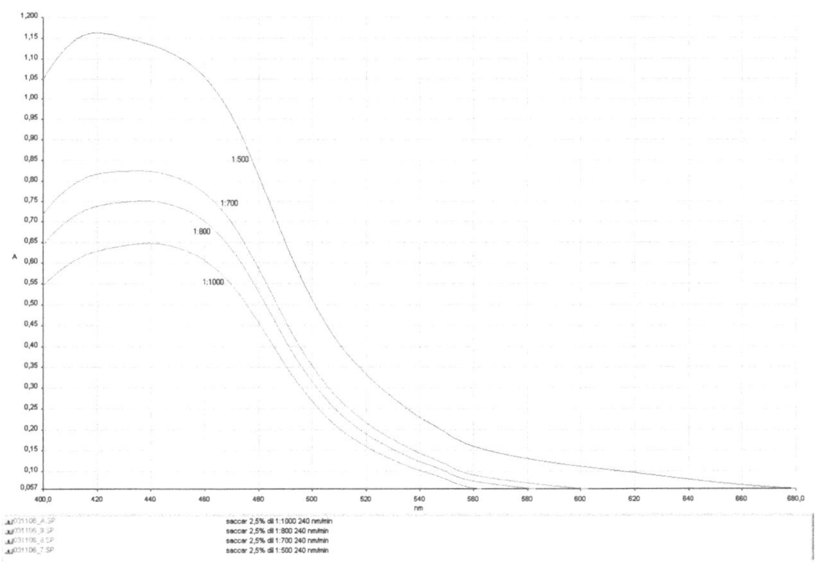

Figure 24.8. Overlay of spectra of sucrose solutions at different concentrations. Spectra of aqueous solutions of sucrose at different concentrations have shown the high absorption and therefore interference of sucrose to wavelength of 564 nm.

is obtained by preliminary SPE using C4 cartridges (200 mg) to retain CRM-bound saccharide and then ultrafiltration to remove the sucrose.

24.3.4 Meningococcal Characterization in Tetravalent $MenACYW_{135}$ Vaccine Using Capillary Zone Electrophoresis

The molecular separation of different glycoconjugates in MenACYW vaccine is obtained by CZE. Capillary electrophoresis (CE) methods are based on the differential migration of charged molecules in an electric field (similar to traditional polyacrylamide gel electrophoresis), allowing a separation selectivity similar (and often complementary) to HPLC. In CE, electrophoretic migration and electroosmotic flow are combined for the separation of charged molecules. In CZE, the simplest form of CE, the separation mechanism is based on differences in the charge-to-mass ratio.

One major application of this analytical tool is the characterization of glycoprotein variants. As shown in Figure 24.9 it is possible to identify the glycoconjugates comprising the MenACYW vaccine by a single CZE analysis. No sample pre-treatments were performed before injection in the CE system. Overlaid electropherograms of MenACYW vaccine and the concentrated bulk of each glycoconjugate are shown in Figure 24.10. The concentrations of glycoconjugates in the bulk and MenACYW vaccine are different.

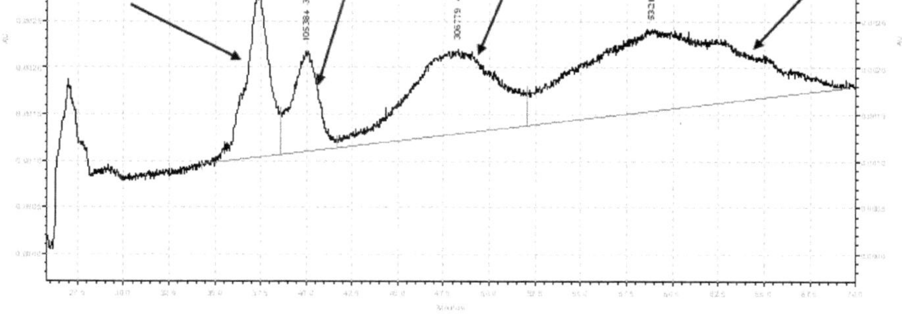

Figure 24.9. CZE analysis of MenACYW vaccine: UV detection 200 nm. It is possible to identify the single glycoconjugates that comprise the MenACYW vaccine by capillary electrophoresis analysis. No sample pre-treatments were performed on the sample before injection in the CE system.

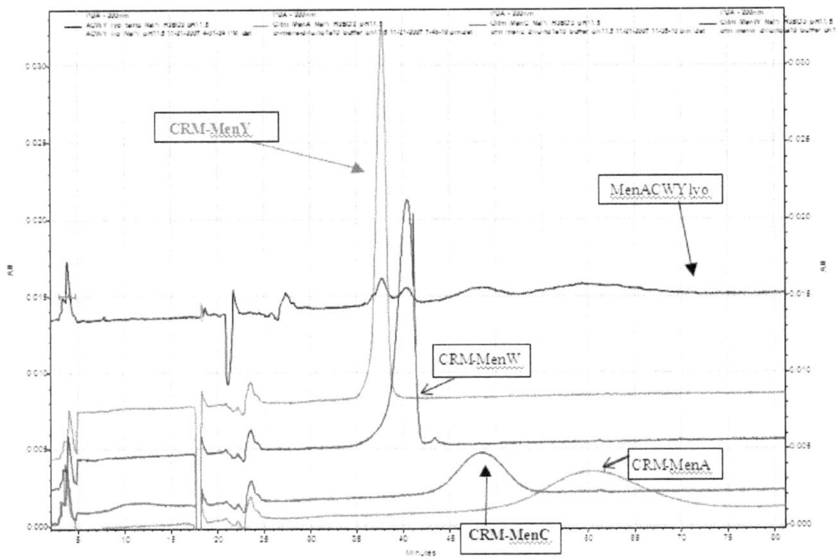

Figure 24.10. Overlay of electrophoregrams obtained from MenACYW vaccine and single concentrated bulks. The single glycoconjugates show the same migration time in the MenACYW formulation.

24.4 CONCLUSION

Different analytical methods are used to monitor the integrity and stability of the MenA, C, Y, and W_{135} meningococcal conjugates vaccines both in the bulk and in the tetravalent MenACYW formulation. The data showed that HPLC-SEC coupled with PDA, RI, and MALLS detection can be used for the dimensional characterization of the conjugate vaccines. In particular the SEC analysis is useful for monitoring the stability of the meningococcal conjugate vaccines and the consistency of their production. A HPAEC-PAD method was developed and validated to monitor the total and free saccharide resulting from the MenA-CRM component in tetravalent $MenACYW_{135}$ vaccine. There was no detectable change in free saccharide content in the final MenACYW vaccine when stored at the recommended temperature for more than a year. A modified Svennerholm colorimetric method was used to determine the total and free sialic acid resulting from MenC-CRM, $MenW_{135}$-CRM, and MenY-CRM components in tetravalent MenACYW135 vaccine. The molecular separation of the different glycoconjugates in $MenACYW_{135}$ vaccine is obtained by CZE. The applicability of this method for a quantitative determination of each single meningococcal glycoconjugate in the final $MenACYW_{135}$ vaccine is under evaluation.

REFERENCES

1. D'Ascenzi S, Vaccines: Characterisation and analytical tools. Vaccines Quality Control, Scale Up and Manufacture. Informa Life Sciences Meeting. 5–6 December 2006 Hamburg Germany.
2. Mentasti E, Saini G. Analisi Chimica Cromatografia. Padova: Ed. Piccin; 1990;part II:15–197.
3. Ricci S, Bardotti A, D'Ascenzi S, Ravenscroft N. Development of a new method for the quantitative analysis of the extracellular polysaccharide of Neisseria meningitidis serogroup A by use of high–performance anion–exchange chromatography with pulsed–amperometric detection. Vaccine 2001;19(15–16):1989–1997.
4. Carinci V, Magagnoli C, Capannoli G, Giannozzi A, D'Ascenzi S. Dimensional Studies on Meningococcal Vaccines A, C, Y, W135 by HPGPC and SEC–MALLS. Poster, IX Convegno sulla Chimica dei Carboidrati. 21–23 Giugno 2004 Certosa di Pontignano Siena Italy.
5. Jumel K, Ho MM, Bolgiano B. Evaluation of meningococcal C oligosaccharide coniugate vaccines by size–exclusion chromatography/multi–angle laser light scattering. Biotechnol. Appl. Biochem. 2002;36:219–226.
6. Bardotti A, Averani G, Berti F, Berti S, Carinci V, D'Ascenzi S, Fabbri B, Giannini S, Giannozzi A, Magagnoli C, Proietti D, Norelli F, Rappuoli R, Ricci S, Costantino P. Physicochemical characterisation of glycoconjugate vaccines for prevention of meningococcal diseases. Vaccine 2008;26(18):2284–2296.
7. Wyatt Technology Corporation, Astra for Windows User's Guide, Software Manual (1998).
8. Wyatt Technology Corporation, Light Scattering University, DAWN Course Manual (1999).
9. Kratochvil P, Classical Light Scattering from Polymer Solutions. Amsterdam: Elsevier Press; 1987.
10. Svennerholm L. Quantitative estimation of sialic acids. II. A colorimetric resorcinol–hydrochloric acid method. Biochim Biophys Acta 1957;24(3):604–611.

PART V

APPENDICES

Appendix 1

DISSOCIATION CONSTANTS (pK_a) OF ORGANIC ACIDS (AT 20°C)

Compound	pK_1	pK_2	pK_3
Acetic	4.56		
Acetoacetic	3.58		
Aconitic (trans-propane-1,2,3-tricarboxylic)	2.80	4.46	
Acrylic (propenoic)	4.26		
Adipic (hexanedioic)	4.26	5.03	
Anisic (4-methoxybenzoic)	4.48		
Ascorbic	4.03	11.34	
Azelaic (nonanedioic)	4.39	5.12	
Benzenesulfonamide	10.1		
Benzoic	4.00		
Bromoacetic	2.72		
Butanoic	4.63		
Caproic (hexanoic)	4.85		
Carbonic	6.37	10.25	

(*continued*)

Applications of Ion Chromatography for Pharmaceutical and Biological Products, First Edition.
Edited by Lokesh Bhattacharyya and Jeffrey S. Rohrer.

(Continued)

Compound	pK_1	pK_2	pK_3
Chloroacetic	2.68		
Citraconic (*cis*-methylbutenedioic)	2.20	5.60	
Citric (2-hydroxypropane-1,2,3-tricarboxylic)	2.87	4.35	5.69
Crotonic (*trans*-but-2-enoic)	4.69		
Cyanoacetic	2.63		
Dichloroacetic	0.87		
Diglycolic (2-(carboxymethyloxy)acetic acid)	2.79	3.93	
Dithiotartaric (2,3-dimercaptobutanedioic)	2.71	3.48	8.89
Ethanethiol	10.6		
Fluoroacetic	2.59		
Formic	3.55		
Fumaric (*trans*-butenedioic)	2.85	4.10	
Galaturonic	3.23	11.42	
Gentistic (5-hydroxysalicylic)	2.70		
Glutaric (pentanedioic)	4.13	5.03	
Glyceric (di-2,3-dihydroxypropanoic)	3.52		
Glycolic (hydroxyacetic)	3.52		
Guanidine	13.54		
2-Hydroxyisobutyric	3.72		
4-Hydroxybenzoic	4.10	9.96	
Hippuric (N-benzoylglycine)	3.50		
Iodoacetic	2.98		
Isobutyric (2-methylpropionic)	4.63		
Isocitric (1-hydroxypropane-1,2,3-tricarboxylic)	3.02	4.28	5.75
Isovaleric (3-methylbutanoic)	4.58		
Itaconic (methylelebutandioic)	3.68	5.14	
α-Ketoglutaric (2-oxopentanedioic)	1.85	4.44	
Lactic (*d*2-hydroxypropanoic)	3.66		
Maleic (*cis*-butenedioic)	1.75	5.83	
Malic (1-hydroxybutanedioic)	3.24	4.71	
Malonic (propanedioic)	2.65	5.38	
Mandelic (1-phenylhydroxyacetic)	3.19		
Mellitic (benzehexacarboxylic)	0.70	2.21	3.52
3-Mercapto propanoic	4.34	10.84	
Mesaconic (*trans*-methylbutene)	2.61		
Mucic	3.08	3.63	
Nitroacetic	1.46		
p-Nitrophenol	7.2		
Octanoic	4.89		
Orotic (Uracil-6-carboxylic)	1.96	9.34	

(*Continued*)

Compound	pK_1	pK_2	pK_3
Oxalic	1.04	3.82	
Phthalic (benzene-1,2-dicarboxylic)	2.75	4.93	
Peracetic (ethaneperoxoic)	8.2		
Phenol	10.0		
Picric (2,4,6-trinitrophenol)	0.3		
Pimelic (heptanedioic)	4.31	5.08	
Pivalic (2,2-dimethylpropanoic)	4.83		
Propanoic	4.67		
Pyruvic (2-oxopropanoic)	2.26		
Quinic (1,3,4,5-tetrahydroxycyclohexanecarboxylic)	3.36		
Salicylic (2-hydroxybenzoic)	2.81	13.40	
Squaric (3,4-dihydroxy-3-cyclobutene-1,2-dione)	0.40	3.10	
Succinic (butanedioic)	4.00	5.24	
Succinimide (Pyrrolidine-2,5-dione)	9.6		
Tartaric (d-2,3dihydroxybutanedioic)	2.82	3.95	
Thioglycolic (mercaptoacetic)	3.42	10.11	
Thiolactic (2-mercaptopropanoic)	3.48	10.08	
Thiomalic (mercaptobutandioic)	3.30	4.60	10.38
Thiophenol	6.6		
Trichloroacetic	0.66		
Trifluoroacetic	0.0		
2,2,2-trifluoroethanol	12.4		
Trimellitic (benzene-1,2,4-tricarboxylic)	2.40	3.71	5.01
Tropolone (2-Hydroxy-2,4,6-cycloheptatrien-1-one)	6.7		
Uric (2,6,8-trihydroxypurine)	5.61		
Valeric (pentanoic)	4.64		
Vinylacetic (but-3-enoic)	4.12		

Appendix 2

DISSOCIATION CONSTANTS (pK_a) OF COMMON SUGARS AND ALCOHOLS

Compound	pK_a at 25°C
Glucose-1-phosphate	1.10, 6.13
Glucose-3-phosphate	0.84, 5.67
Glucose-4-phosphate	0.84, 5.67
Glucose-6-phosphate	0.94, 6.11
Fructose-6-phosphate	0.97, 6.11
N-acetylneuraminic acid (NANA)	2.60
Glucuronic acid	3.20
Galacturonic acid	3.48
Maltose	11.94
Lactose	11.98
Fructose	12.03

(continued)

Applications of Ion Chromatography for Pharmaceutical and Biological Products, First Edition.
Edited by Lokesh Bhattacharyya and Jeffrey S. Rohrer.

(Continued)

Compound	pK_a at 25°C
Lyxose	12.11
Ribose	12.11
Mannose	12.08
Xylose	12.15
Glucose	12.28
Galactose	12.39
Arabinose	12.43
Sucrose	12.62
Raffinose	12.74
Dulcitol	13.43
Sorbitol	13.60
α-Methylglucoside	13.71
Methanol	15.54
Ethanol	16.0
Glycol	14.22
Glycerol	14.15

INDEX

Acclaim Trinity P1, 222, 224–7, 230, 232–3
acesulfame-K, 108
acetate, 261, 265
acetylcholine, 238
acid rain, 38
activated-pulsed amperometric detection (APAD), 52
adduct formation, 384
advanced applications of IC-PED, 73–6
 fingerprinting of bioproducts, 73–4
 bacterial polysaccharides, 74
 in-vitro microdialysis, 74–6
agmatine, 213
alkaloids, 108, 117–18
alkanolamines, 195, 204–5, 210
alkypolyamine, 196
aluminum adjuvant, 399
amikacin, 179, 190
amikacin sulfate, 179
Aminex A7, 314
Aminex Fast, 147, 315–6, 319
Aminex HPX-87H, 151, 314–6, 318–9, 411
Aminex HPX 87N, 315–6
AminoPac PA10, 352–3, 355, 357–9, 361–2
amino acids, 351–63
5-amino-1-pentanol, 204, 209
2-(2-aminoethoxy)ethanol, 204, 209
aminoglycoside antibiotics, 140–1, 262, 339, 346
 amikacin, 141
 kanamycin, 141
 neomycin, 140–1
aminomethylpropanol, 71
AminoTrap, 331–2, 344, 346, 403
3D amperometry, 63
ammonium, 413–18
amylamine, 196
anabolic 380
anion micro membrane suppressor, 407
anion self-regenerating suppressor (ASRS), 384
anion trap column, 407
anomeric hydroxyl, 340
antibiotics, 140–3
 erythromycin, 142
 tetracyclines, 141–2
anticoagulant solutions, 411
antihistamine, 195
aspartame, 108

bag-in-a-bottle, 96
barium, 197, 206, 217–18
Beer–Lambert Law, 41
beta blockers, 148–9
 acebutolol, 149
 alprenolol, 149
 atenolol, 148–9
 metoprolol, 148–9
 oxprenolol, 149
 propranolol, 148–9
beta-agonists, 149
 clorprenaline, 149
 clenbuterol, 149
 fenoterol, 149
 salbutamol, 149
betaine, 238
bethanechol, 150, 152, 217, 237, 238, 240–2, 250, 253–6
bile acids, 70
bininchonic acid procedure, 412
biogenic amine, 72, 196, 209–10, 213
biological product, 395

Applications of Ion Chromatography for Pharmaceutical and Biological Products, First Edition.
Edited by Lokesh Bhattacharyya and Jeffrey S. Rohrer.

bis(2-aminopropyl)amine, 215
bisphosphonates, 143–7
  alendronate, 141–2, 144–7
  clodronate, 143, 145–6, 153–4
  etidronate, 145–7, 152
  olpadronate, 146
  pamidronate, 144, 146–7
  risedronate, 147, 152
biuret procedure, 412
blood, 81
Bradford dye binding method, 412
bromate, 111, 261
brucine, 119
butoconazole, 147–8
butylamine, 217
butyrate, 261

cadaverine, 210, 213, 216
cadmium, 218
caffeine, 140, 150–1
calcium, 403
capillary HPAEC-IPAD-MS, 388
capillary ion electrophoresis, 407
capillary zone electrophoresis 437,443–4
capsular polysaccharide, 342
captopril -122
carbachol, 150, 217, 237–8, 240–3
carbocisteine, 108, 151
carbohydrate membrane desalter (CMD), 384
carbohydrate removal on line, 355
carbohydrates 380, 351–63
carbomyl-kanamycin, 184
carbomyl-tobramycin, 184
CarboPac MA1, 141, 155, 179, 332, 336, 345, 355, 357, 359–60, 362–3, 399, 406, 408, 424–33
CarboPac PA1, 75, 79, 80, 141, 176–7, 179–80, 185, 189, 190, 262, 296, 330–1 341–2, 345–6, 355, 359, 381, 397, 401, 403–4, 407–8, 410, 424–8, 430, 433
CarboPac PA10, 69, 155, 330–1, 341, 345–6, 398–9, 403–4, 442–3
CarboPac PA20, 305, 331, 340–1, 344–6, 355, 358–61, 375
CarboPac PA100, 333, 344–5, 408–9
CarboPac PA200, 334, 345, 369, 375, 381, 383, 387–8
catabolic, 380
cation exchange, 92, 94, 97
cefepime, 216–17
cell culture, 362
charged aerosol detection, 194
chlorate, 111
chlorite, 261
chlorophenol, 114
chlorothiazide, 120, 122, 126
choline, 213–14, 217, 238, 240–4
chromate, 111, 261
chromium, 218
cis-aconitate, 261
citrate acid/citrate, 151–2, 197, 222, 225, 249–52, 261, 264, 411
cobalt, 218
cocaine, 119
coffee, 108
column switching, 355
combination vaccines, 399
contact lens solution, 241
conventional chemical mode, 95–6
copper, 218
c-polysaccharide, 408
CR-CTC, 200, 208–9, 212–13, 215–17
creatine, 82
creatinine, 82, 231
CRM197, 424
cross-ring fragmentation, 386
current efficiency, 98
cyclamate, 108
cyclohexylamine, 196, 204, 209
cystamine, 215

DC amperometry, 101
1,10-decanediamine, 216
deoxycholate precipitation, 403
2-deoxyglucose, 345
2-deoxystreptamine, 184–5
deoxystreptamine-kanosamide, 184
desalter, 383
detection, 15–20
  conductivity detection, 15–20
  mechanism of suppression, 18–20
  non-suppressed conductivity detection, 16–20
  suppressed conductivity detection, 16–20
  pulsed amperometric detection, 20
deuterated glucose, 341–2
dextromethorphan, 264
2,3-diaminopropionic acid, 198
diclofenac, 122, 125–6, 225, 227–30
diethanolamine, 197, 205, 212
diethylamine, 197, 213
2-diethylaminoethanol, 204–5
diethylenetriamine, 215
diethylethylenediamine, 215
diethylhydroxlamine, 195
differential refractometry, 411
dihydrocodeine, 119
dihydromorphine, 119
5,7-dihydroxytryptamine, 231

5,7-dihydroxytryptamine creatinine sulfate, 229, 231
dimethylamine, 203, 209–12, 215, 217
1-dimethylamino-2-propanol, 217
3-dimethylamino propylamine, 196, 215
1,1-dimethylbiguanide, 225–6
dimethylhydroxlamine, 195
1,2-dimethylpropylamine, 208
2,4-dinitrobenzene, 177
diphenhydramine, 122, 126
diptheria, pertussis and tetanus, 399
dipyridamole, 122, 126
disaccharides 382
displacement chemical regeneration (DCR), 95–6
disposable electrodes, 101
disposable gold working electrode, 178
dissolution test, 152–3
distribution coefficient, 436, 438–40
donnan exclusion, 23, 42
donnan forbidden, 94
donnan membrane, 11–12, 24–6, 29, 32, 34
donnan potential, 94
doxepin, 122, 126
drug formulation, 259
DryLab, 123

electroanalysis at noble metal electrodes, 52–8
  amines, 55–8
  carbohydrates, 52–5
  sulfur-containing compounds, 55–8
electrolytic water splitting, 97
eluent generator, 108, 109, 114, 208, 213, 215–17, 261, 267, 344, 346
enzyme characterization, 76–7
ephedrine, 119, 149–50
ethanolamine, 71, 205, 209, 212, 214, 217
ethylamine, 208, 215, 217
ethylenediamine, 196, 215–16, 222
2-ethylhexylamine, 196
ethylhydroxlamine, 195
European Directorate for the Quality of Medicines and Healthcare (EDQM), 310
European Pharmacopoeia (*EP*) 309
evaporative light scattering detection (ELSD), 194, 223–4, 232
evaporative light scattering detector, 132
excipient, 154–5, 224, 230–2 409–11
external water, 94. 98

fermentation, 184
fiber suppressor, 92
fentanyl citrate, 249–52
first aid cream, 186, 189
fluorescence detection, 194
Food and Drug Administration (FDA), 167, 230, 287, 347
forensics, 249–58
formate, 111, 261, 265
free saccharide, 429–31
fucosamine, 342
fucose, 340–1, 343–4
fumarate, 261
funosemide, 122
furosemide, 120, 125–6

gabapentin, 79, 80
galactosamine, 304–5, 340, 341, 344–5
galactose, 340, 341, 343–4, 406
galacturonate, 261
galacturonic acid, 407
gas assisted, 98
genotoxic impurity, 193
genotoxins, 224
gentamicin sulfate 177
gentamicin, 190
glassy carbon electrode, 50, 238
glaucoma, 237
glucosamine, 340–5
glucosamine-1-phosphate, 398, 403
glucose, 340–4
glucose-1-phosphate, 397
glucuronic acid, 406–7
glutarate, 261
glycan, 323, 325
glycan fragmentation, 386
glycan mapping, 335–7
glycerol, 411
glycoconjugate, 339, 424–9
glycoconjugate vaccine, 396, 397, 399, 402, 405, 408
glycolate, 261
glycolipid, 339, 345
glycopeptide, 68–70, 345
glycoprotein, 339–40, 343–7
gM1-gangliosidosis, 383
gold, 101
gold working electrode, 52–6, 59, 61, 65, 73, 80, 342, 346
Gouy-Chapman model, 34
gradient elution, 353
gradient programs, 354, 356
group B streptococcus polysaccharide, 423–4

Haemophilus influenzae, 397, 399
HEMA 1000Q, 145
heparin, 304–5, 339
hepatitis B, 399

1,7-heptanediamine, 216
hexanaesulfonate, 111
1,6-hexanediamine, 216
hexoses, 404
hib conjugate vaccine, 399
high performance anion exchange
  chromatography, 326–8
  eluent, 327–8
  borate contamination, 327–8
high performance anion-exchange chromatography (HPAEC), 20, 138, 141, 379
histamine, 213
HPAEC-PAD, 138, 141, 339–47, 366, 368–72, 375–6, 396–410
HPIC-CS3, 150
humatin, 262–3
hydrazine, 193, 196, 204, 209
hydrocortisone, 122, 126
hydrofluoric acid hydrolysis, 407–8
hydroxlamine, 193, 195
4-hydroxycoumarin, 109
2-hydroxypropyltrimethyl ammonium chloride (2-HPTA), 150, 217, 240–2
hydroxyzine, 147, 152

ibuprofen, 122, 126
ICH, 244–5
IC-MS, 194, 200, 203, 205–6, 210, 215
IC-Pac HC, 145–46, 403
IC-Pak Cation, 414
IC-Pak CM/D, 149–50
IC-Pak HR, 144–46, 148
IC-PED, 64–73
  alcohols, 71
  amines, 72–3
  amino acids, 72–3
  amino alcohols, 71–2
  aminoglycosides, 71–2
  amino sugars, 71–2
  carbohydrates, 68–70
  peptides, 72–3
  pharmaceutical and biological samples, 67–73
  polyols, 71
  proteins, 72–3
  sulfur compounds, 73
IgG, 343–4
imidazolidines, 196
impurity, 139, 141–6, 153–4, 165–7, 177–8, 179–87, 243, 380
indirect absorption detection, 60
indirect conductivity detection, 414
indirect detection, 79–80
indirect Pulsed Electrochemical Detection (InPED), 60, 79–80
infant formula, 213
instrumentation, 14–15, 329
integrated Pulsed Amperometric Detection (IPAD), 52, 59–61, 66–7, 72–3
internal standard, 345
inulin 381
iodide, 111
ion chromatography, 309
  columns for, 310
  applications in pharmacopeial monographs, 311–12
  detection: conductivity, amperometric, and refractive index, 312, 317
  detection: spectroscopic, 317
  monographs that use (non-spectroscopic detection), 313–16
  monographs that use (spectroscopic detection), 310–18
ion exchange surface (IES), 92–3
ion–exchange chromatography, 3–11, 16, 18
  effect of flow rate, 10–11
  effect of pressure, 10
  effect of temperature, 10
  eluent, 9
  mechanism, 5–8
  organic solvent, 9–10
ion-exclusion chromatography, 11–14, 139, 151–2, 154, 411
  complexation, 13
  effect of flow rate, 14
  effect of temperature, 14
  eluent, 13–14
  hydrogen bonding, 13
  hydrophobic properties of analytes, 12–13
  mechanism, 11–13
  steric factors, 13
  $\pi-\pi$ interaction, 13
IonPac AS5, 153
IonPac AS5A, 165, 212
IonPac AS7, 143, 145, 147
IonPac AS7A, 145
IonPac AS11, 7, 167, 168, 401–2, 408, 411
IonPac AS11-HC, 249–50, 261–4, 401, 403, 408
IonPac AS14, 264, 266
IonPac AS15, 97, 266–7, 403
IonPac AS16, 255
IonPac AS18, 262–3, 265
IonPac AS20, 111–12
IonPac AS4A-SC, 111, 140
IonPac CS12, 198, 412, 416–17
IonPac CS12A, 205–7, 213–14, 410, 416

IonPac CS14, 116, 150, 151–3, 250, 254, 264, 266
IonPac CS16, 200–2, 206–8, 210, 214–15, 417
IonPac CS17, 116, 166, 195, 203–4, 208–10, 215–17, 239–40
IonPac CS18, 203, 205, 208–16
IonPac CS5A, 212
IonPac ICE-AS1, 27–4, 38–9, 155
IonPac ICE-AS6, 38–9
IonPac NS1, 31, 33
IonPac SCS-1, 200–1, 210–12, 214, 217–18
iso-butylamine, 208
isocitrate, 261

Japanese Pharmacopoeia (*JP*) 309

kanamycin A, 184
kanamycin B, 176, 180–1, 184–6, 190
kanamycin sulfate, 179
kanosamine, 185
ketomalonate, 261
Kjeldahl procedure, 412–18
Kuwamoto model, 111

β-lactam antibiotics, 142–3
  amoxicillin, 142–3
  flucloxacillin, 142–3
lactate, 261
lactose, 410
lead, 218
linear solvent strength model (LSSM), 110–13, 115
Lowry method, 412, 414
Luna SCX, 142
lysine, 197

malate, 222, 225, 261
malonate, 261
manganese, 197, 218
mannose, 340–1, 343–4
mannoseamine-6-phosphate, 402
meglumine, 197, 225
melamine, 216
MenA PS, 401
MenA saccharide 437, 440–1
MenC PS, 402
meningococcal glycoconjugate, 435–47
MenW135 PS, 402
MenY PS, 402
mesylate, 225
methenamine, 148
methylamine, 212–13, 217
methylaminoethanol, 205
methyldiethanolamine, 215
α-methyl glucoside, 340
methylhydroxlamine, 195
Metrosep A SUPP 5, 277
Metrosep Carb1, 332
Metrosep Cation 1-2, 149
methylsulfonate, 261
microelectrode, 63–4, 80–82
milk, 76, 213–14, 216
mobile phase selection, 139–40
molybdate, 111
monochloroacetate, 261
monoclonal antibody 346–7
monoisopropanolamine, 195
monosaccahride composition analysis, 329–33
  Aminotrap guard column, 332
  column, 330–1
  dissociation constants, 326
  hydrolysis, 330
  standards, 330
monosaccharide, 68–9, 73, 339–47
morphine, 108, 119
morpholine, 204, 209, 217
MRI, 197

N,*N*-dimethylethanolamine, 195
N-acetyl,403
N-acetylgalactosamine, 345
N-acetylglucosamine, 345
N-acetylmannoseamine phosphate, 401
nanopolymer silica hybrid technology, 222, 225
naphthol, 122
naphtholic acid, 122
naproxen, 225–7
N-butylamine, 196, 208
neamine, 175–6, 180, 184–6, 189
nebramine, 179, 184–5
neomycin, 185, 190
Neisseria meningitidis, 39–7, 401, 435
neomycin A, 175
neomycin B, 185, 189
neomycin C, 177, 186
neomycin sulfate, 185
netilmicin, 190
netilmicin sulfate, 177–8
nickel, 218
N-linked oligosaccharide 380
N-methyldiethanolamine 217
N-methylethanolamine, 195
N-methylmorpholine, 217
N-methylpyrrolidine, 216–17
noble metal, 94
1,9-nonanediamine, 216

nonsuppressed IC, 199, 211
norephedrine, 149–50

O-acetyl, 403–4, 407
Obelisc R or N, 223
1,8-octanediamine, 216
N-octylamine, 196
oleic acid, 232–3
oligosaccharide, 69–70, 73, 343, 345, 408
oligosaccharide analysis, 333–5
OmniPac PAX-100, 7, 144, 150
OmniPac PCX-100, 142
Omnipac PAX 500, 142, 407
o-phthalaldehyde fluorometric, 412, 414
ophthalmic, 237, 241
opium, 108
oxalate, 111, 261
oxidation, 98, 101
oxymorphone, 148

palmitic acid, 232–3
paracetamol, 148
paraquat, 215
paromomycin, 190, 262
Partisil SCX, 148
Peclet number, 29
penicillamine, 56, 58
perchlorate, 111
pet food, 216
phenethylamine, 213
phenol, 122
Phenospehere SAX, 146
phosphate, 403, 410
phosphorylated monosaccharides 381
phosphotungstic acid precipitation, 413
phthalate, 111, 261
$\pi-\pi$ interactions, 25, 31
piperazine, 196–7, 215
plasma, 213
platinum, 101
platinum electrode, 51–3, 56, 59, 61, 65, 73
pneumococcal polysaccharides, 406–7
polyamines, 196, 215
polydispersity 436, 438–40
polyethylene amines, 196
Polypore H, 411
polysaccharides, 69–70, 74
potassium, 410
potential sweep pulsed coulometric detection (PSPCD), 52
procaine, 225
propionate, 261, 407
propranolol, 122, 126
propylamine, 208
protein nitrogen, 412
Protein-Pak SP-5PW, 415
pseudoephedrine, 149–50, 264
pulsed amperometric detection (PAD) 138, 141, 328–9, 379
pulsed coulometric detection (PCD), 51
pulsed electrochemical detection (PED)
  fundamentals, 59–61
  mode I, 59, 64–7
  mode II, 59–60, 64–7
  mode III, 60
  historical perspective, 51–2
pulsed voltammetric detection (PVD), 52, 62, 64
pulsed voltammetry, 54, 57, 64
putrescine, 196, 210, 213, 216
pyrophosphate, 263
pyruvate, 261, 407

quinate, 261
quinovosamine, 342

radiopharmaceuticals, 309–10
recycle, 94, 97–8
refractive index detection, 194, 343
reversed-pulsed amperometric detection (RPAD), 51
Rezex, 154
rhamnosamine, 342
rhamnose, 406, 428–30
ribavirin, 147
ribitol, 399
ribitol-ribose-phosphate (RRP), 397, 399
ribose, 399
ribose-5-phosphate, 400
rodenticide, 109
rubidium, 197, 206

*S. Pneumoniae,*404, 407
saccharin, 108
salicin, 77
salicylic acid, 109
salt converter, 100
SC-CSRS 300, 415–16
Scherzo SM-C18, 223
Schomburg column, 198, 211
SDS-PAGE, 413
SEC-butylamine, 208
selenate, 111
self-regenerating suppressor, 410
SeQuant ZIC-HILIC, 223, 231, 233
serum, 213
sialic acid, 402–4, 437, 442
sialic acid assay, 332–3
SIELC, 223

silver, 101
single column IC, 199–200
single-radial immunodiffusion, 413
size-exclusion chromatography, 44
stationary phase (column) capacity, 26, 29, 31, 37
sodium, 403, 417
solid phase extraction (SPE), 429–30
sorbate, 261
sorbitol, 410
spermidine, 196, 213
spermine, 196, 213
Sphereclone SAX, 146
spongy platform, 100–1
stearate, 230, 232–3
streptomycin, 179, 190
stripper column, 91
strontium, 197, 206, 218
succinate, 111, 261
sucrose, 252–3, 339, 410
sugar phosphate, 339
sulfamethoxazole, 122, 126
suppressed conductivity, 99–101
SuperSep IC Anion, 147
suppressor reactions, 92
suspension, 186

tartrate, 111, 222, 225, 261
tert-butylamine, 196, 208
tetracycline, 117
TFA hydrolysis, 399
theobromine, 150–1
theophylline, 150–1
thiocyanate, 111
thiosulfate, 111
tobramycin, 176, 179–81, 184–5, 190
tolfenamic acid, 122, 126
tosylate, 222
toxicological applications, 77–9
  glucuronides, 77–9
trans-aconitate, 261
triethanolamine, 71, 205, 212
triethylamine, 203, 209–11, 215
triethylenetetramine, 196, 215
trifluoroacetate, 261, 265–6
trimethylamine, 212–13
trimethylamine-N-oxide, 215
tris, 195, 225
tromethamine, 197
tropic acid, 122
tuberculin purified protein derivative, 411
tungstate, 111, 261
two-dimensional separation, 359

uric acid, 82
urine, 78, 82, 213, 388
USP L9, 147, 318–9
USP L12, 148, 313, 318–9
USP L17, 147, 313–6, 318–9
USP L19, 314, 316
USP L23, 314–6
USP L34, 315–6
USP L46, 141, 179, 313, 316
USP L47, 141, 179, 313–4
USP L48, 314, 319
USP L50, 142
USP L52, 318
USP L53, 150, 313
USP L55, 150, 254, 313
USP L58, 315–6
USP L61, 150, 249, 313–6, 411
USP-NF, 4, 15, 21, 237, 241, 243
  general chapters, 310
    Assay for Citric Acid/Citrate and Phosphate <345>, 139, 311
    Biotechnology-derived Articles <1045>, 311
    Biotechnology-derived Articles—Amino Acid Analysis <1052>, 311
    Biotechnology-derived Articles—Peptide Mapping <1055>, 311
    Impurities in Official Articles <1086>, 311
    Ion Chromatography <1065>, 139, 310–11
  monographs, 309
US Pharmacopeial Convention, 310

vacancy ion-exclusion chromatography, 26
vaccine, 70, 339
valerate, 261
vigabatrin, 79–80
virtual column, 111–12, 128
vitamin B2, 352

waveform design, 61–3
wine, 361

xanthene, 108
xylitol, 70

zeta potential, 37
zinc, 197, 218
Zipax SAX, 148
Zorbax 300-SCX, 142, 148

www.ingramcontent.com/pod-product-compliance
Ingram Content Group UK Ltd.
Pitfield, Milton Keynes, MK11 3LW, UK
UKHW020226290926
15470UKWH00001B/404

9 780470 46709